PED-Vol. 55

SENSORS, CONTROLS, AND QUALITY ISSUES IN MANUFACTURING

presented at
THE WINTER ANNUAL MEETING OF
THE AMERICAN SOCIETY OF MECHANICAL ENGINEERS
ATLANTA, GEORGIA
DECEMBER 1–6, 1991

sponsored by
THE PRODUCTION ENGINEERING DIVISION, ASME

edited by
T. I. LIU
CALIFORNIA STATE UNIVERSITY

C. H. MENQ
OHIO STATE UNIVERSITY

N. H. CHAO
AT&T BELL LABORATORIES

THE AMERICAN SOCIETY OF MECHANICAL ENGINEERS
345 East 47th Street • United Engineering Center • New York, N.Y. 10017

ISBN No. 0-7918-0840-8

Library of Congress
Catalog Number 91-58408

FOREWORD

The objective of the Symposium on Sensors, Control, and Quality Issues in Manufacturing, held at the 1991 ASME Winter Annual Meeting in Atlanta, is to create a forum for exchange of information on research and development in this field. With rapid progress in factory automation and increasing demand of high quality product, these topics are crucial for the factory of the future.

This symposium discusses sensors, controllers, and various quality issues in manufacturing processes. Furthermore, the mechanics of these processes are also discussed. The latter is fundamental because it is the basis to select adequate sensing and control approaches and to solve the quality problems. The papers presented in this bound volume are organized along three lines:

- Sensors for Manufacturing
- Control for Manufacturing
- Quality Issues in Manufacturing

We would like to express our gratitude to all of the authors and reviewers who contributed to the high quality of this bound volume. The efforts of the session chairmen are also greatly appreciated. Special thanks are due to the Production Engineering Division for supporting this symposium.

T. I. Liu
C. H. Menq
N. H. Chao

CONTENTS

Sensors, Controls, and Quality Issues
in Manufacturing
ASME 1991

PATTERN RECOGNITION BASED BICOHERENCE ANALYSIS OF VIBRATIONS FOR BEARING CONDITION MONITORING

C. James Li and J. Ma
Department of Mechanical Engineering
Columbia University
New York, New York

B. Hwang and G. W. Nickerson
David Taylor Research Center
Annapolis, Maryland

ABSTRACT

For automatic detection/diagnosis of localized defects in rolling element bearings, a pattern recognition analysis scheme based on features extracted from bearing vibrations using bispectral analysis is developed. Features are extracted using the bicoherence spectrum to detect the sum frequency components of bearing characteristic defect frequencies and their harmonics. Employing these features, a linear discriminant classifier has been established to detect localized defects on a roller and outer-race of a bearing. Experimental results show that the bispectral analysis for the detection of localized defects of rolling element bearings is very effective even when power spectral analysis fails to distinguish the abnormal states from the normal one.

1. INTRODUCTION

Rolling element bearing condition monitoring has received considerable attention for many years because the majority of the problems in rotating machines are caused by faulty bearings. The classical failure mode of rolling element bearings is localized defects, in which a sizable piece of the contact surface is dislodged during operation, mostly by fatigue cracking in the bearing metal under cyclic contact stressing [1]. Thus, failure alarms for a rolling element bearing are often based on the detection of the onset of localized defects.

During the bearing operation, wide band impulses result from the passage of the defect through the rolling element-race way contacts. Defects at different locations of a bearing (inner-race, roller and outer-race) have characteristic frequencies at which impulses are generated. Theoretical estimations of these frequencies are denoted as **characteristic defect frequencies** (Appendix A).

Some vibration modes of the bearing parts and its supporting structures may be excited by these periodic impulses. The leading edge of each resonance comprises a very sharp rise corresponding to the impact of the defect, whose decay is approximately exponential, as the energy is dissipated by internal damping (Fig. 1a). Therefore, the signal of a damaged bearing consists of periodic bursts of exponentially decaying sinusoidal vibrations occurring at the characteristic defect frequency (Fig. 1b).

* This material is based upon the work supported by the Department of Navy under award N00038-84 C-0165 P00019.

The characteristic defect frequency makes it possible to detect the presence of a defect and diagnose in what part of the bearing the defect appears. One may conclude that there are localized defects on the outer-race of a bearing if one sees a series of vibration bursts occurring at a rate close to the theoretical estimation of the characteristic defect frequency for defects on the outer-race.

The difficulty of localized defect detection lies in the fact that the energy of the bearing's signature is spread across a wide frequency band and therefore can be easily buried by noise. This can be seen from Fig. 1 which shows a simulated bearing signature and its spectrum. The signature is characterized by periodic bursts of exponentially decaying sinusoidal oscillations. Its spectrum consists of a family of harmonics of the characteristic defect frequency with the largest harmonic in the neighborhood of the resonance. One can see that the amplitude at the characteristic defect frequency (the first peak) is so small that its presence may not be noticed at all if there is noise.

To solve this problem, time domain methods involving indices that are sensitive to the impulsive oscillations have been developed. Examples are peak level, r.m.s. value [2], crest factor analysis [3], kurtosis analysis [4] and shock pulse counting [5]. These methods would yield satisfactory results as long as one selects a frequency band that includes some of the modes that actually get excited by impacts of rollers on the defect and where there are no other sources of impacts than the bearing localized defect. Unfortunately, it has been shown that there is no good way of predicting vibration modes of a bearing's parts and its supporting structures, and which one of these modes may be excited by the impact on the defect. What makes it even worse is that a mode excited at the onset of a defect may not be excited later when the extent of damage increases.

A number of frequency domain techniques have been developed to detect a train of impulses occurring at some characteristic defect frequency from the waveform of the bearing signal. Examples among these approaches are synchronized averaging [6], cepstrum analysis, sum and difference frequencies analysis [7-8], and high frequency resonance technique (HFRT) [9-12]. Both the synchronized averaging and cepstrum analysis reveal the total energy in the whole family of harmonics of the characteristic defect frequency. Naturally, both of the methods are more sensitive to the presence of the localized defect than the spectral analysis. On the other hand, they are also more likely to be influenced by noise that may be present at the frequency of some of the harmonics.

The most widely accepted technique is the high frequency resonance technique (HFRT). This method requires the selection of a frequency band that covers a few vibration modes of the monitored bearing and its structure that are away from the low frequency mechanical noise band. Signal measured at this band is then amplitude demodulated to reveal possible modulation by the characteristic defect frequencies. If one of the modes falling in this selected band is excited by the periodic impulses due to the defect, amplitude demodulation would manifest the characteristic defect frequency. Again, the success of this method depends on the selection of the correct frequency band.

Pattern recognition schemes based on features extracted by short-time signal processing techniques were developed to process bearing vibrations and acoustic emissions [13,14]. The discriminating power of the pattern classifier and the superior ability of the short-time signal processing technique to extract time varying properties due to localized defects have enhanced the ability to detect the defect at an even earlier stage than other state-of-the-art. However, since the mechanical resonances in the bearing system were utilized by the techniques to enhance the signal, their performance also depend on the selection of the correct frequency band.

The objective of this paper is to present a new bearing localized defect detection technique based on bicoherence analysis. The presence of the characteristic defect frequency and its harmonics are detected by using the bicoherence spectrum. Then, the pattern recognition technique is utilized to classify the condition of a bearing based on its bicoherence. It is shown in the experimental study that this method is so sensitive to the presence of a localized defect that it can detect a defect using a low frequency band which is full of mechanical noise. Consequently, neither guess about the modal frequencies nor about which mode may be excited by the defect is needed and therefore a general scheme is made available.

2. BISPECTRUM/BICOHERENCE AND QUADRATIC COUPLING

Let's denote the Fourier Transform of a random process, x(t), as

$$X(f) = |X(f)| \exp(i\phi) \tag{1}$$

where $|X(f)|$ and ϕ are magnitude and phase angle. Then the spectral density of x(t) can be expressed as

$$P(f) = E[\, X(f)X^*(f) \,] \tag{2}$$

where the asterisk stands for complex conjugate and E[ ] denotes the expected value. Using the same notation, the bispectrum, which is defined to be the double Fourier Transform of the third moment of x(t) [15], can be expressed as

$$B(f_1,f_2) = E[\, X(f_1)X(f_2)X^*(f_1+f_2) \,] \,. \tag{3}$$

It is obvious that the bispectrum is symmetric about the axis $f_1=f_2$, namely, $B(f_1,f_2)=B(f_2,f_1)$. In Addition, when the computation of bispectrum is carried out using discrete samples, only are those $B(f_1,f_2)$ for $f_1+f_2 \le f_N$, where f_N is the Nyquist frequency, meaningful because of frequency folding. Therefore one only has to calculate the bispectrum in the triangle region shown in Fig. 2.

The bispectrum measures the statistical dependence between three different frequency components [16]. To see this, let us first rewrite equation (3), using (1), as

$$B(f_1,f_2) = E\Big[\, |X(f_1)X(f_2)X(f_3)| \exp[i(\phi_1+\phi_2-\phi_3)] \,\Big] \tag{4}$$

where $f_3 = f_1+f_2$. If phase angles ϕ_1, ϕ_2 and ϕ_3 are independently and uniformly distributed over $[-\pi, \pi]$, then so is $(\phi_1+\phi_2-\phi_3)$. Hence the expected value in (4) will lead to zero. On the other hand, when $X(f_3)$ is generated by the quadratic coupling between $X(f_1)$ and $X(f_2)$, ϕ_3 is equal to the sum of ϕ_1 and ϕ_2, and consequently the bispectrum will be just

$$B(f_1,f_2) = E[\, |X(f_1)X(f_2)X(f_3)| \,] \,. \tag{5}$$

Now, let's assume that the quadratic coupling of two pure sinusoidal components $x_1(t)$ and $x_2(t)$ produces a new signal $x_3(t)$:

$$x_3(t) = C_{12}\, x_1(t)\, x_2(t) \tag{6}$$

where C_{12} is the coupling coefficient. Apply the Fourier Transformation to equation (6) and use $X_i(f)$ to represent the Fourier Transform of $x_i(t)$, so

$$X_3(f) = \int_0^\infty C_{12} X_1(g) X_2(f-g)\, dg \,. \tag{7}$$

Since $x_1(t)$ and $x_2(t)$ are pure sinusoidal signals at frequencies f_1 and f_2, their Fourier Transform can be written into

$$X_i(f) = X_i(f_i)\, \delta(f-f_i) \qquad (i=1,2) \tag{8}$$

where δ is the Dirac delta function. Then equation (7) can be easily simplified to

$$X_3(f) = C_{12}\, X_1(f_1)\, X_2(f_2)\, \delta[f-(f_1+f_2)] \,. \tag{9}$$

Since the only nonzero value of X_3 is at $f=f_1+f_2$, we conclude that the quadratic coupling of two frequency components generates a new component at their sum frequency. In addition, it can be shown that quadratic coupling can also generate a new component at the difference frequency by extending the above analysis to the negative frequency domain.

Substitution of equation (8), (9) into (3) for $f=f_3=f_1+f_2$ and dropping the subscript of X yield

$$B(f_1,f_2) = C_{12}\ E\left[\,|X(f_1)X(f_2)|^2\right]. \tag{10}$$

Therefore, the quantity

$$b^2(f_1,f_2) = \frac{|B(f_1,f_2)|^2}{E\left[\,|X(f_1)X(f_2)|^2\right]\ E\left[\,|X(f_1+f_2)|^2\right]} \tag{11}$$

is an indication of the degree of quadratic coupling, where $b(f_1,f_2)$ is called bicoherence which is actually the normalized bispectrum. When all the energy at $f_3=f_1+f_2$ is from coupling of components at f_1 and f_2, eq. (11) gives $b^2(f_1,f_2)=1$. On the other hand, when no energy at $f_3=f_1+f_2$ is from coupling ($C_{12}=0$), then $b^2(f_1,f_2)=0$.

Suppose frequency component at f_3 comes from both an independent source and the coupling between frequency components at f_1 and f_2. Then :

$$X(f_3) = C_{12}\ X(f_1)\ X(f_2) + X_{in}(f_3) \tag{12}$$

where $X_{in}(f_3)$ is the spontaneous component. The total power at f_3, according to (2), is

$$\begin{aligned} P(f_3) &= C_{12}^2\ E\left[\,|X(f_1)X(f_2)|^2\right] + E\left[\,|X_{in}(f_3)|^2\right] \\ &= P_c(f_3) + P_{in}(f_3) \end{aligned} \tag{13}$$

where the subscripts "c" and "in" stand for "coupled" and "independent", respectively. Then,

$$b^2(f_1,f_2) = \frac{P_c}{P_c + P_{in}} \tag{14}$$

tells us that the square of the bicoherence is an exact measure of the fraction of power at $f_3=f_1+f_2$ due to the quadratic coupling.

In most cases, not all the energy at frequencies f_1 and f_2 are involved in coupling. Consequently equation (14) is not valid. Under this condition, frequency components at f_1 and f_2 have only part of their energy involved in the coupling. Let's denote the involved and non-involved parts by "c" and "n" respectively. Thus, one has

$$X(f_i) = X_c(f_i) + X_n(f_i) \qquad (\,i=1,2\,) \tag{15}$$

$$X(f_3) = C_{12}\ X_c(f_1)\ X_c(f_2) + X_{in}(f_3)\ . \tag{16}$$

Using expressions (15) and (16) together with (5), (11), and (13), the bispectrum and modulus of bicoherence are

$$B(f_1,f_2) = C_{12}\ E\left[\,|X_c(f_1)X_c(f_2)|^2\,\right] \tag{17}$$

and

$$b^2(f_1,f_2) = \frac{P_c}{P_c + P_{in}}\ \frac{E\left[\,|X_c(f_1)X_c(f_2)|^2\,\right]}{E\left[\,|X(f_1)X(f_2)|^2\,\right]}\ . \tag{18}$$

Equation (18) says that the magnitude of b^2 depends on not only the percentage of energy at f_3 due to quadratic coupling, but also the fractions of energy at f_1 and f_2 involved in the coupling.

A special case where frequency components f_1 and f_2 are completely uncorrelated will be examined. In such a case, $X(f_1)$ and $X(f_2)$ as well as $X_c(f_1)$ and $X_c(f_2)$ are mutually independent random variable pairs. If

$$P_c(f_i) = E\left[\,|X_c(f_i)|^2\,\right] \qquad (i=1,2) \tag{19}$$

one obtains the following simple relationship:

$$b^2(f_1,f_2) = \frac{P_c(f_3)}{P(f_3)}\,\frac{P_c(f_1)}{P(f_1)}\,\frac{P_c(f_2)}{P(f_2)}\ . \tag{20}$$

This means that b^2 is exactly proportional to the fractions of power involved in coupling at all three frequencies.

Noting that $X^*(f)=X(-f)$, thus $B(f_1,-f_2)=B(f_2-f_1,-f_2)=B^*(f_1-f_2,f_2)$, the treatment of the difference frequency in the bispectral/bicoherence analysis is very similar to the sum frequency. If the coupling of components at f_1 and f_2 creates a component at $f_3=f_1-f_2$, then their phase angles have this relationship $\phi_3=\phi_1-\phi_2$, and then the coupling can be seen as components at f_2 and f_3 coupled to generate a component at the sum frequency f_1. Hence, the value of $b^2(f_2,f_3)$ or $b^2(f_3,f_2)$ quantitatively describes the degree of coupling.

3. PATTERNS IN BICOHERENCE SPECTRUM FOR DAMAGED BEARING

It has been shown in the section of introduction that signatures from a bearing with a localized defect consist of a family of harmonics of the characteristic defect frequency. This family of harmonics are the Fourier components of a periodic signal like the one shown in Fig. 1b and therefore are correlated in their phases. Consequently, bicoherence spectrum should have large values at frequency pairs (f, f), (2f, f), (3f, f) ..., (2f, 3f), (2f, 4f), ..., where f is the characteristic defect frequency.

Another possible phenomenon is that some of the harmonics of the characteristic defect frequency are involved in the generation of the sum and difference frequencies with other spontaneous frequency components such as the harmonics of the rotational speed.

Lastly, the bearing signature like the one in Fig. 1b may be amplitude modulated by the rotational speed because the defect rotates in and out of the loading zone. If this happens, sidebands to each side of the harmonics will be seen in the spectrum. The sidebands are located at one rps (rotations per second) to the right and left of each harmonic.

4. BEARING FAULT DIAGNOSIS:

4.1 Experimental Setup: The tapper roller bearing under test is mounted on a mechanical platform which enables the adjustment of the thrust load, the radial load and the position of the center of the bearing. The setup is driven by a DC variable speed motor. The operating conditions were kept the same for all bearings tested.

Five bearings were used, their conditions are listed in Table 1. All defects were artificially simulated using an electric-discharge-machine (EDM) and a grinding machine. We were not able to impose any defect on the inner race of the bearing because of its inaccessibility. Vibration signals were picked up in the radial direction by a PCB 482A accelerometer and then sampled at frequency of 1,000 Hz after a fourth order low-pass filter with a cutoff frequency of 500 Hz. Because most state-of-art schemes don't work well at a low frequency band due to the presence of mechanical noise, the cutoff frequency was chosen to be low so that if the scheme was proven to work, its superiority would be obvious.

The setup was running at a speed of 870 rpm, i.e. $f_r=14.5$ Hz. The characteristic defect frequencies for this speed are listed in Table 1.

4.2 Pattern Classifier Design: Averaged bicoherence of 32 ensembles were estimated from bearing vibrations. An unambiguous example which is a bicoherence spectrum of bearing 5 (with large outer race defect) is shown in Fig. 3 together with that of bearing 3 (without defect). All three patterns under section 3 are evident in this example. More bicoherence plots for bearings with localized defects of various severity at different locations can be found in reference [17].

Since knowledge required for identifying the first pattern, i.e. self-couplings between the characteristic frequency and its harmonics, is the least among the three patterns, magnitude at frequency pairs of this pattern were examined to find effective features for distinguishing defective bearings from the nondefective ones. In the case of a defective outer-race, one should find features which enable us to separate outer-race defective bearings 4 and 5 from the other three bearings. Due to the possible ambiguity resulted from very low frequency mechanical noise generated by other mechanisms of the system, frequency pairs (f_{or}, f_{or}) and ($2f_{or}$, f_{or}) were discarded. Pairs ($3f_{or}$, f_{or}) and ($2f_{or}$, $2f_{or}$) were chosen as the features. Bicoherence values at these two frequency pairs of all five bearings are plotted in Fig. 4. The diamonds and squares represent the bicoherence values of outer-race defective bearings and nondefective ones, respectively. These two groups of points are distinctly separated from each other and can be separated by a straight line as the one indicated in the figure.

In the context of pattern recognition, a linear function used to separate different clusters is termed as a linear discriminant function. If the pattern vector is defined as $\mathbf{y}$ = (feature 1, feature 2) = (magnitude at ($3f_{ro}$, f_{ro}), magnitude at ($2f_{ro}$, $2f_{ro}$)), a linear discriminant function takes the form $d_i(\mathbf{y}) = \mathbf{a}_i^t\mathbf{y} + a_0$ where $\mathbf{a}_i$ is a weight vector for the discriminant function and a_0 is the additional component of the weight vector that corresponds to the intersecting point between $d_i(\mathbf{y})$ and axis.

4.3 Classifier Training: The problem of classifier training is equivalent to finding the coefficient set of the linear discriminant function using a labeled training set. A labeled training set actually consists of a set of available pattern vectors measured from bearings with known conditions.

The approach for finding such a coefficient set with a given training set will be to define a criterion function J(**a**) that is minimized if **a** is a solution vector. This reduces the problem to one of minimizing a scalar function, a problem that can often be solved by a gradient descent procedure.

Consider now the problem of constructing a criterion function, J, for solving the linear inequality: $\mathbf{a}^t\mathbf{y}+a_0>0$ if $\mathbf{y}$ comes from a damaged bearing and $\mathbf{a}^t\mathbf{y}+a_0<0$ if $\mathbf{y}$ comes from a good bearing. Without losing any generality, one may define a new pattern vector $\mathbf{y}'=[\mathbf{y}, 1]$ and add a minus sign to all the pattern vectors computed from a good bearing. Thus the problem becomes finding a coefficient set for the inequality $\mathbf{a}^t\mathbf{y}'>0$ for all the training data available.

A common choice of the criterion function is the **perception criterion function** [18].

$$J_p(\mathbf{a})=\sum_{\mathbf{y}'\in y}(-\mathbf{a}^t\mathbf{y}') \tag{21}$$

where y is the set of samples misclassified by **a**. (If no samples are misclassified, J_p is defined to be zero.) Since $\mathbf{a}^t\mathbf{y}'\leq 0$ if $\mathbf{y}'$ is misclassified, $J_p(\mathbf{a})$ is never negative and is zero only if **a** is a solution vector.

Since the j th component of the gradient of J_p is $\partial J_p/\partial \mathbf{a}_j$, one can see from Eq. (21) that

$$\nabla J_p(\mathbf{a})=\sum_{\mathbf{y}'\in y}(-\mathbf{y}') \tag{22}$$

and hence the basic gradient descent algorithm becomes

$$\mathbf{a}_{k+1}=\mathbf{a}_k+\sum_{\mathbf{y}'\in y_k}(\mathbf{y}') \tag{23}$$

where y_k is the set of samples misclassified by $\mathbf{a}_k$. Thus, the descent procedure for finding a solution vector can be stated as follows: the next weight vector is obtained by adding some multiple of the sum of the misclassified samples to the present weight vector [19].

This gradient search algorithm was programmed in the Fortran programming language. When it is provided with all the training data shown in Fig. 4, the line in the figure is the linear discriminant function yielded by this algorithm.

5. RESULTS AND ANALYSIS

As explained in section 4.2, Figs. 4 and 5 summarize the results of bicoherence analysis of bearing vibrations for roller defects and outer race defects. In Fig. 4, pattern vectors of outer race defective bearings 4 and 5 are represented by diamonds and those of other bearings are represented by squares. In Fig. 5, bicoherence values of roller defective bearings 1 and 2 are represented by diamonds and those of other bearings are represented by squares.

Both figures show that the two clusters of defective bearings and nondefective ones are linear separable although Fig. 4 shows that one of the pattern vector of bearing 4 came close to the cluster of nondefective ones. This is because the defect on this bearing is very small. In both cases, the cluster of nondefective bearings actually includes bicoherence values of bearings with defects at a different location of the bearing. The fact that a defect at a location is not mistaken for a defect at another location indicates that bicoherence analysis provides features not only effective for defect detection but also for defect diagnosis.

In addition, the distribution of pattern vectors shows that the magnitudes of pattern vectors are correlated with the severity of damage. The more severe the damage is, the farther the pattern vector is from the origin. This is particularly true in the case of outer race defects.

6. CONCLUSION

The bearing condition monitoring technique employing the pattern recognition analysis and bicoherence analysis was developed. The effectiveness of the technique is supported by bicoherence values obtained from bearings with outer-race defects, roller defects and no defects. The advantages of the technique are summarized as follows:

1. It enables the early detection of bearing localized defects by using low-frequency vibration signal.
2. It enables the diagnosis of the location of bearing defects.
3. It can be easily implemented with microprocessor hardware.

Appendix A: Bearing characteristic defect frequencies [1]

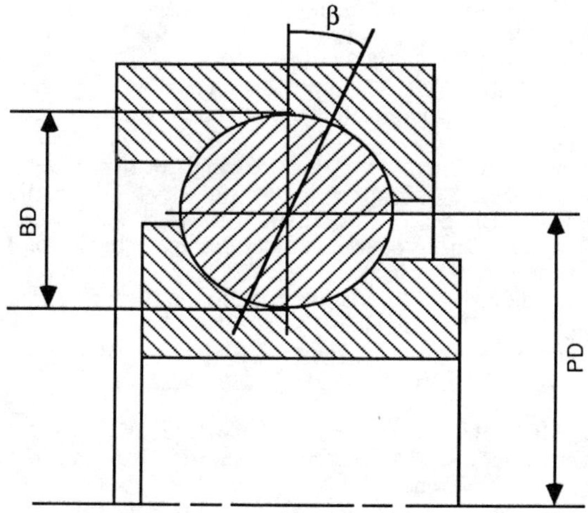

$$f_{or} = \frac{n}{2} f_r \left[1 - \frac{BD}{PD} \cos(\beta)\right] \qquad (A1)$$

$$f_{ir} = \frac{n}{2} f_r \left[1 + \frac{BD}{PD} \cos(\beta)\right] \qquad (A2)$$

$$f_{ro} = \frac{BD}{PD} f_r \left[1 + (\frac{PD}{BD})^2 \cos(\beta)\right] \qquad (A3)$$

where

f_{or} = characteristic defect frequency for defect on outer-race
f_{ir} = characteristic defect frequency for defect on inner-race
f_{ro} = characteristic defect frequency for defect on roller
BD = roller diameter
PD = bearing pitch diameter
f_r = bearing rotating speed
b = contact angle between race and ball
n = number of rollers

REFERENCES

[1] Widner, R. L. and Littmann, W. E., "Bearing Damage Analysis", National Bureau of Standard special publication No. 423, April 1976

[2] Downham, E., and Woods, R., "The Rationale of Monitoring Vibration on Rotating Machinery in Continuously Operating Process Plant", ASME paper No. 71-Vibr-96

[3] Weichbrodt, B., and Bowden, J., "Instrument for Predicting Bearing Damage", GE Co. Report, S-70-1021 AD 869633, March 1970

[4] Dyer, D. and Stewart, R. M.; "Detection of rolling element bearing damage by statistical vibration analysis", J. of Mechanical Design, Vol. 100; 1978, p-229

[5] Gustafsson O. G. and Tallian, T. "Detection of damage in assembled rolling bearings", Trans. ASLE,5,62 1962, p-197

[6] Braun, S. and Datner, B., "Analysis of roller/ball bearing vibrations", J. of Mechanical Design, Jan. 1979, Vol. 101; p-121

[7] Ehrich, F. F.; "Sum and difference frequencies in vibration of high speed rotating machinery", J. of Engineering for Industry, Vol. 94, No. 1, Feb. 1972, p-181

[8] Eshleman, R. L., "The use of sum and difference frequencies in rotating machinery analysis", Proc. of Machinery Vibration Seminar IV, Cherry Hill, N.J., Vibration Institute, Nov. 1980

[9] Mcfadden, P. D. and Smith, J. D., "Vibration monitoring of rolling element bearings by the high frequency resonance technique-a review", Tribology International ,Jan. 1984, p-3

[10] Broderick, J. J., Burchill, R. F. and Clark, H. L., "Design and fabrication of prototype system for early warning of impending bearing failure", NASA CR-123717, Jan. 1972

[11] Burchill, R. F., "Resonant structure techniques for bearing fault analysis", National Bureau of Standards NBSIR 73-252, 1973

[12] Darlow, M. S., Badgley, R. H. and Hogg, G. W., "Applications of high frequency resonance techniques for bearing diagnostics in helicopter gearboxes", USAAMRDL-TR-74-77, 1974

[13] Li, C. James and Wu, S. M., "On-line Bearing Localized Defect Detection by Pattern Recognition Analysis", ASME J. of Engineering for Industry , August 1989

[14] Li, C. James and Li, S. Y., "Acoustic Emission Analysis for Bearing Condition Monitoring", 44th general meeting of Mechanical Failures Prevention Group, Virginia Beach, Virginia, April 1990

[15] Subba Rao, T. and Gabr, M. M., An Introduction to Bispectral Analysis and Bilinear Time Series Models, Springer-Verlag Berlin Heidelberg, 1984

[16] Kim, Y. C. and Powers, E. T., "Digital Bispectral Analysis and Its Applications to Nonlinear Wave Interaction", IEEE Trans. on Plasma Science, Vol. PS-9, No. 2, June 1979, p-120

[17] Li, C. James, Ma, J., Hwang, B. and Nickerson, G. W., "Bispectral Analysis of Bearing Vibrations for Bearing Condition Monitoring," submitted for the 3rd International Machinery Monitoring and Diagnosis Conference, Las Vegas, NV, 1991

[18] Rosenblatt, F., "The Perceptron-A Perceiving and Recognizing Automation," Report 85-

460-1, Cornell Aeronautical Laboratory, Ithaca, N.Y., 1957
[19] Nilsson, N. J., "Learning Machines. Foundations of Trainable Pattern Classifying Systems," McGraw-Hill, New York, 1965

Tab.1 Bearing condition

Bearing	Defect Description	Location	Charact. frequency
#1	1 small hole (dia=0.050 in)	roller	$f_{ro} \approx 90$ Hz
#2	1 slot	roller	$f_{ro} \approx 90$ Hz
#3	no	n/a	n/a
#4	3 small holes (dia=0.050 in)	outer race	$f_{or} \approx 100$ Hz
#5	1 slot	outer race	$f_{or} \approx 100$ Hz

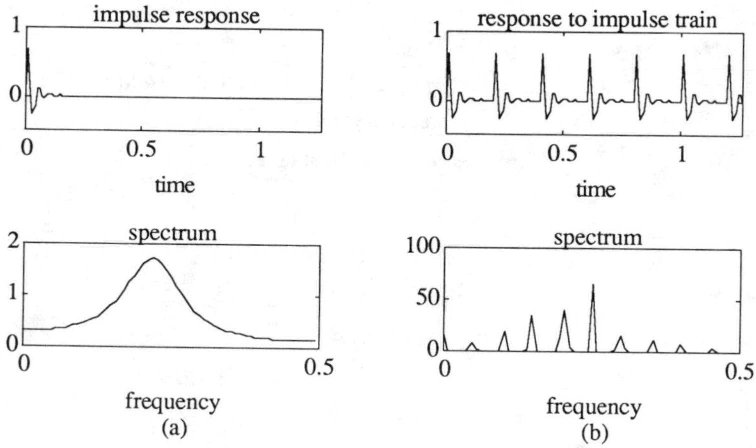

Fig. 1 Structure response to impulse input
(a) Structure impulse response and its spectrum
(b) Structure response to impulse train and its spectrum

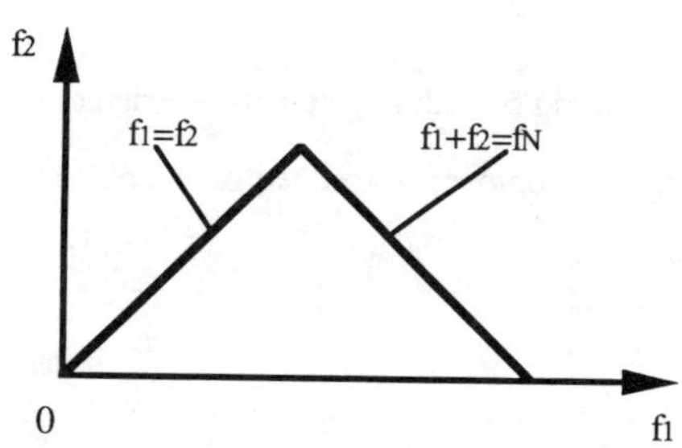

Fig. 2 Area for bispectrum computation

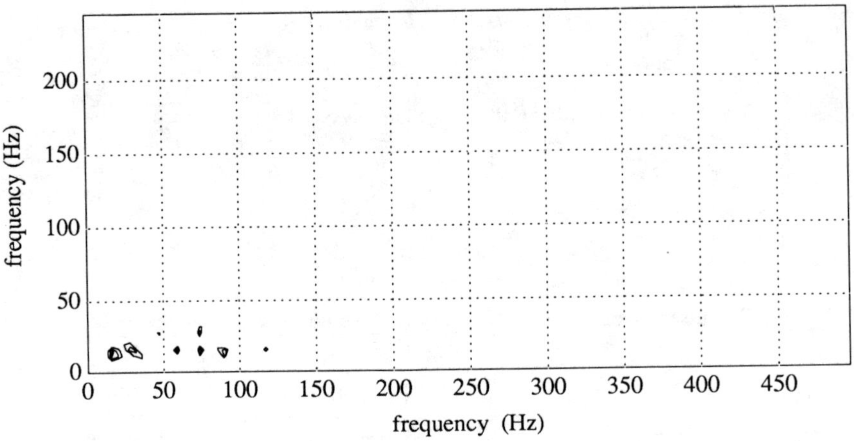

(a) bearing 3 (without defect)

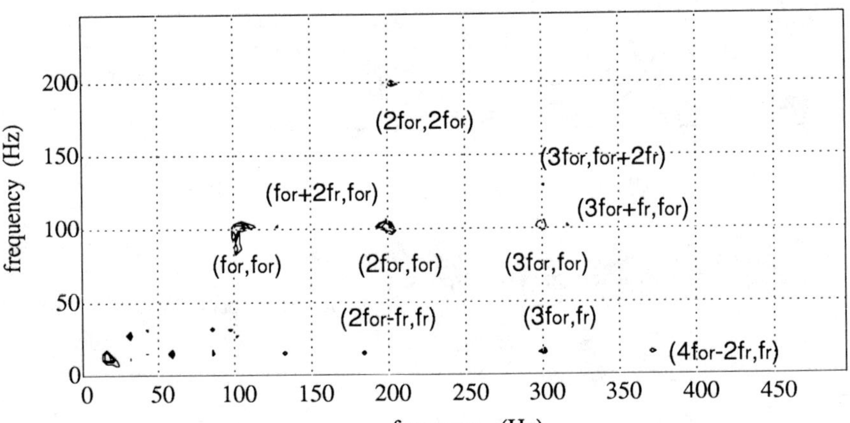

(b) bearing 5 (with large outer-race defect)

Fig. 3 Contours of squared bicoherence

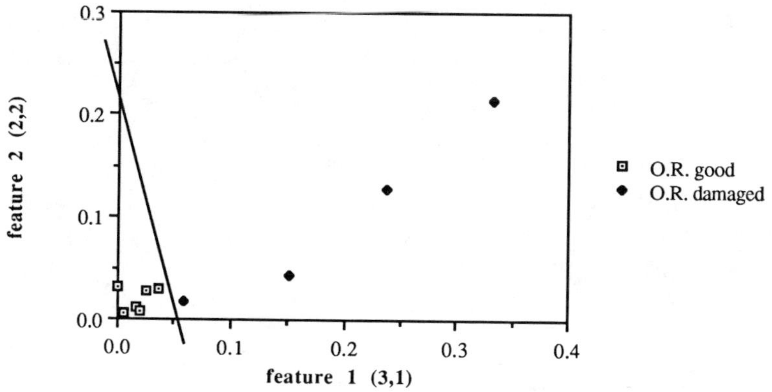

Fig. 4 Pattern vector distribution for outer-race defective bearings

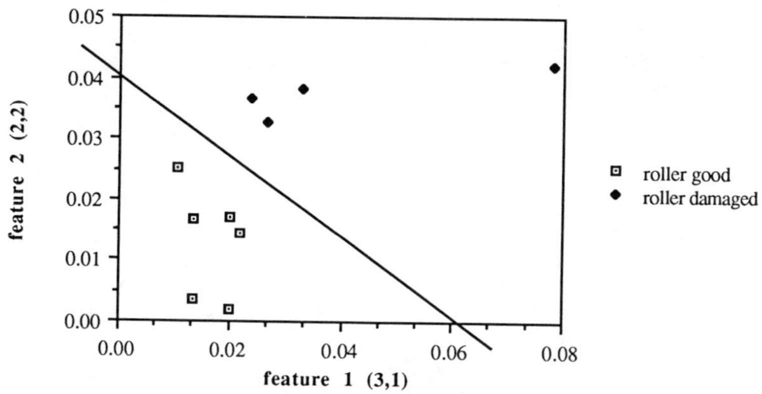

Fig. 5 Pattern vector distribution for roller defective bearings

Sensors, Controls, and Quality Issues
in Manufacturing
ASME 1991

DETECTION OF BALL BEARING CONDITIONS BY AN A. I. APPROACH

T. I. Liu
Department of Mechanical Engineering
California State University
Sacramento, California

J. M. Mengel
Mather Air Force Base
Sacramento, California

Abstract

A 3x12x1 artificial neural network has been used successfully for on-line monitoring of ball bearing conditions. Peak amplitude in the frequency domain, peak RMS, and the power spectrum of vibration signals are used as inputs of the neural network while the outputs indicate the bearing states.

Experimental data were taken to train the neural network in the learning process. Then the trained network was used for on-line detection of bearing states. The neural network can distinguish between the normal bearings and defective bearings with 100% success rate. It can also categorize the bearing conditions into six different states with a success rate of 97%.

1. Introduction

Ball bearings are widely used in the industry. In order to achieve preventive maintenance and automated manufacturing, on-line monitoring of ball bearings is essential.

Vibration analysis was used in this work for the following reasons :

1. It has a record of sufficient accuracy and reliability [1].
2. Different defects produce different vibration patterns, therefore, adequate actions can be taken for a specific bearing defect [2].

A 3x12x1 artificial neural network was used for the on-line monitoring of ball bearing conditions. The peak amplitude, RMS, and power spectrum of vibration signals were used as inputs to the artificial neural network and the outputs were the bearing states [2, 3]. In the first stage, whether the bearing is normal or defective was successfully determined. And in the second stage the neural network was used to classify the type of bearing defect effectively.

2. Artificial Neural Networks

Kohonen defines artificial neural networks as: "...massively parallel interconnected networks of simple adaptive elements. Their hierarchical organizations are intended to interact with the objects of the real world in the same way as biological nervous systems do" [4]. The underlying principle of neural networks is that numerous neurons process the data in a parallel and distributed fashion. The processing units, or neurons, are interconnected and the knowledge is represented by strengths, or weights, of connections between units. The knowledge is usually distributed over a large number of connections so that the operation of these networks may still continue even if part of the connections are destroyed due to malfunction. This parallelism results in high-speed information processing and fault tolerance [5-8].

Artificial neural networks have been successfully used in system learning, process optimization, and tool wear sensing [9-14]. However, so far they have not been applied to bearing condition monitoring. In this work, feedforward neural networks were used.

2-1. Feedforward Neural Networks

A typical feedforward neural network is shown in Figure 1. The feedforward neural networks consist of the input layer, the output layer, and a number of hidden layers in between. The input and the output layers are the two "visible" layers which are located at the bottom and the top of the network, respectively. The information contained in the input is recoded into an internal representation by the hidden layers which perform the mapping from the input to the output. Each node in the network can only send its output to the nodes in the higher layer and receive its input from the nodes in the lower layer [5].

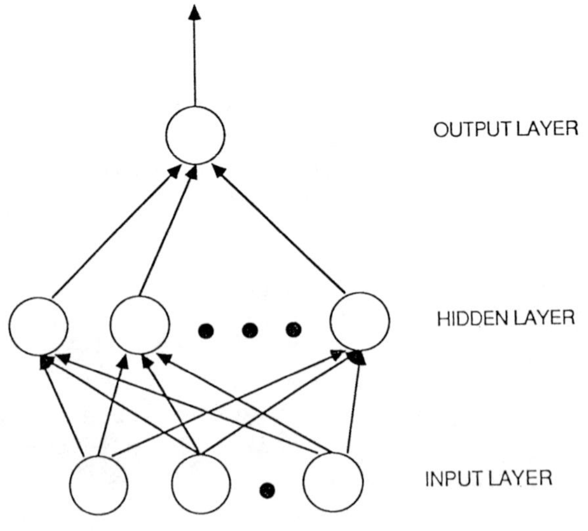

Figure 1 Feedforward Neural Network

At each node two operations are performed. The first is the summation of all the inputs. The following equation is used:

$$s_j = \Sigma w_{ij} x_i \qquad (1)$$

where j represents the layer currently being worked on; i designates the preceding layer; w_{ij} is the connection weight between the nodes; and s_j represents the total summation.

The second operation to occur is the generation of the node output by the application of a nonlinear activation function:

$$f_j = 1 / (1 + \exp\{-[s_j + c_j]\}) \qquad (2)$$

where c_j represents an internal offset value.

From Equation (2), one can see that all of the node output values range between zero and one. Therefore, all the input and output data needs to be normalized before being applied to the neural network. The input and output values are normalized according to the following equation:

$$p_n = [0.8/(p_{max} - p_{min})]\ [p - p_{min}] + 0.1 \qquad (3)$$

where p_{max} and p_{min} are the maximum and minimum values respectively of the data parameter p.

An initial output is obtained by a forward pass through the neural network. This output is quite different from the desired output since the initial values of the weights and offsets are randomly assigned. Therefore, the next step is to go through a learning process so as to adjust the weights and offsets.

2-2. Learning Process

The learning process of neural networks starts by assigning random values to the weights and thresholds. This prevents the neural network from being trapped at a local minimum at the beginning of the process. The input vector is fed to the neural network in order to calculate the output. Because the values of the weights and the thresholds are randomly assigned, the output will be quite different from the desired output.

The back propagation technique, known as the generalized delta rule, uses the difference between the expected output and the derived value to adjust the weights and the thresholds. The weights preceding each output node are updated according to the following equation:

$$w_{ij}(n+1) = w_{ij}(n) + \epsilon\ \delta_j\ x_i \qquad (4)$$

where ϵ is a correction gain factor; w(n+1) represents the updated weighting value; w(n) represents the current weighting value; and δ_j is the correction factor for nodes of the output layer, which is calculated by:

$$\delta_j = y_j\ (1-y_j)\ (d_j-y_j) \qquad (5)$$

wher y_j is the estimated output and d_j is the desired output.

The correction factor used to update weights for all other hidden layers are updated according to following equation:

$$\delta_j = x_j\ (1-x_j)\ \Sigma\ \delta_K\ w_{jK} \qquad (6)$$

where the k applies to the node layer succeeding the one currently being updated. The offset parameter, c_j, is treated as an additional weight factor and updated as follows :

$$c_j(n+1) = c_j(n) + \epsilon \delta_j \qquad (7)$$

This process is repeated until the calculated outputs have converged sufficiently close to the desired output or an iteration limit has been met. After the weights and the offsets have been adjusted for one training vector, additional training vectors can be fed to the neural network so as to further adjust all the weights and offsets. This learning process is finally terminated when the error becomes smaller than a predetermined limit value.

The feedforward neural networks can then be applied to operating conditions which are governed by the same underlying mechanism as the learning process. New input can be presented to neural networks and the output can be estimated and categorized. Since the knowledge is stored in the interconnections of neural networks, they are very suitable for on-line applications.

3. Experimententation

The experimental setup in this work consisted of a spindle driven by an electric motor. Two SKF 1635 DC ground steel ball bearing were placed in a housing with a Bruel and Kjaer 4384 piezoelectric accelerometer attached to pick up radial accelerations. The test setup is shown in Figure 2.

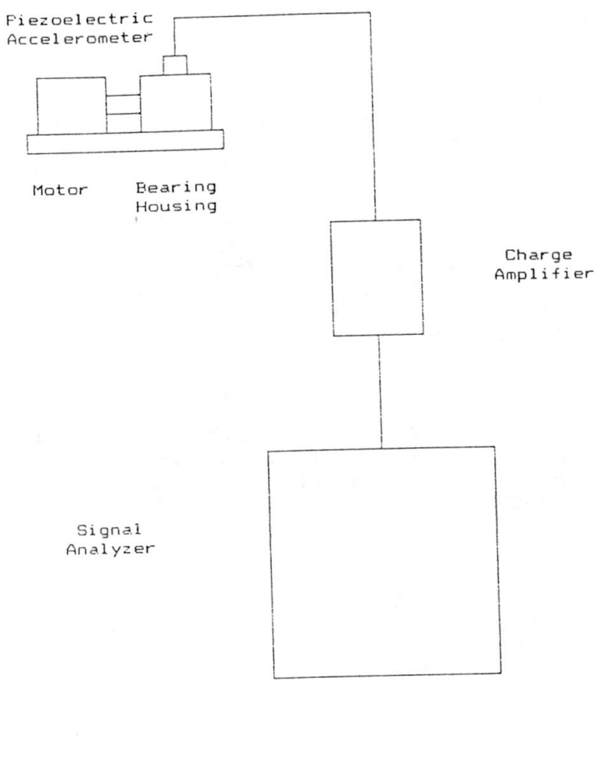

Figure 2 Ball Bearing Test Setup

The output signal from the accelerometer was run through a Bruel and Kjaer 2635 charge amplifier. The 2635 charge amplifier is a four stage device consisting of an input amplifier, low pass filter amplifier, integrator amplifier, and an output amplifier. The amplifier was adjusted for proper filter frequencies (the high pass filter was set to 3 KHz and the low pass filter was set to 2 Hz).

The signal was then processed by a Bruel and Kjaer 2034 Dual Channel Signal Analyzer. The 2034 is a fully self-contained, two-channel, FFT analyzer with 801 lines of resolution. Readings were taken directly from the digital display.

The bearing used for this research were SKF 1635DC ground steel ball bearings with a 44.450 mm (1.750 inch) outer diameter, a 19.050 mm (0.750 inch) inner diameter, a 31.369 mm (1.235 inch) pitch diameter, and eight 19.050 mm (0.750 inch) diameter balls.

Donald Bently pointed out that approximately 90% of all bearing failures are related to either an inner or an outer race flaw. The other 10% are related to either a rolling element flaw or a cage flaw [15,16]. Based on this fact, the following six bearing states were chosen for observation:

1. Normal ball bearing
2. Bearing with one ball defect
3. Bearing with one outer race defect
4. Bearing with one inner race defect
5. Bearing with one ball and one outer race defect
6. Bearing with two inner race defects (180 degrees separation)

The size of the defects were 0.152 mm (60/1000 inch) diameter by 0.127 mm (5/1000 inch) depth, which is typical [17]. The defects were made by Electrical Discharge Machining (E.D.M.).

4. Feature Selection

Several indices were chosen for bearing condition monitoring. The indices chosen were the peak amplitude in the defect region of the frequency domain, percent power in the defect region as compared to the total power in the observed range, and the peak RMS in the defect region.

Peak Amplitude of the Frequency Domain

The first observation taken was the peak amplitude in the defect region of the frequency domain. For the setup in this study, the defect region covers the frequency range from 92 Hz (the frequency of an outer race defect) to 138.3 Hz (the frequency of an inner race defect). The maximum amplitude of the frequency domain in this region was recorded in decibels for each measurement.

Frequency domain information is one of the most common indices used in defect analysis since this information can completely define the vibration. The basic concept behind studying the frequency domain information is that different types of defects produce different characteristic defect frequencies at which impulses are generated. These characteristic defect frequencies are easily calculated as shown in the Appendix. When one sees a series of vibration bursts occurring at a rate which is close to the theoretical estimation of the characteristic defect frequency, one can conclude that there are localized defect [18].

Percent Power

The second observation recorded was percent power of the observation in the defect region as compared to the total power in the observed range. Observation of the power spectrum is another very common method of diagnosing bearing damage [2]. The energy represented at different frequencies is dependent upon the type of defect present [18].

Peak RMS

The final observation recorded was the peak RMS in the defect region. The RMS represents the power content of the vibration [19]. RMS is a very commonly used index for vibration analysis since increasing bearing damage results in increasing RMS [20].

The above indices were chosen because of their documented use in previous work [2,3], their ease of measurement in the observed region, and their ability to be rapidly analyzed by the Bruel and Kjaer 2034 Dual Channel Signal Analyzer.

In this work, a total of 30 readings, five for each bearing states, were taken in the learning process to train the artificial neural network and another 30 readings were taken as on-line tests.

5. Monitoring Bearing Conditions via Neural Networks

In this work, a 3x12x1 neural network was used for bearing condition monitoring. The first number represents the number of inputs (peak amplitude, percent power, and peak RMS), the last number represents bearing states, and the number between them represents the number of nodes in the hidden layer. Both inputs and outputs were assigned normalized values according to Equation (3).

The first step for bearing condition monitoring was to determine whether the bearing was normal or defective. In the learning process, normalized output value of 0.7 was chosen for a defective bearing and 0.3 for a normal bearing. The learning rate was set at 0.1 for this test. The error limit was set to be 0.01.

After the 3x12x1 neural network has been trained, it was used for on-line recognition of the bearing states. Based upon the output values of the neural network, bearings were classified to be in two categories according to the following rules:

Normal bearing:	(0.100 - 0.500)
Defective bearing:	(0.501 - 0.900)

The neural network could distinguish the defective state from the normal state with a success rate of 100%. This is a very favorable success rate when compared to some of the other classification methods currently available. Table 1 summarizes the comparison of these methods [18].

In the second step, bearings were further classified to be one of the following six categories during the learning process:

One ball defect:	0.17
One outer race defect:	0.30
One inner race defect:	0.44
One ball defect and one outer race defect:	0.57
Two inner race defects:	0.70
Normal bearing:	0.83

Once again, the learning rate was set to 0.1 for all networks and the data was trained until the error limit of 0.06 or less was reached.

In the on-line tests, ball bearings were classified one of the six categories according to the following rules:

One ball defect:	.100 - .233
One outer race defect:	.234 - .367
One inner race defect:	.368 - .500
One ball defect and one outer race defect:	.501 - .633
Two inner race defects:	.634 - .767
Normal bearing:	.768 - .900

Table 1 Comparison of Different Techniques for Bearing Condition Monitoring

Monitoring Technique	Success Rate
Kurtosis	62%
Variance	67%
Crest Factor	70%
Peak to Peak	70%
HFRT	77%
Pattern Recognition	91%
Neural Networks	100%

A success rate of 97% was achieved for monitoring bearing states using the 3x12x1 neural network.

6. Conclusions

This research strongly indicates that the artificial neural network is a viable candidate to monitor ball bearing defects. The 3x12x1 neural network used in this work can distinguish between the normal bearing condition and the defective bearing condition with a success rate of 100%. Furthermore, it can categorize the ball bearing conditions into six different states with a success rate of 97%.

ACKNOWLEDGEMENT

This work is partially sponsored by the U.S. Air Force and the Hornet Foundation.

References :

[1] O.G. Gustafsson, T. Tallron, "Detection of Damage in Assembled Rolling Element Bearing," ASLE 5, pp. 197-209, 1962

[2] D. Dyer, R.M. Stewart, "Dectection of Rolling Element Bearing Damage by Statistical Vibration Analysis," Journal of Mechanical Design, Vol. 100, pp. 229-235, April 1978

[3] S. Braun, B. Datner, "Analysis of Roller/Ball Bearing Vibritions," Journal of Mechanical Design, Vol. 101, pp. 118-125, Jan., 1979

[4] T. Kohonen, "An Introduction to Neural Computing," Neural Networks, Vol.1, pp.3-16, 1986

[5] D. Rumelhart, J. McClelland, "Parallel Distributed Processing," Volume 1, MIT Press, Cambridge, MA, 1986

[6] S.Y. Kung, J.N. Hwang, "Neural Network Architectures for Robotic Applications," IEEE Transactions on Robotics and Automation, Vol. 5, No. 5, pp641-656, 1989

[7] R. Lippman, "An Introduction to Computing with Neural Nets," IEEE ASSP Magazine, pp. 4-22, 1987

[8] J.J. Hopfield, D.W. Tank, "Computing with Neural Circuits : A Model," Science, Vol.233, pp.625-633, 1986

[9] T.I. Liu, E.J. Ko, S.L. Sha, "Intelligent Monitoring of Tapping Tools,"ASM Journal of Materials Shaping Technology, Vol. 8, No. 4, 1990

[10] T.I. Liu, E.J. Ko, S.L. Sha, "Dignosis of Tapping Operations Using an A.I. Approach," ASM Journal of Materials Shaping Technology, Vol. 9, No. 1, 1991

[11] T.I. Liu, E.J. Ko, "Learning and Control of Dynamic Systems Using A.I. Techniques," Proceedings of 4th USPS Advanced Technology Conference, Washington, D.C., Nov.5-7, pp.1301-1316, 1990

[12] G. Chryssolouris, M. Guillot, "A Comparison of Statistical and A.I. Approach to the Selection of Process Parameters in Intelligent Machining," ASME Journal of Engineering for Industry, Vol. 112, pp.122-128, 1990

[13] S. Rangwala, D. Dornfeld, "Integration of Sensors via Neural Networks for Detection of Tool Wear States," Proceedings of ASME Winter Annual Meeting Symposium on Intelligent and Integrated Manufacturing Analysis and Synthesis, Boston MA, pp.109-120, 1987

[14] S. Rangwala, D. Dornfeld, "Learning and Optimization of Machining Operations Using Computing Abilities of Neural Networks," IEEE Transactions on Systems, Man, Cybernetics, Vol. 19, No.2, 1989

[15] D. Bently, "Predictive Maintenance Through the Monitoring and Diagnostics of Rolling Element Bearings," Bently Nevada Co., Applications Note, AN044, pp. 2-8, Jan.1989

[16] D. Bently, "Maintenance Practices in the U.S.A.," Bently Rotor Dynamics Research Corporation, Jan., 1990

[17] R.L. Widner, W.E. Littman, "Bearing Damage Analysis," National Bureau of Standards Special Publication, No. 423, 1976

[18] C.J. Li, S.M. Wu, "On-Line Detection of Localized Defects in Bearings by Pattern Recognition Analysis," ASME Journal of Engineering for Industry, Vol.111, pp.331-336, 1989

[19] W.T. Thompson, "Theory of Vibrations with Applications," Prentice Hall, Englewood Cliffs, New Jersey, 1988

[20] J.T. Broch, "Mechanical Vibrations," Bruel and Kjaer, 1984

[21] P.G. Wheeler, "Bearing Analysis Keeps Downtime Down," Plant Engineering, No. 25, pp. 87-89, 1968

APPENDIX

The formulas of bearing characteristic frequencies are listed as follows :

Defect on outer race (Ball pass frequency outer)

$$= \frac{(n)}{2}\,\frac{(RPM)}{60}\left(1-\frac{Bd}{Pd}\cos\phi\right) \tag{A1}$$

Defect on inner race (Ball pass frequency inner)

$$= \frac{(n)}{2}\,\frac{(RPM)}{60}\left(1+\frac{Bd}{Pd}\cos\phi\right) \tag{A2}$$

Ball Defect (Ball spin frequency)

$$= \frac{(Pd)}{2Bd}\,\frac{(RPM)}{60}\left[1-\left(\frac{Bd}{Pd}\right)^2\cos^2\phi\right] \tag{A3}$$

Fundamental train frequency

$$= \frac{1}{2}\,\frac{RPM}{60}\left(1-\frac{Bd}{Pd}\cos\phi\right) \tag{A4}$$

where

Pd = Pitch diameter
Bd = Ball diameter
n = Number of balls
ϕ = Contact angle

Sensors, Controls, and Quality Issues in Manufacturing
ASME 1991

ACOUSTIC EMISSION FROM PLASTIC DEFORMATION OF A CRYSTALLINE ALLOY

Jin-Bok Chung and Elijah Kannatey-Asibu
Department of Mechanical Engineering
University of Michigan
Ann Arbor, Michigan

ABSTRACT

Acoustic emission during plastic deformation is analyzed by considering dislocation interaction with obstacles such as impurities and grain boundaries. Dislocation motion under solute drag is modeled in the same way as that of motion in a pure crystal. As the solute density increases, the dislocation velocity decreases due to solute drag, which reduces the AE output. However, dislocation motion during unpinning from solute atoms and grain boundaries is modeled as a step velocity function. The intense AE from the unpinning process is due to an increase in velocity after breakaway as a result of a change in drag mechanism from solute to phonon drag. The stress state significantly affects AE from impure crystals. An increase in compressive hydrostatic stress decreases the AE energy by reducing the unpinning velocity and event rate. The presence of a grain boundary reduces dislocation velocity. Even though the number of AE sources may increase as the average grain size decreases, the corresponding acoustic emission is diminished because the velocity effect then dominates the radiation wave energy.

1. INTRODUCTION

Acoustic emission during plastic deformation is attributed to acceleration of dislocations during harmonic and initial motion. The presence of obstacles such as impurities and grain boundaries significantly affect the AE generation by changing the mechanism of dislocation motion.

It has been shown experimentally that AE is reduced by adding alloying elements [2, 3], and this is attributed to the dragging of dislocations by impurities. However, the acoustic emission activity is increased by the impurities when dislocation motion is not governed by solute dragging but by the unpinning from solute dragging which results in the serrated or Portevin-Le Chatelier strain. Acoustic emission generation during serrations, stretched yield or the Portevin-Le Chatelier deformation is relatively large due to the abrupt change in velocity compared to that during strain-hardening deformation, as shown in Figure 1. In this case, an increase in impurities raises the AE energy at yield [4], as a result of the repeated locking and unlocking of dislocations. The motion of dislocations during breakaway from solute dragging or a pinning condition is modeled as a step function of velocity because of the abrupt motion.

The stress state significantly affects acoustic emission generation for materials having inclusions. The difference in AE resulting from a change in the stress state has been attributed to inclusion fracture [5]. However, apart from the inclusion fracture, another AE source mechanism related to the compressive hydrostatic stress is analyzed theoretically in this study. It has been shown experimentally that the compressive stress generates new dislocations at the interface between the matrix and an inclusion due to the difference in bulk moduli between them [6-8]. Thus, under compressive hydrostatic stress, AE activity may be reduced at yield due to an increase in initial dislocation density which results in a reduction of dislocation velocity at a given strain rate.

For a solid solution, the effect of hydrostatic stress is also significant as observed by experiments [9] and different from that of an inclusion material. Most of the AE from alloys including solid solutions is attributed to the repeated locking and unlocking of dislocations from solute pinning, which does not appear in pure materials and becomes less noticeable after precipitation [10]. Effect of the hydrostatic stress on AE in a solid solution can thus be obtained by considering the interaction of a moving dislocation with solute atoms under the hydrostatic stress.

Regarding the effect of grain size on AE, it has been observed experimentally by Kiesewetter and Schiller [11] and Scruby et al. [12] that acoustic emission from the deformation of pure material increases with increasing grain size. However, as alloying elements are added to the material, acoustic emission increases with decreasing grain size [13, 14], due to interaction between the grain boundaries and alloying elements. Thus, AE generation cannot be attributed only to the grain boundary effect, since a change in grain size alters the flow stress, which results in an increase in impurity activation with dislocation motion.

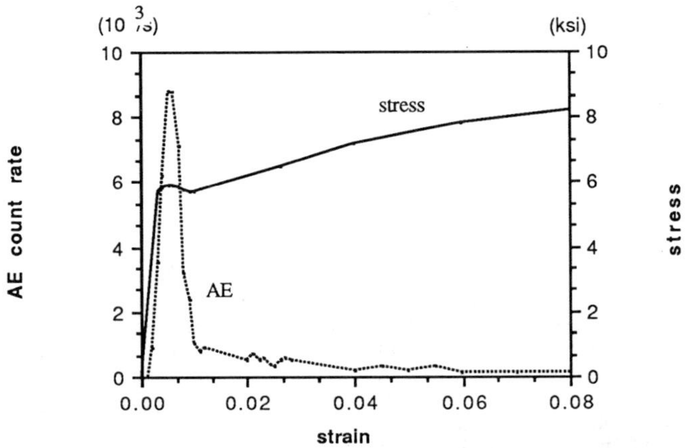

Figure 1 : AE from stretched yield strain of mild steel [1]

2. IMPURITY EFFECTS ON AE DURING PLASTIC DEFORMATION

Impurities, including solid solutions and inclusions in crystalline materials act as barriers to dislocation motion. Since an impurity contains its own strain energy due to the elastic inhomogeneity between it and the matrix, the radiation energy resulting from the release of self strain energy needs to be superimposed on that from dislocation motion. However, in this analysis, only the radiation wave energy from dislocation dynamics will be considered as a source of acoustic emission during plastic deformation of an impure material.

Dislocations can be pinned in a variety of ways, with their motion being interrupted by either a dragging or locking mechanism which affects the flow stress and the mobility of the dislocations. In addition to their effect on flow stress, the obstacles exert a notable influence on the dynamics of dislocation motion. Dislocations moving in a dilute concentration of solute atoms are dragged by solute pinning when the diffusion velocity of the solute is higher than the dislocation velocity. Impurity interactions with dislocations are described as thermal activation, where its microscopic motion is thermally activated over the obstacles. Thus, the velocity can be expressed by the Gibbs free energy of activation. Considering a bowed-out dislocation restrained by pinning points, the average dislocation velocity, υ_s, due to solute drag can be written [15] as

$$\upsilon_s \;\alpha\; \frac{\sigma d_s}{kT} \exp\left(-\frac{w_b}{kT}\right) \tag{1}$$

where d_s is the average spacing of solutes along the dislocation core, w_b is the free energy for a solute diffusion to the dislocation core, T is the absolute temperature, σ is the stress applied to the dislocation, and k is the Boltzman constant. d_s is related to the solute density per unit length of dislocation line, ρ_s, by

$$\rho_s = d_s^{-1} \tag{2}$$

Substituting equation (2) into (1) for a given set of conditions gives

$$\upsilon_s \ \alpha \ \frac{1}{\rho_s} \exp(-w_b) \tag{3}$$

An increase in solute density, ρ_s, or its free energy, w_b, thus reduces dislocation velocity, and increases mobile dislocation density to satisfy Orowan's equation for a constant strain rate. The radiation of wave energy from either harmonic or transient motion of dislocations is proportional to the square of dislocation velocity [16]. In addition, the number of AE sources is inversely proportional to the dislocation velocity when the strain rate is constant. Hence, the acoustic emission energy, E_{AE}, from plastic deformation of a solid solution crystal is related to the solute density ρ_s as follows:

$$E_{AE} \ \alpha \ \frac{1}{\rho_s} \tag{4}$$

When a dislocation breaks away from solute drag, its velocity is then determined by phonon drag instead of solute drag as shown in Figure 2, which results in a sudden increase in velocity. The plastic flow induced by the breakaway from a pinning atmosphere is characterized by Lüder's strain or serrated flow. Thus, the relation of AE to solute density needs to be explained through the rate of unpinning events.

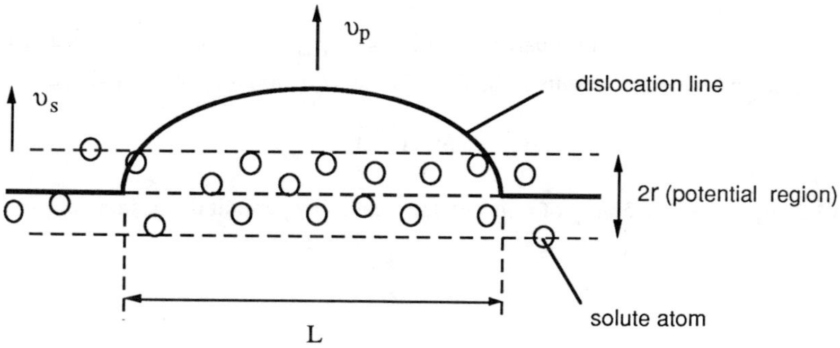

Figure 2 : Unpinning of a dislocation line from solute atoms

For a given external stress σ, a local fraction of the dislocation line is bowed out of the potential region produced by the solute atoms. The stress necessary for gradual bow out from that region is

$$\sigma = \frac{1}{b}\frac{\partial E_f}{\partial r}$$

where b is Burgers vector, r is distance from the dislocation center, and E_f is the free energy of the solute atoms per unit dislocation length and is given by

$$E_f = \rho_s w_b$$
$$= \pi r^2 c_s w_b \tag{5}$$

where r is the radius of the potential region, and c_s is the saturation density of solute atoms at the dislocation core which is constant for a given material and temperature, and ρ_s is the total number of solute atoms at a dislocation core per unit length of dislocation. When the free energy is spread over an arbitrary potential range of r, the stress required for bow-out from the pinned condition is then

$$\sigma = \frac{2\pi r c_s w_b}{b} \tag{6a}$$

Taking the potential region r as $r_o (\cong b)$,

$$\sigma = \frac{2\pi r_o c_s w_b}{b} \tag{6b}$$

According to Fisher's analysis [15], the dislocation can break away and unzip continuously from the core dragging by solutes at a critical instance when the activation energy exceeds the free energy per unit increment of bow-out length. The breakaway occurs when the bowed out length, L, reaches a critical length, L_c, which is given by

$$L_c = \frac{c}{\sigma} (\mu E_c)^{0.5}$$

where $c \cong 0.9$, and μ is the shear modulus. The energy, E_c, required for activation of breakaway at a critical length, L_c, per unit dislocation is

$$E_c = E_b - E_a$$

$$= \frac{0.6 \mu^{0.5} E_f^{1.5}}{\sigma} \tag{7}$$

where $E_b = E_f$ is the free energy of a barrier, and $E_a = \frac{5 \sigma^2 L_c^2}{12\mu}$ is activation energy supplied by external work, given by Fisher's results [15]. Substituting equation (6b) into (7) gives

$$E_c = c' r_o E_f^{0.5} \tag{8}$$

where $c' = 1.2b\sqrt{\mu}$. The probability of a successful breakaway per unit dislocation, p_u, is

$$p_u \, \alpha \, \exp\left(-\frac{E_c}{kT}\right) \tag{9}$$

and the density of unpinning dislocations, ρ_u, is

$$\rho_u = \rho p_u \tag{10}$$

where ρ is the overall dislocation density. Substituting equations (8) and (9) into (10) gives

$$\rho_u \, \alpha \, \rho \exp\left(- \frac{c' r_o E_f^{0.5}}{kT} \right) \tag{11}$$

The macroscopic strain rate is simply related to the dislocation velocity as

$$\dot{\varepsilon} = \rho_m b \upsilon \tag{12}$$

where υ is the average velocity of moving dislocations, and ρ_m is the mobile dislocation density. We assume that when dislocations are pinned by solute atoms, plastic deformation can develop only by dislocation breakaway. ρ_m is then characterized by ρ_u, and the velocity is then governed by phonon drag. Thus, equation (12) can be used to express serrated strain deformation by

$$\dot{\varepsilon} = \rho_u b \upsilon_p \tag{13}$$

Substituting equation (11) into (13), the average velocity of unpinning dislocations, υ_p, is obtained as

$$\upsilon_p \;\alpha\; \frac{1}{\rho}\exp\left(\frac{c' r_o E_f^{\,0.5}}{kT}\right) \tag{14}$$

AE energy during serrated strain at a constant strain rate can thus be represented with the density of unpinning dislocations and their average velocity as follows:

$$E_{AE} \;\alpha\; \frac{1}{\rho}\exp\left(\frac{c' r_o E_f^{\,0.5}}{kT}\right) \tag{15}$$

3. HYDROSTATIC EFFECT ON ACOUSTIC EMISSION

3.1 Solid solutions

Let us consider a given crystalline alloy having solutes distributed along the dislocation. A solute atom inserted in the matrix interacts with the hydrostatic stress σ_m due to the dislocation self stress as a result of the volume difference $\delta V = | V_s - V_m |$, where V_s is the volume of a solute atom and V_m is that of a matrix atom. Considering the reversible work $\sigma_m \delta V$ as a change of the free energy, the solute concentration c is given [15] by

$$c = c_o \exp\left(\frac{\sigma_m \delta V}{kT}\right) \tag{16}$$

where c_o is the solute concentration at a reference state with only the hydrostatic stress due to the strain field of an edge dislocation being present without an external hydrostatic stress, and the concentration of solid solutions are distributed by σ_m.

Now, consider the effect of an external hydrostatic stress σ_{me} on the solute concentration. If there is an external stress on an isotropic continuous elastic material, the hydrostatic stress due to the external stress is uniform. Hence, a change of interaction energy is the same for all solute atoms regardless of their location. However, the presence of an edge dislocation produces an elastic discontinuity, providing a void Δabc as shown in Figure 3. Using the core cut-off of a dislocation center modeled by Volterra for simplicity, the void can be regarded approximately as being of the same order as the Burgers vector located in the half region $\pi < \theta < 2\pi$ near the dislocation core.

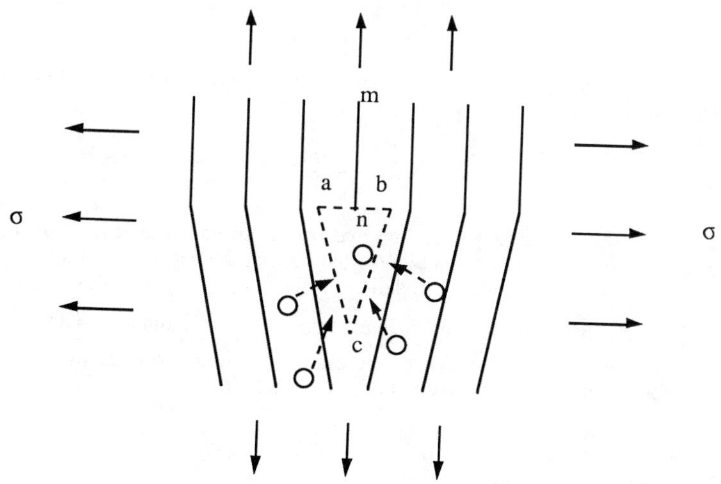

Figure 3 : Edge dislocation and solute atoms under hydrostatic stress

Consider a round void of radius r_o equivalent to the void region Δabc inside an elastic continuum material. The stress field under plain strain conditions with an external stress σ_{me} is written [16] as

$$\sigma_r = -\frac{R^2\sigma_{me}}{R^2-r_o^2}\left(\frac{r_o^2}{r_c^2} - 1\right)$$

$$\sigma_\theta = \frac{R^2\sigma_{me}}{R^2-r_o^2}\left(\frac{r_o^2}{r_c^2} + 1\right)$$

where R is the distance from the void to the external boundary, and r_c is distance from the dislocation center. For an external hydrostatic stress σ_{me}, the stress in the z-direction should be added to that of the r and θ directions.

$$\sigma_z = \sigma_{me}$$

Since r_o is of the order of atomic dimensions, $R \gg r_o$. Thus, the stresses can be written approximately as

$$\sigma_r = -\sigma_{me}\left(\frac{r_o^2}{r_c^2} - 1\right)$$

$$\sigma_\theta = \sigma_{me}\left(\frac{r_o^2}{r_c^2} + 1\right)$$

$$\sigma_z = \sigma_{me}$$

The radial displacement u is

$$u = r_c\varepsilon_\theta$$

and

$$\varepsilon_\theta = \frac{1}{E}\left(\sigma_\theta - \nu(\sigma_r + \sigma_z)\right)$$

$$= \frac{\sigma_{me}}{E}\left((1+\nu)\frac{r_o^2}{r_c^2} + 1 - 2\nu\right)$$

Substituting r_c with r_o, the radius of the void area around the core region, r, under the hydrostatic stress is thus obtained as

$$r = r_o\left(1 + \frac{(2-\nu)\sigma_{me}}{E}\right) \tag{17}$$

where $r_o \doteq b$ is the radius under zero external hydrostatic stress. The voids around a core provide the space for the solute atoms which can diffuse by thermal activation or stress applied during pre-yield strain. The core radius r_o and solute density ρ_s in Figure 2 vary with σ_{me}. Considering that $\frac{(2-\nu)\sigma_{me}}{E} \ll 1$ at yield, then, substituting r_o in equations (5), (14), and (15) with r_1 in equation (17) gives the velocity of a dislocation and acoustic emission energy from breakaway of dislocations during plastic deformation in terms of hydrostatic stress as follows:

$$\upsilon' \propto \upsilon\left(1 + \frac{\alpha_s(2-\nu)\sigma_{me}}{E}\right) \tag{18}$$

and

$$E_{AE}' \propto E_{AE}\left(1 + \frac{\alpha_s(2-\nu)\sigma_{me}}{E}\right) \tag{19}$$

where $\alpha_s = \frac{2b(\pi c_s w_b)^{0.5}}{kT}$. υ and E_{AE} are the velocity and acoustic emission energy at zero external hydrostatic stress, and υ' and E_{AE}' are the velocity and acoustic emission energy under the external hydrostatic stress. A tensile hydrostatic stress provides the short-range diffusion of solute atoms to the dislocation core as shown in Figure 3. But, the atomic layer m-n may move down to fill the void rather than the solute diffusing to the void when the tensile hydrostatic stress exceeds a critical value. A compressive hydrostatic stress, on the other hand, induces the diffusion of solutes out of the core region, which results in a change in the density of solute atoms around a pinned dislocation. Above a critical compressive stress, the atomic layer m-n may be pushed upward.

3.2 Inclusions

The effect of hydrostatic stress on AE generation is only obtained for a crystalline material containing solid solution atoms and interstitials in equation (19). However, for second phase particles such as inclusions, there are two different materials in contact, and the diffusion of inclusions into the dislocation core cannot be expected because of their size. Hence, the effect of hydrostatic stress on a material with inclusions is explained differently from that on a solid solution alloy. The local stress and strain at the inclusion-matrix interface due to a hydrostatic stress, σ_m, have been obtained by Eshelby [17] and Ashby et al. [7], and it has been shown experimentally that dislocations are generated at the matrix-precipitate interfaces under a critical hydrostatic stress [7, 18, 19]. Thus, volume misfit under hydrostatic pressure due to the bulk moduli difference between matrix and particle can be related to local plastic deformation which results in a generation of dislocations. The volume change due to a compressive hydrostatic stress σ_{mc} is written [7] as

$$\Delta V = \Delta(V_p - V_m)$$

$$= -\frac{\sigma_{mc}}{3}\left(\frac{1}{K_p} - \frac{1}{K_m}\right)V_p$$

where K is the bulk modulus, and subscript m is for the matrix material and p for particles. When the hydrostatic stress exceeds some critical stress, the volume change is filled by local plastic deformation. Thus, ΔV determines the amount of local plastic deformation and dislocation generation. Assuming that an increase in dislocation length, ΔL_d, is proportional to the amount of deformed volume,

$$\Delta L_d \;\alpha\; \left|\frac{\sigma_{mc}}{3}\left(\frac{1}{K_p} - \frac{1}{K_m}\right)V_p\right|$$

and, for a unit matrix volume,

$$\Delta\rho \;\alpha\; \left|\frac{\sigma_{mc}}{3}\left(\frac{1}{K_p} - \frac{1}{K_m}\right)\delta_p\right|$$

where L_d is the length of dislocation, $\Delta\rho$ is the increase in dislocation density, and δ_p is the volume fraction of the particle. Thus, the initial dislocation density, ρ_i, at initiation of yield is changed by the hydrostatic stress and becomes

$$\rho_i = \rho_{io} + \Delta\rho$$

$$= \rho_{io} + \frac{\sigma_{mc}}{3}\left(\frac{1}{K_p} - \frac{1}{K_m}\right)\delta_p$$

The AE energy from dislocation motion which is composed of initial and harmonic motion is proportional to the square of dislocation velocity [20]. Considering equation (12) at a given strain rate, there is an inverse relationship between the mobile dislocation density and velocity, and the AE energy is then related to initial dislocation density as follows;

$$E_{AE}' \;\alpha\; \rho_i^{-1}$$

or

$$E_{AE}' \;\alpha\; \left(\rho_{io} + \frac{\sigma_{mc}}{3}\left(\frac{1}{K_p} - \frac{1}{K_m}\right)\delta_p\right)^{-1} \tag{20}$$

Using a series approximation since $\Delta\rho$ due to the hydrostatic stress is relatively small compared to ρ_{io}, equation (20) becomes

$$E_{AE}' \; \alpha \; \left(\rho_{io} - c'\sigma_{mc}\right) \tag{21}$$

where $c' = \left| \frac{1}{3}\left(\frac{1}{K_p} - \frac{1}{K_m}\right)\delta_p \right|$.

Thus, the acoustic emission energy from plastic deformation of either a solid solution or a material with inclusions has a linear relationship with the compressive hydrostatic stress, viz..

$$E_{AE}' = E_{AE}(1 - C\,\sigma_{mc}) \tag{22}$$

where E_{AE} is given at zero hydrostatic stress, and C is a constant determined from equations (19) and (21). Consequently, the presence of a compressive hydrostatic stress σ_{mc} increases the initial dislocation density at the interface between inclusions and the material and reduces the pinning strength, which also results in a decrease in the dislocation velocity and AE energy during plastic deformation. It has been shown experimentally that stretched yield strain or a region of Lüder's strain of impure materials are reduced by compressive hydrostatic stress [6].

4. GRAIN BOUNDARY EFFECT ON AE

Changing the grain size may change the dislocation motion and thus the acoustic emission generation. For plastic deformation at a constant strain rate, the dislocation velocity can be obtained from the plastic compatibility requirement that $\rho_m \upsilon$ = constant. The finer grain size provides more sources for dislocation motion. Since the dislocation glide distance is confined by a grain diameter, the mobile dislocation density due to a grain boundary source, ρ_{mg}, can be expressed as

$$\rho_{mg} = \frac{a}{d_g}$$

where a is a constant. And considering the mobile dislocation density in a single crystal as ρ_o, the total mobile dislocation density becomes

$$\rho_m = \rho_{mg} + \rho_o$$

or

$$\rho_m = \frac{a}{d_g} + \rho_o \tag{23}$$

where d_g is the average grain diameter. And the dislocation velocity,

$$\upsilon \; \alpha \; \frac{1}{\frac{a}{d_g} + \rho_o} \tag{24}$$

However, the velocity function should be obtained to analyze the radiation energy. The interaction force in the x direction, F_x, between a dislocation and a grain boundary composed of edge dislocations is given [15] by

$$F_x = \sigma_{xy} b$$

$$= \frac{\pi \mu b^2 X(\cosh 2\pi X \cos 2\pi Y - 1)}{D(1-\nu)(\cosh 2\pi X - \cosh 2\pi Y)^2}$$

And the interaction force with an array of screw dislocations is

$$F_x = \sigma_{yz} b$$

$$= \frac{\mu b^2 \sinh 2\pi X}{2D(\cosh 2\pi X - \cos 2\pi Y)}$$

where σ_{ij} is the i-th component of stress on a plane normal to the positive j direction, and x and y are distances from the grain boundary, with x=XD, y=YD, and D is the average distance between dislocations in the array. F_x is shown in Figure 4, and as the figure indicates, the interaction forces decrease rapidly with distance from the grain boundary to zero for edge dislocations and to $\frac{\mu b^2}{2D}$ for screw dislocations. Thus, it can be assumed that there is no long range interaction force between a dislocation and grain boundary since the average grain diameter d_g is usually much greater than D. Thus, the velocity function of one discrete motion through a single grain can be approximated by a step function rather than a harmonic function.

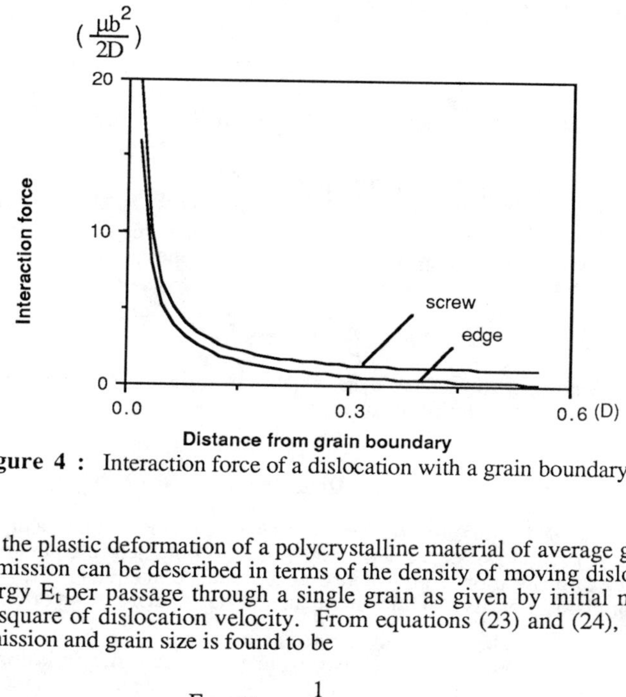

Figure 4 : Interaction force of a dislocation with a grain boundary

Considering the plastic deformation of a polycrystalline material of average grain size d_g, the resulting acoustic emission can be described in terms of the density of moving dislocations and the radiation wave energy E_t per passage through a single grain as given by initial motion which is proportional to the square of dislocation velocity. From equations (23) and (24), the relationship between acoustic emission and grain size is found to be

$$E_{AE} \propto \frac{1}{\frac{a}{d_g} + \rho_o} \qquad (25)$$

In the fine grain region where $\frac{a}{d_g} \gg \rho_o$, equation (25) becomes

$$E_{AE} \propto d_g \qquad (26)$$

As the grain structure approaches a single crystal, i.e., d_g tends to ∞, equation (25) becomes

$$E_{AE} \propto \frac{1}{\rho_o} \qquad (27)$$

Thus, increasing the average grain size raises the acoustic emission energy as found in equation (25).

Considering the AE during deformation of a polycrystalline alloy, the acoustic emission is governed not only by the grain boundary but also by the impurities, where the AE is composed of AE from dislocation motion governed by grain boundaries, $E_{AE(g)}$, and AE from dislocation interaction with impurities, $E_{AE(i)}$.

$$E_{AE} = E_{AE(g)} + E_{AE(i)}$$

From the Hall-Petch relation, reducing the grain size increases the external stress σ required for deformation as follows:

$$\sigma = \sigma_o + k d_g^{-0.5} \tag{28}$$

The higher stress in a finer grained crystal increases dislocation velocity and results in an increase in breakaway events from solute drag. The rate $\dot{n}_u$ of successful breakaways is

$$\dot{n}_u \;\alpha\; \exp\left(-\frac{E_b - E_a}{kT}\right) \tag{29}$$

For a given material, E_b is constant and E_a varies with grain size since flow stress is determined by grain size. Substituting σba^* for E_a in equation (29) gives

$$\dot{n}_u \;\alpha\; \exp\left(\sigma ba^*\right) \tag{30}$$

where a* is activation area. And from equation (28), we have

$$\dot{n}_u \;\alpha\; \exp\left(c d_g^{-0.5}\right) \tag{31}$$

where $c = kba^*$. Hence, the event rate of AE source mechanism due to interaction with impurities is influenced by grain boundaries as shown in equation (31), and the resulting AE energy becomes

$$E_{AE(i)} \;\alpha\; \exp\left(c d_g^{-0.5}\right) \tag{32}$$

As the grain size of an impure crystalline material decreases, $E_{AE(g)}$ decreases as shown in equation (25) while $E_{AE(i)}$ increases as shown in equation (32).

5. Discussion

There are two acoustic emission sources during plastic deformation of a crystalline alloy. One is due to dislocation motion associated with solute drag as obtained in equation (4), and the other source is unpinning of dislocations from solute drag as obtained in equation (15). Acoustic emission from unpinning of dislocations is much greater than that due to dislocation drag, which explains the large AE amplitude during Lüder's strain or serrated deformation.

For the dislocation motion associated with solute drag, acoustic emission generation during deformation of an alloy is found from equation (4) to be inversely proportional to the solute density, and since the rate of AE energy is proportional to the square of RMS voltage, the RMS is inversely proportional to the square root of the solute density. This analysis is compared with experimental results obtained by Kuribayashi [3], Figure 5, where the broken line is from equation (4) and the solid line from experiments. The trend in experimental results is found to correlate well with the theory, the AE energy being inversely proportional to solute density. The unpinning of dislocations from solute atoms is not considered to have a significant influence, since no serrated strain was observed during the strain. The reduction of AE signal due to the addition of solutes is explained by the drag mechanism which reduces the dislocation velocity.

However, a dislocation can break away from a Snoek or Cottrell atmosphere under certain conditions which depend on stress, temperature, solute density and free energy of the solute atom as indicated in equation (11). Most of the acoustic emission generated during serrated strain resulting from breakaway is attributed to the high dislocation velocity. Thus, only the breakaway event is considered in estimating AE, neglecting other mechanisms. As the strain increases, the solute concentration near a dislocation core decreases due to solute diffusion to newly created dislocations, resulting in an increase in breakaway events as given by equation (11). However, the intensity of the breakaway mechanism is diminished because of the reduction in solute concentration in a dislocation core, which results in a decrease of dislocation velocity and a change in the mechanism of dislocation motion from a breakaway event to ordinary initial motion. Thus, AE from the serrated or Portevin-Le Chatelier yielding is expected to decrease after reaching a maximum, and only occurs within a limited range of strain.

Free dislocations are generated at the interfacial zone between an inclusion and matrix by hydrostatic stress due to the difference in bulk modulus between matrix and particle, which results in an increase in initial dislocation density and a reduction of dislocation velocity at a given strain. Thus, the generation of free dislocations due to a compressive hydrostatic stress reduces the AE energy. The generation of free dislocations cannot be expected under a tensile hydrostatic stress since the tensile hydrostatic stress produces decohesion rather than local plastic deformation at the interface.

For solid solution alloys, the hydrostatic stress affects the strength of pinning due to the local rearrangement of solute atoms near the dislocation core by short-range diffusion resulting from a change in free energy due to the hydrostatic stress. Thus, a tensile hydrostatic stress increases the AE signal due to solute diffusion to dislocation core and a resulting increase in free energy. On the other hand, the compressive hydrostatic stress reduces the AE signal due to solute diffusion out of the dislocation core. Variation of the upper yield strength $\Delta\sigma_y$ due to a hydrostatic stress can be also explained by the variation of solute density around the core. Substituting equation (17) into (6a) indicates that the yield stress has a linear relationship with hydrostatic stress. The yield drop phenomenon is eliminated by a compressive hydrostatic stress and enhanced by a tensile hydrostatic stress, which correlates with experimental results [6], Figure 6.

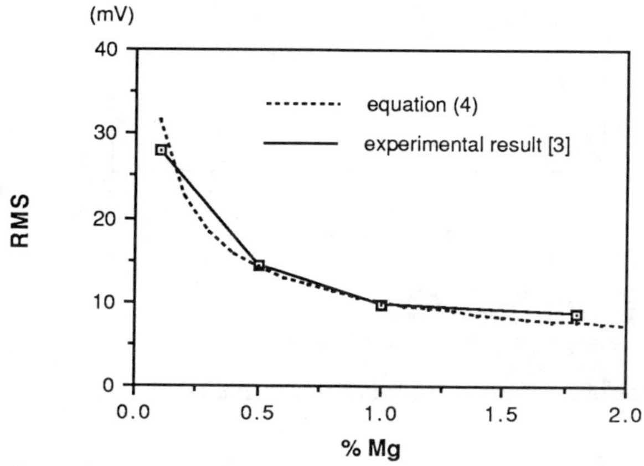

Figure 5 : RMS of AE signal from Al-Mg alloy in terms of solute density

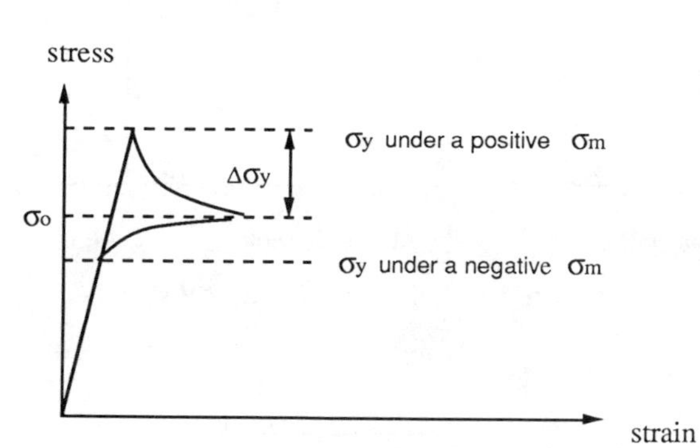

Figure 6: Variation of yield drop due to a hydrostatic stress [6]

The grain boundary influences acoustic emission generation since it affects dislocation motion during strain. For a pure material, as the grain size increases, the resulting AE signal generation also increases, equations (25) and (26), and eventually converges to that of a single crystal as given by equation (27), and this correlates with experimental observations [12], Figures 7 and 8(a). However, as an alloying element is added, the fraction of acoustic emission due to interaction with solute atoms increases and the acoustic emission signal reversely decreases as grain size increases as given by equation (32) and shown in Figure 8(b). This is contrary to that of a pure polycrystalline material. It has been shown experimentally that Lüder's strain and AE are more active during deformation of a finer grain structure for a Silicon-Iron [21] and for an Al-Mg alloy [12].

As explained in the previous section, the breakaway from solute drag or pinning condition is a dominant AE source because of the high velocities involved. As the grain size decreases, acoustic emission energy is expected to increase since AE from the interaction with impurities is more significant than that from the interaction with grain boundaries. Thus, the acoustic emission signal from a polycrystalline alloy depends on the relative amounts of AE from interaction with grain boundaries and with impurities, and can have a maximum value at an intermediate grain size as shown in Figure 8(c).

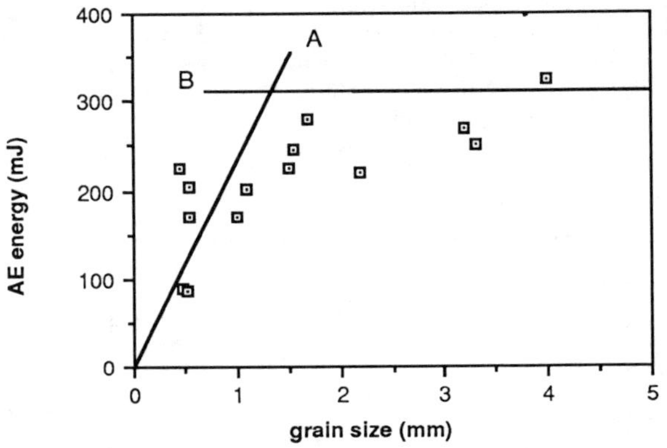

Line A : equation (26)
Line B : equation (27)
Dots : experimental results of pure Al [12]

Figure 7 : Variation of AE from a pure material with grain size

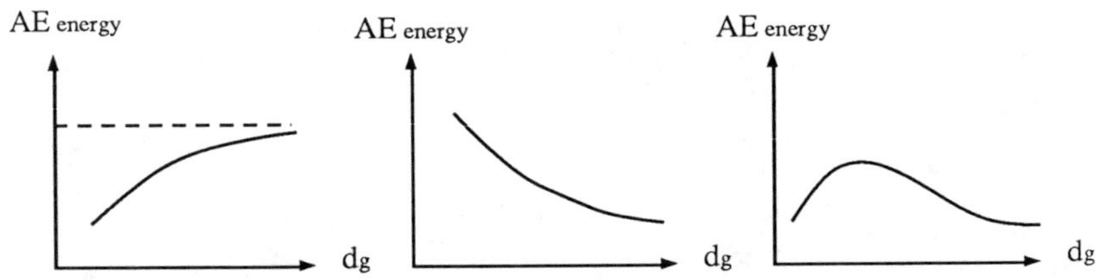

(a) AE from grain boundary (b) AE from impurity (c) AE from both sources

Figure 8 : Grain size effect on AE signal

6. Conclusions

If there are no breakaways from a solute pinning atmosphere, an increase in solute density reduces the intensity of acoustic emission signals as a result of the reduction in dislocation velocity by the dragging mechanism. When breakaway of dislocations from solute dragging or locking occurs, the solute atoms enhance the acoustic emission signal.

Acoustic emission from plastic deformation of a solid solution alloy or one with inclusions has a linear relationship with the hydrostatic stress. AE energy decreases linearly as compressive hydrostatic stress increases.

Considering only the effect of grain boundaries, acoustic emission increases linearly with grain size and asymptotically approaches that of a single crystal. For a polycrystalline alloy, since acoustic emission from the interaction of dislocations with impurities decreases with grain size, the overall acoustic emission in terms of grain size depends on the fraction of AE from interaction with impurities and that from grain boundaries.

ACKNOWLEDGEMENT

Support of this work by the National Science Foundation under grant No. DMC-8607198 is gratefully acknowledged.

REFERENCES

[1] Dunegan, H. L. and Green, A. T., *Acoustic Emission*, ASTM STP 505, pp. 100-113 (1972).

[2] Hatano, H., *J. Appl. Phys.*, Vol. 48, No. 10, pp. 4397-4399 (1977).

[3] Kuribayashi, K., and Kishi, T., *Mater. Sci. Eng.*, Vol. 33, pp. 159-163 (1978).

[4] Higgins, F. P., and Carpenter, S. H., *Acta Metal.*, Vol. 26, pp. 133-139 (1978).

[5] Heiple, C. R., and Carpenter, S. H., *J. Acoust. Emission,* Vol. 6, No. 3, pp. 177-204, and pp. 215-237 (1987).

[6] Bullen, F. P., Henderson, F., Hutchison, M. M., and Wain, H. L., *Phil. Mag.*, Vol. 9, pp. 285-297 (1964).

[7] Ashby, M. F., Gelles, S. H., and Tanner, L. E., *Phil. Mag.*, Vol. 19, pp. 757-771 (1969).

[8] Redfern, B. A. W., Evans, R. A., and Wronski, A. S., *J. Mater. Sci.*, Vol. 5, pp. 784-789 (1970).

[9] Hadjicostis, A. N., and Carpenter, S. H., *Mater. Eval.*, Feb., pp. 19-23 (1980).

[10] Cottrell, A. H., *Pro. Roy. Soc.*, Vol. 44, No. 355, pp. 829-832 (1953).

[11] Kiesewetter, N. and Schiller, P., *Phys. Stat. Sol. (a)*, Vol. 38, pp. 569-576 (1976).

[12] Scruby, C. B., Wadley, H., and Sinclair, J. E., *Phil. Mag.*, Vol. 44, No. 2, pp. 249-274 (1981).

[13] Hatano, H., *J. Acoust. Soc. Am.*, Vol. 57, No. 3, pp. 639-645 (1975).

[14] Kim, H. C. and Kishi, T., *Phys. Stat. Sol. (a)*, Vol. 55, pp. 189-195 (1979).

[15] Hirth, J. P. and Lothe, J., *Theory of Dislocations*, Wiley-Interscience, New York (1982).

[16] Chung, J., and Kannatey-Asibu, Jr., E., " Acoustic Emission from Plastic Deformation of A Pure Single Crystal," J. Applied Physics in press.

[17] Timoshenko, S. P., and Goodier, J. N., *Theory of Elasticity*, McGraw-Hill, New York (1982).

[18] Eshelby, J. D., *Pro. Roy. Soc.*, Vol. 241, pp. 376-396 (1957).

[19] Ball, A., Bullen, F. P., and Wain, H. L., *Mater. Sci. Eng.*, Vol. 3, pp. 283-288 (1969).

[20] Mellor, H. G. and Wronski, A. S., *ACTA Metall.*, Vol. 18, pp. 765-773 (1970).

[21] Tandon, K. N., and Tangri, K., *Mater. Sci. and Eng.*, Vol. 20, pp. 47-54 (1975).

48, Sensors, Controls, and Quality Issues
in Manufacturing
ASME 1991

MONITORING OF TOOL BREAKAGE WITH RESTRICTED COULOMB ENERGY TYPE NEURAL NETWORKS

Ibrahim Nur Tansel
Department of Mechanical Engineering
Florida International University
University Park Campus
Miami, Florida

Charles McLaughlin
Department of Mechanical Engineering
Tufts University
Medford, Massachusetts

ABSTRACT

A supervised neural network system is used for detection of tool breakage in milling operations. The cutting force signals presented are Restricted Coulomb Energy (RCE) type neural networks after encoding. The training period is very short compared to backpropogation neural networks. Also, the RCE networks indicate if their previous training to identify signals is satisfactory.

The effectiveness of the encoding and convenience of the RCE networks are evaluated on simulated and experimental cutting force signals. The neural network is trained on simulated cutting force signals. The RCE network correctly categorized more than ninety-eight of the presented data sets after training, which include simulated and experimental cutting force data never before seen.

The performance of the encoding method and the RCE networks are found to be acceptable for many tool breakage monitoring tasks of end milling operations.

I. INTRODUCTION

Monitoring of tool cutting conditions and tool wear is extremely important in automated manufacturing. Many studies have been done on tool breakage detection. In this study, use of Restricted Coulomb Energy (RCE) type neural networks for detection of tool breakage in milling operations is discussed. The RCE type neural network learns much faster than the most commonly used backpropogation type neural network.

Current international competition among manufacturers has prompted the need for improved quality at reduced costs. These economic considerations and continuously reducing microcomputer prices have motivated many manufacturers to automate their production lines and to control their manufacturing operations by on-line processing of sensory data [1]. Many major manufacturers have recently started to use tool breakage detection systems to recognize machining problems; however, capabilities of these systems are quite limited.

Many researchers have contributed to the development of a model to predict cutting forces in end milling operations [2,3,4,5]. An accurate model of the cutting forces in milling can be used to study many aspects of the machining process, including chatter [6,7], controller design for the milling process [8], and tool breakage detection algorithms. Many different time series models have been applied to identify and monitor machining operations. Mouri and Sata [9] used a Kalman filter to estimate the coefficients of an AR model in turning. They model the signal series of the armature current of the DC Motor that drives the spindle and uses estimation error to detect tool breakage. Matsushima et al. [10] also used a similar approach to detect tool breakage in milling.

Autoregressive (AR) models have also been used to detect tool cutting conditions in milling operations by using acoustic emission intensity [11,12] and cutting force [13,14,15,16,17].

The neural network based systems have been applied to pattern recognition problems in many fields [18-31]. Artificial networks refer to a group of architectures that loosely resemble the brain functioning or structure. Many different neural network paradigms [24,25,26] have been developed to achieve different learning and processing speed capabilities [20]. Most of the available paradigms can use parallel processors for very high computational speed. Among the available neural network paradigms, in this decade, engineers have most commonly used the backpropogation technique [31].

Restricted Coulomb Energy (RCE) [20] type neural networks have been developed in the last fifteen years. These neural networks mainly use three layers that consist of input, hidden and output layers. The neural network inspects presented cases during the training session. For each input pattern (case), the neural network first checks if it can accurately classify the case based on the previous training. If it classifies that input pattern correctly, it will not change the parameters of the model. If it classifies it incorrectly based on the previous information, or if it was not satisfactorily trained for the presented case, the neural network modifies the previously selected parameters, and/or assigns a new hidden node, and then selects the parameters. This procedure allows the neural network to learn most of the problems in two to ten reviews of the training cases, while backpropogation requires sometimes thousands, and even more inspections. Also, RCE networks do not generalize the presented cases, since only the parameters of only the related hidden nodes are changed during training. After training, the neural network can classify the given input patterns. For very demanding classification problems, several RCE neural networks can be used together to monitor different sensors.

In this study, various sensors and signal processing approaches were compared. Although the measurement of acoustic emission or acceleration costs less than measurement of cutting forces, in this study, cutting force measurement was preferred for the following reasons:

- Cutting force signals are much less dependent on the structure. Monitored acceleration or acoustic emission signals reflect the response of the machine tool to cutting forces and must be tuned separately for every machine tool structure.
- Cutting force signals can be simulated easily and more accurately than acceleration and acoustic emission signals.

For analyzing the data, the RCE paradigm was selected for the following reasons:

- Cutting force profiles are quite different from each other at different cutting conditions. If the RCE has not seen a similar pattern in a new case during monitoring, the network can indicate that it was uncertain. Backpropogation type neural networks make unpredictable estimations in these cases.
- Training of RCE is much faster than backpropogation. The entire training can be finished in a few minutes with a single processor microcomputer in a machine shop, while backpropogation would need several hours, or a parallel processor.

In the following chapters, background information on the preparation of the simulation data and RCE will be discussed first. Later, the RCE based tool breakage detection mechanism and experimental verification of the proposed procedure will be outlined.

II. THEORETICAL BACKGROUND

In this section, the mathematical model of end milling is outlined, followed by the RCE presentation.

II.1 Modeling of Cutting Forces in End Milling:

Deterministic models for mathematical simulation cutting forces in end milling have been developed by many researchers [2,3,4,5]. Tlusty [2] modeled the dynamics of cutting forces in end milling at a constant depth of cut and width. In this model, the instantaneous cutting force at cutting angle, ϕ, is assumed to be proportional to the chip thickness. The instantaneous tangential cutting force in end milling can be approximated by:

$$F_T = K\,b\,h\,, \tag{1}$$

where b is the axial depth of cut and K is the cutting force coefficient that depends on the material of the workpiece, geometry of the tool, and average chip thickness. h is the instantaneous chip thickness at angle, ϕ, which is shown below:

$$h = s_t \sin\phi \ , \tag{2}$$

where s_t is the feed rate per tooth. In this model, the radial component of the cutting force is assumed to be 30% of the tangential component,

$$F_R 9 = 0.3 \ F_T \ . \tag{3}$$

The total resultant cutting force is the sum of the components summed over the differential rotation angles, $d\phi$.

$$dF_T = K \ s_t \ \sin\phi \ dy \tag{4}$$

and

$$dF_R = 0.3 \ K \ s_t \ \sin\phi \ dy \ , \tag{5}$$

where dy is the differential cutting edge and is determined by:

$$dy = \frac{r}{\tan\beta} \ d\phi \ , \tag{6}$$

where β is the helix angle and r is the tool radius.

The differential components of the cutting force on two fixed perpendicular directions are:

$$dF_x = - \ dF_T \cos\phi \ - \ dF_R \sin\phi \ , \tag{7}$$

and

$$dF_y = dF_T \sin\phi \ - \ dF_R \cos\phi \ . \tag{8}$$

For the case of nonhelical teeth, the cutting forces may be represented by:

$$F_x = F_T \ (\cos\phi \ + \ 0.3\sin\phi) \tag{9}$$

and

$$F_y = F_T \ (\ - \ \sin\phi \ + \ 0.3\cos\phi) \ . \tag{10}$$

Tlusty and Ismail [6] improved this model to study chatter in end milling. This model allowed for tool deflection in both the x and y directions. The tangential cutting force in Eq. (1) is modified to include tool deflection.

$$F_T = K \ b \ (h-z+zmin) \ , \tag{11}$$

where z is the displacement of the tool normal to the machined surface, and z_{min} is the lowest undulation left behind in preceding cuts at angle, ϕ.

The F_x and F_y components of the cutting force excite vibrations in the x and y directions. The x direction displacements at step $n+1$ are calculated from the cutting force data at step n. Simulation of the structural dynamics is outlined in Ref. [6].

Models of cutting forces have also been developed that include the influence of cutter offset on the cutting force [4,5]. The effects of cutter offset can be included by adding a term in determining the normal z displacement, which adjusts for the eccentricity:

$$z = x \ \sin\phi \ + \ y \ \cos\phi \ + \ t_{ec} \ , \tag{12}$$

where t_{ec} adjusts for the eccentricity. For a four flute end mill, the first and third, and the second and fourth flute values of t_{ec} average to one. For a six flute end mill, the first and fourth, second and fifth, and third and sixth flute values of t_{ec} average to one.

II.2. Restricted Coulomb Energy Neural Networks:

Restricted Coulomb Energy (RCE) [20] is a parallel neural network modeled after the human learning and classification process. The architecture of the RCE network is a feed-forward arrangement. This arrangement allows the network to classify pattern signals in real time without any special hardware. The network is composed of three layers of cells: the input layer, the internal (hidden) layer, and the output layer. The feature vectors of the pattern to be learned are given to the input layer. The nodes of the input layer are connected to every node of the internal layer. The nodes in the internal layer are connected selectively to the output nodes during the training process. The output nodes correspond to different pattern classes. The internal connections occur in such a way that the correct output cell will be fired when an appropriate pattern class is given to the system.

To use the neural networks, sensory signals are processed and a set of features are created. These features are presented to the neural network during the training or monitoring of the process. These features are a set of points that can be mapped to a feature space. The pattern class is a set of points in the feature space and each pattern class has its own region in the feature space.

RCE networks use two learning mechanisms. When new patterns are presented to the network, the response of the neural network is tested without any modification at the weight matrix. If the classification of the network matches the required output, the weight matrix is not changed. Otherwise, the influence of the existing nodes are modified and/or a new node will be created.

The global Liapunov, or energy methods have been applied to the analysis of neural networks [27]. The equilibrium point is the stored pattern based on the input pattern. Hopfield [28] used a relaxation model of memory storage and retrieval using a Liapunov function. Bachmann et al. [29] define a dynamical system in which the minima are known (similar to the Hopfield model). They also define an actual total energy of the system using a Liapunov function. The central idea of the Coulomb energy network is the defining of potential energy of a collection of memory sites. Each memory site can either attract or repel other memory sites. Memory sites, $x_1,.., x_m$, can be grouped into pattern classes in the feature space. The electrostatic potential energy of the configuration of a collection of charges with the same and different charges is defined by the equation for Ψ:

$$\Psi = (1/2L) \sum_{i=1}^{M} \sum_{j=1}^{M} Q_i\, Q_j \mid x_i - x_j \mid - L \tag{13}$$

where L is a parameter related to the memory size and M is the basin of attraction. Memory sites of different classes have repulsive potential energy, while memory sites of the same class have attractive potential energies. For supervised learning [20], $Q_i(c)$ is defined as the charge of memory site, i, and class, c, in such a way that the following criteria are satisfied:

$$sign(Q_i(c)) \neq sign(Q_i(c')) \;\; for \;\; c=c' \tag{14}$$

$$sign(Q_i((c)) = sign(Q_i(c')) \;\; for \;\; c \neq c' . \tag{15}$$

For unsupervised learning [30], all patterns are assumed different and the definition of the attractive potential energy must be modified to:

$$sign(Q_i(c)) = sign(Q_i(c')) \;\; for \;\; c \neq c' . \tag{16}$$

A non-linear sigmoidal logistic function is defined as follows:

$$F_n \left[\sum_{m=1}^{K} \omega_{nm} f_{mi} \right] = \left[1 + \exp(-\,(1/kt)) \left[\sum_{m=1}^{K} \omega_{nm} f_{mi} + \Theta \right] \right]^{-1} , \tag{17}$$

where ω_{nm} is the synaptic weight, e_n is the unit vector of the nth cell of activity, F_n and f_i indicate the afferent activity, k and t are parameters that control the steepness of the sigmoid, and Θ is the threshold. Using the logistic function, the vector of the network activity that results from mapping the afferent pattern, f_i, via the weight matrix, ω_{nm}, can be defined as:

$$x_i = \sum_{n=1}^{N} e_n F_n \left[\sum_{m=1}^{K} \omega_{nm} f_{mi} \right] \tag{18}$$

When the synaptic weight matrix changes, motion of the memory sites in the activity space occurs. The electrostatic energy, Ψ, can be minimized by the correct selection of the synaptic weight matrix. To minimize the potential energy, the gradient, with respect to the weight matrix, is determined,

$$\delta\omega_{nm} = - \partial\psi / \partial\omega_{nm} \tag{19}$$

$$\delta\omega_{nm} = 1/2 \sum_{i=1}^{M} \sum_{j=1}^{M} Q_i Q_j \, |x_i - x_j|^{-(L+2)} \, |x_i - x_j| \, \partial / \partial\omega_{nm} \, |x_i - x_j| \, . \tag{20}$$

As the weight matrix develops, the potential energy for a set of M patterns is minimized.

To implement the RCE algorithm in this study, the *Nestor Development System (NDS)* software developed by Nestor Inc. was used.

III. DETECTION OF TOOL BREAKAGE WITH RCE

A neural network can be trained on experimental data and used to monitor machining operations. However, this approach has two main drawbacks:

1) There are many possible cutting conditions and different types of tools. Collection of data with a good and broken tool is very time consuming and expensive at all the possible cutting conditions with different tools.
2) The good and broken tool data of so many different types of tools require very large memory and some of them may conflict with each other.

To eliminate these problems, the best approach is to train the neural networks by considering tool type and possible cutting conditions by using an analytical model. The present analytical milling models can accurately generate simulated cutting force signals of good and broken tools. The simulation results of these analytical models can be used to train the neural networks.

In this study, the cutting force profiles of good and broken tools (only one tooth is broken) were simulated by using the analytical model that was introduced in Section II.1. The effectiveness of the encoding method and the acceptability of the RCE networks were examined on the simulated data first. Generation of the simulated data and the encoding method will be discussed in this section.

III.1. Preparation of Simulated Data for Training and Testing:

The simulated data sets for training and testing were prepared based on previous studies outlined in Section II.1. However, to generate more realistic data, cutter runout and tooth breakage were added to the simulation program.

The tangential cutting force in each flute at each rotational angle was determined first, Eq. (11). Each angle of rotation is measured with respect to the angle of the first cutter. For the first two revolutions of the spindle, the x and y displacements were assumed to be zero. After the second revolution, the F_x and F_y forces were used to calculate the x and y displacements by assuming that the machine tool has one degree of freedom in both directions. The x and y displacements were considered in the calculation of the removed chip thickness, z, in Eq. (11).

Runout was also considered in the simulations. Cutter runout is encountered when the tool is not centered on the spindle and cutting force variations change considerably. The offset side of the cutter is subject to higher than average cutting forces, while the other side of the tool will remove less material.

Based on this fact, the simulation program was modified to consider the position of each particular tooth to calculate the cutting force and surface profile of the machined part.

Tooth breakage was simulated by assuming that a particular tooth was shorter than others. Consequently, the shorter tooth will remove less material and is subjected to the smaller cutting force, and the machined surface profile is modified by considering this fact. In this study, a certain percentage of a particular tooth was assumed to be broken. All the simulation data were prepared by assuming that one tooth was removing 80% less material compare to the other teeth.

The dynamic cutting coefficient was set at 1991.3E6 N/m^2 for all cases. For displacements in the x direction, the following parameters were used: m_x of 10 kg, c_x of 478.1 kg/s, and k_x of 20E7 N/m. For displacements in the y direction, the following parameters were used: m_y of 12.5 kg, c_y of 331.8 kg/s, and k_y of 20E7 N/m. For the four flute end mill, the values of the offsets were Tooth(1) 0.2 R, Tooth(2) 0.9R, Tooth(3) 0.8R, and Tooth(4) 1.1R; for the six flute end mill, Tooth(1) 0.0, Tooth(2) 0.8R, Tooth(3) 0.8R, Tooth(4) 0, Tooth(5) 0.2R, and Tooth(6) 0.2R.

III.2. Encoding Method:

To work with any neural network effectively, the data must be in its most representative form. The representative form should be as short as possible to keep the memory size of the program small and computation time short. The process by which the sensory signals obtain the most representative form of the inputs is called encoding, Fig. 1.

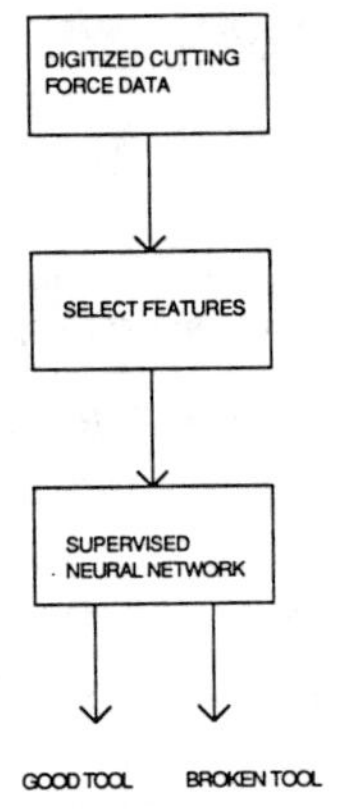

Fig. 1. The diagram of a typical neural network system.

In this study, the cutting force variation of every rotation was represented by 10 values, between 0 to 10, and presented to the RCE neural network. Each of these values was the average of 36 cutting force samples and normalized to work at a very large cutting force range. By averaging 36 cutting force readings, high frequency noise was eliminated and only the cutting force variation profile was obtained.

To obtain the 10 values that represent the cutting force profile of one tool rotation, the cutting forces were digitized at every one-degree rotation of the cutting tool. In total, 720 samples were collected during the two rotations of the cutting tool to start the encoding. The smallest cutting force sample in the first 360 samplings was selected. The smallest cutting force sample and the following 359 samples were used for the encoding. The 360 readings were averaged in groups of 36 samples each, and 10 values were obtained. These 10 values were normalized to have all the values in a range from 0 to 10.

The selected encoding technique has the following advantages:

- Representation of the force profile with 10 samples requires smaller memory size for the parameters of the neural network and our test representations of the force profile with 10 values were found to be reliable.
- Each value represents the sample of 36 readings. This averaging eliminates high frequency noise. In our tests, the experimental noise was minimal.
- The force profile was always initiated with the smallest cutting force reading in one full rotation, and covered the whole rotation. This self-alignment mechanism allows the neural network to inspect the same cutting force profile wherever the cutting tooth is broken.
- The cutting force profile is normalized between 0 and 10. This normalization creates a similar force profile at different depths of cuts and with different tool radii.

IV. EXPERIMENTAL PROCEDURE

The experimental apparatus is presented in Fig. 2. A standard Bridgeport Vertical Milling Machine was used for the experimental study. The cutting force signals in the x (feed direction) and y (perpendicular to the feed direction in the horizontal plane) direction were measured by using a Kistler 9257A 3-Component dynamometer. The dynamometer was mounted on the table of the milling machine. The workpiece was then mounted on the dynamometer. The cutting force signals were measured at fixed rotation angles of the spindle and the resultant force was calculated from the sampled two components. To sample the data with fixed rotation angles, a Litton Model 70 optical encoder was mounted on the spindle. The pulses of the encoder were used to trigger the sampling devices. The resolution of the encoder was 360 cycles per revolution. The output voltage signal of the Charge Amplifier of the dynamometer was collected separately by both an IBM PS-2 50 microcomputer and a Tektronix 2221 digital oscilloscope. The IBM PS-2 50 had a Metrabyte uCDAS-16G A/D board installed in one expansion slot. The UCDAS-16G uses a 12 bit converter in the -10 V to +10 V range to sample up to 16 single end inputs. A Tektronix 2221 digital oscilloscope was also used to sample the data. Again, the TTL compatible output of the encoder was used as an external trigger to sample the data. The 2221 uses an 8 bit converter to represent the data. The data sampled by the Tektronix 2221 was transferred to an IBM PS-2 30 through a GPIB IEEE-488 standard interface bus that was installed in the expansion slot of the PS-2 30. The data sampled by the oscilloscope were then stored in an ASCII file.

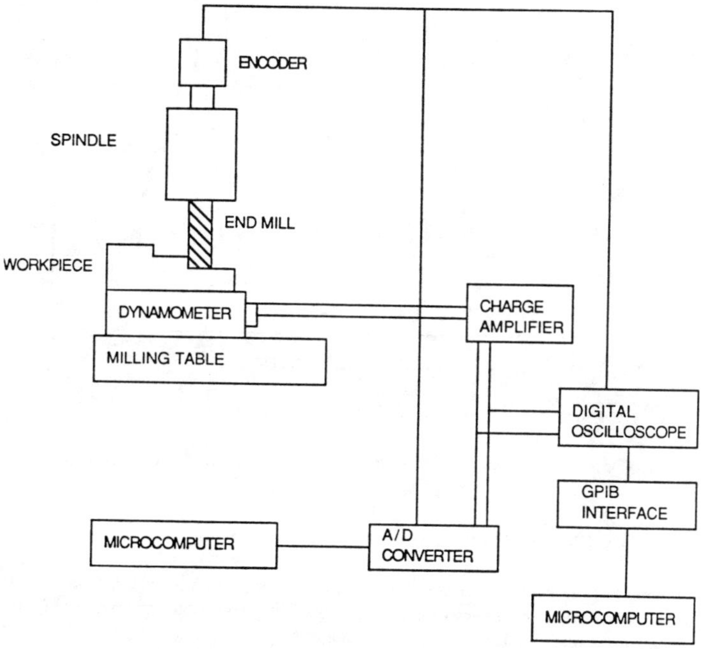

Fig. 2. The experimental setup to measure the cutting forces in end milling operations.

Data were collected by using good and broken tools. The broken tools were prepared by completely grinding off one of the flutes. The typical good and broken tool signals are displayed in Fig. 3. Data were also collected at different cutting immersion angles, including slotting, up-milling, and down-milling.

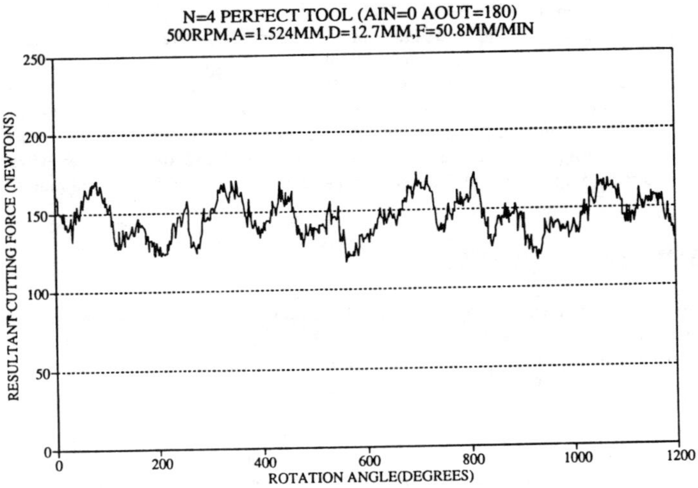

Fig. 3.a. The perfect tool.

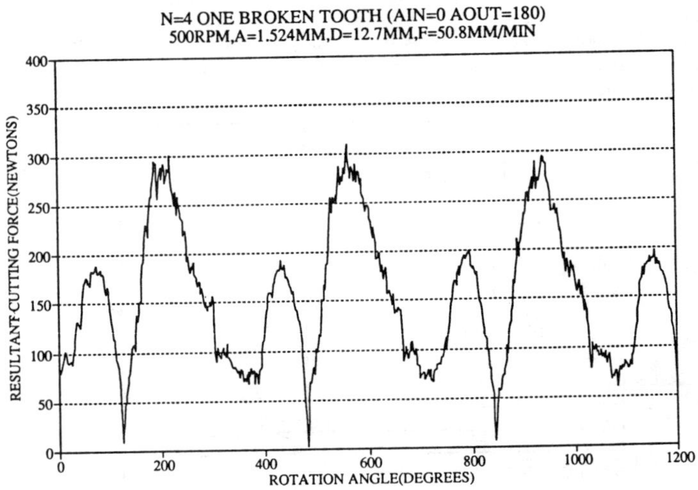

Fig. 3.b. The broken tool.

Fig. 3. Typical experimental resultant cutting force signals.

V. RESULTS AND DISCUSSION

The effectiveness of the developed RCE neural network based tool breakage detection system was tested in two stages. In the first stage, the effectiveness of the encoding approach, number of inputs, and selected paradigms were tested on simulated data. In the second stage, the system was trained on the simulated data and tested on different simulated cutting force patterns and experimental data. These two stages are discussed in detail in the following sections.

V.1. Testing RCE Networks on Simulated Data:

The cutting forces of end milling operations were simulated at twenty different cutting conditions with perfect and broken tools. The simulated data of the broken tool were prepared by considering that one of the teeth removed 80% less material compared to other teeth. The cutting tool was a four flute end mill, moving at 0.0002m per tooth feed rate. In the simulations, the axial depths of cut were selected between 0.000127m and 0.00254m. The cutting force profiles were encoded by following the procedure described in Section III.2. The neural network learned 100% of the 40 cases.

The RCE network was trained on nine different simulation sets. Each set had 200 to 400 simulated cutting force profiles. The neural network learned and accurately identified 100% of all the simulated cases. In these simulations, feed-rates were selected between 0.0001m and 0.001m per tooth. The axial depths of cut ranged from 0.000127m to 0.00254m. The broken tooth removed 80% less material than the other teeth. The other parameters of the simulations are presented in Table 1. The RCE networks assign a prototype for very similar input patterns. The number of prototypes is also presented in Table 1. The RCE network used minimum 2/maximum 5 prototypes to classify the good and broken tool signals. All the prototypes were exclusive. If the prototypes are exclusive, they do not overlap each other. In other words, the neural network will classify the input patterns without any hesitation between two possible classes. The small number of exclusive prototypes indicated that the encoding scheme was very effective and representation of cutting force profiles with ten normalized numbers was proper. Training of RCE networks on a set of 400 input cases took less than ten minutes with an IBM PS2/80 microcomputer. The speed and accuracy of the selected paradigm was found to be excellent on the simulated data.

Table 1

Selected Cutting Conditions for the Simulated Data and the Number of Prototypes the RCE Networks Required

TRAINING OF RCE ON SIMULATED DATA

Number of Teeth	Immersion Angle (Degree)	Tool Diameter	Number of Simulated Cases	Percentage of Correct Identifications	Number of Prototypes
4	0-180	0.50	400	100%	2
4	0-90	0.50	200	100%	5
4	90-180	0.50	200	100%	2
4	0-180	0.75	400	100%	3
4	0-90	0.75	200	100%	5
4	90-180	0.75	200	100%	2
4	0-180	1	400	100%	2
4	0-90	1	200	100%	5
4	90-180	1	200	100%	2

V.2. The Performance of RCE Networks on Experimental Data After Simulation Training:

The RCE network was trained with 400 simulated cases. The training set included 200 perfect and 200 broken tool resultant cutting force patterns. The simulated data were prepared by considering 200 different cutting conditions with a one-inch end mill. The feed rate ranged from 0.0001m to 0.001m per tooth in the simulated cases. The axial depths of cut were between 0.000127m and 0.00254m. One tooth of the broken tools removed 80% less material than the other teeth.

After training on the simulated data, the RCE network was tested on a new simulated data set that it had not seen before. The simulations were prepared by considering 190 different cutting conditions with good and broken tools. The cutting parameters of the test set were in the range of the training data, but they had different values. The broken tooth was assumed to remove 80% less material than the other teeth. 36 experimental, 16 perfect, 16 broken, and 4 with changing depths of cut (Fig. 3) data sets were added to the test set. The cutting conditions of the experiments are presented in Table 2. The test set had a total of 416 cases.

Table 2

Experimental Cutting Conditions

Axial Depth of Cut (mm)	Feed Rate (mm/min)	Spindle Speed (rpm)
1.016	50.8	500
1.016	50.8	500
1.016	50.8	500
1.016	50.8	500
1.524	50.8	500
1.524	101.6	500
1.524	203.2	500
1.524	254	500
1.016	50.8	500
1.016	50.8	500
1.016	50.8	500
1.016	50.8	500
1.524	50.8	500
1.524	101.6	500
1.524	203.2	500
1.524	254	500
0.508-1.016	50.8	500
0.508-1.016	101.6	500
0.762-1.625	50.8	500
0.762-1.625	101.6	500

395 input patterns of the test set were classified "identified correct", which means the network categorized these patterns with total confidence. They were found to be similar to the prototypes in training and the neural network was certain they did not have any similarity to any of the other prototypes. Five input patterns were classified as "identified incorrect", which means the network categorized class was not the same as the real class of the input pattern. Thirteen input patterns were classified as "uncertain correct", which means the network could not confidently select the classes of these input cases; however, the network's estimation was correct. Three input patterns were classified as "unidentified", which means the network was unable to classify the input patterns at all. The throughput of the network using the testing data was 99.3%. The accuracy of the network on the testing data was 98.1%.

VI. CONCLUSION

The resultant cutting force of milling operations was monitored by an RCE type neural network to detect tool breakage. The encoder represented the cutting force profile of one full revolution with ten normalized averages. The RCE neural networks interpreted tool conditions from those ten inputs. The effectiveness of the proposed encoding and selected paradigms was tested on simulated and experimentally collected data.

On small training sets (400 or less) the neural network established two to five prototypes to represent the perfect and broken tool. These results indicated that the proposed encoding approach selected very distinctive characteristics of the good and broken tool cutting force signals for the network by compensating the effects of the cutting force profile variations related to the depth of cut and feed rate.

The RCE type neural networks were able to learn characteristics of the training cases after a few inspections, while the backpropogation approach needed hundreds of presentations. The neural network could identify more than 98% of the presented test data (computer simulations plus experimental signals that it had not seen before) with confidence following computer simulation based training. The learning speed and accuracy was found to be very convenient and helpful for monitoring of milling operations with single processor microcomputers. It was possible to train the system on the spot in less than 10 minutes by using the computer simulations. The RCE networks could also distinguish if they were properly trained to interpret the presented input cases. This feature improved the reliability of the system. However, the network could not update the model during monitoring. The proposed encoding method and RCE type neural networks were found to be convenient and beneficial for detection of tool breakage in milling operations by using single processor computers. The supervised RCE neural networks may classify the inputs "unidentified" if the monitored signal is different than the training set since the cutting conditions drastically change or tool wear level increases. To avoid this problem, unsupervised neural networks with continous learning capability could be used.

REFERENCES

1. Mellichamp, D. A., *Real Time Computing with Applications to Data Acquisition and Control*, Van Nostrand Reinhold Co., 1983.

2. Tlusty, J. and MacNeil, P., "Dynamics of Cutting Forces in End Milling," *CIRP Annals*, Vol. 24, 1975.

3. Gygax, P. E., "Dynamics of Single Tooth Milling," *CIRP Annals*, Vol. 28, 1979.

4. Kline, W. A. and Devor, R. E., "The Effect of Cutter Runout On Cutting Geometry and Forces in End Milling," *International Journal of Machine Tool Design and Research*, Vol. 23, 1983.

5. Sutherland, J. W. and Devor R. E., "An Improved Method for Cutting Force and Surface Error Prediction in Flexible End Milling Systems," *ASME Journal of Engineering for Industry*, Vol. 108, 1986.

6. Tlusty, J. and Ismail, F., "Special Aspects of Chatter in Milling," *ASME Journal of Vibration Acoustics, Stress, and Reliability in Design*, Vol. 105, 1983.

7. Ismail, F. and Bastami, A., "Improving Stability of Slender End Mills Against Chatter," *ASME Journal of Engineering for Industry*, Vol. 108, 1986.

8. Landerbaugh, L. K. and Ulsoy A. G., "Dynamic Modeling for Control of the Milling Process," *ASME PED*-Vol 18, 1985.

9. Mouri, N. and Sata, T., "Automatic Detection of Tool Breakage Using Kalman Filters," *Bulletin of the Japanese Society of Precision Engineering*, Vol. 15, No. 4, 1981.

10. Matsushima, K., Bertok, P., Sata, T., "In-Process Detection of Tool Breakage Monitoring Spindle Motor Current of a Machine Tool, Measurement and Control for Batch Manufacturing," Winter Annual Meeting of ASME, 1982.

11. Diei, E. N. and Dornfeld, D. A., "A Model of Tool Fracture Generated Acoustic Emission During Machining," *ASME PED*- Vol. 18, 1985.

12. Liang, S. Y. and Dornfeld, D. A., "Tool Wear Detection Using Time Series Analysis of Acoustic Emission," *Journal of Engineering for Industry*, Vol. 111, 1989.

13. Lan, M. S. and Naerheim, Y., "In Process Detection of Tool Breakage in Milling," *ASME PED*-Vol. 18, 1985.

14. Altintas, Y., Yellowly, I., Tlusty, J., "Detection of Tool Breakages in Milling," *ASME PED*-Vol. 18, 1985.

15. Tarn, J. H. and Tomizuka, M., "On-Line Monitoring of Tool Cutting Conditions in Milling," *ASME Journal of Engineering for Industry*, Vol. 111, 1989.

16. Altintas, Y. and Yellowly, I., "In-Process Detection of Tool Breakage in Milling Using Cutting Force Models," *ASME Journal of Engineering for Industry*, Vol. 111, 1989.

17. Tansel, I. N., McLaughlin, C., Clements, M., "Identification of Tool Breakage In Milling Operations by Using On-Line Tooth Modelling," presented at the IASTED International Symposium on Robotics and Manufacturing, 1989, and published at *Robotics and Manufacturing* (ed. M. H. Hamza) pp.244-248, 1989.

18. Tansel, I. N., "Neural Network Approach for Representation and Simulation of 3D-Cutting Dynamics," *Trans. of North American Metal Research Institute*, pp. 193-200, 1990.

19. DARPA, *Neural Network Study*, AFCEA International Press, Fairfax, VA, Vol. 23/1, 1988.

20. Scofield, C. L., "Learning Internal Representations in the Coulomb Energy Network," *Proceedings of the IEEE International Conference on Neural Networks*, 1988.

21. Grossberg, S., "Adaptive Pattern Classification and Universal Recording, II: Feedback, Expectation, Olfaction, and Illusions," *Biol. Cybern.*, **23**, 187, 1976.

22. Carpenter, G. A. and Grossberg, S., "ART 2: Self Organization of Stable Category Recognition Codes for Analog Input Patterns," *Applied Optics*, Vol. 26, No. 23, 1987.

23. Carayannis, G., Manolakis, D. G., Kalouptsidis, A., "A Fast Sequential Algorithm for Least-Squares Filtering and Prediction," *IEEE Trans. on Acoustics, Speech and Signal Processing*, Vol. ASSP-31, No. 6, December 1983.

24. McCulloch, W. S. and Pitts, W., "A Logical Calculus of Ideas Immanent in Nervous Activity," *Bulletin of Mathematical Biophysics*, Vol. 5, 1943.

25. Rosenblatt, F., "The Perceptron, a Probabilistic Model for Information Storage and Organization in the Brain," *Psych. Rev.*, **62**, 1958.

26. Widrow, B. and Hoff, M. E., "Adaptive Switching Circuits," *WESCON Convention Record IV*, 1960.

27. Grossberg, S., "Nonlinear Neural Networks: Principles, Mechanisms, and Architectures," *Neural Networks*, Vol. 1, Pergamon Journals Ltd., pp. 17-61, 1988.

28. Hopfield, J. J., "Neuronal Networks and Physical Systems With Emergent Collective Computational Abilities," *Proc. of the National Academy of Sciences*, USA, Vol. 79, pp. 2554-2558, 1982.

29. Bachman, C. M., Cooper, L. N., Dembo, A., Zeitoumi, O., "A Relaxation Model for Memory with High Density Storage," *Proc. Natl. Acad. Sci.*, USA, **21**, pp. 7529-7531, 1987.

30. Scofield, C. L., *Unsupervised Learning in the N-Dimensional Coulomb Network*, to be published.

31. Rumelhart, D. E., Hinton, G. E., Williams, R. J., "Learning Internal Representations by Error Propagation," *Parallel Distributed Processing: Explorations in the Microstructure of Cognition*, Vol. 1, D. E. Rumelhart and J. L. McClelland (Eds.), MIT Press, pp. 319-362, 1986.

Sensors, Controls, and Quality Issues
in Manufacturing
ASME 1991

MONITORING AND DIAGNOSIS OF ROBOT OPERATION THROUGH VIBRATION SIGNAL ANALYSIS

Y. Rong
Department of Technology
Southern Illinois University
Carbondale, Illinois

M. Fang, H. S. Tzou, and W. A. Gruver
Center for Robotics and Manufacturing Systems
University of Kentucky
Lexington, Kentucky

ABSTRACT

Because robots are highly integrated into manufacturing, a methodology for dynamic monitoring and diagnosis of robot operations is needed to prevent failures from disrupting the production process. This research concerns the development of a vibration-analysis based diagnostic monitoring system for robot operation. Changes in robot operation speed, lubricant condition, and belt fit are diagnostically monitored to provide information for preventive maintenance. Effects due to mechanical clearances are also considered. Vibrations of the robot system are described by stochastic models and the state of robot operation state is recognized through cluster analysis. A fuzzy-logic based multi-assessment procedure is applied to improve the reliability of diagnosistic results. A real-time algorithm is developed for on-line diagnostic monitoring. Experimental results demonstrate the applicability of the technique.

I INTRODUCTION

In modern manufacturing systems, industrial robots have important roles. The production in a flexible manufacturing cell (FMC) or a flexible assembly system (FAS) is usually organized through operations of one or more robots. As robots become more highly integrated into manufacturing systems, robot failures must be on-line, automatically detected, predicted and diagnosed for the purpose of preventive maintenance or fault-tolerance control, so that productivity, reliability, and safety of automated production can be ensured [1].

Monitoring and diagnosis techniques have been successfully applied for fault detection of machines and manufacturing systems [2,3]. In the case of robot operation, however, the event to be monitored and diagnosed is not continuous, an event such as metal cutting or bearing vibration. During one cycle of robot motion, various steps are carried out to complete a specific task, resulting in irregular and intermittent operations. Therefore, the detection of robot failures from incomplete measurements of robot motion is a problem of monitoring and diagnosis.

Robot dynamics and control have been extensively studied over the past two decades. It has been established that there is sufficient information in the vibration signal for identification of the state of robot operation [4]. When a robot is programmed to perform a specific task, a vibration pattern can be consistently observed. If this vibration pattern is learned and stored, in advance, in a diagnostic monitoring system, it is possible to detect any variation of robot operation state through analysis of the measured vibration signal. Commonly used, single criterion diagnosistic methods cannot identify changes of robot operation state because of the dynamic complexity of robot motion. This research applies a fuzzy logic-based multi-assessment technique for the diagnostic monitoring of an industrial robot, although the technique is applicable to many types of robots–including mobile robots or telerobots–and other powered machines and material handling devices. In this paper, failures associated with robot operations are discussed. Cluster functions, the multi-assessment method, and the diagnostic monitoring algorithm are introduced. A case study concerning the operation of an industrial robot describes the application of the methodology.

II ROBOT OPERATION FAILURES

Figure 1 shows the top and front views of the IBM 7535 industrial robot. Two DC servo motors drive the primary rotary axes (theta 1 and theta 2), and a stepping motor drives the roll axis. Accelerometers are mounted in the end of the robot arm to measure the vibration signal of the robot during its operation. The state of robot operation may deviate from a normal working condition for many reasons. In this study, the following possible failures are considered: (1) malfunctions of the controller (motor, drive electronics, or software) which result in variation of the operational motion from nominal values; (2) transmission subsystem malfunctions, which could be caused by changes in lubricant conditions (before and after maintenance) and belt fit ; (3) worn out components, which cause enlargement of mechanical clearances. In this study, therefore, the following variables are monitored and diagnosed: (1) operational speed; (2) lubricant condition; (3) belt fit; and (4) clearance variation.

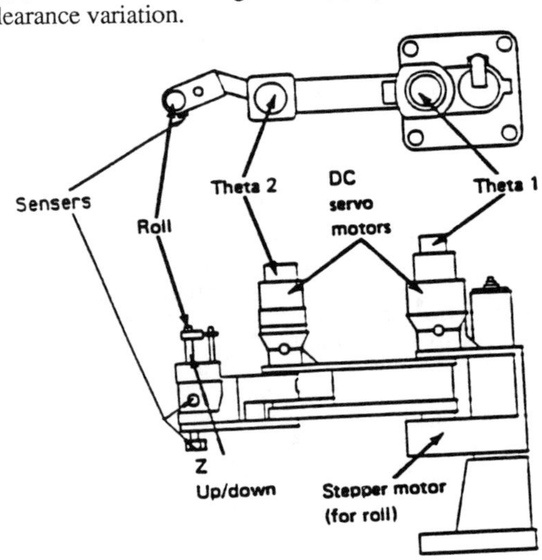

Figure 1. IBM 7535 Robot and Sensor location

III DIAGNOSTIC ANALYSIS OF ROBOT OPERATION

When a measuring system is installed, diagnostic monitoring can be viewed as a decision making process, i.e., given labeled states based on previous robot motions, we classify a new observation into one of the alternative states in an optimal fashion. The cluster analysis can be expressed in terms of the Nearest Neighbor (NN) classification rule [5]: if a system state (denoted by an index i) is defined by a feature vector $\{\xi\}_i$, a new observation $\{\xi\}_o$ should be classified to one of the reference system states $\{\xi\}_r$ if the distance (d) between $\{\xi\}_o$ and $\{\xi\}_r$ is least compared with other samples, i.e.,

If $d(\{\xi\}_o, \{\xi\}_r) < d(\{\xi\}_o, \{\xi\}_i)$, for i=1, 2, ..., n, and $i \neq r$;
then classify $\{\xi\}_o$ to $\{\xi\}_r$.

The NN rule is a widely used method for pattern classification because of its convenience to use and its clear physical meaning. A proper distance measure needs to be defined for a successful application of the NN rule.

Cluster Functions

There are many cluster (distance) functions developed for dissimilarity measures. In this research, a time-series based stochastic model is used to represent the dynamics of the robot operation. When parameters of the stochastic model are estimated from the measured vibration data, the following cluster functions are applied to determine the dissimilarity of the robot operation states. The cluster functions are defined as follows [6]:

Improved Euclidean Distance.

$$d_{ed}(\{\psi_t\}, \{\psi_r\}) = (\{\psi_t\} - \{\psi_r\})^T [W] (\{\psi_t\} - \{\psi_r\}) \tag{1}$$

where $\{\psi_t\}$ and $\{\psi_r\}$ are model parameter vectors of the tested and reference models; and [W] is a weighting matrix.

Cross-Entropy Distance.

$$d_{ce}(\{\psi_t\}, \{\psi_r\}) = \ln\left[\frac{\sigma_t^2}{\sigma_r^2}\right] + \frac{\sigma_{r,t}^2}{\sigma_t^2} - 1 \tag{2}$$

where σ_t and σ_r are standard deviations of the residuals of model fit under the test and reference conditions; $\sigma_{r,t}$ is the standard deviation of the residuals of the reference model fit with the test data, defined as:

$$\sigma_{r,t}^2 = \frac{1}{n-1}\{\psi_r\}^T [X_t]^T [X_t] \{\psi_r\}$$

and $[X_t]$ is the matrix of test data of the vibration signal.

Characteristic Frequency and Damping Ratio. The differences of characteristic frequencies and damping ratios can be used to measure the dissimilarity of robot operation states, which can be estimated through stochastic models [7],

$$\omega_{nk} = \frac{1}{\Delta}[\ln(\lambda_k)\ln(\lambda_k^*)]^{1/2} \tag{3a}$$

$$\zeta_k = \frac{-\ln(\lambda_k \lambda_k^*)}{2[\ln(\lambda_k)\ln(\lambda_k^*)]^{1/2}} \tag{3b}$$

where λ_k and λ_k^* are the kth characteristic roots of the stochastic model.

Normalized Residual Sum of Squares (NRSS).

$$NRSS = \frac{\sigma_t^2}{\Omega_0(X_t)} \tag{4}$$

where $\Omega_0(X_t) = E\{[X_t - \bar{X}_t]^2\}$, is the expected value of vibration data variance.

The improved Euclidean distance and cross entropy distance measure the dissimilarities between system states by calculating the distance between the stochastic models which is estimated using the vibration data with different system (test and reference) conditions. The characteristic frequencies and damping ratios test the variations of the system eigenvalues affected by the robot operation state. The NRSS is a check of the fitness of the stochastic model. If the NRSS becomes large, the tested vibration data are no longer represented by the reference model, i.e., the state of robot operation has changed.

Fuzzy-Logic Based Multi-Assessment

Although there are many cluster functions that can be used for the classification of system states, each has limitations for certain application conditions. When the parameters of a stochastic model are estimated from experimental data, uncertainties always exist in the cluster analysis. If the uncertainty ranges overlap, it is difficult to classify the robot operation state by a single-critical cluster analysis (Figure 2). If more than one cluster function is simultaneously applied to a fuzzy condition, conflicting conclusions may be generated because of the uncertainties. Thus, a multi-assessment method is needed to increase the reliability of the diagnostic decisions. To describe distorted system patterns under ambiguous conditions, fuzzy logic can be applied. Between two extreme situations (where the rate of success, or failure detection, is 1, or 0), we define intermediate values. When one of the cluster methods is used in a fuzzy situation, a diagnosis can be only obtained with a certain level of confidence. By assigning a confidence degree matrix (membership function) to each of the cluster methods, a more confident decision can be obtained from the multi-assessment procedure. The membership functions can be defined through a sensitivity analysis of each cluster method, based on an analysis of the robot operation.

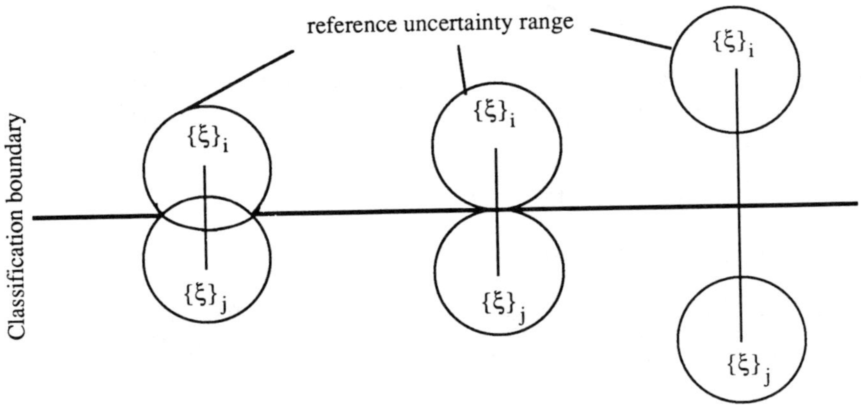

Figure 2. Classification uncertainty and boundary

A feature vector of the robot operation, measured by the cluster functions listed above, is defined $\{U\} = \{u_1, u_2, ..., u_n\}$; and a system state vector of robot operation is $\{V\} = \{v_1, v_2, ..., v_m\}$. When the feature u_j appears in the vibration signal, the probability of the system state v_i is defined by a fuzzy number, r_{ij} $(0 \leq r_{ij} \leq 1)$. Therefore, we obtain a matrix of fuzzy membership for the features $\{U\}$ and states $\{V\}$ of the robot operation, $[R] = (r_{ij})_{nxm}$. When a diagnostic analysis is implemented, the normalized dissimilarity distances between the new observation and reference states are defined and calculated as the vector $\{A\} = \{a_1, a_2, ..., a_n\}^T$ to represent the degree of confidence of the effectiveness and the priority of each cluster method for a specific monitoring variable. The values of r_{ij} and a_i are obtained from a sensitivity analysis for each cluster method through an off-line experimental self-learning process, and stored in an updated knowledge base. Therefore, when a new observation is measured, the diagnosistic decision of the multi-assessment can be expressed as follows:

$$\{B\} = [R] \bullet \{A\} \tag{5}$$

where • denotes an assessment operation [8].

The result of the fuzzy analysis generates recognition of the pattern, i.e., when $b_i \geq \lambda_i$, the system state v_i is true. In this way, the observed robot operation state can be classified into one of the reference states, and the failure modes can be localized and identified.

Diagnostic Monitoring Algorithm

A two-step approach is proposed to implement the diagnostic monitoring procedure in a real-time environment. In the first step, a fast monitoring algorithm is designed for on-line detection, where only the relative standard deviation $s_{n,t}$ is calculated to find deviation of the system state from the normal state. In the second step, when an abnormal system state is detected, the diagnosis is carried out through a cluster analysis and a fuzzy-logic based multi-assessment by using the information stored in the knowledge base. The computational speed in second step is also important because a real-time decision may be necessary to achieve a control action after the diagnosis. A reference is obtained by off-line learning. Figure 3 shows a diagram of the diagnostic monitoring system.

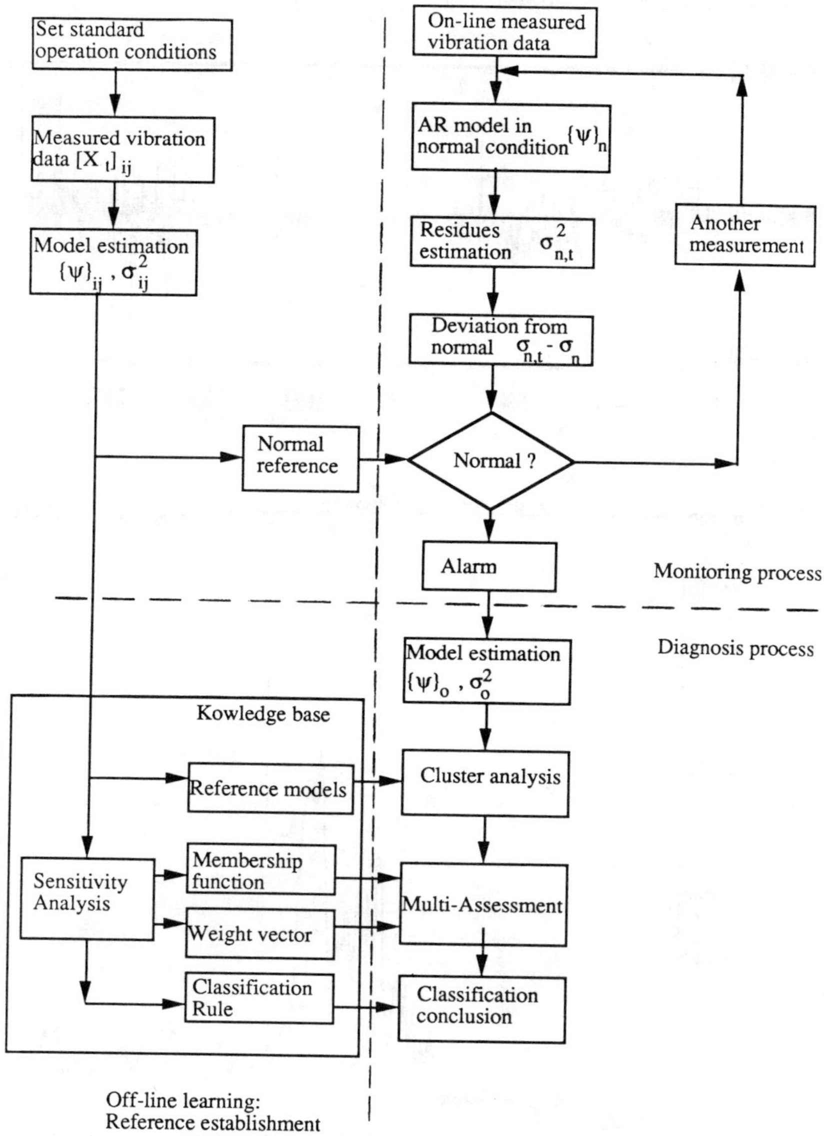

Figure 3. Diagnostic monitoring algorithm

IV EXPERIMENTAL RESULTS AND ANALYSIS

An IBM Model 7535 industrial robot was programmed to perform a pick-and-place operation in a laboratory environment. The vibration signals of the robot drive motors were measured. Figure 4 shows a typical vibration signal and its power spectrum. The cluster analysis methods listed in Section 3 were defined as the feature vector {U} and the possible failures described in Section 2 were defined as the robot operation state vector {V}. The effects of the monitoring and diagnosis variables were studied to establish the diagnosis reference and to build up a knowledge base of robot operational data. After all of the effects for characterizing the monitoring and diagnosis variables were obtained, the matrix of membership function of fuzzy-logic assessment [R] and the weight vector {A} were evaluated.

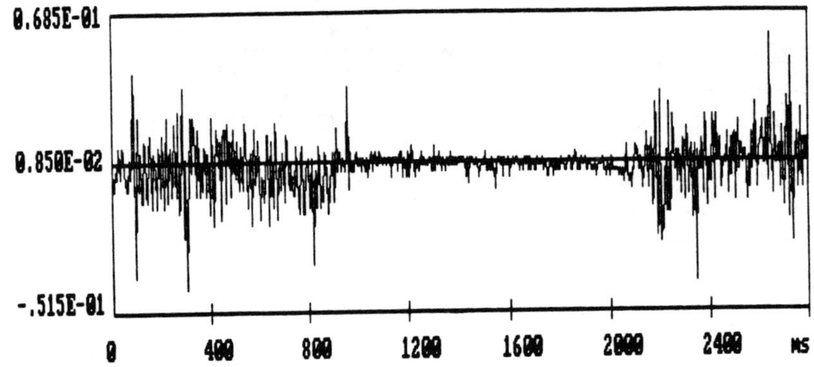

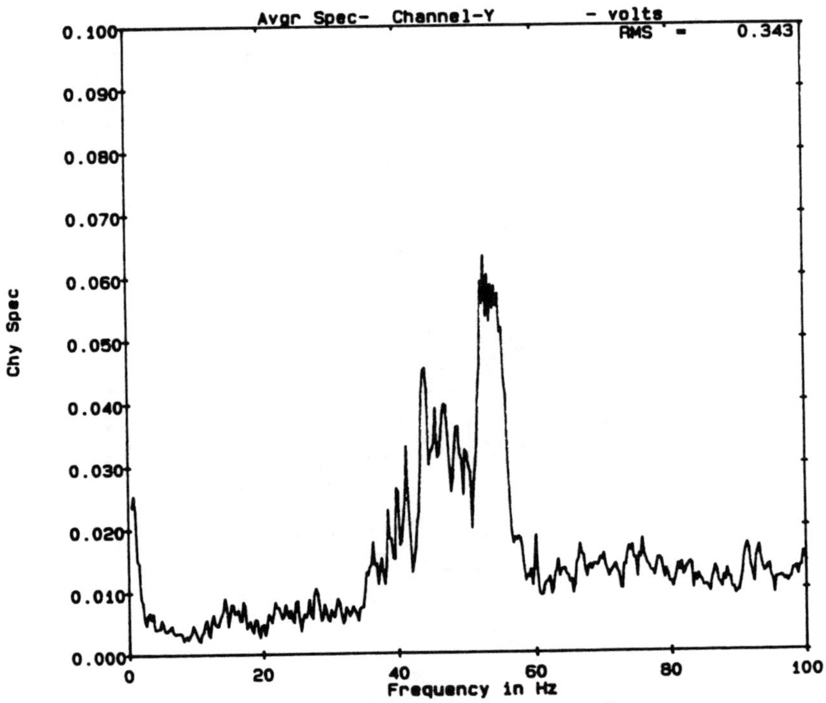

Figure 4. Robot vibration signal in time domain and in frequency domain

Diagnostic Analysis

Operational Speed Effects. The speed adjustment was given in a PAYLOAD command where a small value denotes slow motion. PAYLOAD 5 was considered to be the nominal robot motion. Deviations from this nominal pattern were investigated. Figure 5a shows the normalized intra- distance and inter-distance between the robot operation states. Figure 5b shows the deviations of the robot operation state from normal operational speed. When the robot operational speed was changed, either increased (PAYLOAD 7) or decreased (PAYLOAD 3), deviations where detected through the improved Euclidean distance, cross-entropy distance, NRSS, and characteristic frequency. Malfunctions of the robot controller were therefore identified.

Cross-entropy distances

payload	3	5	7
3	0.781	1.779	4.438
5		1.000	2.598
7			0.981

Improved Euclidean-distances

payload	3	5	7
3	0.914	4.668	7.844
5		1.000	1.738
7			1.213

NRSS

payload	3	5	7
3	0.986	1.713	2.330
5		1.000	1.630
7			1.059

Characteristic Frequency F3

payload	3	5	7
3	0.833	1.962	2.643
5		1.000	1.762
7			0.983

Figure 5.a Normalized Intra- and Inter- "dissimilarity distances"

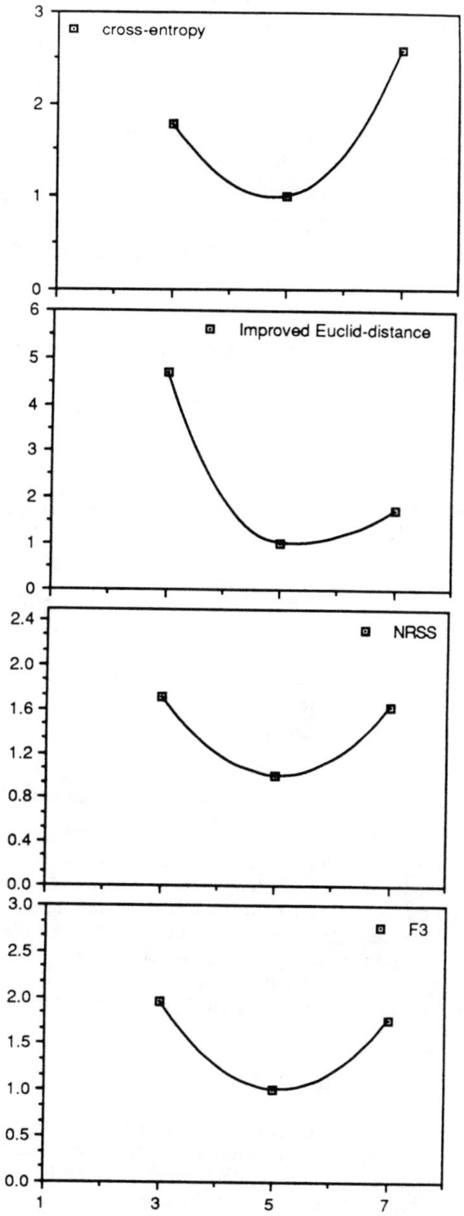

Figure 5.b Illustration of robot operation speed effects

Lubricant Effects. When the robot lubricant condition is changed (either poor lubricant condition before maintenance or an over lubricated condition after maintenance), the most sensitive features are the characteristic damping ratio and the NRSS, as shown in Figure 6. This information is particularly important for preventive maintenance.

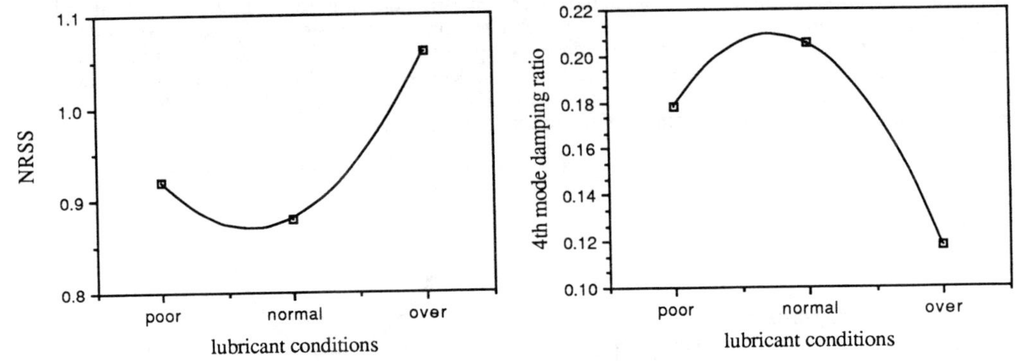

Figure 6. Lubricant effects

Belt Fit. Loose transmission belts potentially affect the performance of robot operation. Changes of belt fit can be tested through diagnostic monitoring. Figure 7 shows the effect of a loose belt, where the most sensitive cluster features are the characteristic damping ratio and the NRSS.

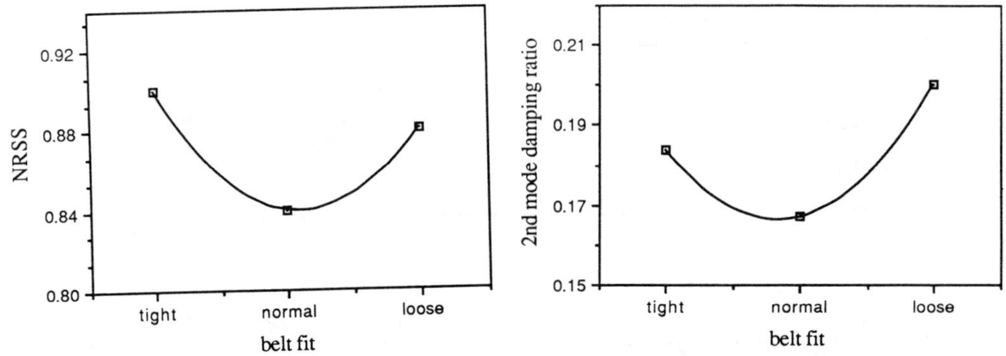

Figure 7. Belt-looseness effects

Clearance Effects. Mechanical clearances will be enlarged as the robot mechanical elements wear out. The clearance in the wrist (see Figure 1) was examined. It has been established that a variation of mechanical clearances affects the system eigenvalues due to dynamics of the joint contacts [9]. The clearances affect robot operations as shown in Figure 8, measured by the characteristic frequency and the NRSS, which changed not only in their average value but also in their range of distribution. In addition, mechanical clearances lead to ambiguous classification of robot operation states. Figure 8 is an example showing diagnostic results of robot speed classification.

Multi-Assessment Analysis

An example of identifying robot operation states is given in this section. Due to different sensitivities of each cluster function for identifying a specific robot operation state (see Figure 5), a membership function and a weight vector need to be assigned to represent different degrees of confidence for each cluster. For the diagnostic analysis, a multi-assessment vector for the robot motion was defined by $\{U\} = \{d_{ed}, d_{ce}, F_3, NRSS\}$ and a system state vector by $\{V\} = \{3, 5, 7\}$, where F_3 is the third mode frequency, a chosen characteristic frequency, and the coordinates of

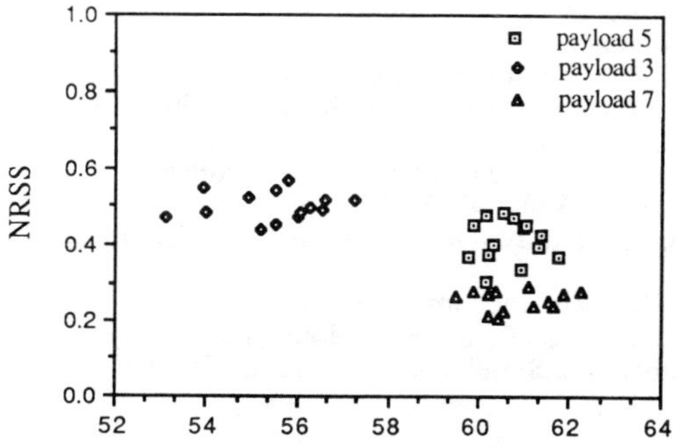

Figure 8.a Diagnosis of robot operation speed without clearance effects

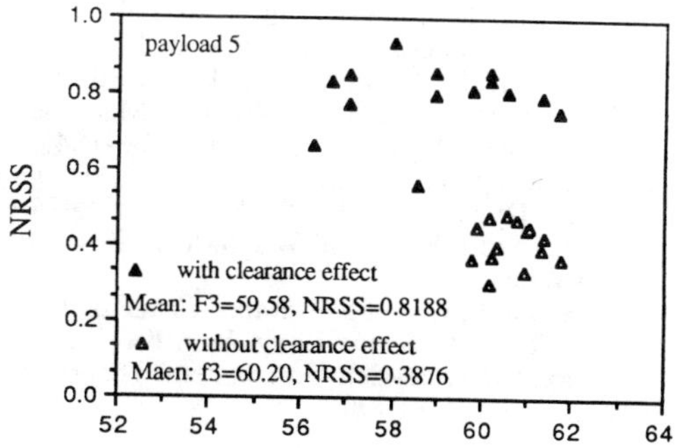

Figure 8.b Diagnosis of robot operation speed with clearance effects

{V} denote the payload. Table 1 shows the membership function matrix, which is based on a sensitivity analysis [6, 8]. The value 0.2 is assigned to the most improbable situation and 0.8 is assigned to the most probable case.

The reliability of the diagnostic decisions was significantly improved by using the fuzzy-logic based multi-assessment procedure. Table 2 shows a comparison of the Rate of Success (ROS) for diagnostic monitoring using single criterion diagnosis and multi-assessment analysis, based on experimental tests. The improvement was significant. Choosing F3 and NRSS as two-criterion cluster functions, the degree of confidence is large (90.8%) when the clearance effect is not significant (Figure 8a). This situation does not occur, however, with the identification of other robot operation states, e.g., lubricant conditions.

Table 1. Membership functions for multi-assessment

payload \ assessment	d_{ed}	d_{ce}	F3	NRSS
payload 3	0.2	0.2	0.2	0.2
payload 5	0.2	0.8	0.5	0.3
payload 7	0.8	0.2	0.5	0.7

Table 2. Comparison of the Rate of Success

assessment	d_{ed}	d_{ce}	F3	NRSS	Multi-
ROS (%)	68.8	70.5	42.7	37.3	97.3

V SUMMARY

This paper studied on-line monitoring and diagnosis of robot operation, which is important for preventive maintenance in automated production environments. A fuzzy-logic based multi-assessment technique using four different cluster functions was developed to provide a more reliable diagnostic decision. Because of the use of fuzzy-logic and multi-assessment approaches, the rate of success of the diagnosis is significantly improved, as indicated by experimental results.

VI REFERENCES

1. Moore, T. N., 1986, "Automatic Diagnostics for Robot-based Assembly Systems," ASME WAM, *Modeling, Sensing and Control of Manufacturing Processes*, PED-Vol. 23, DSE-Vol. 4, pp. 323-331.

2. Tonshoff, H. K., Wulfsberg, J. P., Kals, H. J., Koning, K. and van Luttervelt, C. A., 1988, "Developments and Trends in Monitoring and Control of Machining Processes, *Annals of the CIRP,* Vol. 37/2, pp. 611-622.

3. Li, C. J. and Wu, S, M., 1989, "On-line Detection of Localized Defects in Bearings by Pattern Recognition Analysis," *ASME J. of Engineering for Industry,* Vol. 111, No. 4, pp. 331-336.

4. Luo, R., Chen, J. and Yang, J. C. S., 1984, "Identification of the Dynamic Characteristics of Robotic Systems," SME, Proc. of *Robots 8 Conference*, pp.103-119.

5. Gersch, W. Brotherton, T. and Brown, S., 1983, "Nearest Neighbor-Time Series Analysis Classification of Faults in Rotating Machinery," *ASME J. of Vibration, Acoustics, Stress, and Reliability in Design,* Vol. 105, pp. 178-184.

6. Rong, Y. and Tzou, H. S., 1991, "Diagnostic Monitoring and Sensitivity Analysis of Contact Dynamics in Jointed Structures," *AIAA Journal,* Vol. 29, No. 7.

7. Pandit, S. M. and Wu, S. M., 1978, *Time-Series and System Analysis with Applications,* John Wiley.

8. Yu, P. L., 1985, *Multiple-Criteria Decision Making,* Plenum Press.

9. Tzou, H. S. and Rong, Y., 1991, "Contact Dynamics of a Spherical Joint and a Jointed Truss-Cell Unit System: Theory and Stochastic Simulation," *AIAA Journal,* Vol. 29, No. 1, PP. 81-88.

Sensors, Controls, and Quality Issues
in Manufacturing
ASME 1991

IN-PROCESS MEASUREMENT OF ROUNDNESS AND CYLINDRICITY

K. F. Ehmann and C. Lin
Department of Mechanical Engineering
Robert R. McCormick School of Engineering and Applied Science
Northwestern University
Evanston, Illinois

ABSTRACT

Measurements performed with respect to imperfect references, characterizing most in-process measurement situations, must be corrected to account for the error caused by this fault. A theoretical model expressing the influence of axis error motions on the measurement error is formulated for an arbitrary measuring task, and also for typical cylindricity and roundness measuring configurations. To be able to correct for axis error motions a laser-based sensor system for the measurement of rigid body motions of an axis of rotation is proposed. A detailed theoretical analysis of the sensor is performed including the influence of manufacturing errors, component tolerances and sensitivities. Based on this general purpose device specific in-process sensors are discussed for roundness and cylindricity measurements.

1. INTRODUCTION

Accurate measurements of dimensions, form and surface geometry require well defined frames of reference, which closely approximate the ideal or theoretical frames. This requirement has imposed the prevalent approach to dimensional, form and surface metrology which relies on the use of measuring devices, separated from the actual manufacturing environment, which are capable of providing measuring references of sufficient accuracy. This need is obvious since any error in any of the basic references directly influences measurement accuracy. Instruments built on these principles, however, have a number of disadvantages, namely they lack adaptability for use in the workshop environment, they are not easy to use and also not suited for in-process measurements.

On the other hand, existing approaches to in-situ measurements, on the production machine itself, are based on the assumption that the reference frames of the machine are sufficiently precise to be adopted as measurement reference frames. The validity of this assumption is, however, limited to relatively routine measurements with average tolerance requirements. The accuracy of in-process reference frames for performing high precision in-situ measurements is often orders of magnitudes lower from the accuracy that would be normally required of measuring machines performing commensurate measurements. Thus, to perform in-situ measurements on production machinery it is necessary, by some suitable means, to account for the imperfection of the references with respect to which measurements are made.

The fundamental principle used in dimensional metrology is based on the establishment of a measure of the relative position (in this context position applies to both linear and angular measures) between the workpiece and the gage or equivalently between their respective frames. Therefore, any deviation between the nominal relationship between the workpiece and gage frames will be reflected in a measurement error. Furthermore, since repeated relative measurements necessitate some kind of motion between the workpiece and gage, deviations from ideal motions will also result in error. Additional error sources include thermal drifts, structural error motions, etc. In generalized terms the relationship between the reading of the gage, G, the actual deviation of the workpiece feature being measured from its ideal theoretical value, W, and the unavoidable error in the ideal relative position between the workpiece and gage frames, E, is given by:

$$G = W + E \tag{1}$$

This relationship can be used to expresses, in the simplest terms, the philosophies related to metrology on dedicated instruments as opposed to the in-situ approach. Measurement machines are built such as to reduce the magnitude of E to a level at which it can be neglected so that the gage reading is equivalent to the workpiece deviation. In the case of in-situ measurements on production machinery E is usually not negligible. Consequently, to recover the true workpiece deviation, W, from the gage output, G, E has to be determined. This paper explores the feasibility of measuring the error between the gage and workpiece frames and formulates principles of in-process metrology for rotationally symmetric features of machined workpieces.

First, in Section 2, the geometric and kinematic relationships between the actual machine frames, the tool and/or gage frames and the ideal theoretical reference frames is established. The widely adopted Denavit-Hartenberg (DH) approach is followed. Under the assumption of small deviations a linear error propagation model of a general multi-axis machine is formulated. Second, the physical realization of independent reference frames is considered in Sections 3 and 4. A laser based sensing device is proposed for the measurement of small rigid body motions of axes of rotation and translation respectively of machines used for the production of rotationally symmetric features. The governing optical equations of the sensor facilitate its calibration and resolution of the rigid body motion components of the axes. The inherent inaccuracies in the manufacture of the sensor and the inaccuracies of the optical components themselves are also considered in the modeling of the sensor. Finally, preliminary results on the implementation of sensors based on the proposed principles for the measurement of roundness and cylindricity are presented in Section 5. It has been shown that the proposed approach can provide an independent frame of reference which is considerably more accurate than the machine's own frames.

2. ERROR MODEL OF A MULTI-AXIS MEASURING MACHINE

The purpose of this section is to lay the foundation for the determination of the error quantity E in Eq. 1. as a function of error motions in the active degrees of freedom of motion and the imperfect configuration of the machine. In complex measuring situations the establishment of the desired relationship between the workpiece and gage in general involves motions (both rotary and translatory) in multiple axes resulting in the accumulation and interaction of errors associated with each axis of motion. Under the assumption of rigid body motions, a mathematical expression for the deviation of the gage frame with respect to the workpiece frame from the nominal can be expressed by following the concepts established by Reshetov et al (1988), Lin (1989) and others as briefly summarized below.

By expressing the nominal relative relationship between the individual adjacent axes by a homogeneous DH matrix $\mathbf{A}_i^N$ the nominal position and orientation of the gage with respect to the workpiece can be expressed as (Paul, 1982):

$$T_n^N = \prod_{i=1}^{n} A_i^N \tag{2}$$

where n is the total number of axes. In reality due to errors of various origins the actual position and orientation will differ from the ideal one so that in the above relationship the actual transformation $\mathbf{A}_i^C$ should be used to yield the actual relative position $\mathbf{T}_n^C$. The relationship between the actual and nominal transformations for a single axis can be simply expressed as:

$$\mathbf{A}_i^C = \mathbf{A}_i^N + d\mathbf{A}_i = \mathbf{A}_i^N + \delta\mathbf{A}_i\mathbf{A}_i^N \tag{3}$$

where $d\mathbf{A}_i$ depends on the rotational and translational error motions of the axis.

The development of the error propagation model for the whole machine requires the evaluation and substitution of Eq. (3) into Eq. (2) yielding the following error propagation equation for a general multi-axis machine configuration:

$$\mathbf{T}_n^C = \prod_{i=1}^{n} \mathbf{A}_i^N + \sum_{i=1}^{n} \mathbf{A}_1^N \ldots \mathbf{A}_{i-1}^N \delta\mathbf{A}_i\mathbf{A}_i^N \ldots \mathbf{A}_n^N + \ldots = \mathbf{T}_n^N + d\mathbf{T}_n + \ldots. \tag{4}$$

For small error components, corresponding to the usual practical situation, the retention of the first order error matrix (second term on the r.h.s.) results in sufficient accuracy. The first order error matrix has an anti-symmetric form and contains six independent terms three of which correspond to the total differential translation error and the remaining three to the total differential rotation error components. The final form for the first order error term can be written (Lin 1989) as:

$$d\mathbf{T}_n = \begin{bmatrix} \Sigma_t^e \mathbf{R}_n & \Sigma_t^e \mathbf{p}_n + \delta_t^e \\ 0^T & 0 \end{bmatrix} \tag{5}$$

where the total differential translation vector is given by

$$\delta_t^e = \sum_{i=1}^{n} \{\mathbf{p}_{i-1} \, x(\mathbf{R}_{i-1}\,\mathbf{M}_{1i}) + \mathbf{R}_{i-1}\mathbf{M}_{2i}\}\mathbf{E}_{1i}$$

and the three components of the differential rotation matrix, Σ_t^e, arranged in the standard anti-symmetric form, by the differential rotation vector

$$\varepsilon_t^e = \sum_{i=1}^{n} \mathbf{R}_{i-1}\mathbf{M}_{3i}\mathbf{E}_{2i}$$

and where

$$\mathbf{E}_{1i} = [\Delta\Gamma_i \; \Delta\Psi_i \; \Delta\Phi_i \; \Delta\theta_i \; \Delta\alpha_i \; \Delta x_i \; \Delta y_i \; \Delta z_i \; \Delta h_i \; \Delta a_i\,]^T$$

$$\mathbf{E}_{2i} = [\Delta\Gamma_i \; \Delta\Psi_i \; \Delta\Phi_i \; \Delta\theta_i \; \Delta\alpha_i\,]^T$$

$$\mathbf{M}_{1i} = \begin{bmatrix} 1 & 0 & 0 & 0 & C\theta_i & 0 & 0 & 0 & 0 & 0 \\ 0 & 1 & 0 & 0 & S\theta_i & 0 & 0 & 0 & 0 & 0 \\ 0 & 0 & 1 & 1 & 0 & 0 & 0 & 0 & 0 & 0 \end{bmatrix}$$

$$\mathbf{M}_{2i} = \begin{bmatrix} 0 & 0 & 0 & 0 & -h_iS\theta_i & 1 & 0 & 0 & 0 & C\theta_i \\ 0 & 0 & 0 & 0 & h_iC\theta_i & 0 & 1 & 0 & 0 & S\theta_i \\ 0 & 0 & 0 & 0 & 0 & 0 & 0 & 1 & 1 & 0 \end{bmatrix}$$

$$\mathbf{M}_{3i} = \begin{bmatrix} 1 & 0 & 0 & 0 & C\theta_i \\ 0 & 1 & 0 & 0 & S\theta_i \\ 0 & 0 & 1 & 1 & 0 \end{bmatrix}$$

and x denotes the vector cross product. The elements of error vectors $\mathbf{E}_{1i}$ and $\mathbf{E}_{2i}$ are the angular error motion components (yaw, pitch and roll) ($\Delta\Gamma_i$, $\Delta\Psi_i$, $\Delta\Phi_i$), translatory error motion components (Δx_i, Δy_i, Δz_i), and errors in the customary kinematic parameters (Paul, 1982) of the axis, i.e., ($\Delta\theta_i$, $\Delta\alpha_i$, Δh_i, Δa_i), while $\mathbf{R}_{i-1}$ and $\mathbf{p}_{i-1}$ represent the rotational matrix and position vector of the nominal transformation up to the (i-1)th frame.

2.1. Error Model of Cylindricity Measurements

The above error propagation equation also applies to cylindricity measuring devices regardless whether the measurements are performed on a special purpose instrument or on the production machine itself. The fundamental principle of cylindricity measurements involves a rotary and translatory motion of the gage relative to the cylinder to be measured (Kakino and Kitazawa, 1978, and Murthy and Rao, 1981), and hence, is influenced by errors associated with the rigid body motions of the corresponding axes. The equation describing the total error of the gage is given by Eq. (5) evaluated for n=2 with the appropriate kinematic parameters. In explicit form the error propagation equation, for the cylindricity measurement configuration with a rotating sensitive direction, shown in Fig. 1a, is given by:

$$\begin{aligned} d\mathbf{T}_2 = & \begin{bmatrix} 0 & 0 & 0 & 0 \\ 0 & 0 & -1 & -h_1-h_2 \\ S_2 & C_2 & 0 & a_2 S_2 \\ 0 & 0 & 0 & 0 \end{bmatrix} \Delta\Gamma_1 + \begin{bmatrix} 0 & 0 & 1 & h_1+h_2 \\ 0 & 0 & 0 & 0 \\ -C_2 & S_2 & 0 & -a_2 C_2 \\ 0 & 0 & 0 & 0 \end{bmatrix} \Delta\Psi_1 \\ & + \begin{bmatrix} -S_2 & -C_2 & 0 & -a_2 S_2 \\ C_2 & -S_2 & 0 & a_2 C_2 \\ 0 & 0 & 0 & 0 \\ 0 & 0 & 0 & 0 \end{bmatrix} \Delta\Phi_1 + \begin{bmatrix} 0&0&0&1 \\ 0&0&0&0 \\ 0&0&0&0 \\ 0&0&0&0 \end{bmatrix} \Delta x_1 + \begin{bmatrix} 0&0&0&0 \\ 0&0&0&1 \\ 0&0&0&0 \\ 0&0&0&0 \end{bmatrix} \Delta y_1 + \begin{bmatrix} 0&0&0&0 \\ 0&0&0&0 \\ 0&0&0&1 \\ 0&0&0&0 \end{bmatrix} \Delta z_1 \\ & + \begin{bmatrix} 0 & 0 & 0 & 0 \\ 0 & 0 & -1 & -h_2 \\ S_2 & C_2 & 0 & a_2 S_2 \\ 0 & 0 & 0 & 0 \end{bmatrix} \Delta\Gamma_2 + \begin{bmatrix} 0 & 0 & 1 & h_2 \\ 0 & 0 & 0 & 0 \\ -C_2 & S_2 & 0 & -a_2 C_2 \\ 0 & 0 & 0 & 0 \end{bmatrix} \Delta\Psi_2 + \begin{bmatrix} -S_2 & -C_2 & 0 & -a_2 S_2 \\ C_2 & -S_2 & 0 & a_2 C_2 \\ 0 & 0 & 0 & 0 \\ 0 & 0 & 0 & 0 \end{bmatrix} \Delta\Phi_2 \\ & + \begin{bmatrix} 0&0&0&C_2 \\ 0&0&0&S_2 \\ 0&0&0&0 \\ 0&0&0&0 \end{bmatrix} \Delta x_2 + \begin{bmatrix} 0&0&0&-S_2 \\ 0&0&0&C_2 \\ 0&0&0&0 \\ 0&0&0&0 \end{bmatrix} \Delta y_2 + \begin{bmatrix} 0&0&0&0 \\ 0&0&0&0 \\ 0&0&0&1 \\ 0&0&0&0 \end{bmatrix} \Delta z_2 \end{aligned} \tag{6}$$

where S_i and C_i, i=1,2, stand for $\sin\theta_i$ and $\cos\theta_i$ respectively.

For the sake of simplicity in Eq. (6) only error motion components are retained. To include errors associated with errors in the customary DH kinematic parameters defining the mutual relationship between the two axes eight additional terms would have to be added. The reason for omitting these terms is that they can be relatively easily accounted for, since they remain constant for a given configuration, by suitable calibration methods (see Wu 1986). On the contrary error motions are in general not repeatable.

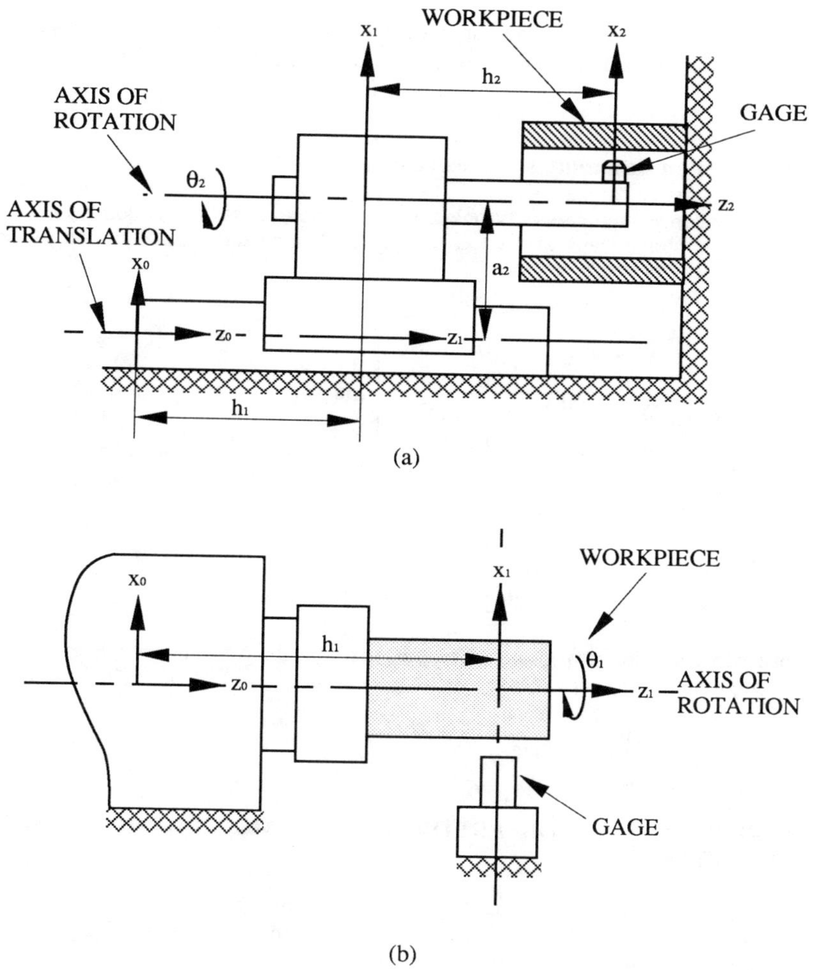

Fig. 1. Plausible Configurations for Cylindricity (a) and Roundness Measurements (b)

From the inspection of Eq. (6) and of the measurement configuration it is clear that the contributions of the various error motions to the output of the gage are not of the same significance. Since in this case measurements are primarily influenced by components which lead to deviations in the sensitive direction (ANSI/ASME, 1986), components which result in errors in other directions can be neglected. For the case under consideration deviations take place in the x and y directions so that the magnitude of the error motion in the sensitive direction is given by:

$$E_{cyl} = \sqrt{E_x^2 + E_y^2} \tag{7}$$

where the total component error motions, from Eq. (6), are:

$$E_x = (h_1 + h_2)\Delta\Psi_1 - a_2 S_2 \Delta\Phi_1 + \Delta x_1 + h_2 \Delta\Psi - a_2 S_2 \Delta\Phi_2 + C_2 \Delta x_2$$
$$E_y = -(h_1 + h_2)\Delta\Gamma_1 + a_2 C_2 \Delta\Phi_1 + \Delta y_1 - h_2 \Delta\Gamma_2 + a_2 C_2 \Delta\Phi_2 + S_2 \Delta x_2 + C_2 \Delta y_2$$

Equation (7), therefore, represents the fundamental equation expressing the contribution of imperfect motions on the readings of the gage. Thus, this quantity is directly related to E in Eq. (1).

2.2. Error Model of Roundness Measurements

In roundness measurements only one axis of rotation causes error motions (Fig. 1b). A simple derivation as above leads to the following complete fundamental equation of the error motions for roundness measurements:

$$dT_1 = \begin{bmatrix} 0 & 0 & 0 & 0 \\ 0 & 0 & -1 & -h_1 \\ S_1 & C_1 & 0 & 0 \\ 0 & 0 & 0 & 0 \end{bmatrix} \Delta\Gamma_1 + \begin{bmatrix} 0 & 0 & 1 & h_1 \\ 0 & 0 & 0 & 0 \\ -C_1 & S_1 & 0 & 0 \\ 0 & 0 & 0 & 0 \end{bmatrix} \Delta\Psi_1 + \begin{bmatrix} -S_1 & -C_1 & 0 & 0 \\ C_1 & -S_1 & 0 & 0 \\ 0 & 0 & 0 & 0 \\ 0 & 0 & 0 & 0 \end{bmatrix} \Delta\Phi_1$$

$$+ \begin{bmatrix} 0 & 0 & 0 & 1 \\ 0 & 0 & 0 & 0 \\ 0 & 0 & 0 & 0 \\ 0 & 0 & 0 & 0 \end{bmatrix} \Delta x_1 + \begin{bmatrix} 0 & 0 & 0 & 0 \\ 0 & 0 & 0 & 1 \\ 0 & 0 & 0 & 0 \\ 0 & 0 & 0 & 0 \end{bmatrix} \Delta y_1 + \begin{bmatrix} 0 & 0 & 0 & 0 \\ 0 & 0 & 0 & 0 \\ 0 & 0 & 0 & 1 \\ 0 & 0 & 0 & 0 \end{bmatrix} \Delta z_1 \qquad (8)$$

In addition since only two error motion components contribute to the error E in the sensitive direction the approximate equation can be written, based on Eq. (8), in the following form:

$$E_{rnd} = h_1 \, \Delta\Psi_1 + \Delta x_1 \qquad (9)$$

3. IN-PROCESS SENSOR FOR ERROR MOTION MEASUREMENTS FOR AXES OF ROTATION

The evaluation of the error motions of axes of rotation given by Eq. (5) in general and the evaluation of Eqs. (7) and (9) in particular requires the knowledge of the error motion components. Existing solutions limit measurements to the sensitive direction only (Vanherck, 1973, and ANSI/ASME, 1986). For the simultaneous measurement of all error motion components a laser based in-process sensor will be proposed and analyzed in this section.

3.1. Sensor Configuration

To measure the six customary error motion components of an axis of rotation a sensor is needed which can provide six independent readings which are a) directly proportional to the error motion components, or b) from which the error motion components can be resolved. The configuration of a laser based optical system capable of providing the six independent readings as a function of the error motion components is given in Fig. 2.

A laser source is attached to the rotating part i. The beam is projected on a beam splitter which splits the beam into beams **bd** and **bg**. Beam **bg** is reflected into beam **gh** and then **hc** by the mirrors g and h respectively. A second beam splitter splits beam **bd** into **df** and **de**. Because the axi-symmetric mirror h is placed on the rotating part any error motion of part i along the z_{i-1} axis will affect the intersection point of beam **gh** and mirror h and, therefore, be recorded by the dual axis lateral photodetector 1. By treating beam **af** as a rigid link it is easy to see that the readings of photodetector 3 are a function of the remaining two translatory and three rotary error motions. To obtain enough equations to solve for the six error motion components, one additional detector, photodetector 2, is provided. The functional relationships between photodetector readings and error motion components are addressed in the two subsequent subsections.

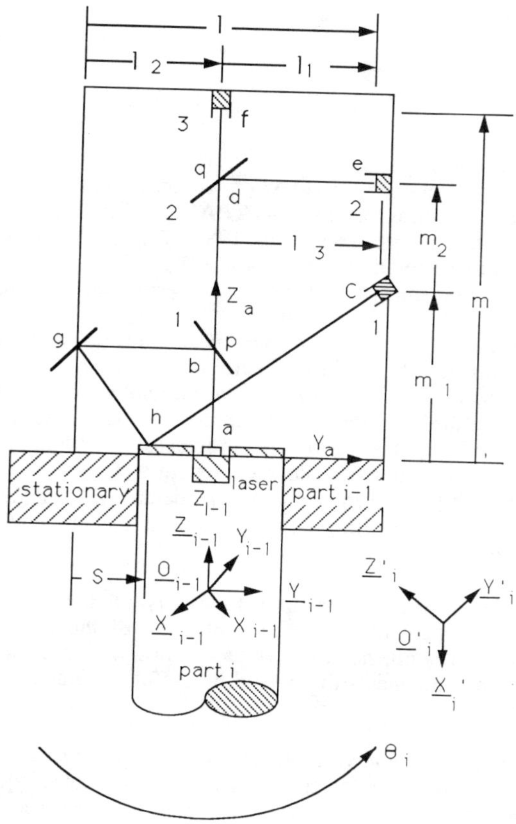

Fig. 2. Error Motion Measurement Sensor for an Axis of Rotation

3.2. Theoretical Analysis

For establishing a model for the sensor it is convenient to assume that the stationary part (i-1) and rotating part i of the axis and also the reference point with respect to which translational and angular error motions are measured are fixed with respect to frame $(xyz)_{i-1}$. Coordinate frame $(\underline{xyz})_{i-1}$ is defined by rotating $(xyz)_{i-1}$ about the z_{i-1} axis so that the $(yz)_{i-1}$ plane becomes parallel to the ideal light traveling plane. In the absence of error motions, a third coordinate frame $(\underline{xyz})_i'$ attached to the part i always coincides with coordinate frame $(\underline{xyz})_{i-1}$ In the presence of spindle error motions, frame $(\underline{xyz})_i'$ will deviate from frame $(\underline{xyz})_{i-1}$ and move to a new position and orientation. Schematically, this deviation corresponds to the vector $\underline{o}_{i-1}\underline{o}_i'$ in Fig. 2. In the derivation of the expressions for the sensor readings, all coordinates of the optical components and the homogeneous transform matrices will be referred to frame $(xyz)_{i-1}$.

Besides the six error motions the readings of the sensors will be also affected by the imperfect mounting of the optical components of the measurement system. These mounting errors will result in a constant (systematic) error in the readings and must be included into the mathematical model of the measurement system. The deviations from the ideal conditions of an arbitrary optical component R, which will result in systematic errors, can be denoted by $(\Delta x_R, \Delta y_R, \Delta z_R)$ and $(\Delta\Gamma_R, \Delta\Psi_R, \Delta\Phi_R)$, for translatory and angular misalignment. For convenience a coordinate frame $(xyz)_R$ is also attached to every optical component. The transformation matrix establishing the relationship between coordinate frames $(\underline{xyz})_{i-1}$ and

$(xyz)_R$, which has the same orientation as $(xyz)_{i-1}$ under ideal conditions and is embedded into component R (R also represents the origin of $(xyz)_R$) is given by the general transformation

$$^{i-1}A_R = \mathbf{TRANS}\,(\mathbf{P}_{xR},\mathbf{P}_{yR},\mathbf{P}_{zR})\ \mathbf{RPY}\,(\Delta\Phi'_R,\Delta\Psi'_R,\Delta\Gamma'_R) \qquad (10)$$

where $(\mathbf{P}_{xR},\mathbf{P}_{yR},\mathbf{P}_{zR})$ is the position vector of frame R depending on the nominal position (x_R,y_R,z_R) and translatory errors $(\Delta x_R',\Delta y_R',\Delta z_R')$, $(\Delta\Phi_R',\Delta\Psi_R',\Delta\Gamma_R')$ are angular deviations of optical component R from its ideal location, and TRANS and RPY designate a translatory and a yaw-pitch-roll transformation respectively (Paul, 1982). Depending on whether component R is mounted on the stationary or rotating part, $(\Delta\Phi_R',\Delta\Psi_R',\Delta\Gamma_R')$ is a function of the imperfect placement only, or the sum of imperfect placement and error motion respectively.

In order to measure the error motions of one frame with respect to another the coordinates of the centroid of the laser beam on the active surface of the dual axis photodetectors have to be established. Utilizing the principles of geometric optics this can be accomplished by deriving the equations of the appropriate laser beams and determining their intersection with the active plane of the photodetectors. Although these derivations are relatively straight forward they are of considerable length and, therefore, will not be given here. The interested reader is referred to Lin (1989) for the details of these derivations. In the sequel a qualitative discussion of the resulting equations and of the procedure of deriving the final usable form will be given.

The resulting equations are implicit functions of the parameters of the optical configuration, the misalignment of the components, and of the error motions. They are, in addition, of a very complex nonlinear form which can not yield a closed analytical solution in terms of the error motion components that would be useful for practical implementation in real time on-line situations.

The fact that in reality the error motions and misalignments are very small leads to the possibility of using first order Taylor series expansions of these equations about the ideal theoretical configuration of the sensor system. The resulting linearized equations contain terms which are a function of misalignment variables and of the relative angular position of the moving relative to the stationary part only, and also terms which strictly depend on the nominal geometric parameters and on the error motion components (Lin (1989)). In the first order approximation only the misalignment of the laser beam results in first order error contributions while all other misalignments contribute to second and higher order errors.

3.3. Governing Equations

Since the linearized equations explicitly express the relationship between the sensor readings, misalignment, and error motions they constitute the foundation for the real time evaluation of the error motion components. The form of the linearized equations also leads to the possibility of further simplification. If the output of the sensor, at a sufficient number of relative angular positions, is compared to an exact known calibration standard the equations facilitate the resolution of the terms which strictly depend on the laser's misalignment and, subsequently, the determination of the exact value of the misalignment. This, in turn, allows the computation and removal of the contribution of the misalignment error from the sensor readings.

Once the contribution due to misalignment has been removed from the raw photodetector readings the resulting relationships reduce to the following simple matrix relationship between the photodetector outputs and error motion components:

$$\begin{bmatrix} 1 & 0 & 0 & 0 & c_1 & 0 \\ 0 & -c_7 & -1 & c_2 & 0 & 0 \\ 1 & 0 & 0 & 0 & c_3 & 0 \\ 0 & 1 & 0 & -c_4 & 0 & 0 \\ 1 & 0 & 0 & 0 & c_5 & 0 \\ 0 & 1 & 0 & -c_6 & 0 & 0 \end{bmatrix} \begin{bmatrix} \Delta x_i \\ \Delta y_i \\ \Delta z_i \\ \Delta\Gamma_i \\ \Delta\psi_i \\ \Delta\Phi_i \end{bmatrix} = \begin{bmatrix} \Delta X_{p1} \\ \Delta Y_{p1} \\ \Delta X_{p2} \\ \Delta Z_{p2} \\ \Delta X_{p3} \\ \Delta Y_{p3} \end{bmatrix} \qquad (11)$$

or in compact form:

$$\mathbf{C\Xi} = \mathbf{S} \qquad (12)$$

where

$$c_1 = \left(2l_2 + m_1 + l_3 + z_a + \sqrt{2}S\right) + \frac{1}{\sqrt{2}}\left(l_2 + l_3 + m_1 - S - t_h\right)$$

$$c_2 = \frac{3S + l_2 - l_3 - m_1 + t_h}{2} + \frac{l_2 + S + t_h + z_a}{\sqrt{2}}$$

$$c_3 = l_1 + m_1 + m_2 + z_a - \frac{M - n}{\sqrt{2}M}t_s$$

$$c_4 = l_1 + m_1 + m_2 + z_a - \frac{n^3 - M^3}{\sqrt{2}M^3}t_s$$

$$c_5 = m + z_a + \frac{\sqrt{2}(n - M)}{M}t_s$$

$$c_6 = m + z_a + \frac{\sqrt{2}\left(n^3 - M^3\right)}{M^3}t_s$$

$$c_7 = \frac{1}{\sqrt{2}}$$

and where

$$M = \sqrt{2n_1^2 - n}$$

and n_1 and n are refraction indexes of the beam splitters and air, respectively, and t_s and t_h are the thickness of the beam splitter and mirror h. The vectors Ξ and **S** represent the vector of error motion components and raw photodetector readings respectively.

From Eq. (11) it can be seen that the last column is a zero vector implying, as it could be expected, that the analyzed configuration cannot measure errors about its nominal axis of rotation. This, however, is not a significant drawback since typically this quantity is anyhow measured by the axis feedback transducer. The zero column also implies that not all six equations are independent. Therefore, the proposed sensor configuration is capable of rendering three translational error motion components and two rotational error motion components.

The elements of the square matrix in Eq. (11) or (12) can be also interpreted as sensitivities of the corresponding readings with respect to a particular error motion component. It is evident that the sensitivities to translational errors are always integers and that the sensitivities with respect to rotational errors, defined by c_i, i=1,...6, are functions of the geometric disposition of the sensor. The functionality of this measurement system clearly depends not only on the sensitivity of the readings but also on the resolution of the photodetectors used. The design variables should be selected in such a way that the product of the sensitivity coefficients with the desired overall resolution of the error motion measurements exceeds that of the utilized photodetectors.

The solution of Eq. (12), Ξ, can now be used to compute E in Eq. (1). The relationship between Ξ and E clearly depends on the configuration of the particular measuring setup.

4. IN-PROCESS SENSOR FOR ERROR MOTION MEASUREMENTS FOR AXES OF TRANSLATION

The measurement of error motions in axes of translation can be accomplished by following the same principles as above. However, since the details of such sensors can be found in the literature, e.g. Ni, et al (1988), Ehmann (1990), etc., deliberations on this subject will be omitted here. It will suffice to mention that the relationship between the sensor readings and error motion components is given by the same general relationship expressed by Eq. (12) with the only difference that the translation sensor can not resolve errors in the direction of translation.

5. REALIZATION AND IMPLEMENTATION OF SIMPLIFIED SENSORS

The utilization of the sensors described above in a general cylindricity or roundness measuring instrument or in-situ on a production machine would result in the capability to remove the influence of the error motions from the readings of the gage and the determination of the exact workpiece deviations. This approach would result in the most complete and accurate means of evaluating roundness or cylindricity measures since one could account for both first and higher order effects. In customary gaging configurations, however, it can be shown that the contribution of second order errors can be neglected implying that the proposed sensor configurations could be conceivably simplified. The simplification would primarily account for error motions that lead to significant errors in the sensitive direction only. In this sense in the following two subsections two simplified sensor configurations implemented with and without a laser for in-process measurement of cylindricity and roundness respectively will be revisited and analyzed in the light of the above derivations.

5.1. Cylindricity Measurement

Figure 3 shows the configuration of a simplified measurement system for evaluating the cylindricity of a bore in a cylindrical boring operation which corresponds to the sensing principle given in Fig. 1a. It represents a combination of the two generalized sensors discussed above. The principal idea is the evaluation of the difference between the nominal helical path the gage should follow and the actual measured trace. The measurement of the actual trace is performed by the use of a stationary laser beam which constitutes a straight reference and which is also used to establish the nominal circular reference. The beam after passing through the opening at the center of a mirror, inclined by 45° with respect to the beam, is reflected twice by the roof prism which is offset with respect to the axis average line and which is attached, in the close proximity of the gage, inside the hollow boring bar. The mirror deflects the returning beam onto the surface of a single dual axis lateral photodetector. In the absence of deformations in the structural loop between the prism and gage, the displacements of the gage can be considered equal to the deflections of the prism.

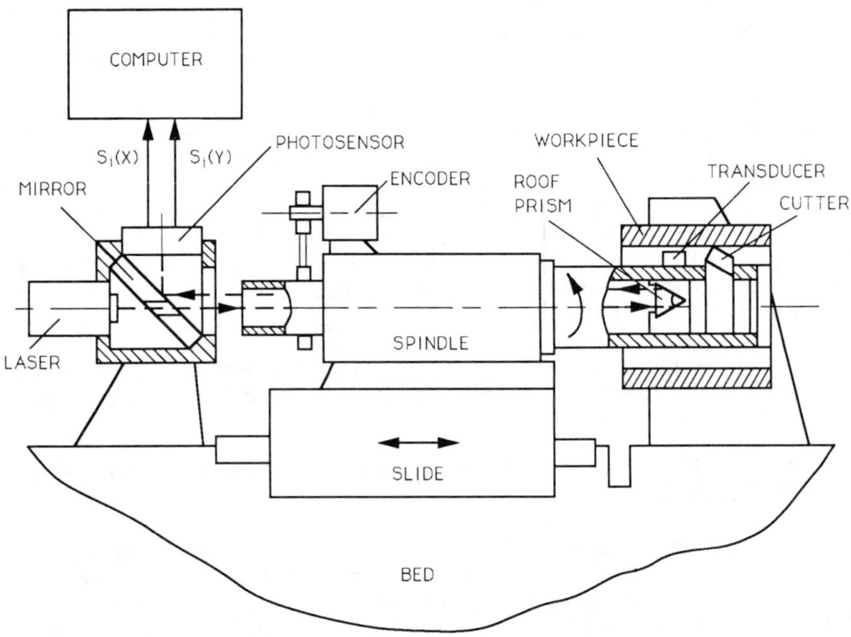

Fig. 3. In-process Measurement of Cylindricity

Under ideal conditions, i.e., ideal linear motion of the spindle along the slide, ideal rotation of the spindle, absence of elastic deflections, parallelism between the axis of rotation and direction of linear motion, and perfect alignment of both optical components of the sensor the centroid of the laser beam on the surface of the photodetector traces a perfect circle. Since, however, the position of the gage (or equivalently roof prism) relative to the reference beam is affected by the imperfect rotation of the boring bar, its deflection, error motion of the linear slide, etc., the trace will deviate from its ideal circular shape. In other words the returning beam is modulated by the various error motion components which lead to angular and translatory displacements of the gage. In addition, the sensor readings will also contain a component which is contributed by the imperfect alignment of the sensor's components. This influence, can be, however, removed by calibrating the system. This contribution will remain constant for a given configuration. For the sensor shown in Fig. 3 a detailed analysis of the influence of component misalignment and a simple calibration procedure was given by Kim et al (1987).

The relationship between a reading at a given axial and angular position on the cylinder's surface (by analogy to Eq. (7)) is given by the following simple relationship:

$$E = \frac{\left(\sqrt{S_x^2 + S_y^2} - 2L\right)}{2} - M \tag{13}$$

where S_x and S_y are the Cartesian coordinates of the centroid of the beam on the surface of the photodetector for a given angular position of the spindle as measured by an encoder, L is the offset of the roof prism with respect to the axis average line, and M the influence of the systematic error due to misalignment.

5.2. Roundness Measurement

For roundness measurements only the error motion components that result in first order error contributions in the sensitive direction need to be considered, i.e., only one translatory and one angular component. A workable configuration is depicted in Fig. 4 in which two non-contact displacement transducers tracing two master cylinders are shown. The output of these two sensors allows the determination of the resulting error motion at say an arbitrary point C.

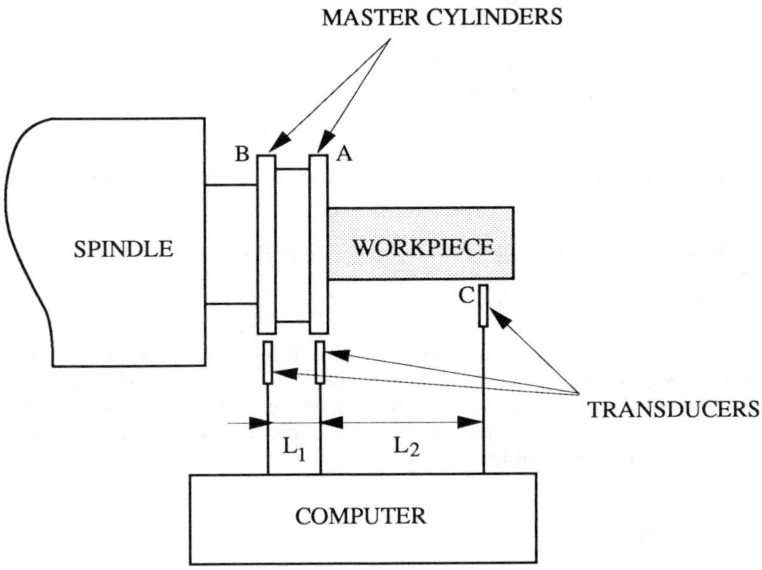

Fig. 4. In-process Measurement of Roundness

Based on the readings by the two sensors at A and B the spindle error motion at point C can be expressed as:

$$E_C = \frac{L_2}{L_1}(E_A - E_B) + E_A \qquad (14)$$

where E_A and E_B are the true spindle radial error motions at locations A and B. Equation (14) is clearly in direct correspondence to Eq. (9). Since the sensor readings at A and B are influenced by the out of roundness of the master cylinders and their eccentric mounting these influences have to be taken into account. The true spindle error motions are related to the transducer readings by:

$$E_q = V_q - U_q - M_q \qquad (15)$$

where q=A,B, V_q is the transducer output, U_q the fundamental component, and M_q the master out of roundness. Methods for determining the fundamental component are given in (ANSI/ASME, 1986, and Kim et al 1987), while the master out-of-roundness deviations can be stored in the computer and removed from the transducer output.

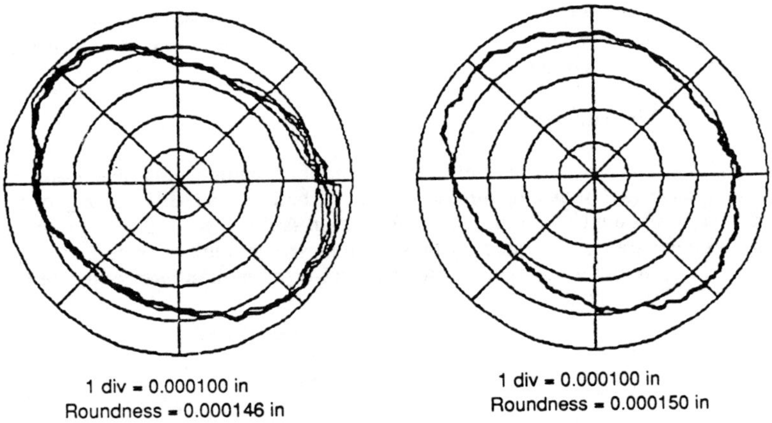

Fig. 5. Five Consecutive Roundness Traces of a Ground Component

The workpiece roundness can now be evaluated as based on Eq. 1 as:

$$W = G - E_C \qquad (16)$$

where G is the signal measured by the sensor scanning the workpiece at the given axial and radial position. An example of the repeatability of this sensor, reproduced from Kim et al (1987), as implemented on a cylindrical grinding machine, is shown in Fig. 5. The repeatable traces for five consecutive revolutions vividly express the ability to in-situ eliminate the error motions from the raw sensor readings.

6. CONCLUSIONS

Principles of in-process measurements in the presence of imperfect references have been established. The following conclusions are drawn:

1. Error propagation relationships can be established for general measurement tasks which facilitate the assessment of the influence of various disturbances on the readings of the gage.

2. In this paper specific error propagation relationships have been developed for roundness and cylindricity measurements influenced by axis error motions.

3. A general sensor for the measurement of five rigid body error motion components for an axis of rotation is formulated and analyzed. The same principles are easily extendable to axes of translation.

4. Two sensors for the in-situ measurement of cylindricity and roundness have been revisited within the generalized framework established in this paper.

REFERENCES

Ehmann, K. F., 1990, "Solution Principles for a New Generation of Precision Self-Correcting Multi-axis Machines," Journal of Robotics & Computer-Integrated Manufacturing, Vol. 7, No. 3/4, pp.357-364.

Kakino, Y., Kitazawa, J., 1978, "In Situ Measurement of Cylindricity," Annals of CIRP, Vol. 27, No. 1, pp. 371-375.

Kim, K. H., Eman, K. F., Wu, S. M., 1987, "Development of a Forecasting Compensatory System for Cylindrical Grinding," Trans. ASME: Journal of Engineering for Industry, Vol. 109, pp. 385-391.

Kim, K., Eman, K. F., Wu, S. M., 1987, "Analysis of Alignment Errors in a Laser-Based In-Process Cylindricity Measurement System," Trans. ASME: Journal of Engineering for Industry, Vol. 109, No. 4, pp. 321-329.

Lin, P. D., 1989, "Error Analysis, Measurement and Compensation for Multi-axis Machines," Ph. D. Thesis, Department of Mechanical Engineering, Northwestern University, Evanston, IL.

Murthy, T. S. R., Rao, S. Y., 1981, "A Simple Approach for Evaluation of Cylindrical Surfaces," Annals of CIRP, Vol. 30, No. 1, pp. 441-444.

Ni, J., Zhang, B. L., Wu, S. M., 1988, "On-line Identification of Volumetric Errors of Multi-axis Machines," Proc. Manufacturing International Conference '88, Atlanta, GA, pp. 77-83.

Paul, R. P., 1982, "Robotic Manipulators: Mathematics, Programming, and Control," The MIT Press.

Reshetov, D. N., Portman, V. T., 1988, "Accuracy of Machine Tools," ASME Press.

Vanherck, P., Peters, J., 1973, "An Axis of Rotation Analyzer,"Proceedings of the 14th MTDR Conference, pp. 299-305.

Wu, C. H., 1986, "Robot Accuracy Analysis Based on Kinematics," IEEE Journal on Robotics and Automation, Vol. 1, pp. 171-179.

ANSI/ASME, 1986, "Axes of Rotation - Methods for Specifying and Testing," ANSI/ASME B89.3.4M - 1985 Standard, ASME.

Sensors, Controls, and Quality Issues
in Manufacturing
ASME 1991

DEVELOPMENT OF AN OPTO-ELECTRIC SENSING SYSTEM TO MONITOR THE ABRASIVE WATERJET NOZZLE WEAR

R. Kovacevic
Center for Robotics and Manufacturing Systems
University of Kentucky
Lexington, Kentucky

ABSTRACT

One of the most critical parts that influences the technical and economical performance of an abrasive waterjet cutting system is the abrasive waterjet nozzle. Nozzle wear could be one of the major contributing factors in reduced precision cutting by abrasive waterjet. The detection of nozzle wear is currently performed visually, either through direct inspection of the tip of the nozzle or indirectly when a deterioration of the quality of surface cut and the workpiece dimensions is observed. Sensing nozzle wear is a key aspect in producing high quality parts on a fully automated abrasive waterjet cutting system.

The final shape of the abrasive waterjet and its cutting performances are determined by the outlet of the nozzle, which is subjected to the erosion wear. An indirect tracking method of the increase in the nozzle outlet diameter is proposed. The proposed method is based on monitoring the increase in the abrasive waterjet diameter by employing a machine vision system. The effect of waterjet pressure, nozzle outlet diameter, and abrasive flow rate on spreading of the abrasive waterjet is analyzed.

INTRODUCTION

In today's rapid development of the manufacturing environment, speed and reliability of manufacturing systems have become the most critical factors in determining the competitiveness of a company. As the integration among different manufacturing units and systems increases, more companies start to replace their conventional manufacturing units with newer ones capable of changing machining parameters automatically and producing several products on the same line. In such systems, various functions of manufacturing processes are monitored by automatic control devices and sensors. Although automatic monitoring is used in various aspects of manufacturing, its most significant application is monitoring the wear of cutting tools.

In abrasive waterjet cutting, the nozzle which is the "cutting tool" of the system has a vital influence on the overall performance of the system. Computer control makes abrasive waterjet (AWJ) cutting a prime candidate for application in flexible manufacturing systems. Automated equipment needs the ability to detect nozzle wear early before kerf width exceeds acceptable limits. The methods used to detect AWJ nozzle wear could be categorized into two groups: direct and indirect. Direct methods make an assessment of nozzle wear either by measuring the inside diameter(ID) of the nozzle at its tip, measuring the material loss of the nozzle at its tip, or measuring the material loss of the nozzle by radiometric techniques. During direct measurement of the nozzle ID, the cutting process has to be interrupted and the jet turned off. Obviously, this method is not suitable for on-line nozzle wear measurement. The requirements for special preparation of the nozzle as well as potential hazards due to radioactivity makes the radiometric technique unsuitable for use on a shop floor. Indirect methods are based on the measurement of some parameters that are correlated to the AWJ nozzle wear such as the change of the stream diameter at the nozzle outlet, the change of the workpiece normal force generated by the impinging jet, level of noise, vibration, etc. A critical review of possible sensors for monitoring the AWJ nozzle wear was given by Kovacevic and Evizi (1990).

In order to develop a satisfactory nozzle wear tracking system a number of attempts were done [Kovacevic, 1990; Kovacevic and Chen, 1989]. Most of the work was oriented toward monitoring of workpiece normal force as a function of the increase in the diameter of the nozzle outlet. It is evident that a wider jet stream through a worn nozzle tends to increase the kerf width and decrease the cutting efficiency, thus contributing to an increase in workpiece normal force level. Therefore, the workpiece normal force generated by an abrasive waterjet could be used as the indicator of the nozzle wear intensity. With an increase in the inside diameter of the nozzle, the workpiece normal force will increase, keeping all other cutting variables constant, and thus nozzle wear can be monitored from an early stage.

Recently, a wear sensor system for direct and almost on-line tracking the wear of an abrasive waterjet nozzle was proposed by Kovacevic (1991). The wear sensor is based on conductive loops which are placed on the ceramic's substrate and embedded on the tip of the nozzle. The integrity or wear condition of the AWJ nozzle is indicated by the electrical closed or open circuit condition of each conductive loop. The continuity status of each wear sensor probe is converted to logic one or logic zero digital data, which is sampled periodically by a central processing unit for providing a visual or audible indication in response to a predetermined erosion or wear condition.

Both previously proposed sensing systems are based on contact sensing units, either with a workpiece if monitoring the workpiece normal force, or with a nozzle if monitoring the increase in diameter at the nozzle outlet. The complexity of the AWJ cutting process, the size of the workpiece, and the requirements for special preparation of the nozzle could put limitations on the practical use of the proposed AWJ nozzle wear sensing methodologies. In the competition among different sensory principles, machine vision systems as a non-contact and micro-computer based method is well suited for flexible automation.

The objective of this study is to investigate the possibility and effectiveness of detecting the AWJ nozzle wear by monitoring the increase in the abrasive waterjet diameter using a machine vision system.

ABRASIVE WATERJET CUTTING SYSTEM AND NOZZLE WEAR

The abrasive waterjet cutting technique, which uses pressurized water with or without abrasive material for cutting various metals and non-metals, is one of the most recently introduced non-traditional machining methods. A standard AWJ system basically consists of high pressure water pump, a cutting head, an abrasive delivering system, and a CNC positioning system.

The basic functions of an AWJ system are shown in Fig. 1. High pressure water from an intensifer pump is supplied to the pneumatic valve (1) which provides a carrier medium for the abrasive material. The air valve provides rapid ON/OFF control of the waterjet at the sapphire nozzle and is controlled by the three-way air control valve (2). The abrasive material stored in the hopper (3) is introduced into the waterjet stream at the mixing chamber (4). The abrasive material flow is controlled by the adjustable abrasive valve (5). Once the waterjet stream and abrasive are combined, the abrasive waterjet exits through the tungsten carbide nozzle (6) toward the material being cut. The AWJ nozzle acts mainly to focus the spreading jet and accelerate the abrasive particles which do not penetrate the jet stream.

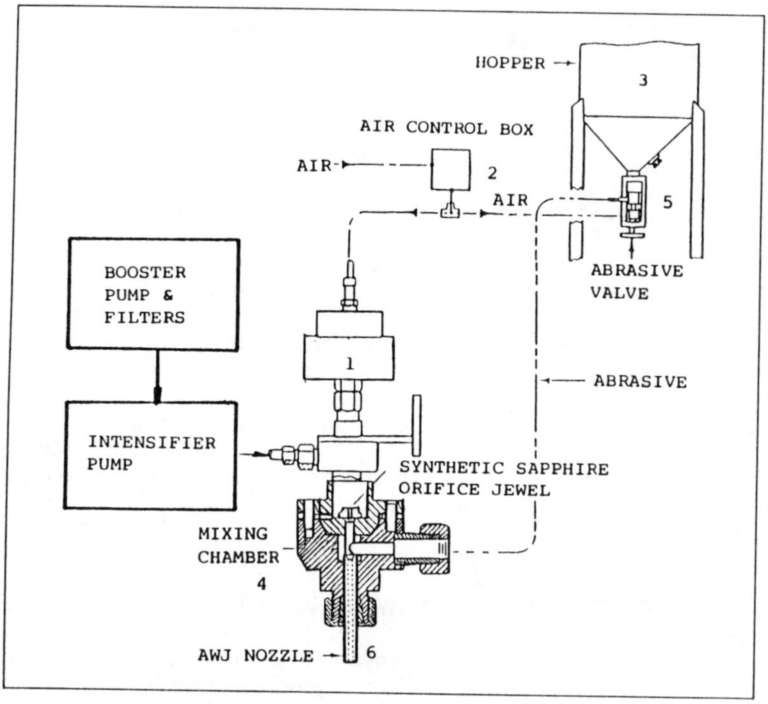

Fig. 1. General concept of abrasive waterjet cutting system

According to Hashish (1987), the nozzle is subjected to both abrasive and erosive modes of wear. When the jet stream first enters the nozzle, the abrasives hit the entrance section of the nozzle at random angles. This form of wear is called erosion wear. Once the jet stream advances through the inner tube of the nozzle, erosive particles travel parallel to the inner surface of the tube. During this movement the particles cause abrasive or shallow impact erosion of the wall.

The outlet of the AWJ nozzle will determine the final shape of the abrasive waterjet and its cutting performances. Fig. 2 shows the effect of an increase of the nozzle-inside-diameter on the depth of the AWJ penetration. Cylinders of 76 mm in diameter and 60 mm in height, made of P20 tool steel, are used as specimens. The only parameter changed was the AWJ nozzle-inside-diameter. From Fig. 2, it is apparent that there is an optimal nozzle-inside-diameter at which the depth of AWJ penetration will reach a maximum. For the selected conditions, the optimal nozzle-inside-diameter is 1.82 mm which is about 62% larger than the original diameter (1.2 mm). Further increase in the nozzle-inside-diameter will cause a sharp decrease in the depth of AWJ penetration, and the surface quality of the kerf will deteriorate. The decrease in the depth of AWJ penetration may be attributed to the decrease in the effective jet diameter which is due to wall friction and a reduction in abrasive impacts caused by rebounding and jet deflection.

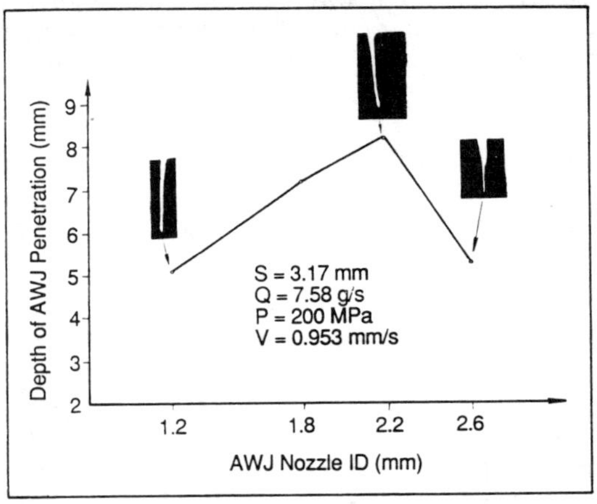

Fig. 2. Effect of nozzle-inside-diameter on depth of AWJ penetration

Nozzle wear in an AWJ cutting operation is undesirable because it severely degrades the quality of cut surfaces and causes undesirable changes in work geometry. Therefore, it is necessary, from a process automation point of view, that a sensing system be devised to detect the extent of AWJ nozzle wear during cutting operations so that the worn nozzle can be identified and replaced in time.

DESCRIPTION OF THE PROPOSED AWJ NOZZLE WEAR SENSING SYSTEM

The rate of change of the inside-diameter at the nozzle outlet, which is used to quantify the nozzle wear, could be indirectly monitored by tracking the increase in the abrasive waterjet diameter. The proposed AWJ nozzle wear monitoring procedure is based on the video capture system. The video capture system, based on the PC, consists of three main components: a CCD matrix-array camera, a frame grabber, and a video interface board. A block diagram of the AWJ diameter monitoring system is shown in Fig. 3. Matching between the object and the sensor is done by the camera. Once this matching is achieved the optical image is then converted into digital signals by using the MV-1 board. The TV monitor is used to make the visual display of the waterjet. The computer is used for overall controlling of the wear monitoring system and performing a quantitative analysis of the image data for extracting information.

The proposed system for monitoring the abrasive waterjet diameter functions in the following way. Because of the complexity of the AWJ cutting process the measurement of the abrasive waterjet diameter will be performed periodically by

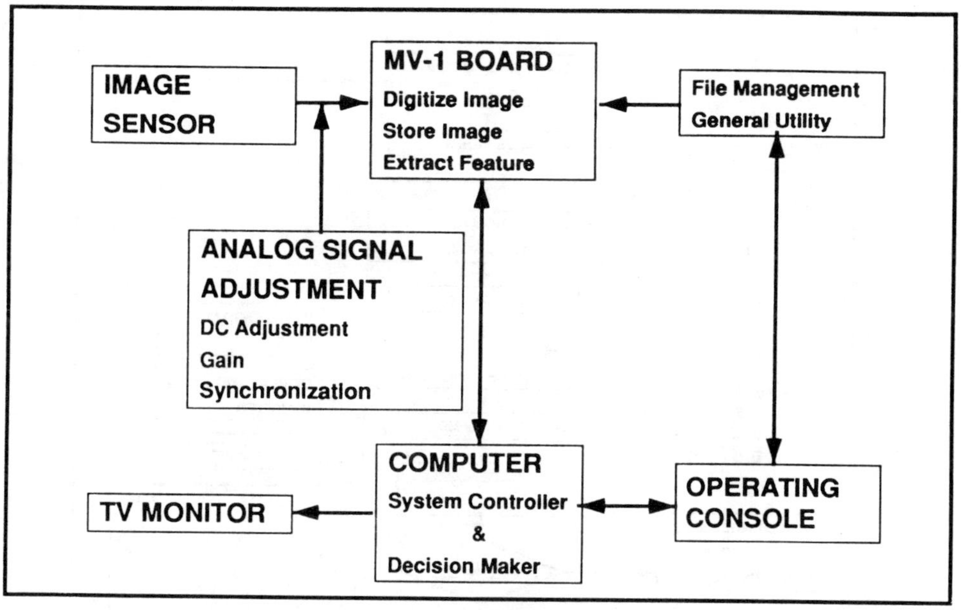

Fig. 3. Block diagram of video capture system to monitor AWJ nozzle diameter

interrupting the cutting process. The abrasive waterjet will be positioned with respect to the CCD camera which is placed at an appropriate position with regard to image acquisition needs. The CCD matrix-array camera monitors the abrasive waterjet which appears as a black area in the whole image. In order to minimize measuring errors the image of the abrasive waterjet is magnified by a zoom lens. The monitoring process of the abrasive waterjet diameter is automatically carried out by a program written in C language. The program is formed from two main sections. In the first section, the image acquisition and digitization steps are carried out by using the built-in functions of MV-1 software. In the second section, a quantitative analysis is performed on the digitized image in order to determine the abrasive waterjet diameter.

Fig. 4 illustrates the procedure supported by the proposed system. The first stage of the procedure is the initialization of the system level functions of the MV-1 software. As soon as the system is initialized, the program automatically switches to the camera mode so that the live image of the object within the camera's scope can be seen on the screen. At this stage, the user should check the monitor to be sure that the image is sufficiently clear. The image acquisition process can then be carried out. Based on the scheduled time, concerned with the particular working cycle, the cutting process will be interrupted and the abrasive waterjet will be automatically positioned with respect to the CCD camera. The section of the abrasive waterjet stream image at the outlet of the nozzle is captured. After converting the input image to a binary image, the next step is to detect the abrasive waterjet edges. In order to reduce the amount of data to be processed and to diminish the influence of the jet spreading on the accuracy of measurement, the window was defined on the screen. The boundary pixels of the jet are identified in order to determine the jet edges by using the edge detecting algorithm of MV-1 software. A routine in the software package is dedicated to computing the number of vertical lines which are the digital representation of the jet image. The output of this routine multiplied with the scaling factor obtained during the camera

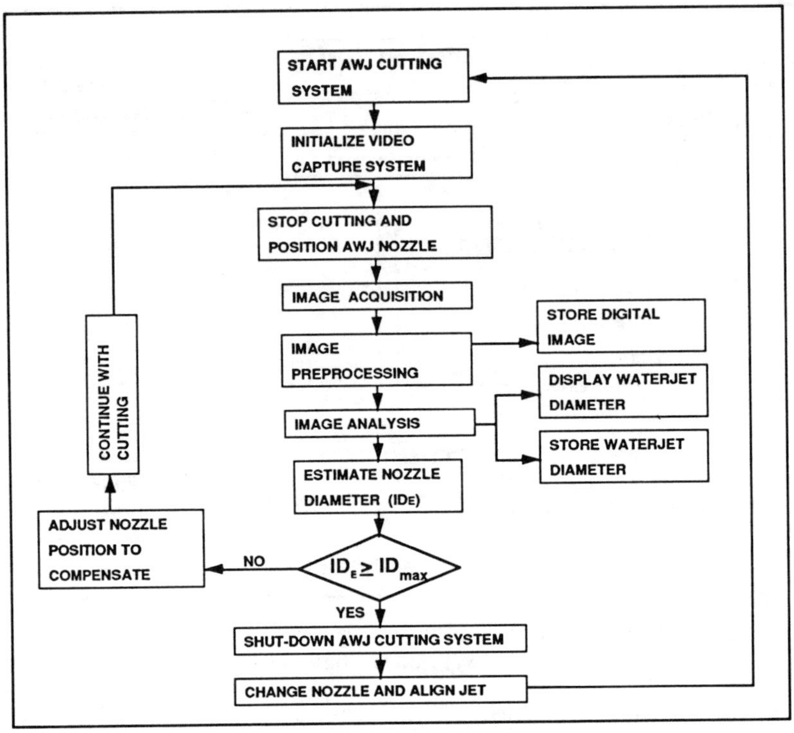

Fig. 4. Flow chart of nozzle wear monitoring system

calibration will determine the actual value of the jet diameter. The actual jet diameter is compared with an allowed value of the jet diameter, representing a life-end value for the used nozzle. If the actual jet diameter is greater or equal to the maximum allowed jet diameter the computer sends a command signal to the AWJ control system to close the pneumatic ON/OFF valve or/and to shut down the whole system. The replacement of the nozzle and an alignment of the jet can take place. In case the actual jet diameter is less than the predetermined threshold value, the nozzle with the cutting head is positioned with respect to the workpiece in order to continue the cutting process. If the positioning control system is equipped with a feed-back loop then the position of the nozzle with respect to the workpiece cutting contour could be adjusted in order to compensate for the increase in the abrasive waterjet diameter.

EXPERIMENTAL TESTS

A number of experimental tests have been carried out to verify the system performance. The measurement's precision and the reliability of the image acquisition are two main potential problem areas. The reliability of the image acquisition is related to nozzle repositioning with respect to the CCD camera. In the proposed set-up the camera is constrained in a fixed position and the cutting head with an AWJ nozzle is moved toward the camera by a CNC controlled x-y positioning system with a positioning accuracy of 0.025 mm. It is assumed that the repositioning of the nozzle will not significantly influence the accuracy of the measured jet diameter.

Before taking measurements of the abrasive waterjet diameter a calibration of the video capture system is performed. Calibration is done using strips of known dimensions (0.5, 1, 2, 3, 4 mm). The width of each strip is expressed in terms of the number of pixels in order to define a scaling factor. The scaling factor will depend on a focusing length of camera, distance between object (waterjet or strip) and camera, and light intensity. The scaling factor and the maximum allowed increase in the AWJ nozzle-inside-diameter are input into the video capture system during the initialization procedure.

In order to access accuracy and reproducibility of the jet diameter measurements a set of experimental tests was performed. The four abrasive waterjet nozzles with inside-diameters of 1.15, 1.5, 1.9, and 2.25 mm were prepared in advance. As was already stated, the rate of change of the nozzle-inside-diameter at the nozzle outlet is used to quantify the nozzle wear. It is evident that the outlet of the AWJ nozzle will determine the final shape of the abrasive waterjet. Thus, by measuring the diameter of the abrasive waterjet close to the nozzle outlet using the proposed system, the nozzle outlet diameter could be estimated indirectly.

However, the spreading of the jet will depend on the waterjet pressure, AWJ nozzle-inside-diameter, abrasive flow rate, abrasive grain size, etc. In the scope of the verification of the video capture system's performance the influence of the waterjet pressure and abrasive flow rate on the spreading of the jet diameter is performed. Three waterjet pressures of 207, 242, and 175 MPa and three abrasive flow rates of 0.45. 0.68, and 0.91 kg/min are selected. Garnet with mesh #80 is used as the abrasive material. The measurements of the jet diameter by the video capture system under the same conditions are repeated three times. It is evident from Table 1 that the jet diameter is larger than the actual diameter of the nozzle

Table 1. Estimated AWJ diameter versus the actual nozzle ID

Abrasive Flow Rate (kg/min)	Nozzle ID (mm)	**Estimated AWJ Diameter (mm)** Waterjet Pressure (MPa) 207			241			275		
0.45	1.15	1.18	1.16	1.19	1.22	1.20	1.23	1.25	1.28	1.24
	1.50	1.58	1.55	1.59	1.59	1.60	1.63	1.63	1.68	1.68
	1.90	1.98	2.01	2.02	2.08	2.05	2.04	2.08	2.14	2.07
	2.25	2.34	2.37	2.40	2.42	2.45	2.48	2.50	2.52	2.54
0.68	1.15	1.16	1.19	1.20	1.24	1.23	1.21	1.25	1.30	1.29
	1.50	1.60	1.59	1.57	1.64	1.59	1.61	1.65	1.70	1.69
	1.90	2.01	1.99	2.03	2.03	2.07	2.09	2.10	2.16	2.09
	2.25	2.41	2.35	2.38	2.49	2.45	2.48	2.51	2.53	2.56
0.91	1.15	1.19	1.20	1.21	1.26	1.25	1.25	1.25	1.30	1.30
	1.50	1.58	1.61	1.61	1.63	1.63	1.60	1.65	1.71	1.70
	1.90	2.02	2.04	1.99	2.05	2.10	2.06	2.17	2.13	2.08
	2.25	2.35	2.40	2.42	2.47	2.49	2.49	2.57	2.55	2.54

outlet. The difference between the measured jet diameter and actual diameter of nozzle outlet increases as the actual diameter of the nozzle outlet and the waterjet pressure increases. The increase of the abrasive flow rate has a negligible effect on this difference. Since tracking the increase in nozzle diameter at constant pressure is the objective, Table 1 indicates that the ratio of estimated to measured nozzle diameter is approximately constant. Thus, if the system is only calibrated before use when the nozzle diameter is known, any subsequent measurements can be simply corrected. The increased wear of the AWJ nozzle makes the clearance between waterjet and AWJ nozzle larger. Obviously, the greater the diameter of the AWJ nozzle's outlet, the smaller the probability of abrasive particles getting into the water due to the lower transverse particle velocity. Increased clearance between the AWJ nozzle and the waterjet will cause incomplete mixing of the abrasive particles with the waterjet. The result of this is the increased spreading of the abrasive waterjet. The structure of the abrasive waterjets expelled from the new and the worn-out nozzle under the same conditions is shown in Fig. 5. From Fig. 5, it is evident that the new nozzle forms a long coherent jet with focused abrasive particles. On the contrary, the worn-out nozzle produces a spreading jet.

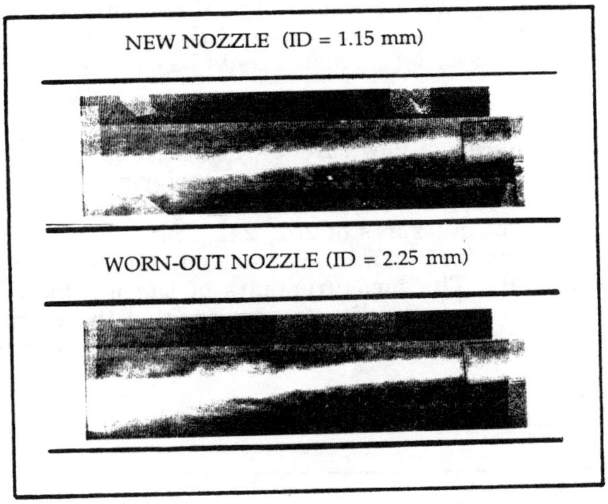

Fig. 5. Abrasive waterjets inssuing from nozzle (waterjet pressure P = 207 MPa, abrasive flow rate Q = 3.7 g/s, exposure time t = 10 μs)

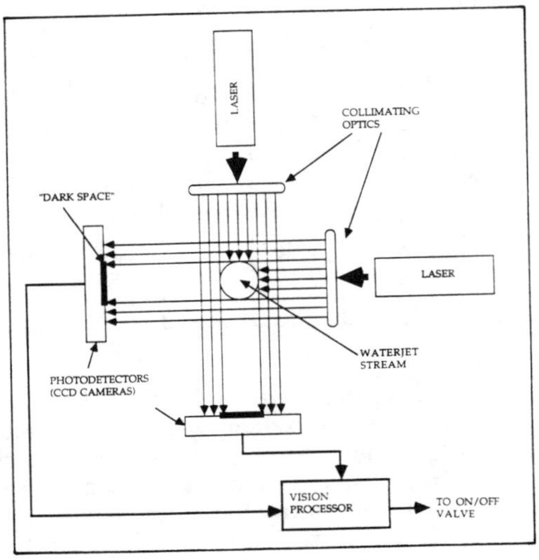

Fig. 6. A concept of the laser-camera assembly

Since nozzle wear is not always concentric, the measurements of the abrasive waterjet diameter should be performed in two perpendicular directions. One of the possible lighting source—camera assemblies is shown in Fig. 6.

CONCLUDING REMARKS

An inexpensive AWJ nozzle wear monitoring system is proposed. AWJ nozzle wear is monitored indirectly by monitoring the increase in the abrasive waterjet diameter close to the nozzle outlet. A proposed machine vision system based on a PC consists of a CCD matrix-array camera, a frame grabber, and a video interface board. The machine vision system is used to acquire the images of the abrasive waterjet periodically. By employing the digital image processing principles the jet diameters are obtained and compared with the actual AWJ nozzle outlet diameters. Through the experimental verification of the system's performance it was shown that the spreading of the abrasive waterjet will influence the accuracy of the estimation of the actual nozzle outlet diameter. The increase in the nozzle-inside-diameter and the waterjet pressure will cause the largest influence on the jet spreading. The proposed optical measuring system could be successfully applied to track the increase in the abrasive waterjet diameter. Thus, the change of the AWJ nozzle outlet diameter could be directly monitored with high accuracy.

ACKNOWLEDGEMENT

The author would like to thank Mr. Mehmed Gulsen for his cooperation in the execution of the experiments.

REFERENCES

Kovacevic, R. and Evizi, M., "Nozzle Wear Detection in Abrasive Waterjet Cutting Systems," Materials Evaluation, Vol. 48, No. 3, pp. 348-353.

Kovacevic, R., 1990, "Estimation of Nozzle Wear Under Varying Cutting Conditions," Proceedings, ASME Winter Annual Meeting - Monitoring and Control for Manufacturing Processes, Dallas, TX, PED-Vol. 44, pp. 9-16.

Kovacevic, R. and Chen, G., 1989, "A Workpiece Reactive Force as a Parameter for Monitoring the Nozzle Wear in Turning Operation by Abrasive Waterjet," Proceedings, ASME Winter Annual Meeting - The High Energy Beam Manufacturing Technologies, San Francisco, CA, PED-Vol. 41, pp. 43-49.

Hashish, M., 1987, "Wear in Abrasive Waterjet Cutting Systems," Proceedings, International Conference on Wear of Materials, Houston, TX, April 5-9, pp. 769-776.

Kovacevic, R., 1991, "A New Sensing System to Monitor Abrasive Waterjet Nozzle Wear," Invention Disclosure, University of Kentucky, Lexington, KY.

Sensors, Controls, and Quality Issues
in Manufacturing
ASME 1991

DESIGN OF A FORCE PROBE WITH INTEGRATED SENSORS

Kenneth A. Hair and Chi-Tsen Lin
California State University, Northridge
Northridge, California

ABSTRACT

Many robotic processes require sensors to transduce contact force information for use in loop closure and data gathering functions. Contact sensors, wrist sensors, tactile force probes and other devices are used in applications such as parts assembly, tool evaluation, and inspection. The California State University Northridge (CSUN) development program focused on the design, fabrication and testing of a contact force probe.

A three degree-of-freedom (DOF) robot manipulator was fabricated to perform the function of identifying and storing the outline coordinates of unknown objects placed before it. The principal means of identification would be a "feel" system using force feedback from a contact probe; much like a human finger being used to trace the outline of an object.

After careful examination of process accuracy and design requirements, the force probe designed herein was developed. As a result, testing of the probe indicated good performance and supports conclusions derived from system simulation.

INTRODUCTION

Many industrial functions are automated with the use of robots to perform activities such as assembly, inspection and other duties. The functions of identifying, segregating and selecting parts for assembly was often performed by human labor; such as loading individual parts into specific trays of bins from which the robot can select. It would be ideal, in most cases, if the human element could be eliminated from the object recognition task. A University design team, headed by a faculty advisor and comprised of multi-disciplined graduate and undergraduate students, conducted a development project with the goal of constructing a robot capable of this task.

The robot was configured to provide three axes of motion about a fixed base. A cylindrical operating envelope was chosen for its minimal coupling and nonlinear effects with respect to joint motion. Attached to the robot arm is a force probe used to detect the surface contour of an object. Figure 1 is a pictorial drawing of the robot's construction.

Figure 2 is a block diagram showing the control system designed to operate the manipulator. The IBM PC-AT (25 MHz) is used to process the search trajectory, and hybrid force/position loop closure.

The robot manipulator responds to commands by repositioning any or all of its three joints. The principal means of control is a hybrid force and position feedback loop. Position information is transduced from encoders located at each of the robot's three principal motion joints. A microprocessor based controller card (which resides within the IBM PC host computer) derives position, rate and acceleration information from the encoders and digitally closes each loop. A proportional plus integral algorithm was employed along with command feedforward (Wells et al. 1990 and Seraji 1989).

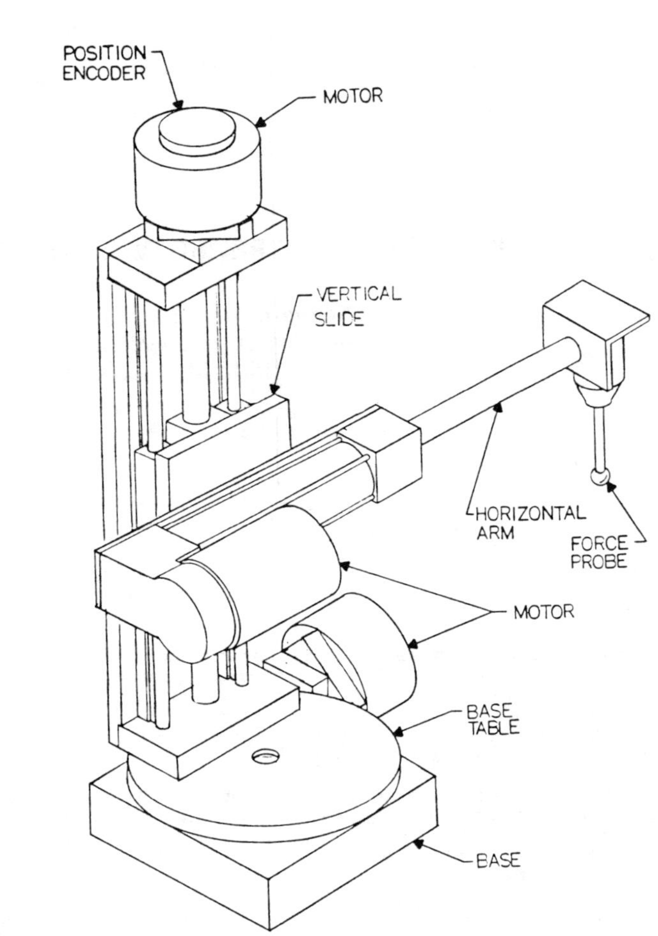

Figure 1: Robot Manipulator.

In addition to closed-loop position control, the robot also features a force control loop which enables it to determine the shape and physical features of the test specimen. Force control is accomplished through the use of the finger-like probe mounted as an end effector. When the probe makes contact with an obstacle (during the spiral search trajectory), the IBM PC reconfigures the control algorithm to bypass the search trajectory and subsequently follow a new trajectory based on maintaining a constant normal force contact with the object. Force information is read through an analog-to-digital converter board also resident inside the PC. Figure 3 is a photograph of the complete system including manipulator, host computer, drive electronics set and force probe.

Tactile sensors are expanding their roles in robotic systems as a recently maturing technology. The traditional approach has been to sense cartesian force information from either a probe or wrist sensor. More recent developments have included membranes on a probe tip with hundreds of digital encoded sensors (typically made from electrically conductive plastics), piezo electric devices, and research into laser technology (Harmon 1982).

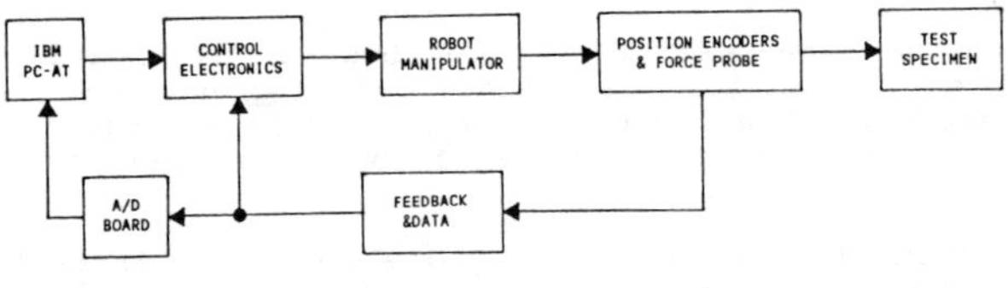

Figure 2: Control Block Diagram.

The challenge facing the CSUN team was to develop a sensor which minimized noise, error and cross-talk in sensory data; yet provided good low-signal sensitivity (threshold) and accuracy. Once completed, the probe will be integrated with a two dimensional vision system for object recognition. It is accepted that both vision and tactile information will be necessary to identify complex object shapes (Allen 1988).

Figure 3: Robot Arm with Force Probe.

FORCE PROBE DESIGN

The principal element in the hybrid force control scheme is the probe. This device sends back relative force data between the probe tip and the robot arm. Figure 4 is a schematic of the system which includes the outer force loop. The two control loops (position and force) are operated exclusively depending on the operating mode (i.e. whether in free space or in contact with an object), and thus constitute a hybrid control system.

During the search mode, the position command is sent by the PC to each actuator joint. The force command is switched out (i.e. set equal to zero) during this mode. Therefore, the control loop follows a programmed spiral search trajectory until the presence of an object is detected.

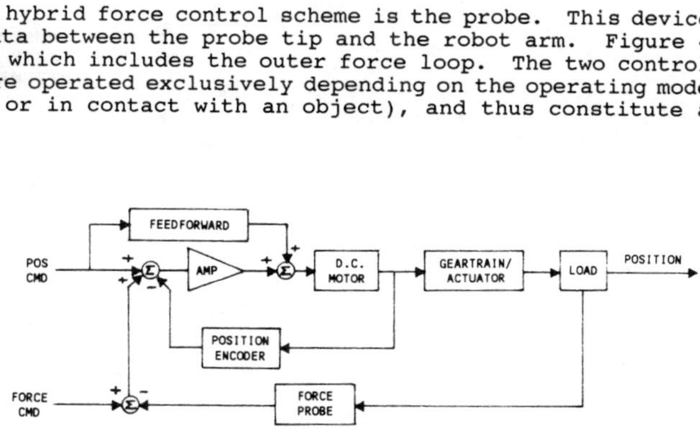

Figure 4: Control Loop Schematic.

Upon detection of an object, as determined by a normal force signal from the probe which exceeds a specified limit, the robot switches to force control. In this mode, the position command is held fixed and a control algorithm which maintains a constant normal force command is imparted. This algorithm also includes a search trajectory which drives the robot to move slowly along the tangential path to the normal force. Thus, the robot attempts to trace the shape of the object or feature (i.e. hole, slot, radius, etc.).

The probe provides three axes of force information; tangential force at the end effector which is seen as torque at the rotating base joint (F_T), radial force along the horizontal actuator (F_R), and upward force along the vertical actuator (F_Z) as shown in Figure 5. The former two are bi-directional; that is they read both retracting and extending forces. The latter vertical probe sensor detects only upward forces along the probe shaft (Z axis).

A tungsten carbide tooling ball is used as a precision "feeler" which makes contact with the object being searched. Forces from this contact are transmitted through the stiff shaft to the sensing elements located in the cylindrical housing. The tooling ball has a surface roughness of less than 2 micro inches which helps reduce

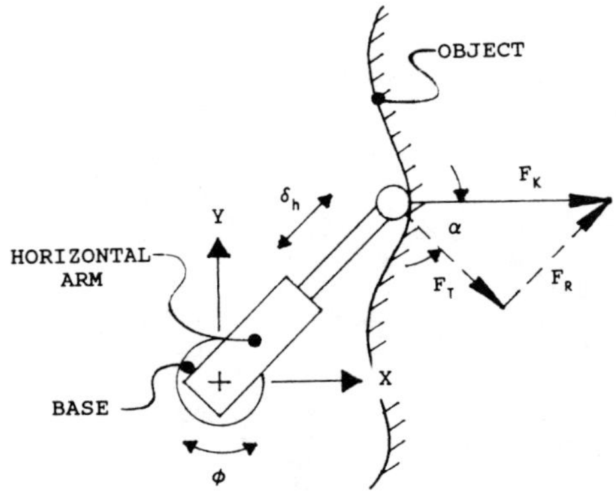

Figure 5: Force Probe Geometry.

the erroneous friction forces of the sliding ball from dominating the contact force readings. Figure 6 depicts the major elements of probe construction.

Force sensing is accomplished with two different types of sensors. For the tangential and radial force components (F_T and F_R), a pair of linear variable differential transformers (LVDT's) are used to transduce the displacement of springs; which is proportional to the corresponding force. For the vertical component (F_Z), a uni-directional linear force transducer reads the shaft compressive force directly.

Figure 7 is a pictorial schematic of the flexure assembly used to determine F_T and F_R. The flexure assembly as shown is a cylindrical aluminum tube with machined orthogonal cantilever spring flexures. The assembly is similar to a "force bridge" where one flexure provides a mechanical resistance (spring rate) in one direction but is sufficiently stiff in the orthogonal direction. Alternately, the other flexure (shown rotated 90° in

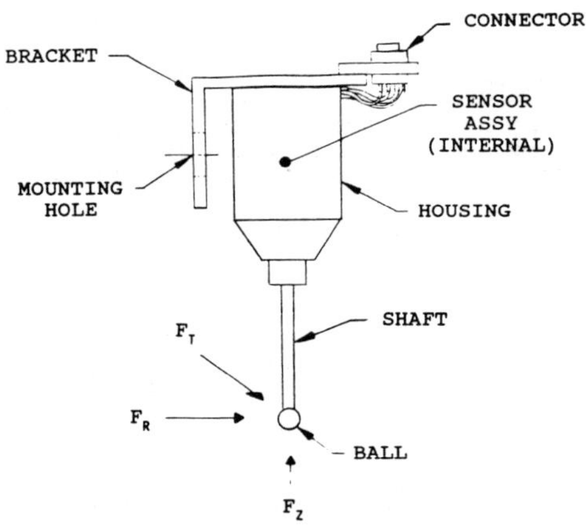

Figure 6: Force Probe Configuration.

the figure) provides a tuned mechanical resistance (spring rate) in the latter direction only. The relationship of force to displacement Δ_i is as follows:

$$F_i = \frac{\Delta_i K}{L_i} \qquad \{1\}$$

where:
- F_i = Force along ith direction [lbs]
- Δ_i = Displacement of flexure assembly at LVDT attach point [in]
- K = Bending spring rate of flexure [in-lb/in]
- L_i = Length from flexure to probe ball tip [in]
- i = Force direction "T"angential or "R"adial [ ]

The equation assumes a linear output for small angular displacement (i.e. probe flexure strokes less than ±5° about each axis) about the sensor's centerline.

As previously mentioned, the displacements Δ_T and Δ_R are transduced by a pair of

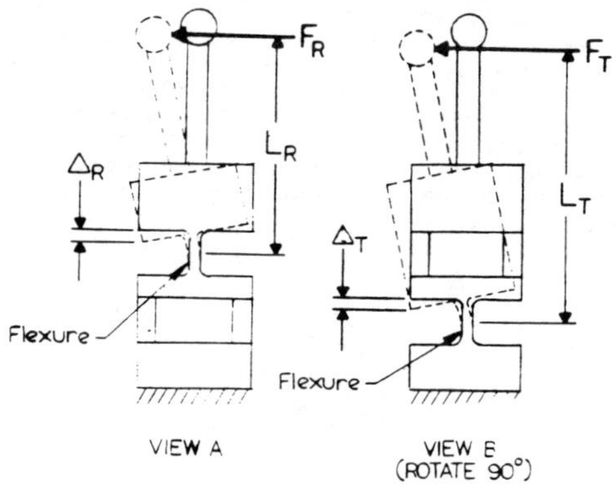

Figure 7: Flexure Assembly Diagram.

LVDT's. Figure 8 is an exploded view of the internal components of the flexure assembly. Table 1 describes the major components.

Table 1: Principal Probe Components.

ITEM #	DESCRIPTION
1	Mounting Base
2	Flexure
7	LVDT Coil
8	LVDT Armature
11	Force Transducer (Compressive Strain Gage Type)
16	Shaft
17	Tungsten Carbide Ball

As previously mentioned, the tangential force (F_T) is read directly from the LVDT (refer to Figure 8, Items # 7 & 8) sensor positioned lateral to the horizontal arm in the probe assembly. The radial force (F_R) is transduced by the LVDT mounted longitudinal to the arm.

These are direct readings with no need for coordinate conversions. From Figure 5 it can be seen that the two commanded force components are related to the two dimensional (X - Y plane) normal force as shown in equations {2} and {3}.

$$F_T = F_K \cos\alpha \qquad \{2\}$$

$$F_R = F_K \sin\alpha \qquad \{3\}$$

where: F_K = Normal contact force [lbs]
α = Normal contact angle [rad]

Consequently, if the two force components F_T and F_R are known from the probe, the angle α can be determined by taking the ratio of {3} over {2}:

$$\alpha = \tan^{-1} \frac{F_R}{F_T} \qquad \{4\}$$

The detailed control diagram for the force control scheme is presented in Figure 9. Since the test object search trajectory is designed to be slow and smooth, it eliminates the need for high response (i.e. rate and acceleration) type control systems as were needed for position control. Thus, the fundamental architecture selected is a proportional force control loop with force and force rate (impulse) feedbacks. The principal design challenge is dealing with the stiffness of the dc motor driven joint actuators in contrast to the flexible (i.e. vibration prone) force probe assembly. The underlying consideration was to select system gains low enough to keep the joint stiffness soft while at the same time providing adequate stiffness to overcome system nonlinearities such as friction. The reduced joint stiffness helps protect the probe from entering a limit cycle or being damaged during sudden force transients.

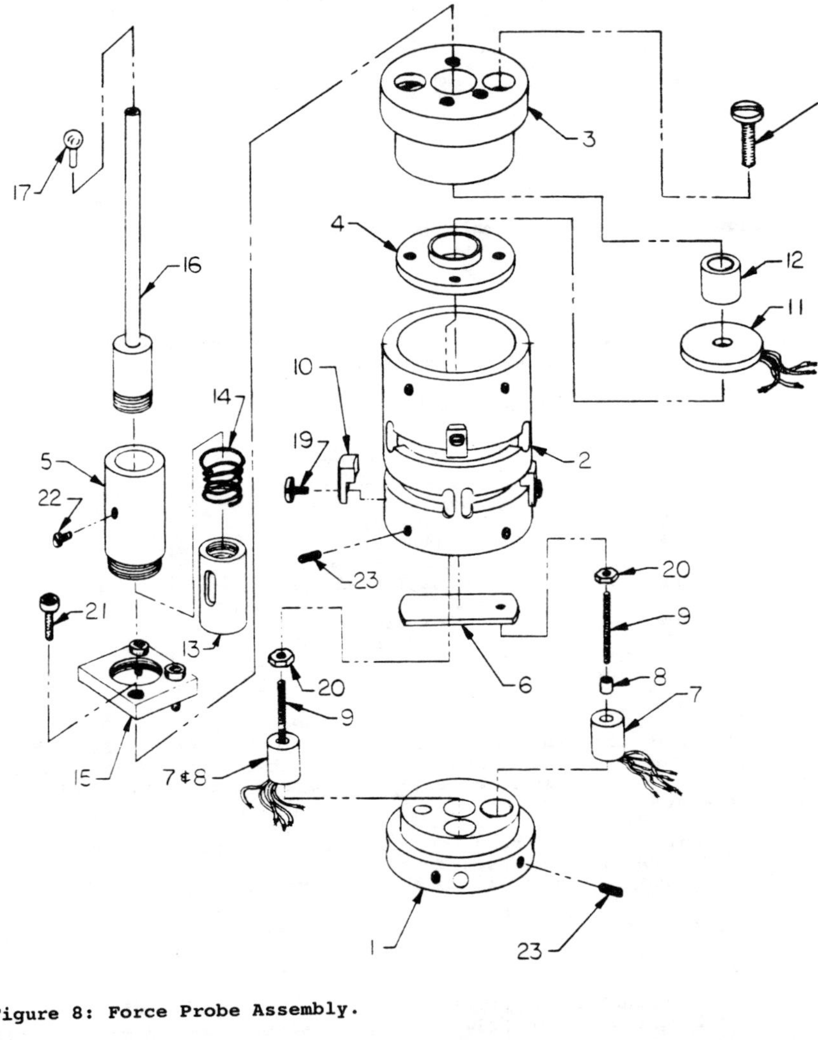

Figure 8: Force Probe Assembly.

The predominant system dynamics are included in the G(s) term; which is the position loop closure transfer function featuring P+I with feedforward control. Of concern here are the associated non-linearities of joint and actuator friction, and external disturbances such as centripetal, coriolis and gravitational forces. By using position control inside the force loop, the affects of these disturbance torques/forces are minimized (i.e. they do not effect the accuracy of the control loop as a parasitic error) as the system tries to satisfy $\phi_c(s)$ and $\delta h_c(s)$ commands via the high performance P+I plus feedforward control system. A pure force control system would have used joint torques and forces as feedbacks instead. As a result of the selected inner-loop position control approach, external joint disturbances are minimized by the high stiffness (i.e. high loop gain) characteristics of the position loop, while the contact force loop is adequately soft via the low gain probe force and force-rate feedback loops.

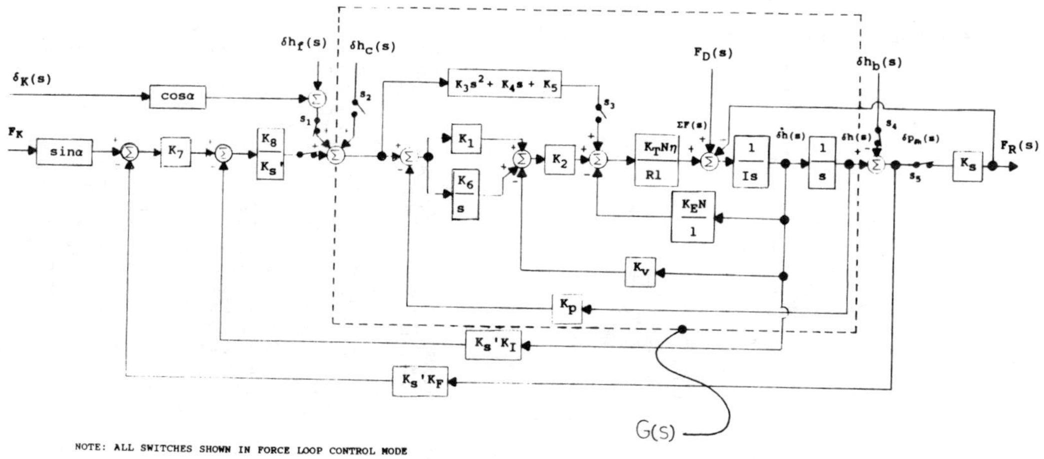

Figure 9: Position and Force Control Block Diagram.

The drawback to the selected approach is that the system commands a probe tip position error $\delta_T(s)$ or $\delta_R(s)$, instead of a force error directly to the joint motors. The force is approximated by the relation:

$$F_i(s) = \delta_i(s) \times K_s \qquad \{5\}$$

where: δ_i = Probe tip displacement [in]
K_s = Effective linearized flexure spring rate at probe tip [lb/in]

Thus, the hybrid force control scheme is actually an additional position control loop, wrapped around the existing manipulator position control loop G(s). This position control loop is scaled to represent force by virtue of the linearized probe spring rate K_s. In order to derive position commands from force information, Figure 9 showed the use of the term K_s'in the force feedback paths, which is the electrical approximation to the actual spring rate K_s (i.e. volts/volt instead of lbs/in). The accuracy of the "force" loop depends on how well the linear K_s' terms represent the real spring. The K_s' terms were determined from a calibration procedure established by applying known forces (i.e. weights) to the probe tip, with the data presented in the TEST RESULTS section.

This approach was selected for several important reasons:

- The desire to maintain a constant normal contact force on the profile of a test specimen is only for convenience, the actual information needed to describe the profile is position (i.e. directly control the variable that is most important to the problem).
- The existing position loop G(s) minimized the destablizing effects of system inertias ($J_b(s)$, I_v and I_h), which would have had the greatest effect on performance if a "pure" joint torque/force control scheme had been chosen.
- Elimination of the need for joint torque/force transducers greatly reduced system complexity and cost.

FORCE CONTROL SIMULATION

The proportional control scheme for the force loop was designed using computer simulation. All of the dynamic effects are included in the position control loop (i.e. inertia, friction, damping, gravity, coriolis, centripetal, etc.). The probe was modelled as a simple spring rate K_s, with probe tip mass and damping assumed to be very small and thus neglected.

For the simulation, the boundary of the test specimen was fixed at a 45° angle (α) at the global position (0,10,6) as shown in Figure 10. This corresponds to the case where the robot initially is touching the surface of the object, yet the force at the probe is zero. The subsequent result is a step response to the force control loop as the manipulator attempts to establish a constant normal force. Additionally, the simulation sets δ_K, the force trajectory displacement for tracing the specimen's surface, equal to zero (i.e. the trajectory is turned off). This latter condition was implemented to constrain the model of the specimen's boundary to a point in space along the 45° surface, and not complicate the simulation by requiring a detailed model of the specimen's shape as a function of position and time. The force loop step response is most useful in determining controller performance and evaluating loop gains since damping, stability and rise time are readily apparent.

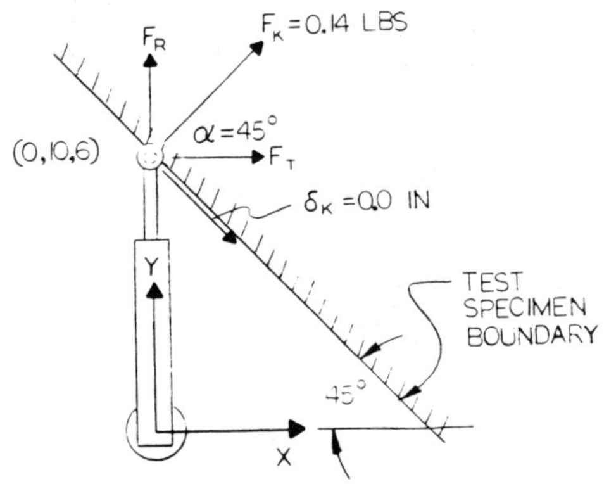

Figure 10: Computer Simulation Scenario.

The force model (configured with the simple point boundary at α = 45°) will attempt to maintain a constant normal force with respect to the commanded value F_K. If F_K is set equal to approximately 0.14 pounds, the corresponding radial force F_R and tangential force F_T required is approximately 0.10 lbs at the 45° angled surface.

Figure 11 presents the step response for both forces as a function of the commanded 0.10 lbs. As indicated, the response is adequately damped and results in a steady-state force error as expected from the proportional control. This error is less than 0.005 lbs. The force error manifests itself as a position error when divided by K_s', and is used to provide a small dc voltage and current to the motor in order to counter and hold the contact force reacted through the probe assembly.

The base and horizontal actuators themselves provide joint torques and forces necessary to respond and maintain the steady-state probe forces shown in the previous figure. The response part of the motion requires large forces to move the predominant system inertias, yet then decay to a finite level to maintain probe contact. The design effort centered around the desire to select force loop gains which provided a smooth response with a damping factor near 1.0 or 1.1. This was achieved by iterative selection of gains and witnessing the corresponding force step response from the non-linear model. The response was analyzed and the gains iterated until the

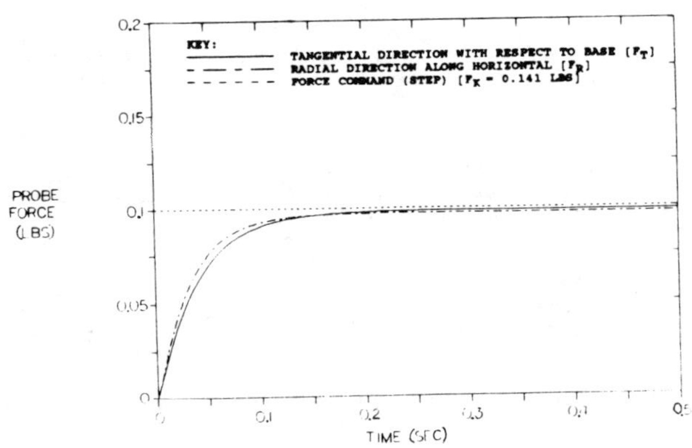

Figure 11: Simulated Force Loop Step Response.

appropriate response was achieved.

TEST RESULTS

Testing was performed on the force probe in order to verify the original design requirements. Testing consisted of the procedures listed in Table 2.

The probe was mounted in the vertical position and calibrated spring-scale force gauges were used to determine the applied load level. A large range scale of ±1 lb and a smaller range scale of approximately ±45 grams were used to take force data. The primary sensor elements are LVDT's in the tangential (F_T) and radial (F_R) directions. The sensors are operated at approximately 10 kHz excitation frequency and each is demodulated by an ASIC (application specific integrated circuit) which are located in the drive electronics chassis. The output of the ASIC's is a DC voltage which represents the output of the probe.

Table 2: Verification Tests.

Procedure	Description
Linearity	Apply a calibrated load from maximum negative force to maximum positive force. The maximum deviation from the least squares fit through the data points in output voltage is the linearity (expressed as a percentage when divided by full output voltage).
Hysteresis	Apply a calibrated load from zero to maximum to minimum and back to zero force. The maximum width of the band of output voltages is the hysteresis (expressed as a percentage when divided by full output voltage).
Null Offset	Apply a calibrated load from 10% maximum load to 10% minimum passing through the zero load point. The voltage deviation from zero when the load is zero is the null offset (expressed as a percentage when divided by full output voltage)
DC Gain	Defined as the mean slope of the linearity curve expressed as volts per pound force.
Repeatability/ Accuracy	Apply a calibrated load repeatedly for a minimum of 30 times. Defined as the statistical deviation at 95% confidence of indicated output voltages (expressed as a percentage when divided by full output voltage).
Null Repeatability	Same as above except for repeated loading from 50% maximum and minimum force back to zero.
Threshold	Apply an increasing then decreasing load of approximately 0.2 Hz, while increasing amplitude from zero force. Record the first output level where both an increasing and decreasing voltage is detected (expressed as a percentage when divided by full output voltage).
Cross-axis Sensitivity	Perform a linearity test while recording the output of the orthogonal axis' force sensor. The maximum deviation from 0 Vdc is the sensitivity (expressed as a percentage when divided by the output of the principal axis' force sensor.

Data was gathered by applying the displacement to the force scale and reading the scale force versus the probe's output voltage. Figure 12 is a schematic of the test set-up. For the input versus output data, the full scale range was broken into 30 segments and data taken at each load point. For the repeatability test data, the calibrated loads were applied and reapplied a minimum 30 times, and the statistical mean and standard deviation computed from the indicated output signals (Lipson and Sheth 1973).

Figures 13 and 14 are output versus input plots for both the tangential and radial force components of the probe. The saturation limits at the extreme ranges of force are due to mechanical stops built into the flexure assembly. These are adjustable and are used to protect the cantilever springs from overstroke and fatigue. The gain of the F_R curve is 5.63 vdc/lb while the F_T curve yields about 7.25 vdc/lb. These scale factors are adjustable within both the ASIC assemblies and the PC for proper loop gains. For verification tests both were left at the factory pre-set values for convenience.

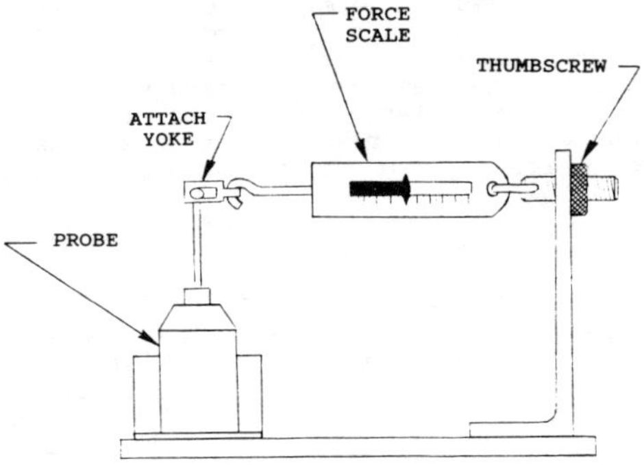

Figure 12: Force Probe Test Set-Up.

Linearity for both curves was almost identical and resulted in a value of about ±3.2% full scale. The hysteresis band was slightly greater for the radial component but both were less than 4.5% total hysteresis. The final indicator of input/output performance is repeatability or accuracy. Repeatability is a statistical measurement of the output when continuously subjected to the same load input. The parameters affecting repeatability are both mechanical (such as material creep and wear) and electrical (heating and circuit stability). Tests showed the repeatability/accuracy to be a maximum of 5.5% within a 95% confidence interval. This represents the minimum accuracy guaranteed for the full scale of all axes. Performance usually exceeded this level as indicated by applying 0.2 lbs to the tangential component which resulted in the best accuracy of about 1.3%. All of these numbers, however, indicate that good static performance has been achieved with the selected probe design.

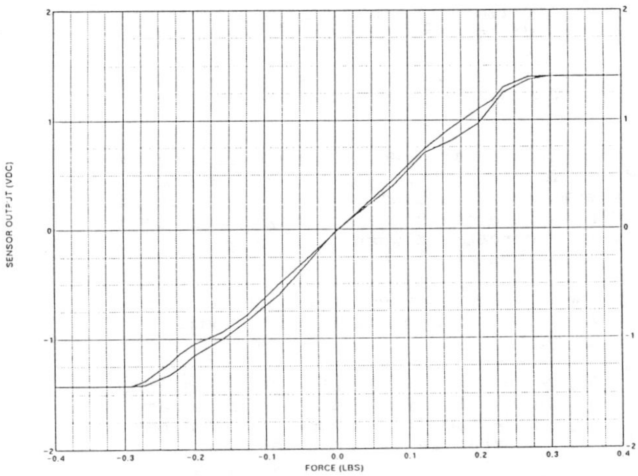

Figure 13: Static Performance - Tangential Force.

Another important factor is the performance of probe with very small force inputs. This is indicated by the sensitivity and threshold of the device. Figures 15 and 16 and curves depicting the performance of the probe with inputs less than ±25 grams. As shown, the performance appears satisfactory with good linearity and a small hysteresis band. Threshold testing of the probe showed a discernable output was achieved at less than 0.8% full scale.

The final static characteristics are concerned with performance about the neutral or null position. Null offset for both components is adjusted by turning the threaded armature shaft (refer to Figure 8 Item # 9) to raise or lower the LVDT armature within the coil. For both force components of the probe, the LVDT's were nulled within 1.5% of full scale output.

Null repeatability is a measure of the probe's ability to maintain the neutral setting after repeated inputs. Again, a statistical representation of performance results from measuring the deviation between recurring zero load conditions. Null repeatability for both components of force was less than 0.6%, within a 95% confidence interval.

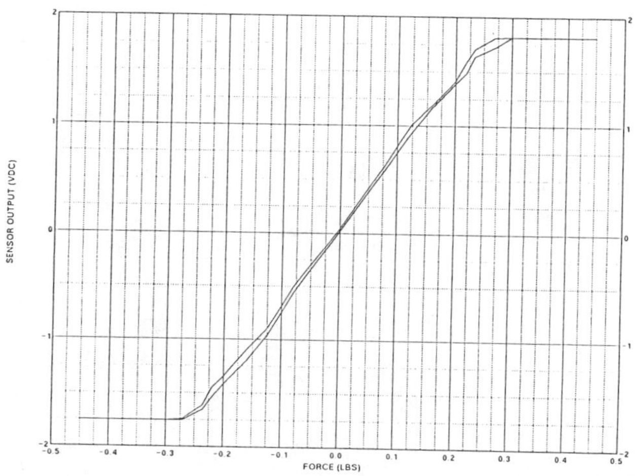

Figure 14: Static Performance - Radial Force.

The last measure of performance was to observe the cross-axis sensitivity of the probe. Specifically, a load is applied in one direction and the output of the orthogonal component is measured. From testing, the radial force component F_R was the least sensitive to the application of tangential forces F_T. The maximum radial component output observed was more than -40 db down from the applied tangential load. Conversely, the maximum tangential component was only slightly greater than -30 db down, but this still represents less than 3% coupling between the two directional axes.

CONCLUSIONS

A tactile force probe has been developed to provide orthogonal force information for use in a robotic system. The probe exhibited linear characteristics and accuracy when subjected to test loads. In addition, the probe provides good sensitivity and range (threshold less than 1 gram with a dynamic range greater than 125). Using an LVDT as the principal sensing element minimized temperature effects and installation complexity when compared to strain gage type sensors. The resulting design was also compact, lightweight and rugged.

When combined with the hybrid force control system, the probe's minimal cross-axis sensitivity and accuracy are essential for deriving the normal contact force and angle. The LVDT's also allowed for direct probe tip displacement measurements; which is the underlying information needed to describe the coordinates of a test object's surface. Consequently, the force probe developed satisfactorily met the objectives of the CSUN design team.

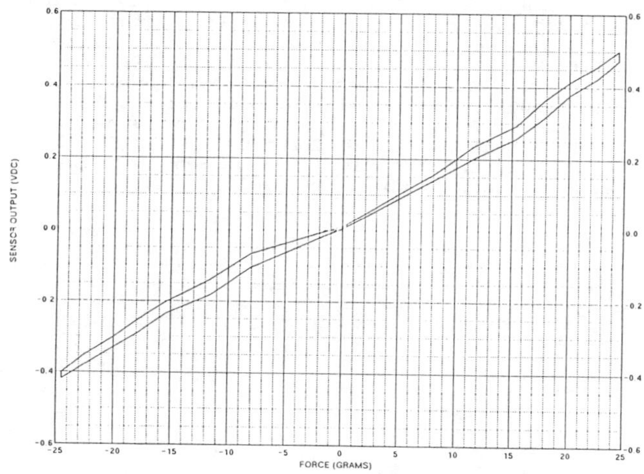

Figure 15: Small Signal Performance - Tangential Force.

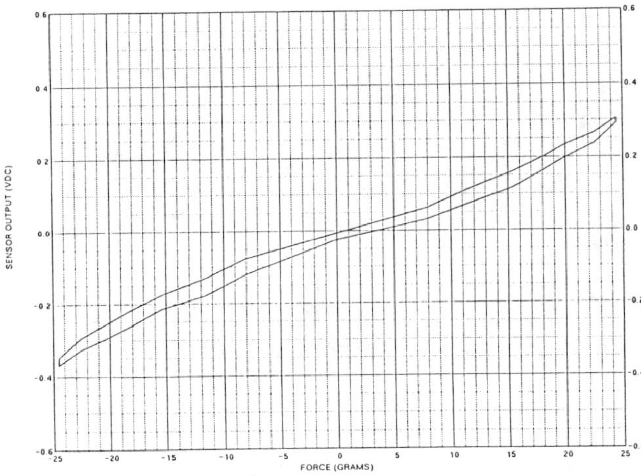

Figure 16: Small Signal Performance - Radial Force.

ACKNOWLEDGEMENT

This work was supported by the California State University Northridge Foundation, CSU Awards for Research, Scholarship for Creative Activity, and the Mechanical Engineering Department of CSUN.

REFERENCES

1. Craig, J., "Introduction to Robotics", Addison-Wesley Publishing Co., Reading, MA, 1986.
2. Wells, R.L., Schueller, J.K., and Husty, J., "Feedforward and Feedback Control of a Robotic Arm", IEEE Spectrum Periodical, NY, 1990.
3. Seraji, H., "Simplified Linear Multivariable Control of Robots", NASA Tech Brief, Vol 13, No. 4, Item #89, April, 1989.
4. Lipson, C., and Sheth, N.J., "Statistical Design and Analysis of Engineering Experiments", McGraw-Hill Co., New York, 1973.
5. Allen, P.K., "Integrating Vision and Touch for Object Recognition", The International Journal of Robotics Research, Vol. 7, No. 6, pp. 15-33, 1988.
6. Harmon, L., "Automated Tactile Sensing", The International Journal of Robotics Research, Vol. 1, No. 3, pp. 3-32, 1982.

Sensors, Controls, and Quality Issues
in Manufacturing
ASME 1991

LINEAR MODEL BASED FAULT DETECTION AND ISOLATION FOR A SCREW COMPRESSOR

C. James Li and Taehee Kim
Department of Mechanical Engineering
Columbia University
New York, New York

G. W. Nickerson
David Taylor Research Center
Annapolis, Maryland

Abstract: This paper presents a model based fault detection and isolation (FDI) method for a screw compressor. Mathematical models which describe the causal relationships linking measurable system variables are derived from physical laws and identified through system identification techniques. These mathematical models are in the form of linear ordinary differential equations and difference equations. The parameters of the models are estimated using Least Square estimators under normal and faulty conditions. It is shown that the use of model parameters, which frequently are more directly influenced by faults, has enabled early detection and accurate recognition of faults.

I. INTRODUCTION

To meet future manufacturing tasks, manufacturing equipments will necessarily become increasingly complicated. Concurrently, the pressure to reduce the cost of the manufacturing will necessitate reduced manning. Above all, the quality of the products has to be assured by the manufacturing processes. These driving forces work in opposition to each other and a new method of maintenance, other than planned maintenance system (PMS) for the upkeep of manufacturing machinery, must be developed. Mechanical system condition based maintenance (CBM) has been recognized as the solution. The benefits of a CBM are improved maintenance procedures and scheduling, increased machinery operational readiness, reduced logistics support cost and better product quality.

In a fielded system, an effective CBM system shall detect faults of a machine well before getting to the failure stage. Thus, monitoring systems must be in place to measure parameters of interest. Diagnostic algorithms must be available to convert the data obtained from the monitoring system into variables which signify the state of operation of the machine.

Whereas previous methods of condition monitoring only permitted recognition of the state of an equipment using measurable signals or their derivatives, an attempt is now being made to develop a more sensitive and reliable method by the use of model based indirect variables as well [1,2]. Previous methods can identify the change of the state of an equipment only after some measurable variables have been considerably affected. The use of indirect variables, which frequently are more directly influenced by faults, is likely to enable earlier and more accurate recognition of a change of the state due to a fault. The approach is described as follows.

Measurable variables of a process are linked by causal relationships which can be described by mathematical models. For instance, the best well known mathematical models are differential/difference equations. In general, these mathematical models have the following form:

$$Y = f\{U,N,\theta,X\} \tag{1}$$

where U(t) and Y(t) are vectors of the measurable input and output variables, N generally represents a vector of nonmeasurable disturbance signals, θ is a vector of constant or slow time varying non-

* This material is based upon work supported by Dept. of Navy under award N00038-84-C-0165.

measurable process parameters, and X is a vector of partially measurable and partially nonmeasurable internal variables.

Knowing the mathematical model, indirect variables can be derived based on the following parameters:

(1) Nonmeasurable part of variables X.

(2) Nonmeasurable process parameters θ.

(3) Nonmeasurable characteristic quantities $\eta = g\{U, Y, \theta\}$.

Frequently, the basic process/equipment model can be derived from physical laws which govern the process/equipment behavior. Hence, the structure of the models is known and the parameters have to be estimated. The nonmeasurable internal variables X and process parameters θ can be obtained by state estimation and parameter estimation schemes described in ref.[3-7].

Using model based indirect variables for fault isolation has the following advantages [1];

- The indirect variables X(t) or $\Theta(t)$ are in many cases closer to the process faults. Hence the process faults may be detected earlier and localized more precisely than just observing U and Y.
- A process fault usually causes changes of several output variables with different signs and dynamics. The model based fault detection now takes into account all these entailed changes and determines the indirect variable which has been changed directly by the fault. Hence, a more significant change in the indirect variable can be extracted and the sensitivity will be improved.
- Closed loops generally compensate for changes of the outputs by changing the inputs. Therefore deviations by faults can not be found with direct measurement of outputs. Model based fault detection methods have no such limitation.
- Fewer sensors will be required. All different sensors are driven by the same dynamics of the system and are therefore functionally related. This functional redundancy can be used to achieve sensor fault tolerance or to reduce the number of sensors.

We have chosen a 20 Hp single screw compressor as a test platform for this new model based fault detection and isolation approach. The reason for this selection is because more and more pneumatic equipment such as robots' endeffectors and tool holder/exchangers are used in today's production line. We will first illustrate the modeling of the compressor and the means to estimate the parameters of the model. Then, we will present the experimental results and conclude the paper with what we have learned from the experiments performed at our laboratories.

II. MODELING

The compressor system is regarded as a system consisting of a motor, an air-end, a pipe system and a radiator used as an oil cooler as illustrated in Fig.1 (see Appendix for nomenclature). A model of the compressor is obtained by combining models of these subsystems.

Torque balance on the motor gives:

$$\Theta_M \frac{dw_m}{dt} = T_M - T_{FM} - T_S \tag{2}$$

where, $T_M = C\frac{I_{ms}^2(w-w_m)}{w_m}$ and $T_{FM} = C_{FM0} + C_{FM1}w_m$

Torque balance on the air-end gives:

$$\Theta_C \frac{dw_m}{dt} = T_S - T_{FC} - T_C \tag{3}$$

where, $\Delta T_C = a\,\Delta w_m + b\,\Delta \dot{M} + d\,\Delta P_l$ and $T_{FC} = C_{FC0} + C_{FC1}w_m$

Eq.(2) and(3) can be added together to eliminate T_S. Furthermore, the combined equation is linearized around a steady-state equilibrium. Let us denote by $\bar{w}_m$, $\bar{T}_S$, $\bar{T}_C$, $\bar{M}$, $\bar{I}_{ms}$ and $\bar{P}_l$ as values of measurable variables at equilibrium. Then,

$$(\Theta_M+\Theta_C)\frac{dw_m}{dt} = -\left(C\frac{\bar{I}_{ms}}{\bar{w}_m^2}w + C_{FM1} + C_{FC1} + a\right)\cdot \Delta w_m - b\cdot \Delta\dot{M}$$

$$+ 2C\bar{I}_{ms}\frac{w-\bar{w}_m}{\bar{w}_m}\cdot \Delta I_{ms} - d\,\Delta P_l \tag{4}$$

(see Appendix for details of derivations).

The pressure drop in the pipe system can be obtained from the following equation:

$$\Delta P = \Delta P_v + \Delta P_R + \Delta P_{ac} \tag{5}$$

where, $\Delta P = P_l - P_{atm}$, $\Delta P_v = C_V \frac{\dot{M}^2}{A_V}$

$\Delta P_R = C_R \dot{M}^2$ and $\Delta P_{ac} = \frac{L}{A_F}\frac{d\dot{M}}{dt}$

After perturbation, Eq. (5) can be written as:

$$\frac{L}{A_F}\frac{d\dot{M}}{dt} = -2\left(\frac{C_V}{A_V} + C_R\right)\bar{\dot{M}} \cdot \Delta\dot{M} + \frac{L}{A_F} \cdot \Delta P_l \tag{6}$$

Heat balance equation on the radiator gives:

$$Q_r = Q_{in} - Q_{out} \tag{7}$$

where, $Q_r = \rho_r c_{pr} V_r \dot{T}_r$

$Q_{in} = \dot{m}_{oil} c_{oil}(T_{in} - T_{out})$ and $Q_{out} = a_r h_r (T_r - T_\infty)$

Therefore, Eq.(7) becomes

$$\rho_r c_{pr} V_r \dot{T}_r = \dot{m}_{oil} c_{oil}(T_{in} - T_{out}) - a_r h_r (T_r - T_\bullet) \tag{8}$$

Eq.(4), (6) and (8) give a mathematical model of the compressor:

$$\frac{d}{dt} w_m = a_{11}\Delta w_m + a_{12}\Delta\dot{M} + b_{11}\Delta I_{ms} + b_{12}\Delta P_l \tag{9a}$$

$$\frac{d}{dt}\dot{M} = a_{22}\Delta\dot{M} + b_{22}\Delta P_l \tag{9b}$$

$$\dot{T}_r = C_1(T_{in} - T_{out}) + C_2(T_r - T_\infty) \tag{9c}$$

where, $a_{11} = -\left(C\frac{\bar{I}_{ms}^{\,2}}{\bar{w}_m^{\,2}} w + C_{FM1} + C_{FC1} + a\right)/(\Theta_M + \Theta_C)$

$a_{12} = -b/(\Theta_M + \Theta_C)$

$a_{22} = -2\frac{A_F}{L}\left(\frac{C_V}{A_V} + C_R\right)\bar{\dot{M}}$

$b_{11} = 2C\bar{I}_{ms}\frac{w - \bar{w}_m}{\bar{w}_m}/(\Theta_M + \Theta_C)$

$b_{12} = -d/(\Theta_M + \Theta_C)$

$b_{22} = \frac{A_F}{L}$

$C_1 = \dot{m}_{oil} c_{oil}/\rho_r c_{pr} V_r$

$C_2 = -a_r h_r/\rho_r c_{pr} V_r$

III. PARAMETER ESTIMATION

Eq.(9) is a system of linear differential equations. We need $d(w_m)/dt$, $d(\dot{M})/dt$ and $d(Tr)/dt$ to estimate the parameters of these differential equations. Numerical differentiation can be used to estimate these derivatives from w_m, $\dot{M}$ and T_r. In practice, the order of the numerical differentiation is restricted to second or third order and filtering is used due to the presence of measurement noise [5].

A very effective and well developed parameter estimation method is the Least Square(LS) method. LS method performs well when equation errors can be characterized as uncorrelated white noises. Otherwise, LS method provides biased estimates. If unbiased estimates are desirable, instrumental variable method, generalized LS method, and extended LS method can be used [3,6].

IV. FAULT DETECTION AND ISOLATION

4.1 Radiator Fault Detection and Isolation Possible radiator faults include oil passage clogging and air passage clogging. From Eq.(9c), one concludes that the oil passage clogging should induce a change in the coefficient C_1 as the oil flow is restricted by a narrowed passage. Similarly, the air passage clogging should result in a change in the coefficient C_2.

To estimate both C_1 and C_2 in the Eq.(9c), one would need the measurements of T_{in}, T_{out}, T_∞, and T_r. Two T-type thermocouples are attached to the outside of the inlet and outlet copper tubes of the radiator and thermally insulated from the surrounding atmosphere to measure T_{in} and T_{out}. The third thermocouple is used to measure the ambient temperature, T_∞. The radiator temperature, T_r, is obtained from a thermocouple attached to the radiator.

The outputs of these thermocouples are sampled at 1 Hz while faults of varying severity are introduced to the radiator. The estimated values of the parameters are plotted against time in Fig. 2 and 3. The changes in both C_1 and C_2 signify the moment that the faults were introduced. It is obvious that the changes of both C_1 and C_2 are very well correlated with the severity of oil passage clogging and air passage clogging. The result associated with C_1 is of particular importance because the oil mass flow rate is a nonmeasurable parameter for this system. Through the modeling and parameter estimation, C_1 can be easily assessed now.

4.2 Air-end/Motor Subsystem Fault Detection and Isolation To estimate the parameters of Eq. (9a) and (9b) one would need the measurement of the rotational speed (w_m), the motor current (I_{ms}), the air-end pressure (P_l), and the mass flow rate ($\dot{M}$). A tachometer is used to measure the shaft rotational speed. A current sensor is used to measure the motor current. A piezoelectric pressure sensor is employed to measure the air-end pressure and a nozzle is used to measure the volumetric flow rate from which the mass flow rate can be derived.

The outputs of these four sensors are sampled at 1 kHz while different faults are present in the compressor. Parameters of Eq. (9a) and (9b) are estimated. It was found that the estimates of parameters a_{12}, b_{12} and b_{22} failed to converge. Those parameters do converge behave inconsistently with what one would have concluded from the model, and are very sensitive to changes in operating conditions. Further study of residuals revealed that this model is not adequate. Therefore, finding an adequate model which correctly describes the air-end/motor subsystem becomes the most important task for this study.

Examination of the spectrum of the mass flow rate measurement revealed that it only contains useful information at very low frequencies. It is discarded to reduce the complexity of the modeling task. Thus, only Eq. (9a) remains. Based on Eq.(9a), the following linear difference equation is proposed to model the dynamics of the air-end/motor subsystem.

$$\begin{aligned} & w(t) + a_1 w(t-1) + a_2 w(t-2) + \ldots\ldots + a_n w(t-n) \\ & = b_0 I(t-d_1) + b_1 I(t-d_1-1) + \ldots\ldots + b_m I(t-d_1-m) \\ & + c_0 P(t-d_2) + c_1 P(t-d_2-1) + \ldots\ldots + c_k P(t-d_2-k) + r(t) \end{aligned} \tag{10}$$

where $w(\cdot)$, $I(\cdot)$ and $P(\cdot)$ are discrete samples of Δw_m, ΔI_{ms}, and ΔP_l; t is the sampling instant; a_1, $a_2, \ldots, a_n$, $b_0, b_1, \ldots, b_m$, $c_0, c_1, \ldots, c_k$ are the coefficients of the difference equation; d_1 and d_2 are delays; n, m and k are orders of the difference equation; and r(t) is the equation error.

Apply the z-transform to Eq.(10) and obtain the following equation:

$$\begin{aligned} & w(z) + a_1 z^{-1} w(z) + a_2 z^{-2} w(z) + a_3 z^{-3} w(z) + \ldots\ldots + a_n z^{-n} w(z) \\ & = z^{-d_1}(b_0 I(z) + b_1 z^{-1} I(z) + b_2 z^{-2} I(z) + \ldots\ldots + b_m z^{-m} I(z)) \\ & + z^{-d_2}(c_0 P(z) + c_1 z^{-1} P(z) + c_2 z^{-2} P(z) + \ldots\ldots + c_k z^{-k} P(z)) + r(z) \end{aligned} \tag{11}$$

Alternatively, one could write Eq. (11) as follows.

$$\begin{aligned} w(z) = {} & I(z) \frac{z^{-d_1}(b_0 + b_1 z^{-1} + b_2 z^{-2} + \ldots\ldots + b_m z^{-m})}{1 + a_1 z^{-1} + a_2 z^{-2} + a_3 z^{-3} + \ldots\ldots + a_n z^{-n}} \\ & + P(z) \frac{z^{-d_2}(c_0 + c_1 z^{-1} + c_2 z^{-2} + \ldots\ldots + c_k z^{-k})}{1 + a_1 z^{-1} + a_2 z^{-2} + a_3 z^{-3} + \ldots\ldots + a_n z^{-n}} \\ & + \frac{r(z)}{1 + a_1 z^{-1} + a_2 z^{-2} + a_3 z^{-3} + \ldots\ldots + a_n z^{-n}} \end{aligned} \tag{12}$$

In this form the model is called a transfer function model and is often schematically represented by a block diagram shown in Fig. 4. The transfer functions are represented by H_1 and H_2. They transfer the input I(z) and P(z) respectively to the output. To this transfered input the "noise" represented by $H_3(z)r(z)$ is added, giving the actual output w(z).

Poles and zeros are defined as values of z that nullify the denominator and the numerator of a transfer function. From the poles and zeros of a transfer function, one could derive the natural frequency (f), damping ratio (ζ) and energy dispersion (D) associated with each pair of complex conjugate poles. Let's denote poles and zeros by λ_i and υ_i. Then,

$$f_n = \frac{\pi}{T}\sqrt{[\ln(\lambda\lambda^*)]^2 + 4[\cos^{-1}(\frac{\lambda+\lambda^*}{2\sqrt{\lambda\lambda^*}})]^2} \tag{13}$$

$$\zeta = \frac{\ln(\lambda\lambda^*)}{\sqrt{[\ln(\lambda\lambda^*)]^2 + 4[\cos^{-1}(\frac{\lambda+\lambda^*}{2\sqrt{\lambda\lambda^*}})]^2}} \tag{14}$$

$$D(i) = \left|\sum_{j=1}^{n}\frac{G(i)G(j)}{1-\lambda_i\lambda_j}\right| \tag{15}$$

where

$$G(i) = \frac{\prod_{j=1}^{m}\frac{\lambda_i-\upsilon_j}{\lambda_i}}{\prod_{\substack{j=1\\ j\neq i}}^{n}\frac{\lambda_i-\lambda_j}{\lambda_i}} \tag{16}$$

The task of model identification is to determine the delays, the orders and the coefficients of the model from the measurements. First, the Akaike's Information Criterion (AIC) is minimized to find the delay. Fig. 5 and 6 show AIC values of a model of order (n, m, k) = (20, 19, 19) with respect to d_1 and d_2 respectively. It is obvious that the AIC is minimized for $d_1=8$ and $d_2=50$. Second, the order (n, m, k) is determined by minimizing AIC with delays previously determined. Since a real dynamic system always has m and k smaller than n, we have assumed that (n, m, k) = (n, n-1, n-1). Therefore, only n has to be determined instead of n, m, and k. In addition, n is assumed to be always even to avoid imposing a real pole to the model. Fig. 7 shows AIC versus n, it is obvious that the AIC does not improve much for n larger than 16. Furthermore, the fact that the autocorrelation function of the residual is found to be close to a Kronecker delta function and small cross correlation between the residual and the two inputs also support that 16 is an adequate order. To summarize, (n, m, k, d_1, d_2) = (16, 15, 15, 8, 50). Experiments were conducted on the 20 Hp single rotor screw compressor with three different conditions at two different operating pressures (85 psi and 115 psi). The three conditions include the normal condition and the following two simulated faulty conditions: bearing friction increase and worn gaterotor which is induced by changing the mesh clearance between the rotor and gaterotors. At least three sets of measurements were taken under a particular condition at a particular operating pressure.

Outputs from transducers after appropriate signal conditioning circuits were sampled at 1 KHz. Parameters of the model are estimated using the least square method to minimize the sum of the square of the equation errors.

4.2.1 Fault Detection of Air-end/Motor Subsystem There was no need for separate fault detection and fault isolation scheme in the case of the radiator because the model was developed based on physical laws which govern the radiator's behavior. The coefficients of the model are expressed in terms of physical parameters of the radiator and the flow goes through it. In this case, changes of model coefficients resulting from a fault can be easily predicted through analysis of the model. For example, one would conclude that a restricted oil passage is likely to be the case if a reduction in C_1 is observed while no significant change in C_2 can be found.

On the other hand the model of the air-end/motor assembly is identified through a "Black-Box" approach, there is no way to analyze how the model is going to be affected by the presence of a fault until the fault actually occurs. One would need a learning population which scan every class of faults at different levels of degradation to complete such an analysis. Clearly, this is not a practical approach for the majority of real problems such as the one we have due to prohibitively high cost.

To have a practical and useful fault detection/isolation system, it is imperative to devise a model based scheme which can detect the presence of abnormality without any prior experience of faults. Once an abnormalities is declared by the fault detection device, the effect of the fault on the model can be studied through the analysis of the newly available data.

The scheme we employed for the fault detection can be explained using fig. 8. A normal model identified while the compressor is healthy is used to predict the output W(t) using the measurement of the I(t) and the P(t) from the compressor. This prediction is compared to the actual measurement of W(t) from the compressor and the difference e(t) is computed. Statistical properties of the e(t) are examined to determine the state of the compressor.

The principle of this scheme is that if the compressor's dynamics have been changed by a fault, the normal model would no longer be adequate to describe the behavior of the compressor. Therefore, its prediction will be off and the prediction error will change its characteristic.

Fig. 9 shows the cross correlation between the prediction errors and the input P(t) obtained under different conditions. If the model is an adequate one, the corss correlation should be close to zero for all the lags. However, one can easily see that the cross correlation of a faulty compressor is larger than that of a healthy one. Therefore, an index based on the variance of the cross correlation between lag 0 and lag 100 is established to distinguish abnormal from the normal condition. Fig. 10 shows this index for the normal and for different fault conditions. It is clear that one can easily distinguish abnormal from normal condition.

4.2.2 Fault Isolation of the Air-end/Motor Subsystem The magnitude of $H_1(e^{j\omega})$ and $H_2(e^{j\omega})$ obtained at 115 psi operating pressure are plotted against frequency in Fig. 11. Comparing these frequency responses, it is clear that the first mode (the peak at 0.4 rad.) in both $H_1(e^{j\omega})$ and $H_2(e^{j\omega})$ becomes more prominent compared to other modes as friction is increased. Note that these plots are in db scale and therefore the real changes in relative magnitude of mode one due to friction are much larger than what it appears to be. Similar magnitude changes of the same mode are also observed in the frequency responses identified at a different operating pressure of 85 psi. From the plot, it is also obvious that gaterotor wear would make the 4th mode in $H_1(e^{j\omega})$ the 2nd mode in $H_2(e^{j\omega})$ more prominent. Consistent phenomenon is also found in the frequency responses identified at 85 psi operating pressure.

Although frequency response plot comparison is a natural way for human beings to conduct fault isolation, it can not be easily imitated by a computer. Therefore, dispersion (Eq. 15) of each mode normalized with the total energy of the frequency response is used. Fig. 12 shows the difference of the normalized dispersion of each mode of $H_1(e^{j\omega})$ and $H_2(e^{j\omega})$ between the normal and the faulty compressor. It is clear that the dispersion of mode 1 of both $H_1(e^{j\omega})$ and $H_2(e^{j\omega})$ is changed significantly by the friction, and that the dispersion of mode 4 in $H_1(e^{j\omega})$ and mode 2 in $H_2(e^{j\omega})$ is changed significantly by the gaterotor wear. Therefore, both faults can be diagnosed by observing the change of the dispersion changes.

V. CONCLUSION

We have applied model based fault detection and isolation technique to a single rotor screw compressor. The approach is shown to be very effective in fault detection and isolation. Under all conditions, none of the direct measurements including the motor current, rotating speed, mass flow rate and efficiency has a change of more than two percent in comparison to their normal values. Therefore, the approach has enabled the formation of indices which are more sensitive to faults than measurable variables are. It is also found that the success of such approach depends on the adequacy of the model one used. Therefore, a good modeling and parameter estimation methodology is very critical to the evolving of this relatively new approach to fault detection and isolation. In particular, a practical nonlinear modeling scheme is called for if the model based fault detection and isolation technique is used for highly nonlinear systems. In summary, it is our belief that this technique will compliment the inadequacy of existing signal processing based fault detection and isolation techniques.

References:

1. Isermann, R., "Process Fault Detection Based on Modeling and Estimation Methods --A Survey, Automatica, Vol 20, No 4, pp387-404, 1984.
2. Geiger,G., "Fault Identification of a Motor-pump system using Parameter Estimation and Pattern Classification", IFAC 9th Triennial World Congress, Budapest, Hungary, pp 1783-1788,1984.
3. Ljung, L. , Soderstrom, T. , Theory and Practice of Recursive Identification, The MIT press,

Cambridge, Massachusetts, 1983.
4. Goodwin, G. C. and Sin, K. S., "Adaptive Filtering, Prediction and Control", Prentice-Hall, Inc., Englewood Cliffs, New Jersey, 1984.
5. Young, P. C., "Parameter estimation for continuous-time models-a survey. Automatica, Vol 17, No. 23.
6. Strejc,V., "Least Square Parameter Estimation", Automatica, Vol 16, pp 535-550, 1980
7. Rao,G.P. and Sivakumad,L., "Identification of Deterministic Time-Lag Systems", IEEE Transaction of Automatic Control, AC-21,pp 527-530,1976.
8. Van Ormen Jr., H.P., "Profiles of air-end rotors",Hydraulics and Pneumatics, Mar, 1986.
9. Engelmanm,R.H.,"Static and Rotating Electromagnetic Devices",Marcel Dekker, Inc., 1982

Appendix:

nomenclature

a	air-end torque characteristic coefficient for angular velocity
a_r	surface area of radiator
A_F	sectional area of pipe
A_V	sectional area of valve
b	air-end torque characteristic coefficient for mass flowrate
C	motor characteristic coefficient($= \frac{9n}{8r_r} M_{sr}^2 \sin^2(\alpha-\beta)\ w$)
C_{FM0};C_{FM1}	motor friction coefficients
C_{FC0};C_{FC1}	air-end friction coefficients
C_{oil}	specific heat of oil
C_R	fluid resistance coefficient of pipe system
C_V	valve coefficient
c_{pr}	specific heat of radiator
h_r	convection coefficient of radiator
I_{ms}	motor stator current
L	characteristic length of pipe system
$\dot{m}_{oil}$	oil mass flowrate in the radiator
$\dot{M}$	air mass flowrate
M_{sr}	mutual inductance between stator and rotor of motor
n	number of cage in motor
P_l	air-end outlet pressure
ΔP_P	pressure drop of pipe
ΔP_R	pressure drop due to fluid resistance
ΔP_V	pressure drop due to valve
ΔP_{ac}	pressure drop due to acceleration
r_r	resistance of rotor winding
T_M	motor torque
T_{FM}	motor friction torque
T_S	shaft torque
T_{FC}	air-end friction torque
T_C	air-end torque
T_r, T_∞	radiator and ambient temperature
T_{in}, T_{out}	oil inlet and outlet temperature
V_r	volume of radiator material
w	field rotating speed
w_m	motor speed

ρ_r	density of radiator material
Θ_M, Θ_C	motor and air-end inertia

Torque balance on motor can be expressed as

$$\Theta_M \frac{dw_m}{dt} = T_M - T_{FM} - T_S \tag{A1}$$

For the Cage Induction Motor [9], motor torque can be expressed as the following;

$$T_M = \frac{3}{4} n I_{mr} I_{ms} M_{sr} w \sin(\alpha-\beta) \frac{1}{w_m} \tag{A2}$$

The relationship between I_{mr} and I_{ms} is

$$I_{mr} = \frac{3}{2r_r}(w-w_m) I_{ms} M_{sr} \sin(\alpha-\beta) \tag{A3}$$

Substituting Eq.(A3) into (A2), we have

$$T_M = C \frac{I_{ms}^2 (w-w_m)}{w_m} \tag{A4}$$

where, $C = \frac{9n}{8r_r} M_{sr}^2 \sin^2(\alpha-\beta)\, w$

Friction force on motor T_{FM} is expressed as

$$T_{FM} = C_{FM0} + C_{FM1} w_m \tag{A5}$$

From Eq.(A1),(A4) and (A5), we obtain

$$\Theta_M \frac{dw_m}{dt} = C \frac{I_{ms}^2 (w-w_m)}{w_m} - C_{FM0} - C_{FM1} w_m - T_S \tag{A6}$$

Small deviations around the steady-state ($\bar{w}_m$, $\bar{T}_S$, $\bar{I}_{ms}$) are considered;

$$w_m = \bar{w}_m + \Delta w_m$$

$$T_S = \bar{T}_S + \Delta T_S$$

$$I_{ms} = \bar{I}_{ms} + \Delta I_{ms}$$

After perturbation, Eq.(A6) can be expressed like the following:

steady-state : $0 = - C_{FM0} - C_{FM1} \bar{w}_m - \bar{T}_S$

transient : $\Theta_M \frac{dw_m}{dt} = -\left(C \frac{\bar{I}_{ms}}{\bar{w}_m^2} w + C_{FM1} \right) \cdot \Delta w_m - \Delta T_S$

$$+ 2C\bar{I}_{ms} \frac{w - \bar{w}_m}{\bar{w}_m} \cdot \Delta I_{ms} \tag{A7}$$

Torque balance on the air-end can be expressed as

$$\Theta_C \frac{dw_m}{dt} = T_S - T_{FC} - T_C \tag{A8}$$

This torque is a function of rotational speed, mass flowrate and discharged pressure [8]. Therefore, the following equation can be obtained:

$$\Delta T_C = a \cdot \Delta w_m + b \cdot \Delta \dot{M} + d \cdot \Delta P_l$$

Air-end friction torque:

$$T_{FC} = C_{FC0} + C_{FC1} w_m$$

Substitute ΔT_C and T_{FC} into Eq.(A8) and linearize it, we have

steady-state : $0 = \bar{T}_S - \bar{T}_C - C_{FC0} - C_{FC1} \bar{w}_m$

transient : $\Theta_C \frac{dw_m}{dt} = \Delta T_S - (C_{FC1} + a) \cdot \Delta w_m - b \cdot \Delta \dot{M} + d \cdot \Delta P_l$ (A9)

Eq.(A7) and(A9) can be added together to eliminate T_S.

$$(\Theta_M+\Theta_C)\frac{dw_m}{dt} = -\left(C\frac{\bar{I}_{ms}}{\bar{w}_m^{\,2}}w + C_{FM1} + C_{FC1} + a \right)\cdot \Delta w_m - b\cdot \Delta\dot{M} + 2C\bar{I}_{ms}\frac{w-\bar{w}_m}{\bar{w}_m}\cdot \Delta I_{ms} + d\cdot \Delta P_l \qquad \text{(A10)}$$

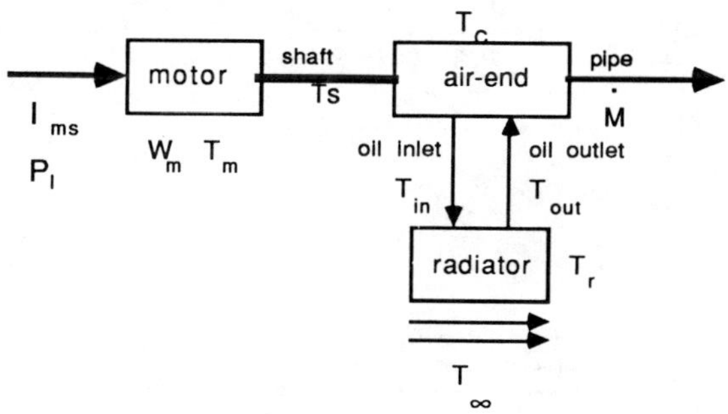

Fig. 1 A model of the compressor

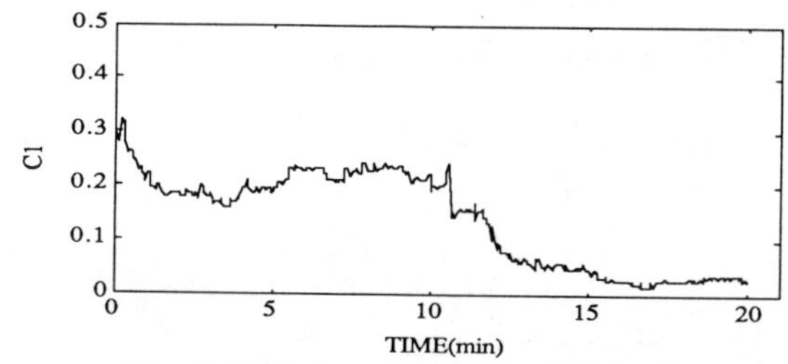

Fig. 2 Variation of C1 due to 50% oil clogging

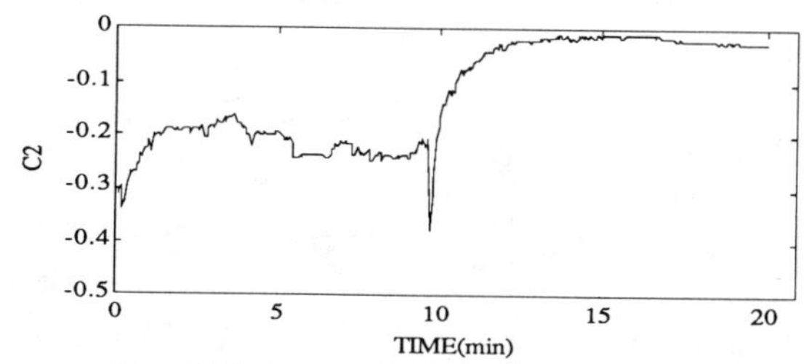

Fig. 3 Variation of C2 due to 50% air passage clogging

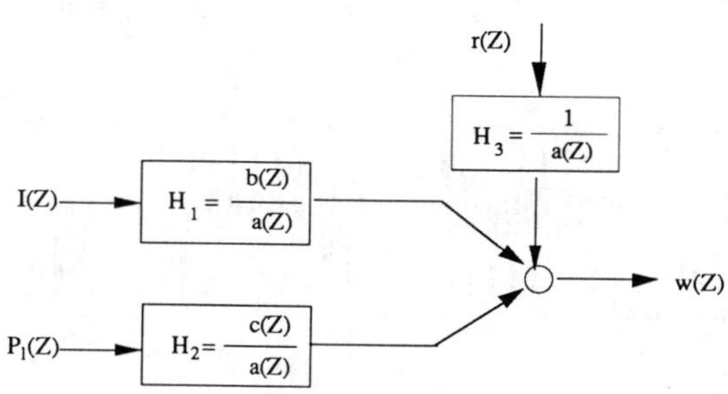

Fig. 4 Block diagram for the transfer function model

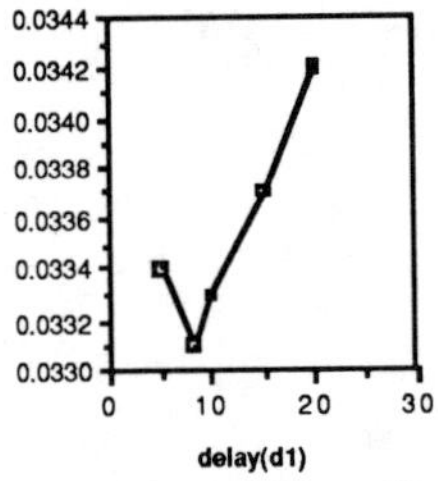

Fig. 5 AIC vs. d1

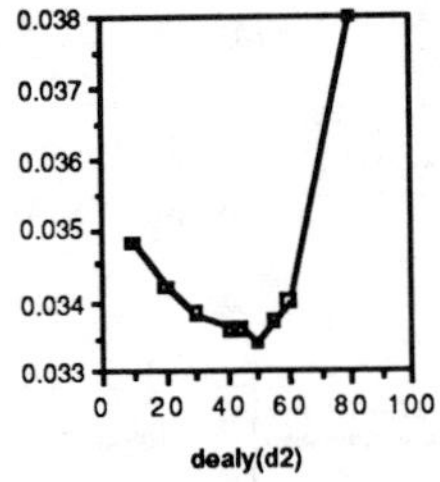

Fig. 6 AIC vs. d2

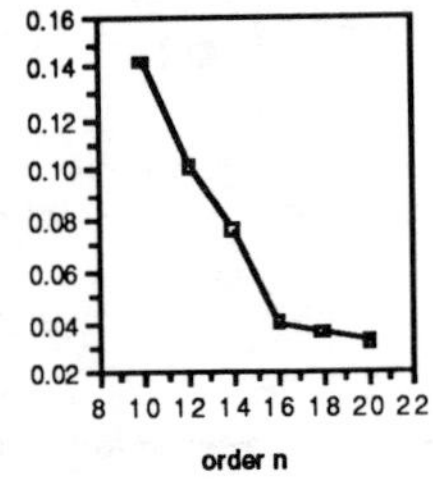

Fig. 7 AIC vs. n

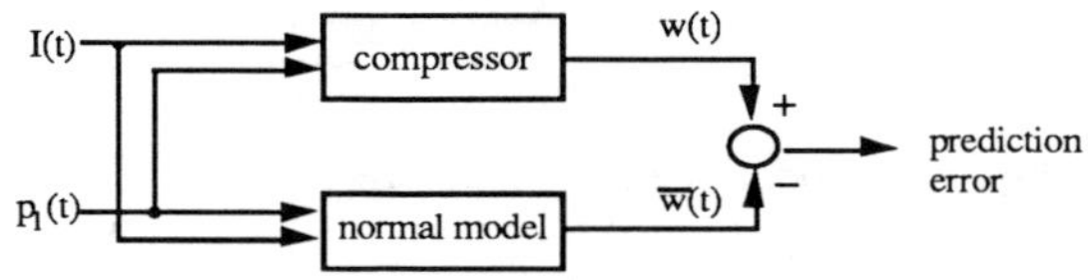

Fig. 8 Fault detection scheme

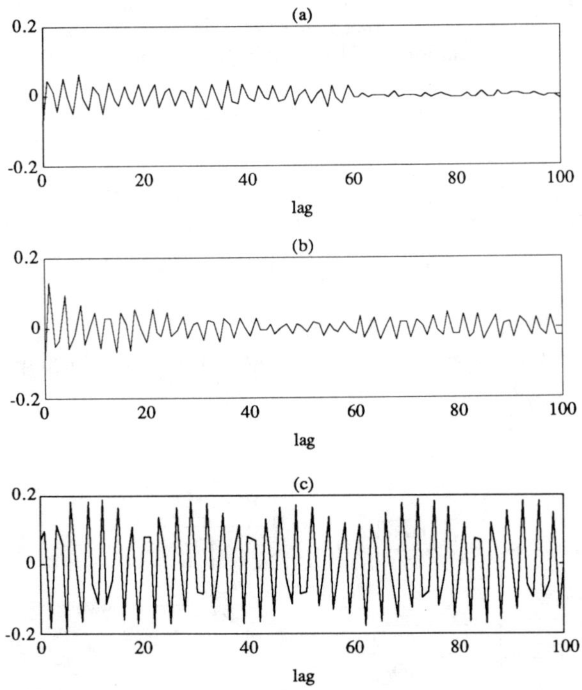

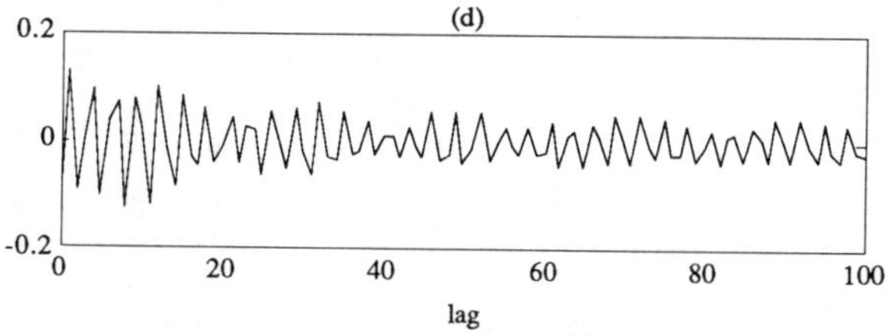

Fig. 9 Cross correlations between prediction error and air-end pressure
(a) model, (b) normal condition, (c) friction, and (d) gaterotor wear

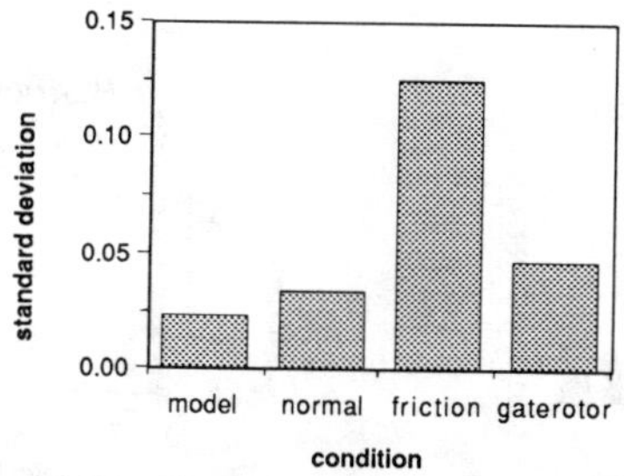

Fig. 10 Standard deviation of the cross correlations

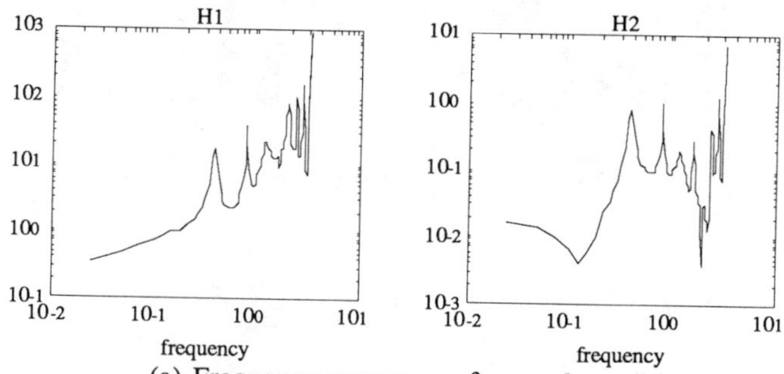

(a) Frequency response of normal condition

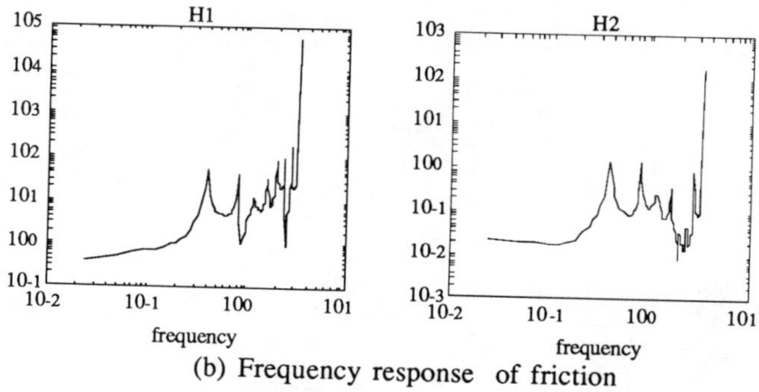

(b) Frequency response of friction

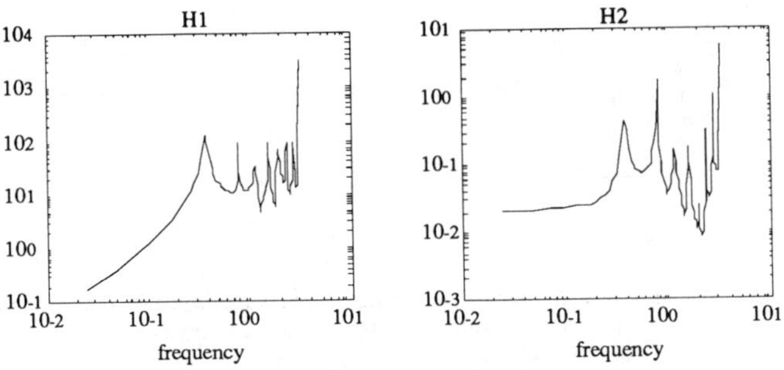

(c) Frequency responses of gaterotor wear
Fig. 11 Frequency response

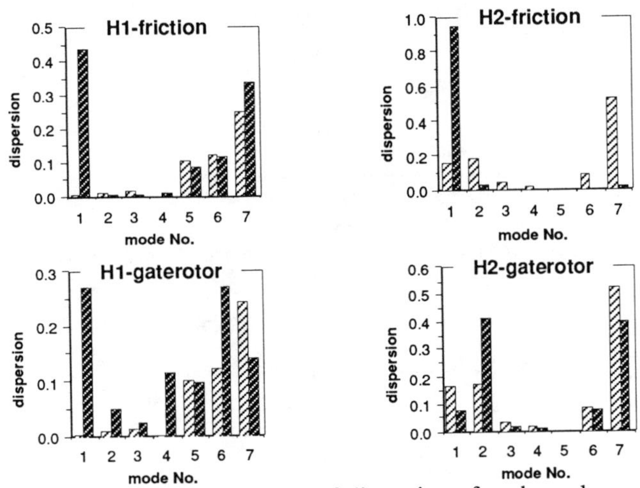

Fig. 12 Comparison of dispersion of each mode

Sensors, Controls, and Quality Issues in Manufacturing
ASME 1991

DESIGN CONCEPT OF AN INTEGRATED VISION SYSTEM FOR COST-EFFECTIVE PART-PRESENTATION

Kok-Meng Lee
George W. Woodruff School of Mechanical Engineering
Georgia Institute of Technology
Atlanta, Georgia

ABSTRACT

Recognizing that the lack of an off-the-shelf software-control generic part-presenter satisfying both cost and functional requirements is a major source of problems which prevent the flexibility of the overall flexible automation to be fully exploited, the design concept of an integrated microprocessor-controlled vision system based on retroreflective vision sensing is developed. In this paper, optic design concepts for collocated illumination for cost effective part-presentation are presented. The uniformity of the image irradiance across the sensor surface, in general, depends on several factors such as the intensity distribution of the light source, the reflectance characteristics of the object and background, and the geometrical relationship between the imaging sensor, the illumination source, and the target. In order to establish design guidelines for effective vision-guided part-presentation, an analytical model which characterizes the influences of these factors on the image characteristics is described, along with the presentation of illustrative applications and experimental results.

1. INTRODUCTION

Machine vision can add considerably to flexibility by simplifying grippers, component feeders, and location tooling, and can reduce the engineering time required to implement it. Elsewhere the author and his co-workers describe the concept of retroreflective vision sensing technique for generic part presentation [1] and its application to flexible part-feeding for assembly automation [2] [3]. Two major problems often associated with the use of machine vision are poor reliability and excessive image processing time, both of which depend on the illumination technique, the complexity of the geometry, and the surface reflectance of both the background and the objects to be handled. In contrast, retroreflective vision sensing has significant potentials in improving the vision reliability, reducing the computational load/time, and lowering the cost in implementation for part feeding since a distinct contrast between the object and the background is obtainable regardless of the surface reflectance.

Perhaps, illumination and its measurements are among the least understood technologies. Discussion and communication of desired operating parameters of optics are complicated by a plethora of nomenclatures for seemingly similar units of measurements. Furthermore, system specifications for illumination are often vague or may be left to the final user and thus expensive digital devices have long been paid for the illumination deficiencies. When illumination deficiencies become apparent, it is then that critical light parameters such as luminous intensity illumination uniformity, spectral output, and image spectral irradiance become an unwelcome part of our vocabulary. Direct symptoms of an unfit light source are not only limited to elongated exposure times, but can also include inaccurate image characteristics due to non-uniform illumination. The uniformity of the image irradiance across the sensor surface depends on several factors, which are the intensity distribution of the light source, the reflectance characteristics of the object and background, and the geometrical relationship between the imaging sensor, the illumination source, and the target.

This paper presents the optic design concepts for flexible part-feeding applications which are referred here as feeding workpiece from pallets using a pre-programmed robot for subsequent processes such as machining or assembly. Most surfaces on objects exhibit a combination of diffuse and specular reflection. The third reflectance known as retroreflection is not commonly found on the surfaces of objects. The ideal retroreflector returns the incident radiation in the direction from which it came. Retroreflective materials have been widely used in traffic control and safety sign on highways and airports [4] [5] [6]. Incidentally, the retroreflector has also been used as a non-contact position sensor of large space structure [7] and more recently, in field tracking of automated guided vehicles (AGV) [8]. The unique feature of the retroreflective materials as an effective means for presentation of parts of which orientation can be determined from the silhouette or by modifying part design to include "landmarks", was conceptualized by Lee et al. [1]. The basic principle of retroreflective vision sensing for part-presentation is to use retroreflective surfaces to maximize the ability of a vision system to reliably and quickly locate parts. One way is to make the surface of a pallet retroreflective so that it can be made to appear much brighter than objects on it. Alternatively, retroreflective landmarks can be intentionally created on objects for location tracking. Since the retroreflective material has a distinctive surface reflectance that is not commonly found in natural or man-made objects, it enables reliable digital images of high object-to-background contrast to be obtained without a prior knowledge of object geometry and surface reflectance. The location and orientation of the part can then be determined with relatively simple, high-speed computation without the need for a detailed reflectance map of the part to be handled.

The remaining of the paper is organized as follows: The fundamental equations of optics for retroreflection are described. Then optic design concepts and the sensitivity of the operational parameters on image characteristics are described and followed by a presentation of illustrative applications. Finally, the conclusions are drawn and further works are suggested.

2. FUNDAMENTAL EQUATIONS OF OPTICS

Consider an optical aperture located at a distance f from the image plane as shown in Fig. 1(a). The solid angle of the cone of rays leading to the patch on the object is equal to that corresponding patch in the image.

$$\frac{\delta O}{\delta I} = \frac{\cos\theta}{\cos\beta}\left[\frac{D'}{f}\right]^2 \qquad (1)$$

where δO and δI are the patch areas on the object surface and the image respectively; the entrance angle β is defined as an angle between a light ray and the surface normal; and the angle between a light ray and the optical axis is denoted as θ.

The amount of light radiated from a surface is called the radiance L. It is the power per unit area per unit solid angle ($W.m^{-2}.sr^{-1}$ - watts per square meter per steradian). Thus, the power of the light originating on the patch and passing through the lens is

$$\delta P = L\ \delta O\ \Omega \cos\beta \qquad (2)$$

where Ω is a solid angle subtended by the lens as seen by the patch.

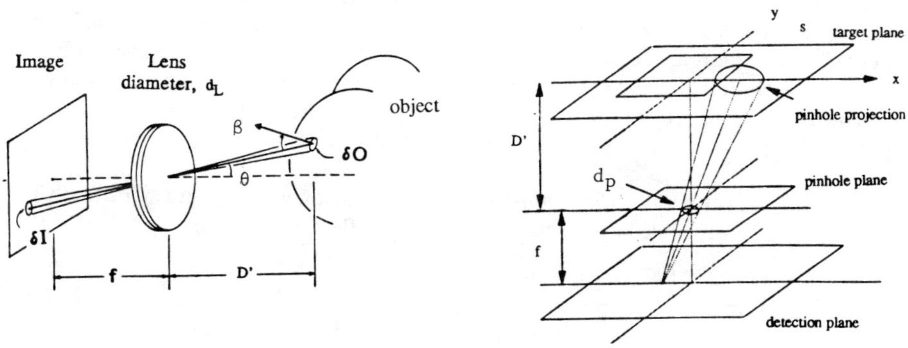

(a) Schematics of Optic Relationship (b) Solid Angle of Pinhole Optics

Fig. 1 Fundamental of Optics

The amount of light falling on a surface is called the irradiance I. It is the power per unit area (W/m^2 - watts per square meter) incident on the surface. The irradiance of the image at the patch under consideration is given by $I = \delta P/\delta I$. Substituting for δO from Equation (1) into Equation (2), the irradiance of the image is given by

$$\frac{I}{L} = \Omega \left[\frac{D'}{f} \right]^2 \cos\theta \tag{3}$$

Equations (1) - (3) are valid for a pin hole as well as a lens as shown in Fig. 1(a) since rays passing through the center of the lens are not affected. The solid angle that the lens of diameter d_L subtends when viewed from the light-emitting surface is given by Equation (4):

$$\Omega = \frac{\pi}{4} \frac{d_L^2 \cos\theta}{(D'/\cos\theta)^2}. \tag{4}$$

Fig. 1(b) illustrates the solid angle for the case of a pin-hole of diameter d_p, i.e. $d_L = d_p$ in Equation (4). Equation (3) leads to

$$\frac{I}{L} = \frac{\pi}{4} \frac{\cos^4\theta}{F^2} \tag{5}$$

where

$$F = \begin{cases} f / d_L & \text{for a lens} \\ f / d_p & \text{for a pin hole.} \end{cases}$$

Scene radiance depends on the amount of light that falls on a surface, the fraction of the incident light that is reflected, and the geometry of light reflection. In other words, the radiance of a surface L will generally depend on the direction from which it is viewed as well as on the direction from which it is illuminated. The radiance of the reftroreflective sheeting is typically characterized by the coefficient of retroreflection $R'(\alpha,\beta)$ [5] which is defined as follows:

$$R'(\alpha, \beta) = \frac{I_r}{E A_r} \tag{6}$$

where I_r is the luminous intensity of the retroreflector in the direction of observation in watts/sr, candela, or lm/sr; E is the illuminance at the retroreflector on the plane perpendicular to the direction of the incident light in watts/m^2 or lm/m^2; A_r is the retroreflecting surface area; and the observation angle α is defined as an angle between the observation axis and an axis of an incident light. The coefficient of retroreflection of a typical retroreflective sheeting is displayed in Fig. 2(b) as a function of observation and entrance angles, where the nomenclature is defined in Fig. 2(a). Within an angle of 0.2° from the principle axis of incident light, typical retroreflective sheetings reflect with over 250 times the intensity of a typical diffuse white surface.

Consider an incident light from a spectral illuminator of wavelength λ falls on a retroreflective sheeting characterized by the coefficient of retroreflection $R'(\alpha,\beta)$. Since

$$L = R'(\alpha, \beta) \frac{E_s(\phi)}{(D/\cos\phi)^2} \tag{7}$$

where $E_s(\phi)$ is the illuminous intensity distribution of the spectral illuminator; ϕ is an angular displacement of an incident light measured from the optical axis of the source; and D is the

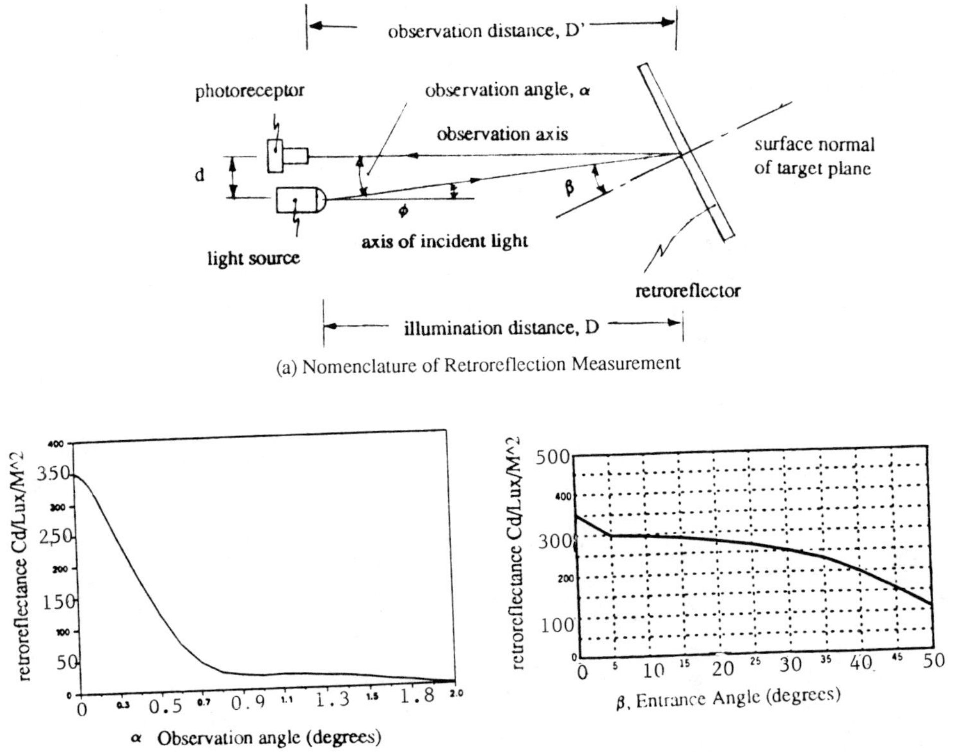

(a) Nomenclature of Retroreflection Measurement

(b) Coefficent of Retroreflection vs. Observation Angles

(c) Coefficient of Retroreflection vs. Entrance Angles

Fig. 2 Typical Characteristics of Retroreflection

illumination distance defined in Fig. 2(a). Upon substituting for L, the image irradiance is written from Equation 5 as

$$I = \frac{\pi}{4} \frac{E_s(\phi)\ R'(\alpha,\beta)}{F^2 \quad (D/\cos\phi)^2} \cos^4\theta \ . \tag{8}$$

The linearity of gray-scale responses of a imaging sensor is commonly specified by an exponent "gamma γ" as

$$A = K\ E^{\gamma} \tag{9}$$

where E is a time integral version of image irradiance; K is a constant, and A is the amplitude of the sensor output signal. A gamma of unity yields a linear response, whereas less than unity compresses the bright end of the response curve and greater than unity compresses the dark end. Many video imaging sensors have gamma of less than unity to compensate for video monitors that exhibit the opposite effect. The output signal of the imaging sensor element is of the form

$$A = K \left[\tau \frac{E_s(\phi)\ R'(\alpha,\beta)}{F^2 \quad (D/\cos\phi)^2} \cos^4\theta \right]^{\gamma} \ . \tag{10}$$

where τ is the effective exposure time.

3. PINHOLE OPTIC DESIGN FOR RETROREFLECTION

Lenses are often used with camera to gather more of the light reflected from (or emitted by) the object, to magnify the field-of-view, and to improve the image quality. Actual lenses, in general, tend to have certain geometrical distortion on the image. It introduces several factors which limit the accuracy of dimensional measurements made with simple machine vision. These factors are the nonlinearity of the lens and the state of focus. The ability of retroreflective vision sensing to generate a high contrast digital image with relatively low illumination power demonstrates the potential of eliminating the needs of lenses for light-gathering. In addition to the simplicity of the pinhole design, pinhole optics offer freedom from geometrical distortion and virtually infinite depth of field.

As illustrated in Fig. 1(c), the intensity at any point P along an axis s on the detector is proportional to the overlapping area of the pinhole projection with radius R on the object plane and the bright area. The highest and lowest intensities, I_{max} and I_{min} are corresponding to the cases when the pinhole projection is completely covered by bright area and dark area respectively. If δO is the area of pin-hole projection and δA is the portion of the object patch which reflects light into the detector.

$$\frac{\delta A}{\delta O} = 1 - \frac{1}{\pi} \left[\cos^{-1} S - S \sqrt{(1 - S^2)} \right] \qquad |S| \leq 1 \tag{11}$$

where $S = \frac{s}{R}$ and $R = \frac{1}{2F} (1 + \frac{f}{D})$.

The normalized distance S can be expressed as a function of β as follows:

$$S = \frac{D}{R} \left(\tan\beta - \tan\beta_o \right) \tag{12}$$

where β_o is the entrance angle defined at the edge on the target plane.

The image irradiance model which describes the normalized output signal of the imaging sensor along the axis s on the detector as a function of the edge location can be expressed as:

$$A = K \left[\tau \frac{E_S(\phi) \quad R'(\alpha, \beta) \quad \cos^4\theta}{F^2 \quad (D/\cos\phi)^2} \frac{\delta A}{\delta O} \right]^\gamma \quad . \tag{13}$$

The constraint imposed by the small lateral distance d limits the physical design and the heat generation of the illuminator. Unlike the general purpose light sources which have a wide angular range of illuminance for uniform lighting on factory floor, the illuminator for the retroreflector is designed to have a relatively narrow angular range of illuminance, generally no more than that required for the field-of-view (FOV). For a given array size of a CCD or CID detector, the FOV is defined as the cone angle of the directions encompassed by the scene being imaged and is given by

$$FOV = \pm \tan^{-1} \frac{1}{2}\frac{W}{D} = \pm \tan^{-1} \frac{1}{2}\frac{w}{f} \tag{14}$$

where W and w are the diagonal dimension of the FOV and the detector respectively. As discussed in the author's past work [1], it is desired that the collocated source, should be chosen to have a spectral characteristic which is distinctive from the general ambient lighting and is optimal to the photodetector response.

Two illumination concepts are simulated for pin-hole optics using the fundamental optic relationship given in Section 2 in order to illustrate the influences of the optic parameters on image characteristics. The simulated target is a flat encapsulated glass-bead retroreflective sheeting. The coefficient of retroreflection as shown in Fig. 2 is given by the following polynomial

$$R'(\alpha, \beta=-4°) = \sum_{i=0}^{10} a_i (\alpha/2)^i , \qquad 0 \le \alpha \le 2° . \tag{15}$$

The coefficients (a_i, i=0, .., a_{10}) are 3.548E3, 2.419E2, –4.023E3, –6.919E4, 5.954E5, –2.131E6, 4.429E6, –5.763E6, 4.642E6, –2.117E6, and 4.172E5. Similarly, the coefficient of retroreflection as a function of entrance angles for $\alpha = 0.2°$ is approximated by

$$R'(\alpha=0.2°, \beta) = 297.97 + 1.1208\,\beta - 0.0882\,\beta^2 \tag{16}$$

where $5° \le \beta \le 65°$

The LED spectral characteristic used in the computation is given in Fig. 3[9]. The luminous intensity ranges from 2.9 to 8.4 candelas for the LED with typical values of 4.0 candelas. The maximum junction temperature of the LED is limited to 110° C.

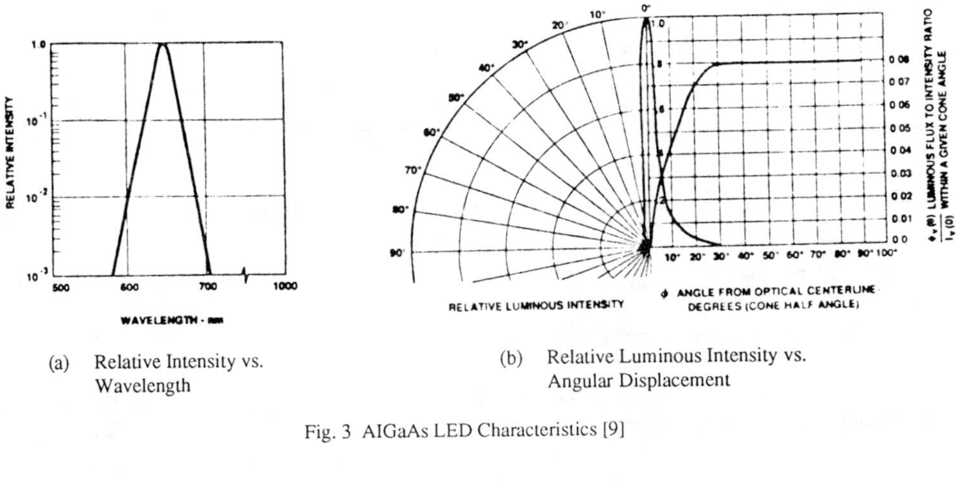

(a) Relative Intensity vs. Wavelength

(b) Relative Luminous Intensity vs. Angular Displacement

Fig. 3 AlGaAs LED Characteristics [9]

3.1 Optic Design I

Fig. 4(a) illustrates a collocated illumination scheme for imaging small components (size less than 12 mm), which combines a beam splitter with a single high intensity LED. A beam splitter is an optical component that has the property of reflecting some portion of the light that strikes it and transmitting the remainder. Beam splitter appear transparent and are generally spectral non-selective. Typical ratio of transmission-to-reflectivity is 50%/50%. Although the effective illumination efficiency of the scheme is 25%, the beam splitter which transmits illumination along the same optic axis allows small parts to be observed by the imaging sensor at a relatively short observation distance.

The design parameters are shown in Fig. 4(b). As shown in Equation (8), the image irradiance is inversely proportional to the square of the illumination distance D. Other parameters which could significantly influence the image characteristics are the angle of mis-alignment

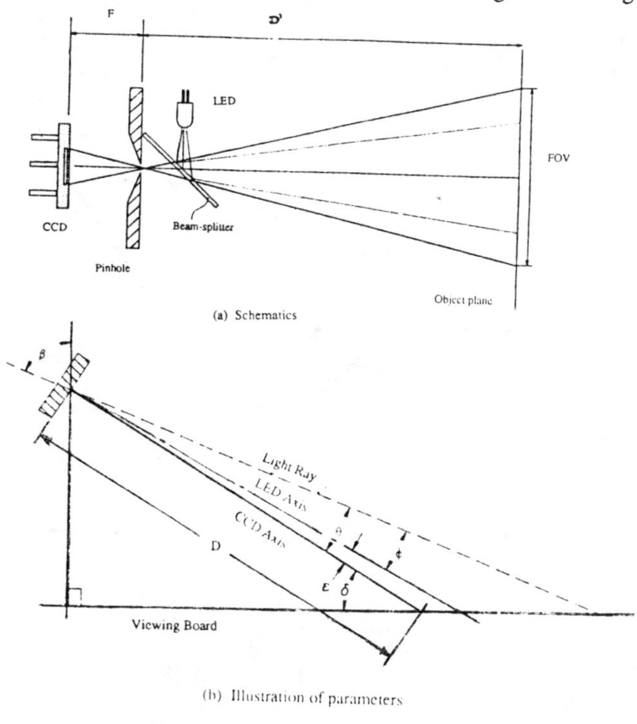

Fig. 4 Optic Design Example 1

between the observation and illumination axes, ϵ, the angular displacement of light ray measured from the illumination axis, ϕ, and the viewing angle δ. The influences of these parameters on image characteristics are illustrated in Figs. (5) to (8). Fig. 5 presents the computed result of Equation (10) for several different gamma values, where the relative sensor outputs are plotted as a function of entrance angles for δ=90°, ϵ=0°. The effects of the non-uniform illumination can be reasonably well characterized by the following model:

$$\frac{A}{A_{max}} = \frac{1}{1 + a\,(\tan^2 \beta)} \quad (17)$$

The value of 'a' was found to be 230 for $\gamma = 0.9$. Since ($\tan \beta$) is linearly proportional to the radius of the illumination projection on the target plane, the model given by Equation (17) indicates that the illumination non-uniformity has a similar effect on grey-scale computation as the geometrical radial distortion introduced by a typical lens.

As shown in Fig. 6, the illumination non-uniformity has significant effects on the threshold selection as well as the image irradiance model which describes the normalized intensity along a path perpendicular to an edge given by Equation (13). For a non-uniform source, the image irradiance model, in general, is location dependent. The effects of the mis-alignment angle ϵ and the viewing angle on illumination non-uniformity are shown in Figs. 7 and 8 respectively.

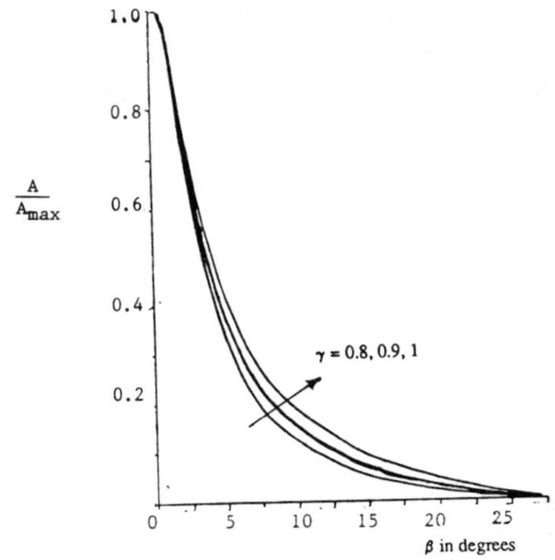

Fig. 5 Effects of Sensory Linearity ($\delta = 90°$, $\epsilon = 0$, $\theta = \phi$)

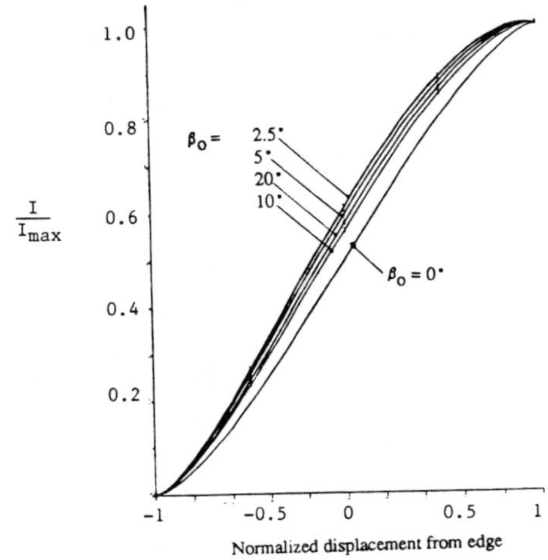

Fig. 6 Effect of Non-Uniformity on Edge Detection ($\delta = 90°$, $\epsilon = 0$)

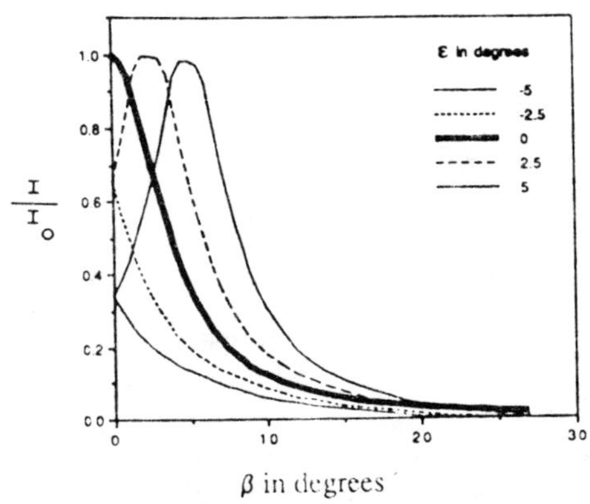

Fig. 7 Effect of Mis-Alignment ($\sigma = 90°$)

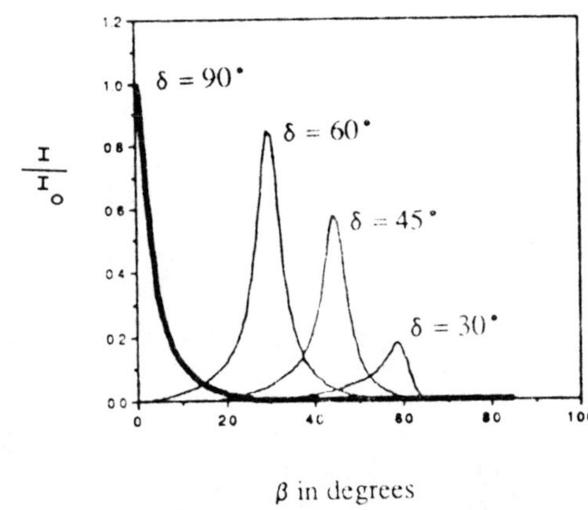

Fig. 8 Effect of Viewing Angles δ ($I_o = I_{max}$ at $\delta = 90°$)

3.2 Optic Design II

For relatively large components (typically component size of 150 mm), the use of a single wide angle LED to achieve a desired image intensity is likely to accompany an elongated exposure time. An alternative scheme using multiple LEDs arranged in circular rings around a pin-hole is shown in Fig. 9(a), along with the geometrical parameters illustrated in three-dimensional space in Fig. 9(b).

The constraint imposed by the small lateral distance, $d = (dx^2 + dy^2)^{0.5}$, limits the maximum number of LEDs which can be effectively packed along the pin-hole. Figs. 10 - 12 present the computed results using a design where twelve LEDs are arranged at the vertex of two regular hexagons with the nominal radii of 6 mm and 12 mm. For a small variation of observation angle about $\alpha = 0.2$, Equations (15) and (16) were used to characterize the targeted retroreflective surface in the simulation.

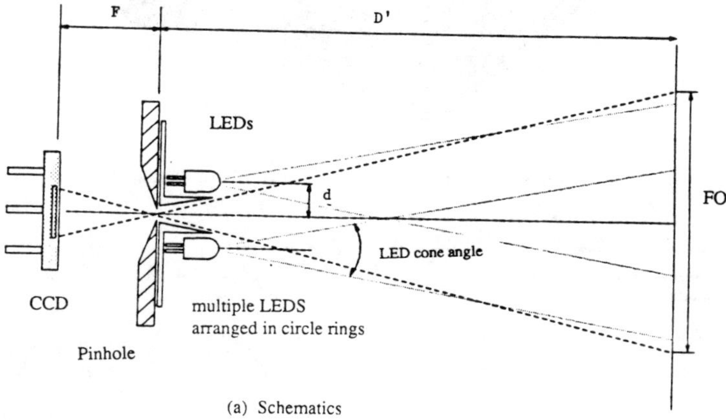

(a) Schematics

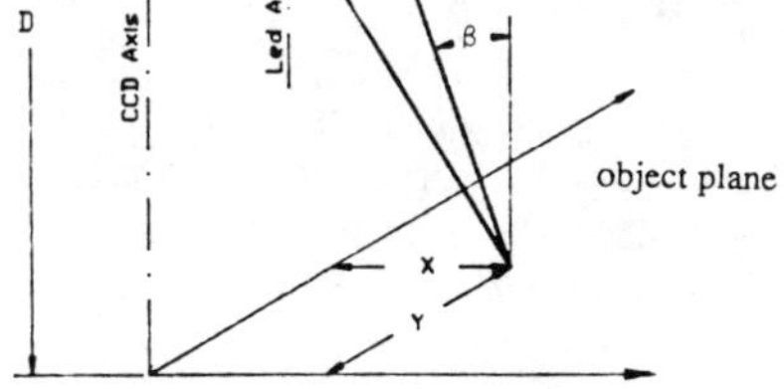

(b) Illustration of parameters

Fig. 9 Optic Design Example II

In Fig. 11, the LED spread factor indicates the relative spacing between the adjacent LEDs from a nominal design characterized by a spread factor of unity. The design trade-off is clearly between the overall uniformity and the absolute intensity, both of which depends the spread factor and the cone angle of the LEDs. However, for a given LED arrangement at a nominal radius, there exists an optimal observation distance with respect to both illumination uniformity and intensity as demonstrated in the simulated results in Fig. 12. The optimal viewing distance for the particular design occurs at 36 inches approximately.

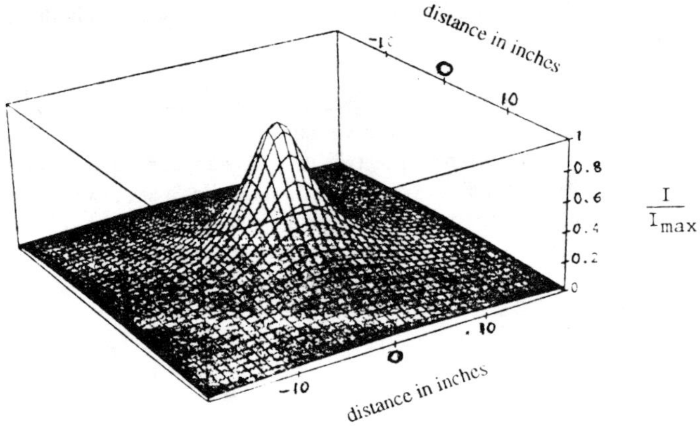

Fig. 10 Effects of LED Illuminous Intensity Distribution

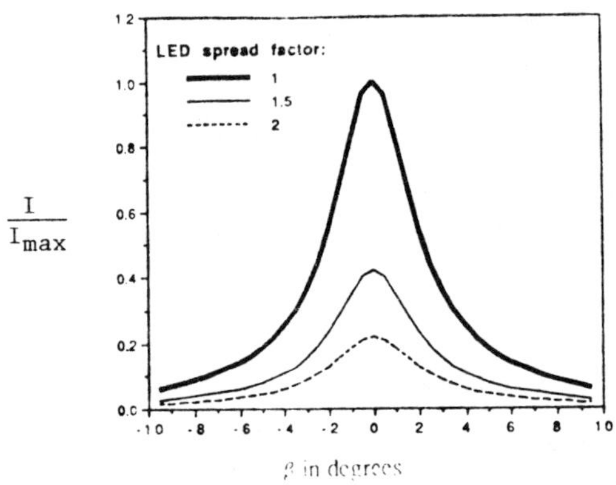

Fig. 11 Effects of Relative LED Spacing

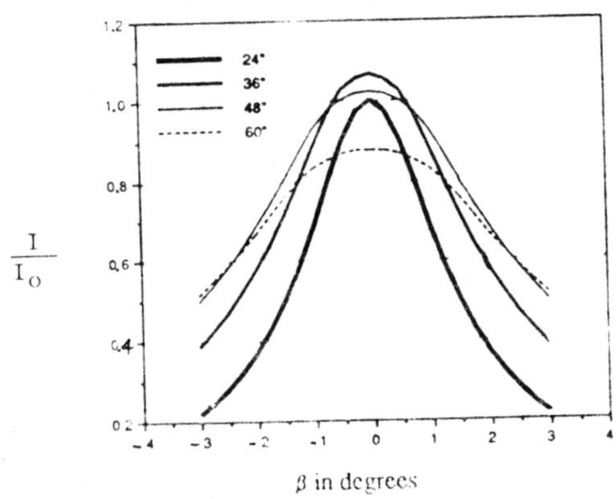

Fig. 12 Effects of Observation Distance ($I_o = I_{max}$ at 24")

4. ILLUSTRATIVE APPLICATIONS

Retroreflective materials can be used as background in part-presentation or as a landmark on parts. The choice clearly depends on the part design and manufacturing process as reported by the author and his co-worker [1]. If the orientation of the parts can be characterized by the two-dimensional object silhouette, retroreflective materials can be used as a background in generic part-presentation. In order to eliminate the needs to apply retroreflective materials on each of individual trays, it is desired that the generic egg-crate-style tray is transparent such that when it is placed on a retroreflective surface, the object will appear as a dark silhouette against a reliable bright-field background. The concept feasibility was demonstrated experimentally using an integrated vision system [3] which allows a Motorola 68000 micro-processor (μP) to have direct control of a Texas Instruments charge-coupled-device (CCD).

4.1 Illustrative Example 1: Small component imaging

Small components (typically less than 10 mm) are often kitted in egg-crate-style waffle pack for assembly. The pack may contain all components necessary to assemble one product. Since the pack can be identified to the cell controller either by a bar code on the pack itself or by a bar-coded label stuck on the base assembly component, the pack may handle many components for several varieties of the same product.

Fig. 13(a) shows an example where small electrical components are placed in hemispherical pockets of a typical egg-crate-style transparent plastic tray. Using the hemispherical pockets (1 inch diameter), the objects can be positioned approximately in the center of the packet due to gravity. As a result, the object can be viewed at a close distance with a low power illumination since each of these small components require only a relatively small field-of-view for imaging.

The optic design I consisting of a 50%/50% beam splitter with a single narrow-angle LED as shown in Fig. 4(a) was used with a 0.1 mm diameter pin-hole to investigate the feasibility for these applications. Example image which was taken with the beam-splitter 6 inches vertically above the object surface are displayed in Fig. 13(b).

4.2 Illustrative Example 2: Machined component imaging

For relatively large components, the imaging sensor must be located at a considerable distance above the target plane in order to have a sufficiently large FOV as well as a reasonably uniform illumination.

Typical examples are machine components as in Fig. 14(a). Common re-usable pallets are used to house the components for several different machine processes. The pallet could be a generic rectangular container with a retroreflective surface on which an egg-crate-style transparent tray can be placed. This pallet design offers several advantages: (1) It reduces the application of retroreflective sheetings on individual trays to a minimum. (2) It allows flexibility to accommodate different part sizes without the needs to re-apply retroreflective materials for new part change-over. (3) It allows different tray configurations to be interchangeable for use with the retroreflective container.

To prevent a prolonged exposure time with pin hole optics, a collocated illumination scheme similar to the optic design II was designed and constructed for concept demonstration. Eighteen narrow-angle LEDs were closely packed around a pin-hole of 0.1 mm diameter. The LEDs were arranged in hexagon patterns on three circular rings of diameters of 6mm, 10mm, and 12mm. Due to highly intensive retroreflectance and the ability to direct control of the integration time, the simple optic allows low-power spectral source to be used to faithfully produce a high object-to-background image for location and orientation computation as displayed in Fig. 14(b) and 14(c), where a relatively simple thresholding technique has been used to detect the outline of the part.

(a) Example Images of Typical Small Components in Transparent Trays

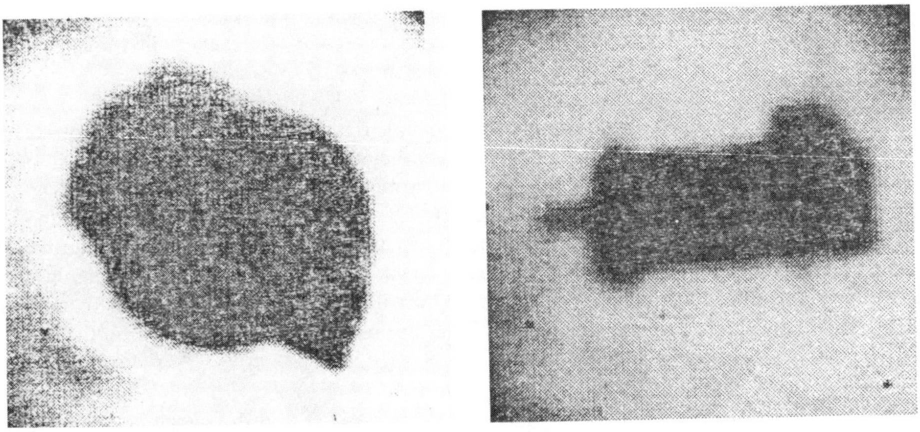

(b) Illustrative Images of Pin-Hole Optics

Fig. 13 Typical Examples of Small Objects

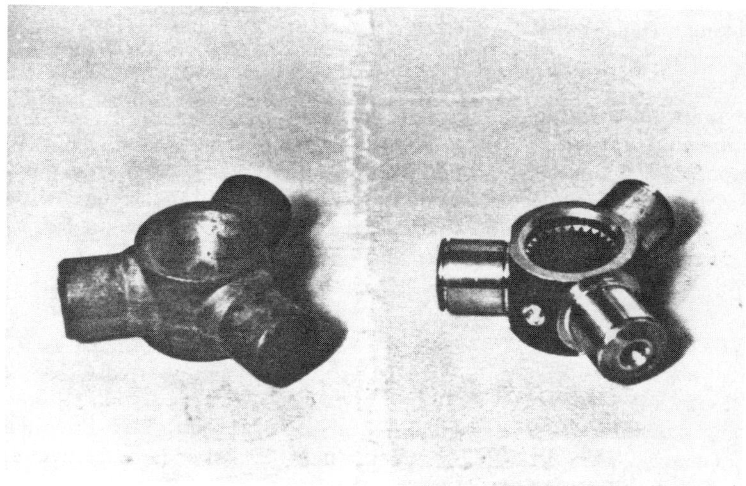

(a) Typical Machine Components

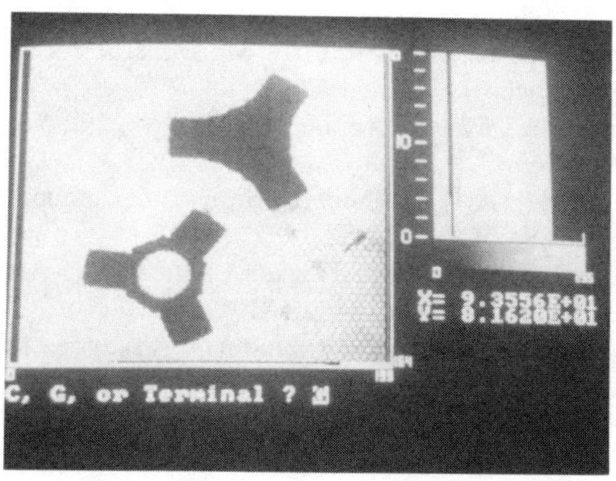

(b) Illustrative Images of Pinhole Optics

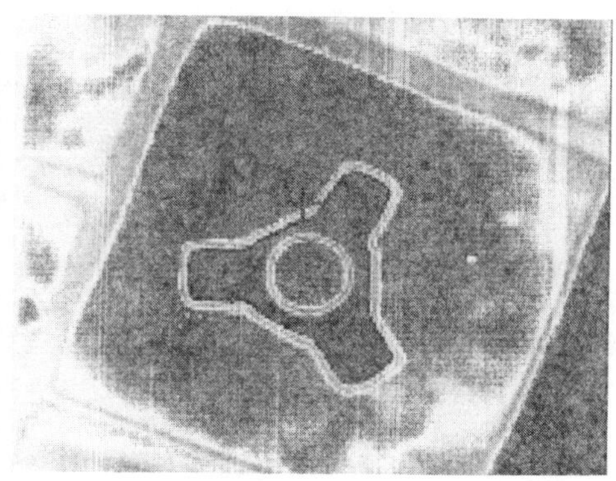

(b) Contour Following of Digital Image Generated by Pin-Hole Optics

Fig. 14 Illustrative Examples of Machine Components

5. CONCLUSIONS

Pinhole optic design concepts for applications of retroreflection for part-presentation were presented along with illustrative applications. The concept of giving retroreflective materials an integral role in vision sensing for generic part-presentation has been proven to have significant potentials in improving the vision reliability, reducing the computational load/time, and lowering the cost in implementation for part-feeding.

The manner in which parts are fed to robots in a FMS depends on the nature of the manufacturing processes, the product design, and the material handling system as a whole. The analytical models which have significant influences on image characteristics are expected to provide insights to the overall manufacturing system modeling and simulation, guidelines for part-presentation equipment standardization, and would aid establishing guidelines for part-design and part-packaging for high-performance part-feeding.

ACKNOWLEDGEMENT

The Research is co-sponsored by the Georgia Tech Material Handling Research Center (MHRC) and the National Science Foundation (NSF) through NSF Presidential Young Investigator Grant No. 8958383. Strong support from Dr. Ira Pence, MHRC Director, Dave Santo of GM, Matt Orlson of Ford, are greatly appreciated.

REFERENCES

1. K-M. Lee and D-R. Li, "Retroreflective Vision Sensing for Generic Part-Presentation," MHRC Technical Report MHRC-TR-89-06. Presented at 1990 IEEE SouthCon/90, Orlando, Florida, March 20-22. Also in Journal of Robotic Systems, February 1991, Vol. 7, Number 1.
2. K-M. Lee, "A Low-cost Flexible Part-Feeding System for Assembly Automation," Proceedings of the ICARCV International Conference on Automation, Robotics and Computer Vision, 18-21 September 1990, Singapore.
3. K-M. Lee, S. Dickerson, and J. Mercurio, "Generic Retroreflective Part Pickup System," Proceedings of the 1991 NSF Design and Manufacturing Systems Conference, January 9-11, Austin, Texas.
4. Federal Test Method Standard 370, Instrumental Photometric Measurements of Retroreflective Materials and Retroreflective Devices, March 1, 1977
5. J. J. Rennilson, "Retroreflective Measurements: A Review," Applied Optics, Vol. 19., No. 8, April 1980. (A collection of five papers)
6. U.S. Department of Transportation, "Retroreflectivity of Roadway Signs for Adequate Visibility: A Guide," Final Report FHWA/DF-88/001, November 1987.
7. F. E. Wargocki, A. Ray and G. E. Hall, "Retroreflector Field Tracker," State-of-the-Art Imaging Arrays and Their Applications, SPIE Vol. 501, 1984, pp. 283-291.
8. W. D. Holcombe, S. L. Dickerson, J. W. Larsen, and R. A. Bolander, " Advances in Guidance Systems for Industrial Automated-Guided-Vehicles," Proceedings of the SPIE - Mobile Robots III, Cambridge, MA, November 9-11, 1988.
9. Optoelectronics Designer's Catalog 1988-1989, Hewlett Packard, pp. 6-28, 6-31.

Sensors, Controls, and Quality Issues
in Manufacturing
ASME 1991

ON ESTIMATING ROTATIONAL SPEED AND TORQUE FROM THE PERIOD OF AN ENCODER

Ciro N. Ramirez and R. Joe Thornhill*
International Business Machines
Austin, Texas

* Department of Mechanical Engineering
University of Texas

ABSTRACT

A method to estimate the speed and torque of rotating machinery is studied. The characteristics of speed estimation using an idealized encoder are examined, followed by a discussion of discrete speed estimation and the inherent aliasing problem. This technique is well suited to milling and drilling equipment where the rotor experiences small transient speed changes during machining and speed fluctuations are harmonically related to the average shaft speed. A technique for speed and torque estimation using an inexpensive magnetic probe is presented. Empirical results from a high speed air turbine machine drilling printed circuit boards are presented. The technique presented is useful for evaluating the dynamic response of any encoder-based system for speed estimation.

NOMENCLATURE

B	rate of linear speed change
$E[\cdot]$	expected value operator
f_N'	Nyquist frequency of resampled data sequence
i	integer sample counter
I	rotor mass moment of inertia about principal axis
j	integer sample counter
k	integer number of encoder slits per revolution
n	p/k, number of torque cycles per encoder cycle
R	radius of rotor
p	number of torque cycles per revolution of rotor
t, t_o	time, initial time
t_s, t_s'	time sampling interval, resampling time interval for speed estimate Ω^*
$T(t)$	rotor torque
T_d, T_o	amplitude of sinusoidal part of total torque, constant part of total torque
$T_M(t)$	motor torque
$v(t)$	position encoder signal
$x_c(t)$	lateral displacement of rotor center perpendicular to the sensor axis
$x_g(t)$	gap between rotor and magnetic probe

x_o	amplitude of $x_c(t)$
$\alpha(t)$	instantaneous angular acceleration due to external torque
$\tilde{\alpha}(t)$	$d\Omega(t)/dt$ = instantaneous angular acceleration
α_d	amplitude of sinusoidally varying part of acceleration
β	arbitrary phase angle
δ	phase angle of $x_c(t)$
$\theta(t)$	rotor angular displacement
τ_e, τ_m	cycle time of encoder signal, period of torque fluctuation
ϕ	arbitrary phase of external torque
ω_m, ω_v	torque modulation frequency, frequency of $x_c(t)$
$\Omega(t)$, $\Omega^*(t)$	true rotor speed, estimated rotor speed

INTRODUCTION

In high speed rotating machinery it is often desired to measure the speed of a rotor or the torque imposed on that rotor. Because of instrumentation difficulties resulting from high operating speeds or machine design, direct torque measurement is sometimes impractical, and torque must be calculated from rotor speed changes. If any one of three factors exist, the measurement is difficult using conventional tachometers. Conventional instrumentation may not operate in the required speed range; added transducer mass may not be balanced well enough to avoid changing the system dynamics; and conventional instrumentation may not be able to sufficiently resolve small speed changes occurring during the observation interval.

A model for the net torque applied to a rotor is posed and used to study a speed estimation technique based on the period of an idealized analog encoder. This is done for the case of (1) no torque, (2) constant torque, (3) a sinusoidally varying torque, and (4) a combination of cases (2) and (3). An examination of the characteristics of this speed estimation technique reveals a speed averaging effect. This averaging operation is the convolution of the shaft speed with a rectangular time window which has unit area but (except for special cases) varying width.

The digital implementation of the technique is also studied and reveals that even if the encoder signal is filtered and sampled in such a manner as to avoid frequency aliasing, aliasing can still occur during subsequent signal processing. Experimental data using this technique is presented. Lastly, a brief discussion of the effect of lateral vibration on the estimation technique is presented.

Of particular interest to the authors is the measurement of the rotor speed and transient cutting torque of high speed machines used to drill small holes in printed circuit boards. Such measurements can be used to detect abnormal operating conditions, such as worn or broken drills. In this application, all of the difficulties described above exist, ruling out conventional transducers. A magnetic probe which is non-contact, simple, and inexpensive functions well if certain precautions are taken.

MATHEMATICAL MODEL

Cutting torque in milling and drilling often consists of a constant part added to a dynamic part which fluctuates one or more times per shaft revolution (Elbestawi, et al., 1991 and König, 1989). In order to study the speed/torque estimation method, a simple model of such fluctuating torque was required.

Assume that a rotor inertia I, is subjected to motor torque T_M, a constant external torque T_o, and a sinusoidal external torque of amplitude T_d and phase ϕ, which cycles p times per rotor revolution. The only restriction on p is that it be non negative. The rotor torque can be written as

$$T(t) = T_M(t) - T_o + T_d \, COS[p\theta(t)+\phi] \tag{1}$$

where $\theta(t)$ is the angular displacement as a function of time.

Assuming that the rotor undergoes rigid body rotation about its longitudinal principal axis, the angular acceleration is given by $\tilde{\alpha}(t) = T(t)/I$. If the motor torque is known, we can write the acceleration due only to externally applied torque as

$$\alpha(t) = \tilde{\alpha}(t) - \frac{T_M(t)}{I} = -\frac{T_o - T_d \, COS[p\theta(t)+\phi]}{I} . \tag{2}$$

So

$$T_o = -\{I\alpha(t) - T_d \, COS[p\theta(t)+\phi]\} . \tag{3}$$

For a given value of p and phase ϕ, the expected value of T_o is

$$E[T_o] = E[-\{I\alpha(t) - T_d \, COS[p\theta(t)+\phi]\}] = T_o = - I \, E[\alpha(t)] . \tag{4}$$

The unknown constant component of cutting torque, T_o, may be estimated from values of $\alpha(t)$. One method of estimating $\alpha(t)$ is to obtain a time record of the rotor's angular speed, $\Omega(t)$, and estimate its slope, since $\alpha(t) = d\Omega(t)/dt - T_M(t)/I$.

The angular speed and position can always be expressed as

$$\Omega(t) = \Omega(t_o) + \int_{t_o}^{t} \alpha(t')dt' + \int_{t_o}^{t} \frac{T_M(t')}{I} dt' \quad \text{and} \quad \theta(t) = \theta(t_o) + \int_{t_o}^{t} \Omega(t')dt' . \tag{5}$$

So, using the form of $\alpha(t)$ given by Equation (2), $\Omega(t)$ can be expressed as

$$\Omega(t) = \left[\Omega(t_o) + \frac{T_o t_o}{I} + \int_{t_o}^{t} \frac{T_M(t')}{I} dt'\right] - \frac{T_o t}{I}$$

$$+ \frac{T_d}{I} \int_{t_o}^{t} COS\left[p\left[\theta(t_o) + \int_{t_o}^{t'} \Omega(t'')dt''\right] + \phi\right] dt' \tag{6}$$

or, letting the constant term and the coefficients of the linear and integral terms be designated Ω_o, B, and α_d, respectively,

$$\Omega(t) = \Omega_o - Bt + \alpha_d \int_{t_o}^{t} COS\left[p\int_{t_o}^{t'} \Omega(t'')dt'' + \beta\right] dt' + \int_{t_o}^{t} \frac{T_M(t')}{I} dt' \tag{7}$$

where $\beta = p\theta(t_o)+\phi$.

Equation (7) is clearly a complicated implicit integral equation with no closed form solution known to the authors. Nevertheless, it is instructive to study special cases of this equation. This is done using an idealized analog position encoder in the next section. In general, T_M must be known in order to use Equation (7) to estimate B (and thus T_o). In the most general case, this may be a difficulty, but for circuit board drilling and other applications, the requirement can be substantially met. For circuit board drilling with a high speed air drill, for example, the time required to drill a hole is quite short, but significant time is required to reposition the board for the next hole. High transient cutting torque is supplied primarily from the angular momentum of the drill spindle, while the low T_M available from the air turbine acts over a much longer time to restore lost momentum. By measuring the spindle moment of inertia and speed changes during drilling and comparing the estimated mean transient torque to the maximum value of T_M estimated from a simple speed torque curve constructed using measured stall torque and no load speed, it was found that their ratio was roughly 100:1. In this case, the effects of T_M can be safely neglected. T_M might also be measured from changes in motor current on electric drilling machines, and its contribution to $\Omega(t)$ accounted for. In the work that follows, T_M is neglected, but its effect can be easily included if T_M is known.

SPEED ESTIMATES USING AN IDEALIZED ENCODER

Consider the idealized encoder system shown in Figure 1.

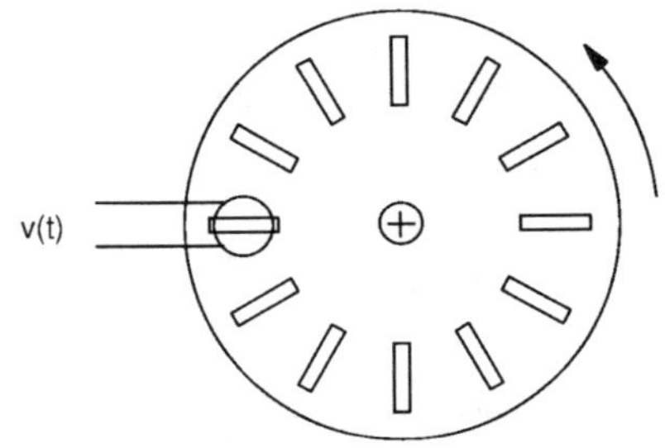

Figure 1. Encoder.

The encoder wheel, which has k uniformly spaced slits, is attached to a rotor rotating at speed $\Omega(t)$ which has an angular displacement $\theta(t)$ defined in Equation (5). The output signal of the encoder system is assumed to be an analog voltage which is a function only of the rotor angular displacement $\theta(t)$, and of the form

$$v(t) = \mathrm{SIN}[k\theta(t)] \quad . \tag{8}$$

<u>Case Of Constant Rotor Speed</u>

If the rotor has a constant speed Ω_o, the angular displacement (setting $t_o = 0$ and $\theta(t_o) = 0$ in Equation (5)), is $\theta(t) = \Omega_o t$. The encoder signal is then $v(t) = \mathrm{SIN}(k\Omega_o t)$. The rotor speed is estimated from the periodic encoder signal, with $\tau_e = 2\pi/k\Omega_o$, which is the time for the encoder wheel to rotate $\Delta\theta = 2\pi/k$ radians. We assume for now that τ_e can be determined continuously. The estimated rotor speed $\Omega^*(t)$, is the same as the true speed, since $\Omega^*(t) = (2\pi/k)(1/\tau_e) = \Omega_o$.

<u>Case Of Linearly Changing, Unmodulated Rotor Speed (Constant Torque Case)</u>

If the rotor has a linear speed variation described by $\Omega(t) = \Omega_o + Bt$ then, if $t_o = 0$,

$$\theta(t) = \theta(0) + \int_0^t (\Omega_o + Bt')dt' = \Omega_o t + (t^2/2)B + \theta(0) \ . \qquad (9)$$

For this case, τ_e is defined to be the value which satisfies the equation

$$\Delta\theta(t) = \theta(t+\tau_e/2) - \theta(t - \tau_e/2) = 2\pi/k = \Omega_o\tau_e + Bt\tau_e = (\Omega_o + Bt)\tau_e \qquad (10)$$

so $\tau_e = 2\pi/[k(\Omega_o+Bt)]$ and

$$\Omega^*(t) = \frac{\Delta\theta}{\tau_e} = \frac{2\pi}{k}\left[\frac{k(\Omega_o+Bt)}{2\pi}\right] = \Omega_o + Bt = \Omega(t) \ . \qquad (11)$$

Therefore, angular speed estimation is also exact for linear speed changes.

Case Of Lightly Modulated Uniform Rotor Speed (Fluctuating Torque Case)

Now consider the specific case where the linear coefficient B is zero and $|[\Omega(t)-\Omega_o]/\Omega_o| \ll 1$ (i.e. $\Omega(t) \approx \Omega_o$). Letting $t_o = 0$ and $\beta = 0$, Equation (7) becomes

$$\Omega(t) \approx \Omega_o + \alpha_d \int_0^t \text{COS}(p\Omega_o t')\, dt' = \Omega_o + \frac{\alpha_d}{p\Omega_o}\text{SIN}(p\Omega_o t) \ . \qquad (12)$$

Define $\Omega_d = \alpha_d/(p\Omega_o) = \alpha_d/(kn\Omega_o)$ and $\omega_m = p\Omega_o = kn\Omega_o$. We can write the approximation for the angular speed as

$$\Omega(t) \approx \Omega_o + \Omega_d \,\text{SIN}(\omega_m t) = \Omega_o + \Omega_d \,\text{SIN}(kn\Omega_o t) \qquad (13)$$

where ω_m is the modulation frequency, $n = p/k$ is the number of torque cycles per encoder cycle, and it is assumed that $\Omega_d \ll \Omega_o$. The only restriction on p is that it be non-negative, so $p\Omega_o = kn\Omega_o$ can represent any frequency.

Using the approximation in Equation (13) and performing the integration of Equation (5) to obtain the angular displacement of the rotor yields

$$\theta(t) = \theta(0) + \Omega_o t - \frac{\Omega_d}{kn\Omega_o}[\text{COS}(kn\Omega_o t) -1] \ . \qquad (14)$$

The encoder signal, using Equation (8) is

$$v(t) = \text{SIN}\left[k\theta(0) + k\Omega_o t - \frac{\Omega_d}{n\Omega_o}\left[\text{COS}(kn\Omega_o t) -1\right]\right] \qquad (15)$$

which is now a frequency modulated sinusoid.

An interesting question is, are there any conditions under which τ_e is a constant? Let us assume that τ_e is a constant and given by $\tau_e = 2\pi/(k\Omega_o)$. The angular displacement at time $t+\tau_e$ must satisfy Equation (14) so,

$$\theta(t+\tau_e) = \theta(0) + (t+\tau_e)\Omega_o - \frac{\Omega_d}{kn\Omega_o} \text{COS}\{[kn\Omega_o(t+\tau_e)] -1\}$$

$$= \frac{2\pi}{k} + \theta(0) + \Omega_o t - \frac{\Omega_d}{kn\Omega_o} [\text{COS}(kn\Omega_o t + n2\pi) -1] \tag{16}$$

If n, the number of torque cycles per encoder cycle, is an integer, we have

$$\theta(t+\tau_e) = \frac{2\pi}{k} + \theta(t) \tag{17}$$

and using Equation (8) we have

$$v(t+\tau_e) = \text{SIN}[k\theta(t+\tau_e)] = \text{SIN}[2\pi+k\theta(t)] = v(t). \tag{18}$$

τ_e is constant if n = 1 , 2, 3, ... This case is of practical interest because for machining processes such as milling and drilling n may be an integer.

The production of the speed estimate signal $\Omega^*(t)$ from the true speed $\Omega(t)$ is illustrated in Figure 2.

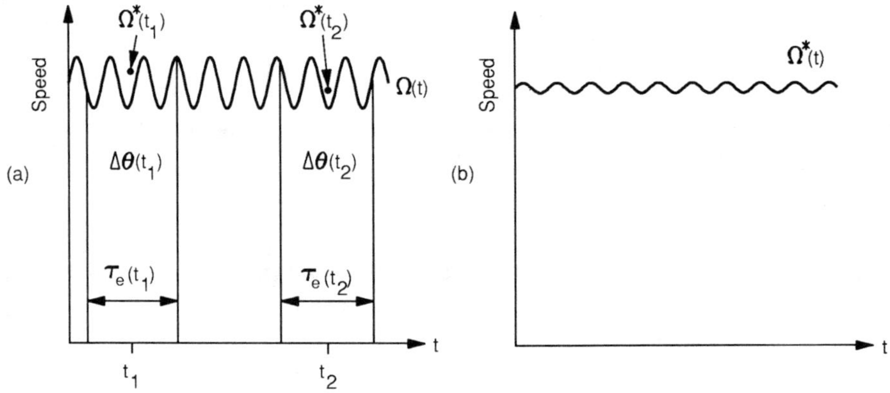

Figure 2. (a) Rotor speed estimation. (b) Smoothed speed estimate.

$\Omega(t)$ is averaged over the encoder period τ_e continuously in time. Because τ_e, in general, is not the same as the period of $\Omega(t)$, the estimate signal $\Omega^*(t)$ is a smoothed version of $\Omega(t)$. The smoothing operation is given by the convolution integral

$$\Omega^*(t) = \frac{\Delta\theta}{\tau_e} = \frac{2\pi}{k\tau_e} = \frac{1}{\tau_e} \int_{t-(\tau_e/2)}^{t+(\tau_e/2)} \Omega(t')dt' \tag{19}$$

The effect of this smoothing can also be discussed in terms of the equivalent filter characteristic. Equation (19) is the convolution of a rectangular smoothing window with the speed signal. The filter characteristic is given by the Fourier transform of this window, which is a SINC function. The window s(t) and its transform $S(\omega)$ are shown in Figure 3 , plotted against $n \approx \omega_m\tau_e/2\pi$. At integer values of n, none of the velocity fluctuations are passed by

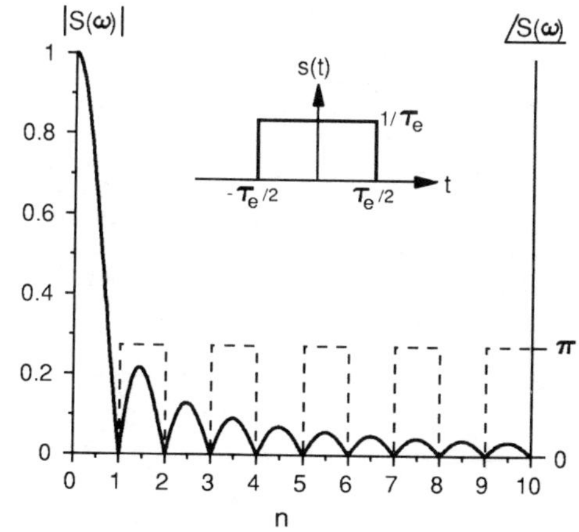

Figure 3. Rectangular time window s(t) and its transform.

the filter. The encoder period is thus constant and not a function of time for these values of n. For this case all speed estimates are equal and are given by

$$\Omega^*(t) = \frac{\Delta\theta(t)}{\tau_e} = \frac{2\pi}{k\tau_e} = \Omega_o \tag{20}$$

For the small speed fluctuation case, modulation at integer multiples of the encoder frequency are thus precisely canceled, leaving only the mean velocity component Ω_o. If Equation (7) included modulation at several integer values of p, the discussion above could easily be modified to include these by superposition, and Equation (13) would contain multiple modulating frequencies, each of which would be filtered separately.

Case of Significantly Modulated Uniform Rotor Speed

When modulation becomes significant, and Ω_d is no longer small compared to Ω_o, the approximations used above cannot be made. In addition, although the convolution of Equation (19) is still valid, the limits of integration are not fixed for all time values, because the period τ_e of the encoder signal will vary with time. In this case, the estimation operation of Equation (19) applied to the time varying speed of Equation (7) is a nonlinear filtering operation. If however, n is an integer, the encoder period will still be an integer multiple of the period of speed fluctuation, and it seems reasonable to assume that speed estimates should still be constant, although not necessarily equal to Ω_o. The proof of this, however, appears to be quite difficult.

In order to evaluate the speed estimation operation for significant modulation, Equations (7) and (19) were input directly to a computer program. The program generated a speed signal which could be modulated at any frequency, and 4th order Runge-Kutta numerical integration was used to generate both the speed and angular position signals. An encoder signal was created from the position signal, and its period determined numerically at 50 intermediate points in each cycle. The speed estimate signal was computed from Equation (19). The program was used to evaluate the estimation response for both small and large speed fluctuations.

For the case under consideration in this section, only fluctuating torque was included, and B = 0. The effective filter response could then be evaluated by taking the ratio of the peak-to-peak values of the actual speed and the estimated speed signals. Runs were made for values of n between 0.1 and 4 for k = 1. In order to study the effect of large speed fluctuations (large modulation), the ratio $Q = \Omega_d/\Omega_o$ was increased until the speed modulation was large enough to cause stalling, and negative speed was observed. This condition was reached when Q = 0.5. The magnitude of the effective filter characteristic is shown in Figure 4. The characteristic matches very well with that of Figure 3 for small values of Q, and is only moderately distorted even for extremely large modulation. Even more significant is the fact that the cancellation effect at integer values of n persists even for extreme modulation. This implies that even large torque components occurring at harmonic multiples of the ratio between rotor rotational frequency and the encoder frequency will be filtered out, and will not appear in the resulting speed estimate signal. If the objective of the speed estimation is to measure only the mean torque, this is a very useful result. Although not shown, the phase response in all cases matched that of Figure 3.

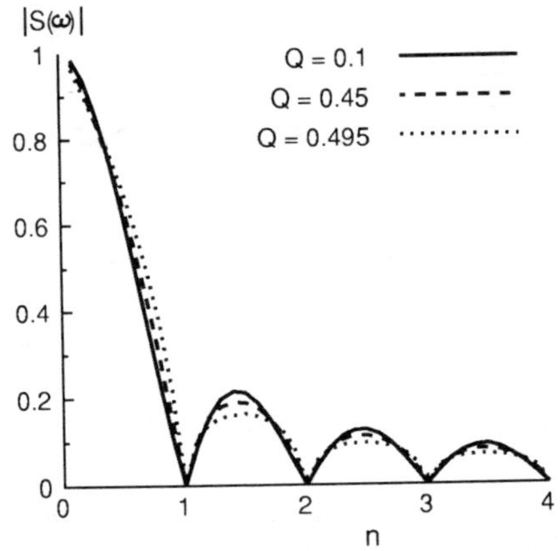

Figure 4. Effective filter characteristic for B = 0, for several values of Q = Ω_d/Ω_0

While speed estimates may not reflect the dynamic fluctuations in torque for the conditions described above, it is quite possible that the mean of such estimates are biased, and do not accurately reflect the mean value of the true speed. In order to evaluate this, the results of the study were analyzed to compare the true mean speed to the mean of the speed estimate. These were computed from

$$\bar{\Omega} = \frac{\Delta\theta}{\tau_m} = \frac{1}{\tau_m}\int_t^{t+\tau_m} \Omega(t')dt' \quad \text{and} \quad \bar{\Omega}^* = \frac{1}{\tau_m}\int_t^{t+\tau_m} \Omega^*(t')dt' \ , \tag{21}$$

where τ_m is the period of the modulation, or torque fluctuation. This was done by numerically evaluating Equations (21) and computing the Normalized Mean Error = $(\bar{\Omega}^*-\bar{\Omega})/\Omega_0$. The results are shown in Figure 5. Despite the presence of what appears to be numerical noise (caused by the large number of data manipulations needed and the digitization errors involved), this error also appears to be least at integer values of n, and is relatively small except for very large modulation levels. Even at high modulation levels, the error at integer values of n appears to be negligible. For $Q < 0.1$, the speed estimation technique appears to be quite accurate for mean speed values.

General Case

The general case of Equation (7) includes both significant periodic torque fluctuation and constant torque (i.e., $B \neq 0$). This case cannot be described by an effective filter characteristic. For small constant torque, the results of the previous section should be applicable, since linear speed changes will be small over any one cycle of the periodic torque. In order to examine the general case, the computer program was modified to include a non-zero B, and run for various cases. Figure 6 shows the results for the case of

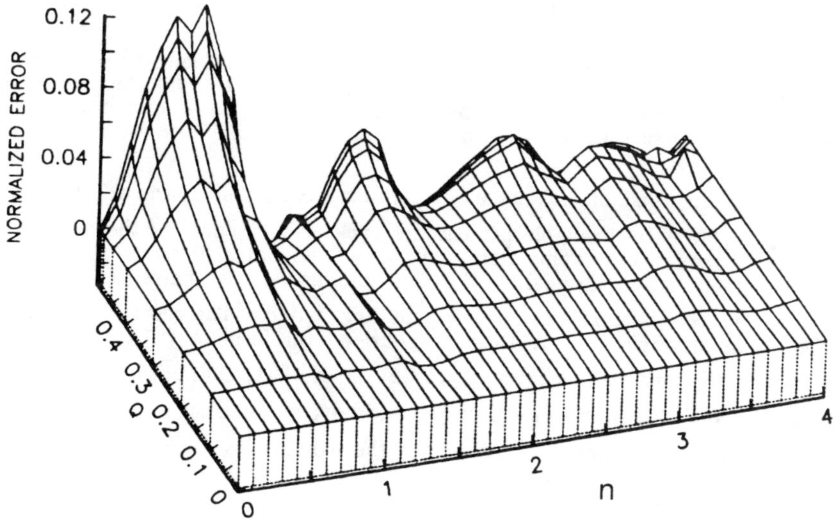

Figure 5. Normalized mean error ($\overline{\Omega}^* - \overline{\Omega}$) $/\Omega_0$ vs. Q = Ω_d/Ω_0

Ω_0 = 100 rad/sec, Ω_d = 10 rad/sec, B = 50 rad/sec^2, and k = 1, for two values of n. The curves shown would correspond to a severe and rapid slowing of the rotor, almost to the stall point. The response of the estimate signal resembles those for the special cases. At n = 1.0, the estimate signal is remarkably insensitive to velocity fluctuations, showing a slight response only when the velocity has dropped to nearly the stall point. This indicates that the conclusions of the previous analyses can be reasonably extended to the general case.

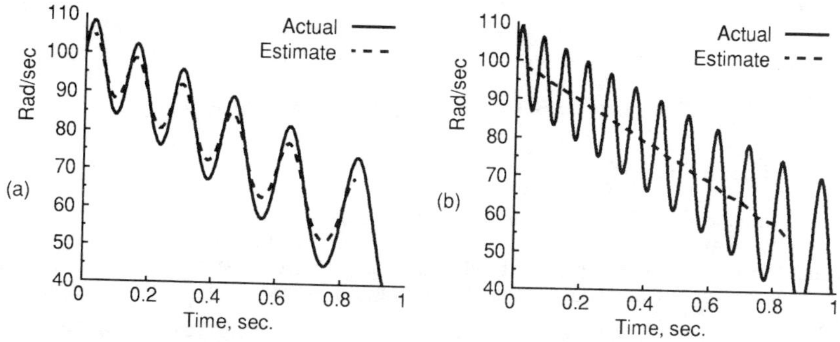

Figure 6. Actual and estimated velocity for Ω_0 = 100, Ω_d = 10, B = 50, k = 1 (a) n = 0.5. (b) n = 1.0.

DISCRETE SPEED ESTIMATION

Although the preceding discussions were presented with the assumption that the encoder period, τ_e, could be determined continuously, encoder periods are most easily measured between zero crossings only. In addition, analog signals are usually sampled digitally for manipulation by computer.

Consider the situation in which the analog encoder signal v(t) is sampled at uniform time intervals t_s to produce $v(it_s)$. Assume that digitization is done after passing the signal v(t) through an analog low pass anti-aliasing filter with the break frequency set appropriately to guarantee that there will be no frequency aliasing as a result of the sampling operation. Because there is no guarantee that a sample falls precisely on a zero crossing of v(t), interpolation of the sampled signal $v(it_s)$ may be required to obtain good estimates of τ_e.

If only zero crossings of v(t) are used to produce values of τ_e and Ω^*, the maximum number of speed estimates available per encoder cycle will be equal to the number of zero crossings per encoder cycle. This finite number of estimates amounts to sampling the idealized, continuous estimate signal $\Omega^*(t)$. The time between estimates, t_s', (which we will consider to be nearly constant) is the new sampling period. The resampled speed signal, $\Omega^*(jt_s')$, is shown in Figure 7. A Discrete Fourier Transform of $\Omega^*(jt_s')$ will produce a spectrum with a Nyquist frequency of $f_N' = 1/(2t_s')$. If $\Omega^*(t)$ contained signal components higher than f_N', the digital spectrum will be aliased, despite the initial anti-alias filtering step. Thus, information about the dynamic content of the signal is lost through the filtering effect and contaminated by aliasing during resampling. These effects are not critical if the objective is the estimation of mean torque or speed.

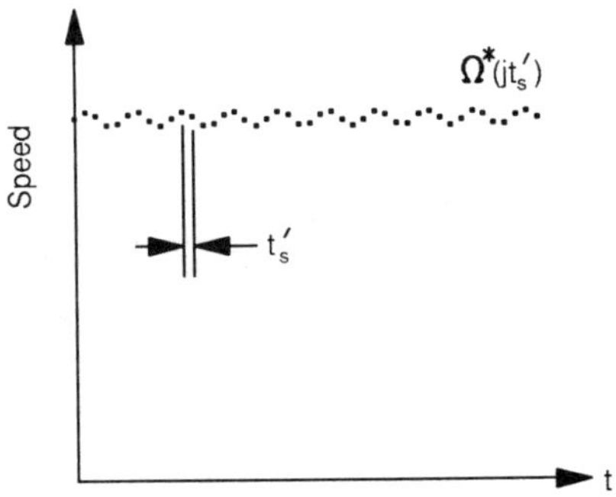

Figure 7. Resampled velocity estimate signal.

EXPERIMENTAL RESULTS

In situations where the rotor speed is quite high and/or a speed sensor (tachometer) which inertially loads the shaft is not suitable, a magnetic probe may be acceptable. This type of sensor, which is shown schematically in Figure 8, is inexpensive and is essentially a non-contact measurement method because the magnetic forces are usually negligible.

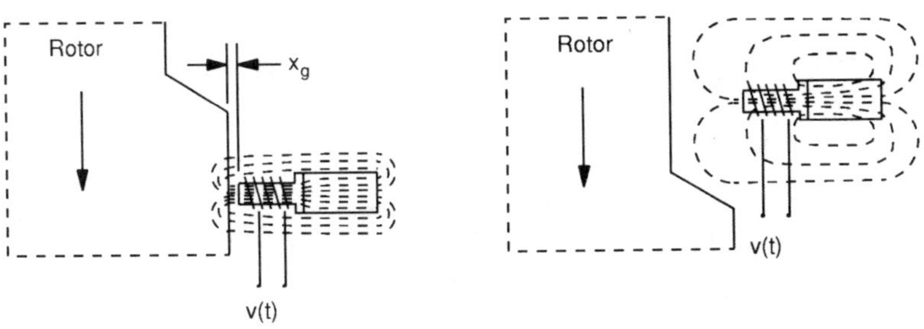

Figure 8. Magnetic probe.

The magnetic sensor consists of a pickup coil, a core with high magnetic permeability, and a permanent magnet. The magnetic flux enclosed by the coil is sensitive to the gap between the core and the target, which must also have high permeability. Assuming that the target geometry changes gradually and eddy current effects are negligible, the flux linkage is a function only of the gap between the sensor tip and the target and of the angular speed and position of the rotor. The voltage will be proportional to the target speed.

A practical application of this type of sensor is found in machines used for drilling small holes in printed circuit boards which often have a rotational speed in the range of 25,000 to 200,000 RPM and low rotational inertia. The rotating shaft may have features such as flats, holes, grooves, etc., which are not necessarily radially or axially symmetric. The sensor signal will have a basic pattern which can be used to determine rotational displacement. In the case of non-symmetric features, this corresponds to having an encoder with $k = 1$. In this case the signal $v(t)$ is not sinusoidal but will repeat every 2π radians.

A magnetic probe of the type described above was installed on a high speed air turbine drilling machine. The rotor had a through hole and two flats which were diametrically opposed and not quite symmetric. A typical magnetic probe signal for this drilling machine, shown in Figure 9, indicates a strong component at four cycles per shaft revolution (3120 Hz) but has a true period of 1.28 ms which corresponds to a speed of 46,800 RPM (780 Hz). Speed estimates can be extracted from this signal by peak detection or RMS averaging over single rotations of the shaft but such estimates will be sensitive to noise and shaft vibration which affects the sensor gap. The preferred procedure is to use the period of the signal to calculate the shaft rotational frequency.

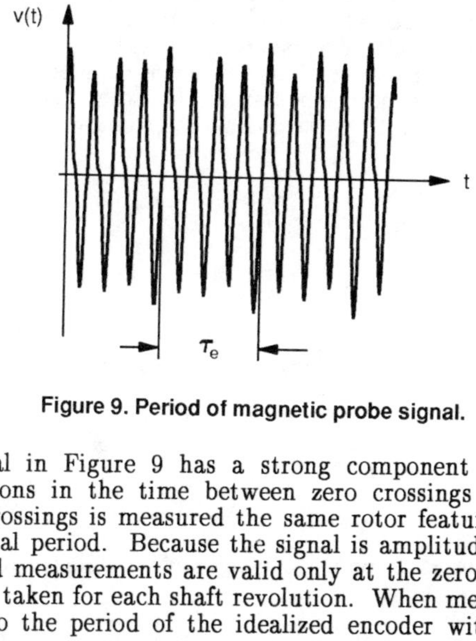

Figure 9. Period of magnetic probe signal.

Although the signal in Figure 9 has a strong component at four cycles per shaft revolution, slight variations in the time between zero crossings can occur. If the time required for eight zero crossings is measured the same rotor features are detected, and this defines the rotor rotational period. Because the signal is amplitude modulated by the shaft velocity $\Omega(t)$, such period measurements are valid only at the zero crossings, and only eight samples of period may be taken for each shaft revolution. When measured in this fashion, the period τ_e is equivalent to the period of the idealized encoder with $k = 1$ and sampled at $t_s' = 1/[8\bar{\Omega}(t)]$ (where $\bar{\Omega}(t)$ is in Hz).

Calculation Of Torque Value

Figure 10 shows the velocity estimate signal computed from the magnetic probe signal for a rotor speed of approximately 575 Hz. The beginning of the hole is clearly detectable and the signal, though showing some contamination, clearly indicates the slowing of the rotor. Although accelerometer measurements indicated the presence of significant vibration at harmonics of the rotor speed, no harmonic torque fluctuations are clearly evident in the velocity estimate. The estimate signal appears capable of resolving speed changes of less than 1% of the rotor speed. Because the torque of the air motor was small and could be neglected, the average drilling torque could be found by a least-squares line fit to the signal, as shown.

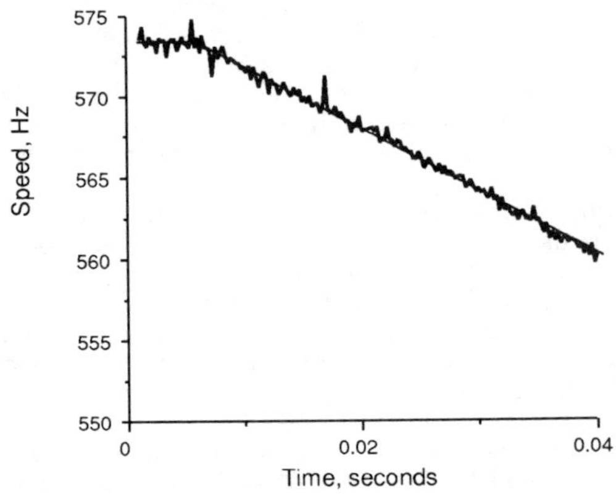

Figure 10. Speed estimate signal and line fit for drilling machine.

Such estimates have successfully resolved mean torque variations of 0.0002 N-m in actual drilling studies.

Effects Of Shaft Vibration

The discussion above on rotor speed examined the effects of torsional vibration on speed estimates. Lateral vibration of the spindle shaft can, however, result in slight variation of the measured period which are not related to shaft speed at all. Figure 11 depicts a smoothly rotating shaft with features which are detected by a magnetic probe, and which is also vibrating laterally in a direction perpendicular to the probe axis. The motion of the shaft center vibrating at frequency ω_v with phase angle δ is given by $x_c(t) = x_0\ \mathrm{SIN}(\omega_v t+\delta)$.

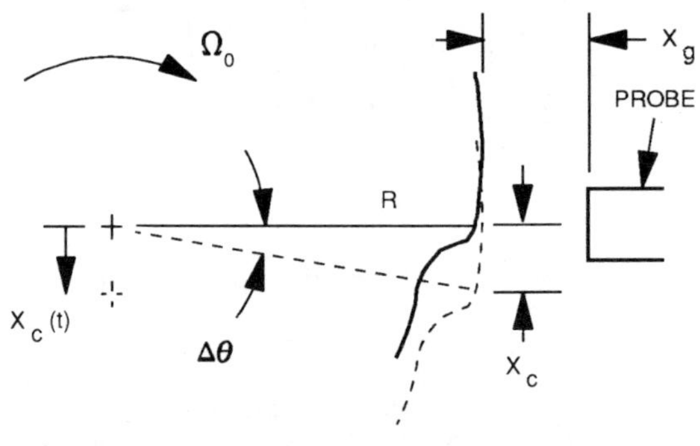

Figure 11. Smoothly rotating shaft, vibrating laterally.

The features of the surface of the shaft also are displaced, so that an apparent angle $\Delta\theta$ is added to the mean rotational angle of the shaft, $\Omega_0 t$. This angle is given by $\Delta\theta = \mathrm{SIN}^{-1}(x_c/R) \approx x_c/R$ for small displacements. The apparent shaft angle and the apparent shaft velocity are given by

$$\theta(t) = \Omega_0 t + \frac{x_0}{R}\mathrm{SIN}(\omega_v t+\delta) \quad \text{and} \quad \Omega(t) = \Omega_0 + \frac{\omega_v x_0}{R}\mathrm{COS}(\omega_v t+\delta) \tag{22}$$

This is the same as Equation (13) with $\Omega_d = (\omega_v x_o/R)$ and $\delta = -\pi/2$.

The observations made for shaft velocity fluctuations can thus be extended to cover lateral shaft vibration. Vibration of the shaft along the probe axis causes vibration in the probe gap x_g, which will amplitude and frequency modulate the probe signal, but its effects on the velocity estimates should be similar to those of lateral vibration. In general, shaft vibration at frequencies which correspond to integer values of n will not affect the velocity estimates obtained. This is important, because the experimental results reported earlier showed that the strongest vibration components present in all axes occurred at precisely these frequencies.

CONCLUSIONS

A rotational speed and torque estimation technique was presented and analyzed using an idealized analog encoder and a specific model of rotor torque appropriate for milling and drilling applications. For the case of no net torque and constant torque the speed estimate is exact. For the case of sinusoidally varying torque, with or without the constant term, the speed estimate does not fluctuate if the modulation period is an integer multiple of the encoder period. Otherwise the estimate is attenuated in a predictable manner based on the SINC function. The SINC function filter model is a good approximation even for large fluctuating torque. Mean speed estimates with large fluctuating torque can be biased, except for integer of n.

Implementation of the speed estimation technique can produce frequency aliasing which results from data processing and not digitization of the analog encoder signal.

Experimental data from a magnetic probe on a high speed electronic card drilling machine was presented. The magnetic probe encoder and the speed estimation technique produced repeatable estimates of the rotor speed and torque in this application.

REFERENCES

Beckwith, T.G, Buck, N. L., and Marangoni, R. D., *Mechanical Measurements*, 3rd Ed., Addison Wesley, 1982.

Brigham, E. Oran, *The Fast Fourier Transform*, Prentice-Hall, 1974.

Carnahan, Brice, Luther, H. A., and Wilkes, James O., *Applied Numerical Methods*, John Wiley & Sons, 1969.

Elbestawi, M.A., Papazafiriou, T. A., and Du, R. X., "In-Process Monitoring Of Tool Wear In Milling Using Cutting Force Signature", *International Journal of Machine Tools and Manufacture*, Vol. 31, No. 1, 1991, pp. 55-73.

König, W., "Quality Definition and Assessment in Drilling of Fibre Reinforced Thermosets", *Annals of the CIRP*, Vol. 38/1, 1989, pp. 119-124.

Oppenheim, Alan V. and Schafer, Ronald W., *Discrete–Time Signal Processing*, Prentice-Hall, 1989.

Sensors, Controls, and Quality Issues
in Manufacturing
ASME 1991

ANALYSIS OF MILLING FORCES VIA ANGULAR CONVOLUTION

J.-J. Junz Wang, Steven Y. Liang, and Wayne J. Book
George W. Woodruff School of Mechanical Engineering
Georgia Institute of Technology
Atlanta, Georgia

Abstract

The measurement of cutting force systems is one of the most frequently used techniques for the monitoring of machining processes. Its wide spread application ranges from tool condition identification, feedback control, cutting system design, to process optimization. To gain fundamental understanding of the force system in machining, this paper presents the work of establishing a closed form expression for the cutting force in end milling as an explicit function of cutting parameters and tool/workpiece geometry. Based on the theoretical local cutting force model, the generation of total cutting forces is formulated as the angular convolution of three uncorrelated cutting process component functions, namely the elemental cutting force function, the chip width density function, and the tooth sequence function. The elemental cutting force function is related to the chip formation process in an elemental cutting area and it is characterized by the chip thickness variation, specific cutting pressure constants, and entry/exit angles. The chip width density function defines the chip width per unit cutter rotation along a cutter flute within the range of axial depth of cut as the function of the angular position of each cutting point. The tooth sequence function represents the spacing between flutes as well as their cutting sequence as the cutter rotates. The analysis of cutting forces is extended into the Fourier domain by taking the frequency multiplication of the transforms of the three component functions. Fourier series coefficients of the cutting forces are shown to be algebraic functions of various tool parameters and cutting conditions. Simulation results are presented in the frequency domain to illustrate the effects of process parameters. A series of end milling experiments are performed and their results discussed to validate the analytical model.

1. Introduction

The measurement of cutting forces is one of the most frequently used techniques for machining process monitoring because the cutting forces are closely related to part dimensional accuracy, surface finish, tool condition, chip morphology, and machining stability. The fundamental understanding of the cutting force systems thus plays an important part in the monitoring, planning, and control of machining processes as well as in its traditional role in the design of machine tool structure and axes servomechanisms. Research in this area dates back to the work of Martellotti [1941, 1945] on the kinematics of the milling process. The basic cutter-chip thickness relationship took the form of

$$t_c = t_x \sin\theta \tag{1}$$

where t_c is the instantaneous undeformed chip thickness, t_x is the feed per tooth in the feed direction and θ is the angular position of the tooth in the cut as seen in Figure 1. In this expression, the tooth trajectory of a milling cutter was assumed to be circular instead of cycloidal, therefore the equation is not an exact solution for the chip thickness, however it is a good approximation when the feed is much less than the cutter diameter, and it has been widely used in the analysis of the milling process. The relation indicated that the cutting force generated by a certain chip thickness depends upon tool edge geometry such as clearance and rake angles as well as properties of the tool and workpiece material.

Simple models have been proposed [Koenigsberger et al. 1961, Sabberwal 1962] to relate the tangential cutting force (f_t) at any location on the cutting edge to the chip thickness in the following form:

$$f_t = K_t\, b\, t_x \sin\theta \tag{2}$$

where b is the width of chip and K_t is the tangential specific cutting pressure constant, which is a function of tool edge geometry and workpiece material properties and is usually evaluated from experiments. It has been shown that K_t is a function of the chip thickness and, for simplicity, its value corresponding to the average chip thickness has been commonly used since the average chip thickness can be easily calculated from the average cutting force.

Based on the local cutting force model, total cutting force models for the milling process were derived by incorporating the considerations of cutting conditions, cutter geometry, feedrate, and axial and radial depth of cut. Closed form integration expressions for the total cutting force on a given flute have been established as a function of the cutter angular position [Koenigsberger et al. 1961, Tlusty et al. 1975]. Using integration models, Koenigsberger and Sabberwal investigated the force pulsation during the milling process and developed quantitative force relationships for slab and face millings. These relationships showed the magnitude of the average force and the ratio of maximum to average forces for different dimensions and positions of the cutter relative to the workpiece. In addition to equation (2), Tlusty and MacNeil related the tangential cutting force (f_t) to the the radial cutting force (f_r) by the radial specific cutting pressure K_r:

$$f_r = K_r\, f_t \tag{3}$$

and developed the cutting dynamics in end milling. It was experimentally observed that the cutting force responding to a sudden feedrate change can often be characterized by a time delay, which may give rise to instability in adaptive control. These cutting force models were given in terms of the cutter angular position and different expressions were needed to evaluate forces at different angular positions. Therefore the integration solutions were segmented into different formulations with different integration boundary points.

Most of the integration expressions of the cutting force system applied to one cutter flute only. For a multi-flute cutter, care has to be taken to determine which flutes are engaged in the cutting and which appropriate integration form is to be used. Gygax conducted a detailed analysis [1979] and a series of experiments [1980] on the dynamics of single-tooth face milling under various cutting parameters. It was indicated that the extension of single-tooth milling to multi-tooth face milling could be achieved by convolution integration in the time domain. This was the first appearance of the convolution modeling concept in the literature. However, no cutting force modeling technique was presented and more complicated end millings were not treated.

Resorting to numerical integration, Kline, DeVor and Lindberg [1982, 1983] developed mechanistic discrete models for end milling in which the cutter is treated as an aggregation of discretized thin disk cutters along the cutter axis. At any angular position, chip load of each disk cutter can be computed as the product of chip thickness and disk thickness. The associated cutting forces can then be found from equations (2) and (3). Summation of forces from all disks yields the total cutting forces. Sutherland and DeVor [1986] improved on the previous model by taking into account the effects of system deflection on the chip load. Following from Kline's model, Fu, DeVor, and Kapoor [1984] developed a cutting force model for face milling which included the effects of spindle tilt and cutter runout. Alternatives to treat the elemental cutting forces differently were reported in the works of Zhou [1983], Ber [1988] and Armarego [1989], however, these formulations still rely on numerical integration to compute the total milling forces.

Up to now there is no documented technique to model the complete cutting force system of milling processes in a single closed form representation. The fact that the existing cutting force models are either inappropriate for multi-flute cutting or rely upon numerical integration for solutions leads to the difficulties in interpreting the cutting force measurements. Therefore their applications in cutting dynamic analysis and process optimization are restricted. On the other hand, there is a lack of milling force model in the frequency domain or transfer function representation. As a result, some of the adaptive control schemes with cutting force as the control constraint had to resort to the parameter identification based on assumed process models [Tomizuka et al. 1983, Lauderbaugh et al. 1988, Fussel et al. 1988]. However, uncertainties in the assumed model structure often raised stability concerns [Rohrs et al. 1985] and limited the scope of milling process controls.

This paper presents the work of establishing a closed form frequency domain expression for the cutting force in end milling as an explicit function of cutting parameters and tool/workpiece geometry through angular convolution modeling. The angular domain analysis is discussed in section 2. The frequency domain formulation along with numerical results are presented in section 3. Section 4 provides the description of experiments and the discussion of analytical model validation based on experimental results.

2. Convolution modeling of Multi-tooth Milling Forces

In the following analysis the complete cutting force system at any instance of cutting is considered as the time-domain convolution of the cutting forces on any elemental point on the cutter, the width of chip produced during unit angular rotation of the cutter, and an impulse train representing the sequence of tooth engagement. These three components of the time-domain convolution can be written as functions of cutter angular position thus the convolution integral can be carried out in the angular domain. The derivation of these components, namely the elemental cutting force function, the chip width density function, and the tooth sequence function, is discussed in detail in this section. Attention will be directed to the cases of ideal cutting in which cutting speed and feedrate are both constant while cutter runout and tilt do not exist, although deviations from ideal cases can be treated as well.

2.1 Elemental Cutting Force Functions

The analysis begins with considering a differential cutting point on any flute of a multi-tooth cutter. As the cutter rotates, the cutting point experiences a pattern of cutting forces as functions of the cutter angular position with respect to the workpiece. These force functions are described in equations (2) and (3), which can be expressed in X and Y coordinates as follows:

$$f_x(\theta) = f_t \cos\theta + f_r \sin\theta \tag{4}$$

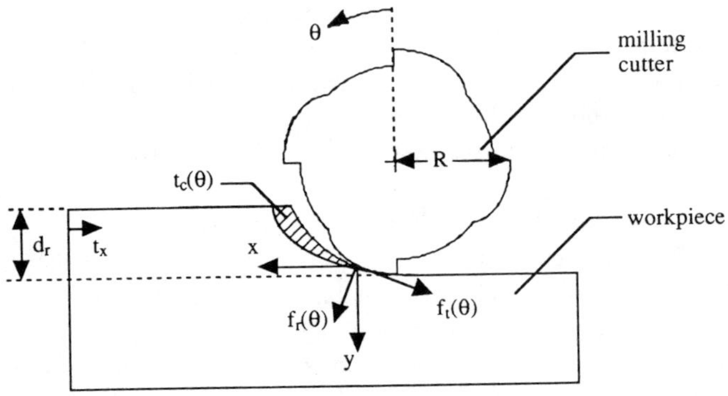

Figure 1. Geometric relationship between undeformed chip thickness (t_c), feed per tooth (t_x), and cutting forces ($f_t(\theta)$, $f_r(\theta)$).

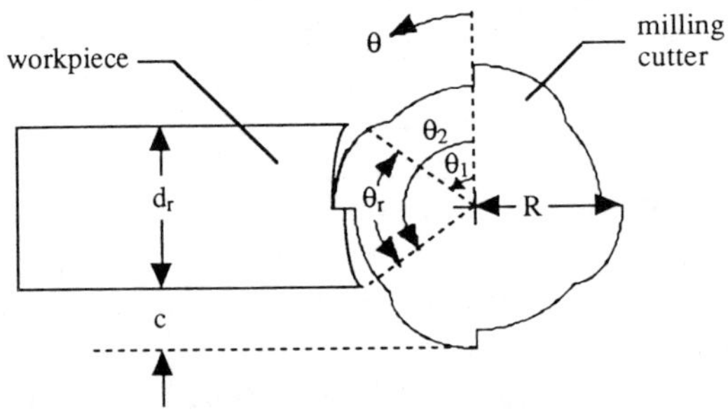

Figure 2. Entry angle (θ_1)and Exit angle (θ_2)of a milling cutter.

$$f_y(\theta) = f_t \sin\theta - f_r \cos\theta \tag{5}$$

Notice that the cutting force expressions in (4) and (5) are applicable only over the range of cutter angular position during cutter-work engagement. This range of angular position is the difference between the cutter entry angle θ_1 and the cutter exit angle θ_2 as shown in figure 2. Denoting the radius of cutter by R, radial depth of cut by d_r , and cutter width excess by c in the figure,

$$\theta_1 = \cos^{-1}\left(\frac{d_r + c}{R} - 1\right) \tag{6}$$

$$\theta_2 = \cos^{-1}\left(\frac{c}{R} - 1\right) \tag{7}$$

By defining a window function $w(\theta)$ as follows

$$w(\theta) = \begin{cases} 1, & \theta_1 \le \theta \le \theta_2 \\ 0, & \text{otherwise} \end{cases} \tag{8}$$

cutting forces in equations (4) and (5) can be expressed in more precise terms as

$$\begin{bmatrix} f_x(\theta) \\ f_y(\theta) \end{bmatrix} = \begin{bmatrix} \cos\theta & \sin\theta \\ \sin\theta & -\cos\theta \end{bmatrix} \begin{bmatrix} f_t(\theta) \\ f_r(\theta) \end{bmatrix} w(\theta) \tag{9}$$

The position and width of the window function are determined by the type of milling process and its cutting conditions. For a peripheral milling, the window depends on the cutter radius and the width of cut, and it generally starts at an angle greater than 90 degrees and ends at 180 degrees for down milling. For a slot milling, it starts at 0 and ends at 180 degrees.

The elemental cutting force functions, p_x and p_y, are defined as the forces acting on an elemental cutting area per unit specific cutting pressure (K_t). That is

$$p_x(\theta) = \frac{f_x(\theta)}{K_t\, b\, t_x} \tag{10}$$

$$p_y(\theta) = \frac{f_y(\theta)}{K_t\, b\, t_x} \tag{11}$$

From equations (4), (5), (9)-(11), it can be shown that

$$\begin{bmatrix} p_x(\theta) \\ p_y(\theta) \end{bmatrix} = \begin{bmatrix} 1 & K_r \\ -K_r & 1 \end{bmatrix} \begin{bmatrix} p_1(\theta) \\ p_2(\theta) \end{bmatrix} \tag{12}$$

where

$$p_1(\theta) = \frac{\sin 2\theta}{2} w(\theta) \tag{13}$$

$$p_2(\theta) = \frac{1-\cos 2\theta}{2} w(\theta) \tag{14}$$

which are graphically depicted in Figure 3.

The elemental cutting force functions are the same for all cutting points anywhere on the cutter since they are defined with respect to a unit cutting area. The contributions of different cutting points on a flute to the total cutting forces differ only in the cutter angular position in which they engage in the cutting. Since points on a flute engage in the cutting in a predetermined sequence as dictated by the cutter geometry, these points produce orderly shifted functions of the same elemental cutting forces. The superposition of these shifted force functions sums to the total cutting forces. Thus, we can treat the elemental cutting process as a linear shift-invariant system and think of the force functions generated at an elemental cutter point, namely p_x and p_y, as finite cutting impulse response functions of duration θ_r.

2-2 Chip Width Density Function

The elemental cutting forces are defined with respect to a unit width of cut while the chip width density function, cwd(θ), discussed herein is the chip width produced during unit angular rotation of the cutter; therefore the chip width density function can be used to weigh each elemental force within the range of the axial depth of cut. In general, for constant or variable helix angle cutters, the chip width density function is determined from

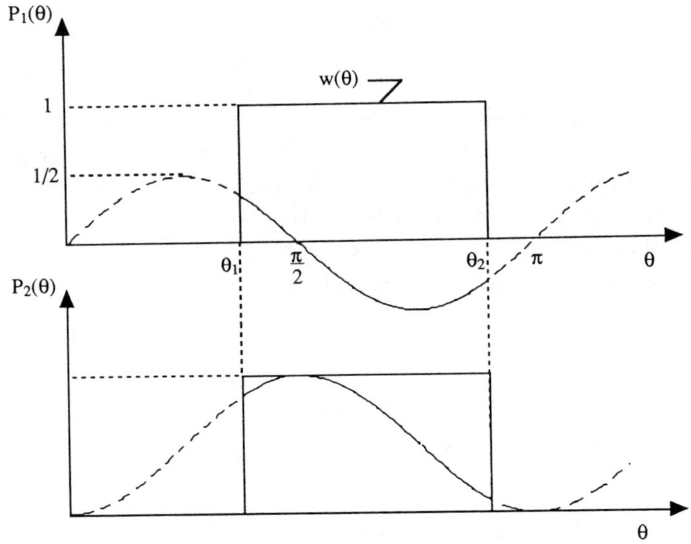

Figure 3. Functions $P_1(\theta)$ and $P_2(\theta)$ in the expression of elemental cutting forces.

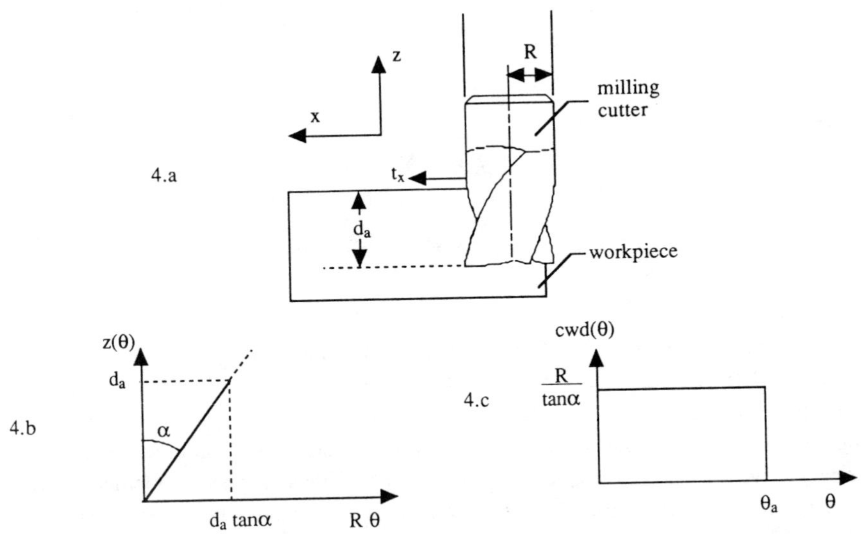

Figure 4. Relationship between the axial depth of cut and chip width density function.

$$\text{cwd}(\theta) = \frac{dz(\theta)}{d\theta} \tag{15}$$

where $z(\theta)$ is the axial position of a point along the cutter flute as a function of the angular position of that point. Shown in Figure 4.b is an unfolded flute, with helix angle α, bounded by the axial depth of cut d_a. Its chip width density function is

$$\text{cwd}(\theta) = \frac{R}{\tan\alpha} \qquad 0 \le \theta \le \frac{d_a \tan\alpha}{R} \tag{16}$$

which is a rectangular pulse function as shown in Fig 4.c. Notice that the total area of the rectangular pulse has a value equivalent to the axial depth of cut. For cutters with a smaller helix angle, the width and height of their chip width density functions are smaller and greater respectively. In the extreme case of a straight teeth cutter (α=0), the function degenerates to an impulse function with an area of d_a.

The chip width density function derived above is applicable to any flute on the cutter. As the milling cutter rotates, each flute will contribute to the total cutting force system through the same chip width density function except that the function is shifted by an angle equivalent to the spacing between flutes. Therefore, the chip width density representation can again be treated as a shift-invariant process with the chip width density function being the impulse response weighed by a tooth sequence function, which represents the continuous cutter rotation and the angular spacing between cutter flutes.

2-3 Tooth Sequence Function

For a cutter with evenly spaced flutes, the tooth sequence function ts(θ) can be determined from

$$ts(\theta) = t_x \sum_{k=0}^{\infty} \delta(\theta - \frac{2\pi}{n} k) \tag{17}$$

where n is the number of flutes on the cutter and δ is the unit impulse function. The function is essentially a train of impulses separated by $\theta_p = \frac{2\pi}{n}$, the angular spacing between any two neighboring flutes.

2-4. Total Cutting Force

The total cutting forces, $\bar{f}_x$ and $\bar{f}_y$, are the angular domain convolution of the elemental cutting force function, the chip width density function, and the tooth sequence function. Denoting convolution operation by "*", cutting forces are expressed as:

$$\begin{bmatrix} \bar{f}_x(\theta) \\ \bar{f}_y(\theta) \end{bmatrix} = K_t\, ts(\theta) * cwd(\theta) * \begin{bmatrix} p_x(\theta) \\ p_y(\theta) \end{bmatrix} \tag{18}$$

Defining the angular convolution of the tooth sequence function and the chip load function as inp(θ), it can be shown that

$$\begin{aligned} inp(\theta) &= ts(\theta) * cwd(\theta) \\ &= \int_0^{\theta} ts\,(\tau)\, cwd(\theta-\tau)\, d\tau \\ &= \int_0^{\theta} t_x \sum_{k=0}^{\infty} \delta(\tau - \frac{2\pi}{n} k)\, cwd(\theta-\tau)\, d\tau \\ &= t_x \sum_{k=0}^{\infty} cwd\,(\theta - \frac{2\pi}{n} k) \end{aligned} \tag{19}$$

which is a periodic function with an angular period of θ_p as shown in Figure 5.

It follows from equations (18) and (19) that

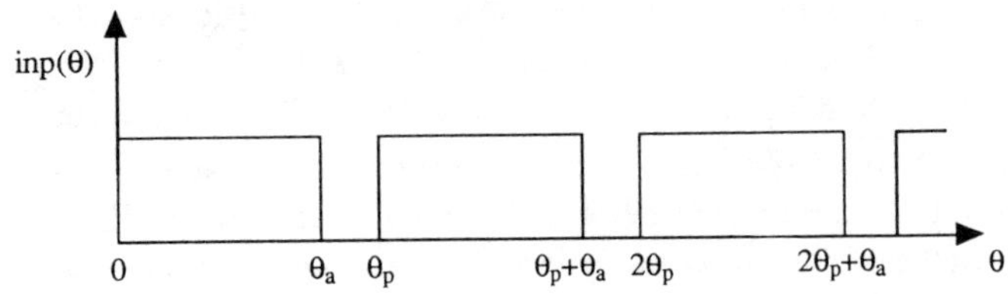

Figure 5. The angular convolution of the tooth sequence function and the chip load function.

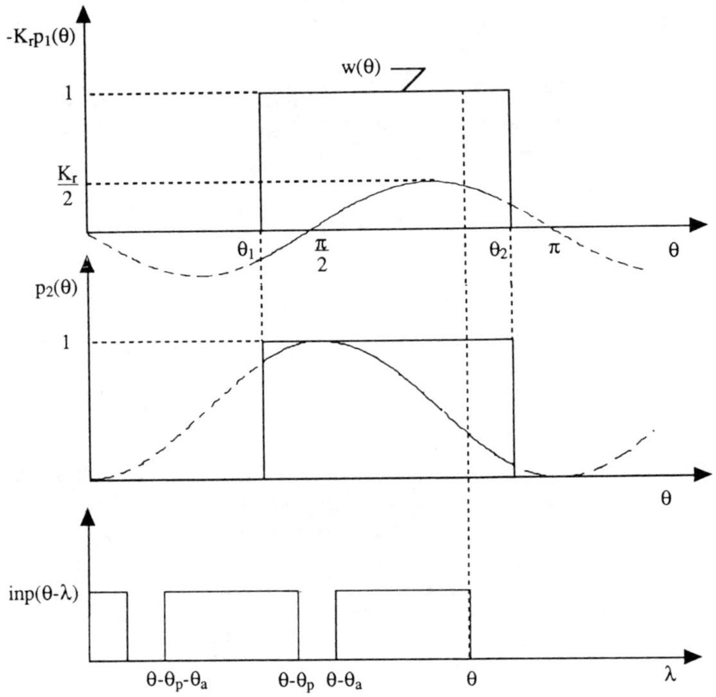

Figure 6. Angular convolution model of total cutting force in y direction.

$$\begin{bmatrix} \bar{f}_x(\theta) \\ \bar{f}_y(\theta) \end{bmatrix} = K_t \, inp(\theta) * \begin{bmatrix} p_x(\theta) \\ p_y(\theta) \end{bmatrix} = K_t \, t_x \int_0^{\theta} \sum_{k=0}^{\infty} cwd(\theta - \tau + \frac{2\pi}{n}k) \begin{bmatrix} p_x(\tau) \\ p_y(\tau) \end{bmatrix} d\tau \tag{20}$$

This is a unified expression for the milling force system in the angular domain. The convolution integral provides a clear insight into the effects of cutting parameters on the milling process. To illustrate this point, a graphical representation of the convolution process for the y component of the cutting force is given in Figure 6. In the convolution integral, the function inp(θ) is rotated with respect to the ordinate and shifted toward the right side. The y cutting force is the area under a function defined as the product of the elemental cutting force $p_y(\theta)$, the rotated and shifted inp(θ), and the specific cutting pressure K_t. Since $p_y(\theta)$ is bounded by the window function w(θ), the cutting force is equivalent to the total area of the product function within the window. As the cutter rotates, the function inp(θ-λ) moves toward right and the total cutting force will experience a periodic variation with a period of θ_p. Three phases of cutting as discussed in [Tlusty et al., 1975] can be clearly visualized as the first cutting edge of the flute enters the window, stays in it, and exit from it. In addition, several comments can be made from the convolution expression regarding the effects of θ_a, θ_p, and θ_r (defined as θ_2-θ_1 as shown in Figure 2) on the characteristics of total cutting forces:

1. If the axial depth of cut (d_a) and the angle of circumferential engagement (θ_r) are both small such that the gap in the inp(θ) is greater than the width of the window function, that is, $\theta_p-\theta_a>\theta_r$, there will be zero cutting force for a duration of $\theta_p-\theta_a-\theta_r$. In this case, at most one flute is engaged in the cutting at any instant of time.

2. If the axial depth of cut and the angle of circumferential engagement are both large such that $\theta_p-\theta_a<\theta_r$, at least two flutes are engaged in cutting for a duration of $\theta_r-\theta_p+\theta_a$.

3. If θ_a is identical to θ_p, one flute engagement will be immediately followed by the engagement of another. The function inp(θ) in this case becomes a DC function independent on time. As a result, cutting force system will assume a constant value at steady state cutting.

4. If the angle of axial engagement (θ_a) is greater than the angular tooth spacing (θ_p), cutting edges will overlap in time. The cutting force system in this situation can be described by that in the previous case superimposed by that in a case with axial engagement angle equals to θ_a-θ_p.

5. The areas under p_x and p_y versus θ curve within the window is an indication of the average cutting force. By appropriate positioning of the cutting window, zero average cutting force could be obtained.

3. Frequency Domain Model

The frequency characteristics of cutting forces can be studied by examining the three model functions in the Fourier domain. In the following discussion, the transforms of these functions from the angular domain to the spectral domain is performed while the frequency variables are normalized with respect to the spindle rotation frequency. In other words a normalized frequency of m in the following analysis is equivalent to m times the spindle frequency.

3-1. Transformation of Elemental Cutting Force Functions

The Laplace transforms of $p_x(\theta)$ and $p_y(\theta)$ in equation (12) are

$$\begin{bmatrix} P_x(s) \\ P_y(s) \end{bmatrix} = \begin{bmatrix} 1 & K_r \\ -K_r & 1 \end{bmatrix} \begin{bmatrix} P_1(s) \\ P_2(s) \end{bmatrix} \tag{20}$$

The Laplace transform of $p_1(\theta)$ can be derived from equation (13) using the complex convolution theorem:

$$P_1(s) = \frac{1}{4\pi j} \int_{q_0-j\infty}^{q_0+j\infty} \frac{2}{q^2+4} W(s-q)\, dq \qquad q_0 > 0 \tag{21}$$

where

$$W(s) = \frac{e^{-s\theta_1} - e^{-s\theta_2}}{s} \tag{22}$$

which is the Laplace transform of $w(\theta)$. By Cauchy's residue theorem, $P_1(s)$ can be shown to be

$$P_1(s) = \frac{1}{2(s^2+4)} \left[e^{-s\theta_1} (s \sin 2\theta_1 + 2\cos 2\theta_1) - e^{-s\theta_2} (s \sin 2\theta_2 + 2 \cos 2\theta_2) \right] \tag{23}$$

Similarly, $P_2(s)$ can be shown as

$$P_2(s) = \frac{e^{-s\theta_1} - e^{-s\theta_2}}{2\,s} - \frac{1}{2(s^2+4)}\left[e^{-s\theta_1}\,(s\cos 2\theta_1 - 2\sin 2\theta_1) - e^{-s\theta_2}\,(s\cos 2\theta_2 - 2\sin 2\theta_2)\right] \tag{24}$$

The Fourier transform can be obtained by substituting $s=j\omega$ into the above equations since $j\omega$ axis is in the region of convergence of $P_1(s)$ and $P_2(s)$. The frequency spectrum of the elemental cutting forces are shown in Figure 7 for various cutting conditions. It is seen in the figures that the position of the workpiece relative to the cutter has significant effects on the dynamics of the cutting force system, especially when the radial depth of cut is small.

3-2 Transform of Chip Width Density Function

The Fourier transforms of chip width density function $cwd(\theta)$ can be derived from Figure 4:

$$CWD(\omega) = \frac{2R}{\tan\alpha}\,\frac{\sin\frac{\omega\theta_a}{2}}{\omega}\,e^{-j\omega\frac{\theta_a}{2}} \tag{25}$$

The Fourier transform assumes the form of a sinc function, which has periodic zeros with a period of $2\pi/\theta_a$. Figure 8 illustrates the effects of tool-workpiece geometry on the frequency spectrum of the chip width density function. It is evident from the figure that the axial depth of cut significantly affects the DC content of the spectrum, while the cutter radius and helix angle dominate the zeros of the spectrum.

3-3. Transform of Tooth Sequence Function

The Fourier transform of the tooth sequence function can be shown from equation (17):

$$TS(\omega) = t_x\, n \sum_{k=-\infty}^{\infty} \delta(\omega - nk) \tag{26}$$

Since $ts(\theta)$ is a periodic impulse sequence function in the angular domain, its Fourier transform is also a periodic discrete impulse function with a period of n. This implies that the cutting forces have non-zero frequency contents only at the harmonics of the tooth passing frequency in addition to their DC components.

3-4. Transform of the Total Cutting Forces

The Fourier transform of the total cutting force is the product of the transforms of the elemental cutting force function, the chip width density function, the tooth sequence function, and the tangential specific cutting pressure constant. That is

$$\begin{bmatrix} \overline{F}_x(\omega) \\ \overline{F}_y(\omega) \end{bmatrix} = \begin{bmatrix} 1 & K_r \\ -K_r & 1 \end{bmatrix} \begin{bmatrix} P_1(\omega) \\ P_2(\omega) \end{bmatrix} .\ CWD(\omega)\ .\ TS(\omega)\ .\ K_t \tag{27}$$

The total cutting force in the angular domain is represented by equation (20) in a convolution integral framework. As an alternative, they can now be expressed in a more useful form using Fourier integral formula:

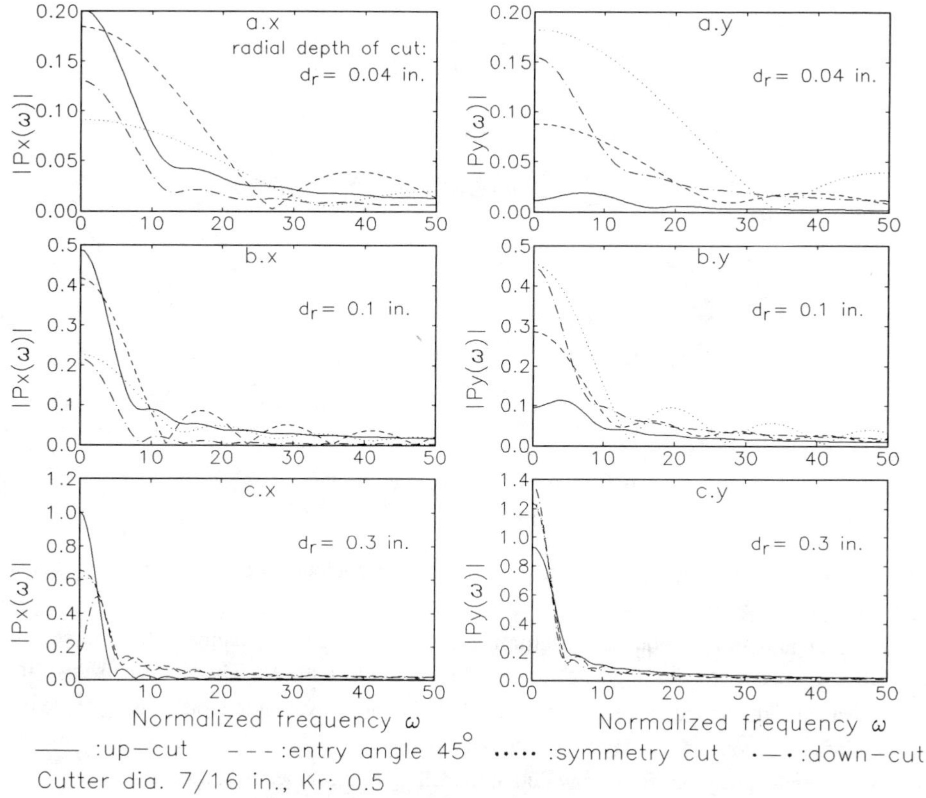

Figure 7. Frequency spectra of elementary cutting forces for various cutting geometry.

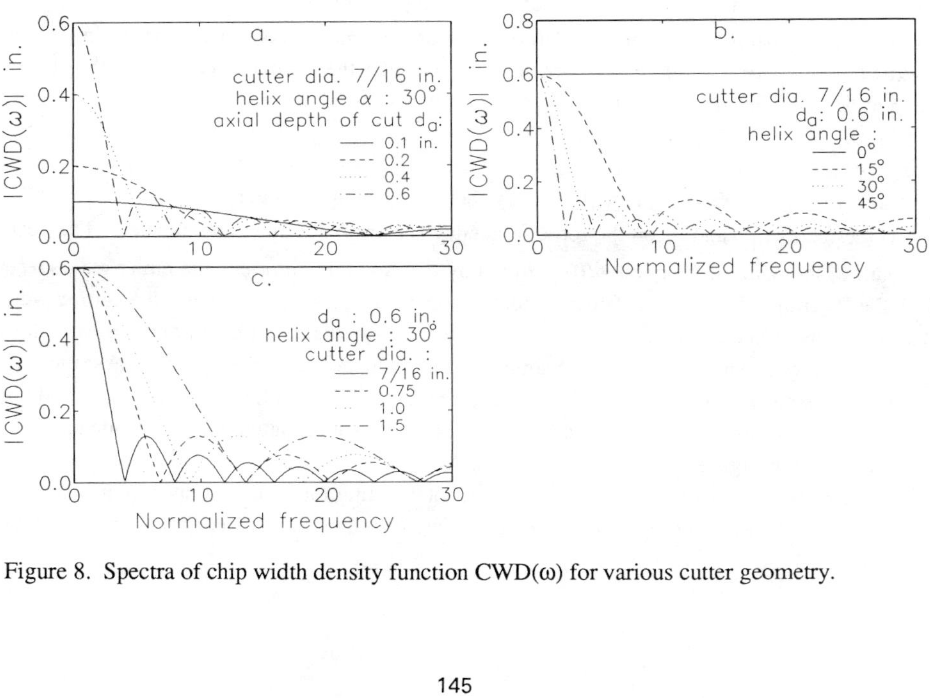

Figure 8. Spectra of chip width density function CWD(ω) for various cutter geometry.

$$\begin{bmatrix} \bar{f}_x(\theta) \\ \bar{f}_y(\theta) \end{bmatrix} = \frac{K_t t_x}{2\pi} \int_{-\infty}^{\infty} TS(\omega)\, CWD(\omega) \begin{bmatrix} 1 & K_r \\ -K_r & 1 \end{bmatrix} \begin{bmatrix} P_1(\omega) \\ P_2(\omega) \end{bmatrix} e^{j\omega\theta}\, d\omega$$

$$= \frac{K_t t_x}{2\pi} \int_{-\infty}^{\infty} n \sum_{k=-\infty}^{\infty} \delta(\omega - nk)\, CWD(\omega) \begin{bmatrix} P_x(\omega) \\ P_y(\omega) \end{bmatrix} e^{j\omega\theta}\, d\omega$$

$$= \sum_{k=-\infty}^{\infty} \begin{bmatrix} A_{xk} \\ A_{yk} \end{bmatrix} e^{jnk\theta} \tag{28}$$

where

$$\begin{bmatrix} A_{xk} \\ A_{yk} \end{bmatrix} = \frac{n\, K_t t_x}{2\pi} CWD(nk) \begin{bmatrix} P_x(nk) \\ P_y(nk) \end{bmatrix} \tag{29}$$

It is noted from equation (28) that A_{xk}'s and A_{yk}'s are the coefficients of the Fourier series expansion of total cutting forces $\bar{f}_x(\theta)$ and $\bar{f}_y(\theta)$.

The total cutting force frequency spectrum is calculated based on equation (29) for three cutting cases each with different axial depths of cut as illustrated in Figure 9. The last case shown in the figure was calculated with a cutter engagement angle θ_a being equivalent to tooth passing angle θ_p. In this particular case a constant cutting force without any dynamic frequency component results, which coincides with the implications from convolution modelling as discussed in section 2-4.

The contribution of tooth sequence function, ts(θ), and chip width density function to the frequency spectrum of total cutting forces can be realized from equation (28). The tooth sequence function determines the frequencies at which dynamic components of the cutting forces will be present. Chip width density function possesses zeros, whose periodicity is a function of θ_a. Thus, it is possible to have a pair of TS(ω) and CWD(ω) such that the dynamic components of cutting forces do not exist regardless of other cutting parameters, in particular the parameters related to the elemental cutting force functions. Figure 9(c) is a depiction of such a cutting process.

On the other hand, tooth sequence function has no bearing on the DC component of cutting forces provided that the feed per tooth is constant. Chip width density contributes to the DC component through a gain of d_a, regardless of the helix angle and the cutter radius.

The effects of P_x and P_y on the force frequency spectrum are more complicated. The position and width of the window function w(θ), defined by the radial depth of cut and cutter radius, affects subtly the spectrum of these two elemental cutting forces. It can be shown that if the cutting window is located such that the elemental force has a symmetric shape, its spectrum will have clearly defined peaks and valleys and a high DC content similar to the frequency response of a low pass filter. If the elemental force has an anti-symmetric shape, its spectrum will have a low DC component and higher dynamic contents similar to a high pass filter response. These statements are evidenced by the spectrum shown in Figure 7.

The effects of elemental cutting forces on the total cutting forces DC component hinge upon the location and width of the cutting window, which is determined by the radial depth of cut, cutter radius and relative position of cutter and workpiece.

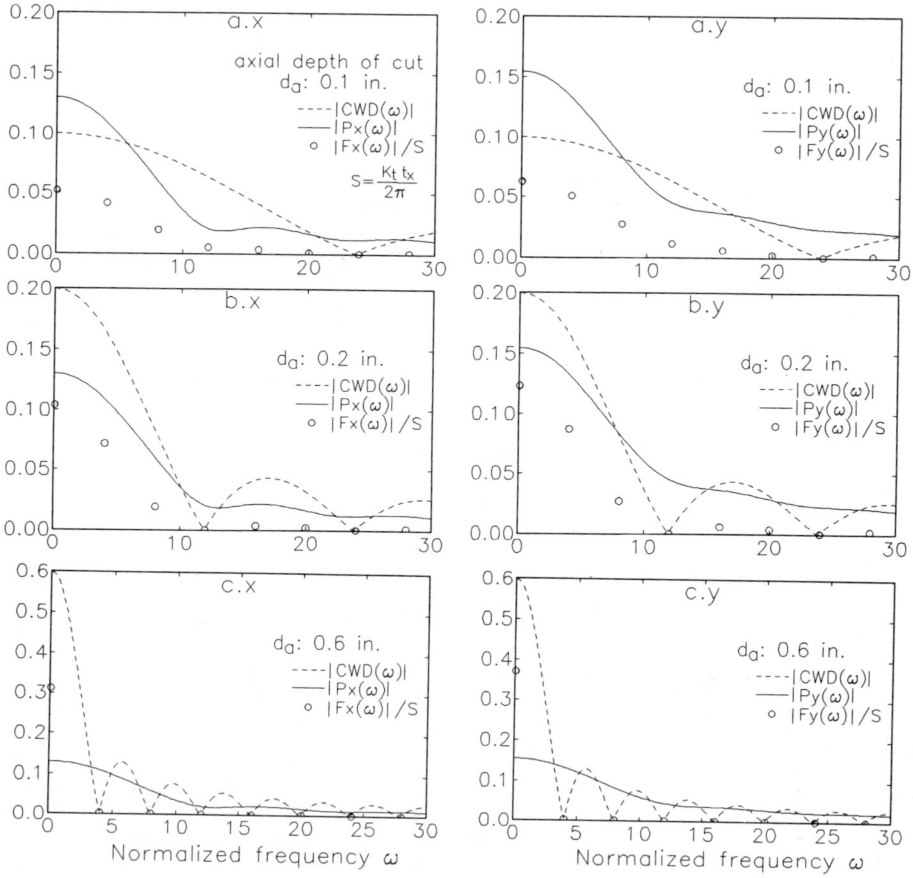

Figure 9. Frequency spectra of the total cutting forces $F_i(\omega)=K_t*TS(\omega)*CWD(\omega)*P_i(\omega)$. Circles represent the areas of the spectral impulses. Assumed cutting conditions are: cutter dia.=7/16 in., helix angle=30°, flute number=4, K_r=0.5, d_r=0.04 in., down-cut.

4. Experimental Verification

A series of end milling experiments was conducted to evaluate the validity of the analytical frequency domain model. The experimental cutting parameters were identical to those used in the three calculation cases shown in Figure 9. Cutting was performed with Aluminum 2024-T4 on a vertical milling machine running at 297rpm and 0.00675inch feed per tooth. The data acquisition system consisted of a two-component dynamometer, signal conditioning unit and a two channel PC-based digitizer with a sampling rate of 1KHz. Digitized cutting force data are transformed into frequency domain using standard FFT software. Since the cutting forces are inherently periodic due to the spindle rotation, the frequency spectrum obtained is discrete with a fundamental frequency equivalent to the spindle frequency. The spectral density function associated with the discrete frequencies are the same as the coefficients in the Fourier series representation of the digitized cutting force data.

Due to the lack of cutting pressure constants K_t and K_r for the specific work material and cutting conditions, a similar approach to that of Kline et al. [1983] was used to obtain these two constants from the x and y DC components of the experimental cutting force data. It can be shown from equation (28) that

Table 1. K_t and K_r calculated from average cutting forces

Cases	d_a (inch)	K_t (lbf/in^2)	K_r
a	0.1	246000	0.593
b.	0.2	240990	0.741
c.	0.6	218070	0.593

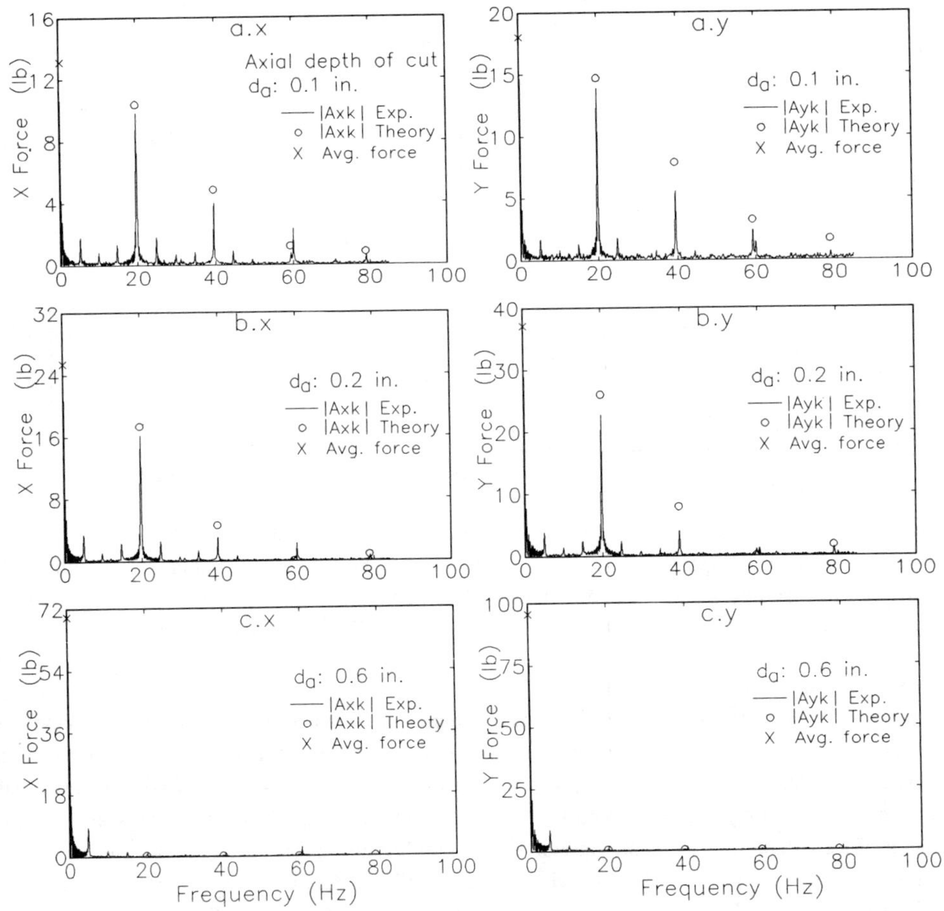

Figure 10. Comparison between the analytically calculated cutting force coefficients and experimental results. Workpiece material was aluminum 2024-T4 and spindle speed was 297 rpm. Other cutting conditions were identical to those of Figure 9 (cutter dia.=7/16 in., helix angle=30°, flute number=4, d_r=0.04 in., down-cut) except that K_t and K_r in Table 1 were used for the calculation of analytical cutting forces.

$$\begin{bmatrix} K_t \\ K_t K_r \end{bmatrix} = \begin{bmatrix} P_1(0) & P_2(0) \\ P_2(0) & -P_1(0) \end{bmatrix}^{-1} \begin{bmatrix} A_{x0} \\ A_{y0} \end{bmatrix} \left[CWD(0) \frac{n\, t_x}{2\pi} \right]^{-1} \tag{30}$$

in which A_{x0} and A_{y0} are the DC components of the measured forces. Constants K_t and K_r calculated in three different cutting conditions are listed in Table 1.

The model-predicted Fourier series coefficients were calculated by substituting K_t and K_r in equations (28) and (29) with the experimentally obtained values. Note that the frequency domain model, (28) and (29), is used herein to estimate the dynamic characteristics of cutting forces based upon given specific cutting pressures. To evaluate the validate of the proposed model, Fourier series coefficients from measurements and those from the model are compared at the first harmonic, the second harmonic, and the third harmonic of the tooth passing frequency (20 Hz) as shown in Figure 10. The experimental spectra at 60 Hz have lost their fidelity due to in-line electrical disturbances. It is seen from the figures that the results from experiments agree well with that from the analytical model in general. Notice that spectrum peaks occur at spindle frequency and at the sidebands of the harmonics frequencies, besides, the model-calculated Fourier coefficients are consistently higher than their experimental counterparts. This point is consistent with the observation reported by Kline [1983] regarding the effects of cutter runout on the frequency contents. He noted that the frequency content at the tooth passing frequency is shifted to the spindle frequency in the presence of cutter runout.

5. Summary

A closed form expression for the cutting forces in end milling was derived as explicit functions of cutting conditions and tool/workpiece geometry. This was achieved by decomposing the cutting process into three uncorrelated component functions, namely the elemental cutting force function, the chip width density function, and the tooth sequence function. The total cutting forces were derived as the angular convolutions of these three component functions. The effects of various cutting parameters on the cutting forces were discussed and illustrated in the context of the convolution model.

The analysis was extended into the Fourier domain by taking the frequency multiplication of the transforms of the three component functions. Fourier series coefficients of the cutting forces are shown to be algebraic functions of various tool parameters and cutting conditions. The illustration of the cutting force mode shapes in relation to the milling process conditions was discussed based on simulation results, which are presented in terms of the power spectra of the cutting component functions as well as that of the total cutting forces in both x and y directions.

A series of end milling experiments were performed to validate the analytical model. The discrete spectral density function of the measured forces agreed well with the calculated Fourier series coefficients at the harmonics of the tooth passing frequencies for different axial depths of cut. The frequency shift in the force spectrum from tooth passing frequency to spindle rotation frequency as a result of cutter runout was also observed.

References

Armarego, E. J. A., and Despande, N. P., "Computerized Predictive Cutting Model for Forces in End-Milling Including Eccentricity Effects," Annals of the CIRP, Vol. 38, pp. 45-49, 1989.

Ber, A., Rotberg, J. and Zombach, S., "A Method for Cutting Force Evaluation of End Mills," Annals of the CIRP, Vol. 37, pp. 37-40, 1988.

Fu, H. J., DeVor, R. E., and Kapoor, S. G., "A Mechanistic Model for the Prediction of the Force System in Face Milling Operations," ASME J. of Engineering for Industry, Vol. 106, pp. 81-88, Feb. 1984.

Fussel, B. K. and Srinivasan, K., "On-line Identification of End Milling Process Parameters," Proceedings of the USA-Japan Symposium on Flexible Manufacturing, Minnesota, pp. 967-974, 1988.

Gygax, P. E., "Dynamics of Single-Tooth Milling," Annals of the CIRP, Vol. 28, pp. 65-70, 1979.

Gygax, P. E., "Experimental Full Cut Milling Dynamics," Annals of the CIRP, Vol. 29, p.p. 61-66, 1980.

Kline, W. A., DeVor, R. E., Lindberg, J. R., "The Prediction of Cutting Forces in End Milling with Application to Cornering Cut," International Journal of Machine Tool Design and Research, Vol. 22, pp. 7-22, 1982.

Kline, W. A. and DeVor, R. E., "The Effect of Runout on Cutting Geometry and Forces in End Milling," International Journal of Machine Tool Design and Research, Vol. 23, No 2/3, pp. 123-140, 1983.

Koenigsberger, F., and Sabberwal, A. J. P., "An Investigation into the Cutting Force Pulsations During Milling Operations," International Journal of Machine Tool Design and Research, Vol. 1, pp. 15-33, 1961.

Lauderbaugh, L. K., and Ulsoy, A. G., "Dynamic Modeling for Control of the Milling Process," ASME J. of Engineering for Industry, Vol. 111, pp. 367-374, November 1988.

Martellotti, M. E., "An Analysis of the Milling Process," Transaction of. ASME, Vol. 63, pp. 677-700, 1941.

Martellotti, M. E., "An Analysis of the Milling Process, Part 2: Down Milling," Transaction of ASME, Vol. 67, pp. 233-251, 1945.

Rohrs, C., Valavani, L., Sthans, M., and Stein, G., "Robustness of Continuous-time Adaptive Control Algorithms in the Presence of Unmodelled Dynamics," IEEE Transaction on Automatic Control, Vol. 30, pp. 881-889, 1985.

Sabberwal, A. J. P., presented by F. Koenigsberger, "Chip Section and Cutting Force During the Milling Operation," Annals of the CIRP, 10, pp. 197-203, 1961/62.

Sutherland, J. W., and DeVor, R. E., "An Improved Method for Cutting Force and Surface Error Prediction in Flexible End Milling Systems," ASME J. of Engineering for Industry, Vol. 108, pp. 269-279, Nov. 1986.

Tlusty, J., and MacNeil, P., "Dynamics of Cutting Forces in End Milling," Annals of the CIRP, Vol. 24, pp. 21-25, 1975.

Tomizuka, M., Oh, J. H., and Dornfeld, D. A., "Model Reference Adaptive Control of the Milling Process," Proceedings of the Symposium on Control of Manufacturing Processes and Robotic Systems, D. Hardt ed., ASME Winter Annual Meeting, pp. 56-63, 1983.

Zhou, R., and Wang, K. K., "Modelling of Cutting Force Pulsation in Face-Milling," Annals of the CIRP, Vol. 32, pp. 21-26, 1983.

Sensors, Controls, and Quality Issues
in Manufacturing
ASME 1991

COORDINATED MOTION GENERATION OF FIVE-AXIS CNC/CMM MACHINES IN CAD/CAM INTEGRATION

Jui-Jen Chou
Department of Agricultural Machinery Engineering
National Taiwan University
Taipei, Taiwan, Republic of China

D. C. H. Yang
Department of Mechanical, Aerospace and Nuclear Engineering
University of California
Los Angeles, California

ABSTRACT

This paper presents an analytical study on the command generation for five-axis CNC machining or CMM measurement. In the integration of CAD and CAM, it is necessary to relate machine tool kinematics and control in a CAM process to the geometrical data in a CAD model. The data stored in a CAD model is usually static in nature and represented by unitless parameters. Yet, in machine tool motion and control, the data should be transformed into a time dependent domain. In this paper, a general theory on the conversion from desired paths to motion trajectory is analytically derived. The geometrical properties of a desired path, including position, tangent, and curvature are related to the kinematics of coordinated motion including feedrate, acceleration, and jerk. As a result, the motion commands used as control references to track arbitrary space curves for five-axis computer-controlled machines can be generated in a rather straight-forward as well as systematic way.

1. INTRODUCTION

The integration of computer-aided design and manufacturing can be considered as the achievement of an automatic production process from the design specifications in a CAD data base to the manufactured part yielded from a CNC machine tool. In order to do so, the information contained in the CAD data base needs to be transferred and some times even transformed in the various stages of the integration.

Roughly speaking, the CAD/CAM integration can be divided into three stages: 1. CAD model to path conversion; 2. cutter path to motion trajectory conversion; and 3. motion trajectory realization. Stage 1 deals with problems such as feature recognition, geometry intersecting and offsetting, etc. Stage 3 deals with control theory and controller design. Stage 2 which is the main concern of this paper, also referred to as "command generation", involves kinematics of coordinated motion, machine dynamics and interpolators design.

Modern machine tools usually have the capability of command generation for simple curves of primitive geometry, such as straight lines, circles and parabolas via techniques of interpolation [1 – 6]. The most frequent type of interpolation techniques are: 1. Search Step technique; 2. Digital Differential Analyzer (DDA) technique; and 3. Direct Function Calculation technique (also referred to as the contouring technique). Both the search-step and DDA techniques are used to perform very simple motions such as straight lines and circular arcs. Machines adopting search-step method moves only one axis at a time (a step pattern motion), whereas those using DDA technique can provide coordinated motion. For a machine to track an arbitrary curve, currently the direct-function technique is used. This method is to first perform a rough interpolation to the curve, then to perform finer linear interpolation between points of rough interpolations.

All these three above mentioned methods are for generating position references for computer control. By our awareness, there is no general theory in this regard which has the capability for command generation of five-axis machines for tracking arbitrary space curves. Some work has been done on the trajectory design of industry robots [7 – 10]. In this investigation, a different approach is taken, the geometrical properties of a tool path, including position, tangent and curvature are analytically related to the desired coordinated motion of the machine tool, including feedrate, acceleration, and jerk. The conversion of the parameter and time between CAD and CAM data is also derived.

2. COORDINATION OF TRANSLATIONAL MOTION

Let $\underline{p}(= (p_x, p_y, p_z))$ be one of a set of data points which specify a desired path, and $\underline{o}(= (o_x, o_y, o_z))$ be the corresponding approach direction, where $\underline{o}$ is a unit vector and its components are direction cosines. Both $\underline{p}$ and $\underline{o}$ are defined with respect to the workpiece. The workpiece is mounted on the table of the machine and has translational displacement $\underline{w}(= (w_x, w_y, w_z))$ with respect to the fixed coordinate system. The fixed coordinate system XYZ, as shown in Fig. 1, has the three axes aligned with the three translational axes of the machine tool and its origin is at Σ. The position Σ coincides with the tool tip C when the machine tool is at its nominal position, i.e., when all displacements are equal to zero.

To formulate the entire coordinated motion, the coordinated translational motion and rotational motion need to be treated separately. We first start with the translational motion. The desired path $\underline{P}(= (P_x, P_y, P_z))$ with respect to the fixed coordinate system Σ can be found by applying a translational transformation to the $\underline{p}$ as

$$P_x = p_x + w_x; \quad P_y = p_y + w_y; \quad P_z = p_z + w_z$$

This sequence of desired position data is interpolated into an nth order parametric form as

$$P_x(u) = a_{nx}u^n + a_{n-1x}u^{n-1} + \ldots\ldots + a_{1x}u + a_{0x} \tag{1}$$

$$P_y(u) = a_{ny}u^n + a_{n-1y}u^{n-1} + \ldots\ldots + a_{1y}u + a_{0y} \tag{2}$$

$$P_z(u) = a_{nz}u^n + a_{n-1z}u^{n-1} + \ldots\ldots + a_{1z}u + a_{0z} \tag{3}$$

where u is the parameter. For Eqs.(1), (2) and (3), composite parametric cubic curves are recommended for the reason of simplicity and less global propagation of change.

2.1 Desired Velocity for Tracking a Tool Path

In a previous study [11], the coordinated motion of three-axis machines for translation has been derived. Let $\underline{V}(u)$ be the feedrate of the tool for tracking a path, and $\dot{P}_x(u)$, $\dot{P}_y(u)$, and $\dot{P}_z(u)$ be its three Cartesian components, where

$$\underline{V}(u) = \frac{dP_x(u)}{dt}\underline{i} + \frac{dP_y(u)}{dt}\underline{j} + \frac{dP_z(u)}{dt}\underline{k} = \frac{dP_x}{du}\frac{du}{dt}\underline{i} + \frac{dP_y}{du}\frac{du}{dt}\underline{j} + \frac{dP_z}{du}\frac{du}{dt}\underline{k} \tag{4}$$

The function of $\frac{du}{dt}$ which is yet to be identified defines the relationship between the static information in the CAD model and the dynamic information needed to control the machine. The magnitude of feedrate can be obtained by using

$$V = \sqrt{\dot{P}_x^2 + \dot{P}_y^2 + \dot{P}_z^2} = \sqrt{\overset{u}{P}{}_x^2 + \overset{u}{P}{}_y^2 + \overset{u}{P}{}_z^2}\,\frac{du}{dt} \tag{5}$$

where $\overset{u}{P}$ denotes $\frac{dP}{du}$ Therefore, we have

$$\frac{du}{dt} = \frac{V}{\sqrt{\overset{u}{P}{}_x^2 + \overset{u}{P}{}_y^2 + \overset{u}{P}{}_z^2}} \tag{6}$$

Let

$$\underline{V}(u) = V(u)\,\underline{T}(u)$$

where $\underline{T}$ represents the unit tangent of the path, and

$$\underline{T}(u) = t_{px}(u)\underline{i} + t_{py}(u)\underline{j} + t_{pz}(u)\underline{k} = \frac{\overset{u}{\underline{P}}(u)}{|\overset{u}{\underline{P}}(u)|} \tag{7}$$

where, t_{px}, t_{py}, and t_{pz} are the direction cosines of $\underline{T}$. Then we have

$$\dot{P}_x(u) = t_{px}(u)\, V(u) \quad where \quad t_{px}(u) = \frac{\overset{u}{P}_x}{\sqrt{(\overset{u}{P}_x^2 + \overset{u}{P}_y^2 + \overset{u}{P}_z^2)}} \tag{8}$$

Similar expressions of Eq.(8) are also true for P_y and P_z components.

If the travelling speed V is specified, then Eq.(8) provides the desired velocity for the machine to exactly track the given tool locations $\underline{P}(u)$ in a coordinated manner (however with no concern of approach direction yet).

2.2 Desired Acceleration for Coordinated Motion

The second derivative of $P_x(u)$ with respect to time is

$$\ddot{P}_x = \frac{d}{dt}(\dot{P}_x) = \frac{d}{dt}(t_{px}\, V) = V\frac{dt_{px}}{ds}\frac{ds}{dt} + t_{px}\frac{dV}{dt} = \frac{dt_{px}}{ds}V^2 + t_{px}\, A \tag{9}$$

where $A = \frac{dV}{dt}$ is the tangential acceleration which is meant the change rate of speed and s represents the arc length.

From differential geometry, the curvature of a curve [12 – 13], denoted by $\underline{K}$, is defined by

$$\underline{K}(u) = \kappa_{px}\underline{i} + \kappa_{py}\underline{j} + \kappa_{pz}\underline{k} = \frac{d\underline{T}(u)}{ds}$$

therefore,

$$\kappa_{px} = \frac{dt_{px}}{ds} \tag{10}$$

Consequently, equation (9) can be written as

$$\ddot{P}_x = \kappa_{px}\, V^2 + t_{px}\, A \tag{11}$$

The first term on the right hand side represents the x component of centripetal acceleration due to the tangential velocity; the second term represents the x component of tangential acceleration along the path. By similar derivation, we can find the desired acceleration for y and z components.

The three components of curvature $\underline{K}(u)$ for a curve can be obtained by the following equation.

$$\underline{K}(u) = (\overset{uu}{\underline{P}}(u) - \frac{\overset{uu}{\underline{P}}(u)\cdot\overset{u}{\underline{P}}(u)}{|\overset{u}{\underline{P}}(u)|^2}\overset{u}{\underline{P}}(u))\frac{1}{|\overset{u}{\underline{P}}(u)|^2} = \frac{(\overset{u}{\underline{P}}(u)\times\overset{uu}{\underline{P}}(u)\times\overset{u}{\underline{P}})}{|\overset{u}{\underline{P}}(u)|^4} \tag{12}$$

3. COORDINATION OF ROTATIONAL MOTION

Two types of orientating structures are commonly used in five-axis machine tools, namely the Euler-angle type and the rotary-table type. In this paper, we use the Euler-angle type orientating structure for developing the theory. The result can be easily applied to the rotary-table type with some minor modification [14]. The arrangement of three rotary joints to form an Euler-angle structure is shown in Fig. 1.

3.1 Euler Angle Structure for Orientating

To access a given position $\underline{P}$ from the desired direction $\underline{O}$, the direction of the tool, axis Y_2, shown as BC in Fig. 1, should be in this direction $\underline{O}$. The direction cosines of the vector defined by BC can be obtained by rotational transformations of first a Θ_1 angle about the Z_1 axis and then a Θ_2 angle about the Z_2 axis. Let coordinate systems $X_1Y_1Z_1$ and $X_2Y_2Z_2$ be attached to links AB and BC, respectively. Successful machining demands the tip of the cutter follow the path $\underline{P}$ as close as possible. Since the tool is symmetrical, represented as S in the figure, the rotation of the spindle will not affect the direction of approach. Consequently, BC is considered as one link in this derivation and five (instead of six) degrees of freedom are enough to perform this task of space curve tracking.

After these two rotations, the direction BC becomes $\underline{Y}_2 = (\sin\theta_2 \sin\theta_1,\ -sin\theta_2 \cos\theta_1,\ cos\theta_2)$ Because the motion of the machine table is only translational, it does not affect the orientation of the approach direction, i.e., $\underline{O} = \underline{o}$.

To orientate the tool BC to the desired direction of approach $\underline{O}$, we can equate $\underline{Y}_2$ and $\underline{O}$ as:

$$\underline{Y}_2 = (\sin\theta_2 \sin\theta_1,\ -\sin\theta_2\cos\theta_1,\ \cos\theta_2) = (O_x,\ O_y,\ O_z) = \underline{O} \tag{13}$$

Equation (13) can be solved by using

$$\begin{aligned} \theta_1 &= ATAN2(O_x,\ -O_y) && 0 \le \theta_1 \le 2\pi \\ \theta_2 &= ATAN2(-SIGN(O_y)\sqrt{O_x^2+O_y^2},\ O_z) && -\frac{\pi}{2} \le \theta_2 \le \frac{\pi}{2} \end{aligned} \tag{14}$$

3.2 Coordination of Rotational Motion

The angular displacement of the two rotary axes for orientating the tool direction to access each data point is provided in Eq.(14). To systematically derive the coordinated motion, the entire sequence of the angular displacement of each joint needs also to be interpolated into composite parametric spline curves as

$$\theta_1(u) = b_{nx}u^n + b_{n-1x}u^{n-1} + \ldots\ldots + b_{1x}u + b_{0x} \tag{15}$$

$$\theta_2(u) = b_{ny}u^n + b_{n-1y}u^{n-1} + \ldots\ldots + b_{1y}u + b_{0y} \tag{16}$$

3.2.1 Desired Angular Velocity for Tracking the Direction of Approach

To match correctly the orientation to the position, the parameter u in the position funciton $\underline{P}(u)$ of Eqs.(1)-(3) and the orientation function $\theta(u)$ of Eqs.(15)-(16) should be the identical. Similarly, the parameter/time conversion, i.e., $du/dt's$, for both position and orientation need also to be the same. The first derivatives of θ_1 and θ_2 with respect to time are:

$$\dot{\theta}_1(u) = \frac{d\theta_1(u)}{du}\frac{du}{dt} = \overset{u}{\theta}_1(u)\frac{V}{\sqrt{\overset{u}{P}{}_x^2 + \overset{u}{P}{}_y^2 + \overset{u}{P}{}_z^2}} \equiv t_{\theta_1}V \quad where \quad t_{\theta_1} \equiv \frac{\overset{u}{\theta}_1(u)}{\sqrt{\overset{u}{P}{}_x^2 + \overset{u}{P}{}_y^2 + \overset{u}{P}{}_z^2}} \tag{17}$$

$$\dot{\theta}_2(u) = \frac{d\theta_2(u)}{du}\frac{du}{dt} = \overset{u}{\theta}_2(u)\frac{V}{\sqrt{\overset{u}{P}{}_x^2 + \overset{u}{P}{}_y^2 + \overset{u}{P}{}_z^2}} \equiv t_{\theta_2}V \quad where \quad t_{\theta_2} \equiv \frac{\overset{u}{\theta}_2(u)}{\sqrt{\overset{u}{P}{}_x^2 + \overset{u}{P}{}_y^2 + \overset{u}{P}{}_z^2}} \tag{18}$$

Equations (17) and (18) ensure a coordinated motion in approach direction along the path. These results can be used for the command generation to the controller for velocity controls.

3.2.2 Desired Angular Acceleration for Coordinated Motion

The angular accelerations of the two rotary joints are derived as:

$$\begin{aligned} \ddot{\theta}_1(u) = {} & \overset{uu}{\theta}_1(u)\frac{V^2}{\overset{u}{P}{}_x^2 + \overset{u}{P}{}_y^2 + \overset{u}{P}{}_z^2} + \overset{u}{\theta}_1(u)\frac{A}{\sqrt{\overset{u}{P}{}_x^2 + \overset{u}{P}{}_y^2 + \overset{u}{P}{}_z^2}} \\ & - \overset{u}{\theta}_1(u)\frac{V^2(\overset{u}{P}_x\overset{uu}{P}_x + \overset{u}{P}_y\overset{uu}{P}_y + \overset{u}{P}_z\overset{uu}{P}_z)}{(\overset{u}{P}{}_x^2 + \overset{u}{P}{}_y^2 + \overset{u}{P}{}_z^2)^2} \equiv \kappa_{\theta_1}V^2 + t_{\theta_1}A \end{aligned} \tag{19}$$

$$\begin{aligned} \ddot{\theta}_2(u) = {} & \overset{uu}{\theta}_2(u)\frac{V^2}{\overset{u}{P}{}_x^2 + \overset{u}{P}{}_y^2 + \overset{u}{P}{}_z^2} + \overset{u}{\theta}_2(u)\frac{A}{\sqrt{\overset{u}{P}{}_x^2 + \overset{u}{P}{}_y^2 + \overset{u}{P}{}_z^2}} \\ & - \overset{u}{\theta}_2(u)\frac{V^2(\overset{u}{P}_x\overset{uu}{P}_x + \overset{u}{P}_y\overset{uu}{P}_y + \overset{u}{P}_z\overset{uu}{P}_z)}{(\overset{u}{P}{}_x^2 + \overset{u}{P}{}_y^2 + \overset{u}{P}{}_z^2)^2} \equiv \kappa_{\theta_2}V^2 + t_{\theta_2}A \end{aligned} \tag{20}$$

where

$$\kappa_{\theta_1} \equiv \frac{\theta_1''(u)}{P_x'^2 + P_y'^2 + P_z'^2} - \frac{\theta_1'(u)\,(P_x'P_x'' + P_y'P_y'' + P_z'P_z'')}{(P_x'^2 + P_y'^2 + P_z'^2)^2} \tag{21}$$

and

$$\kappa_{\theta_2} \equiv \frac{\theta_2''(u)}{P_x'^2 + P_y'^2 + P_z'^2} - \frac{\theta_2'(u)\,(P_x'P_x'' + P_y'P_y'' + P_z'P_z'')}{(P_x'^2 + P_y'^2 + P_z'^2)^2} \tag{22}$$

Equations (19) and (20) provide desired reference values of acceleration to the machine controller.

4. COMBINED COORDINATED MOTION OF FIVE-AXIS MACHINES

Equations derived in Sections 2 and 3 can not be combined directly, for changing the orientation of the tool axis will affect the position of the tool tip. To include this coupling effect, an additional transformation is needed. The transformation is slightly different. It depends on the type of translational structure of a machine tool. In general there are three types of translational structure, namely, fixed-bed, gantry and knee-and-column. In this paper, a theory is developed on the fixed-bed type of translational structure (see Fig. 2). The result can be easily applied to other types [14] if so desired.

For the fixed-bed type of machine with Euler-angle structure, the position of tool tip caused by the rotation of θ_1 and θ_2 is of $(d_2 \sin\theta_2 \sin\theta_1,\ -d_2 \sin\theta_2 \cos\theta_1,\ -d_2 + d_2 \cos\theta_2)$. To track the path $\underline{P}$, the tables of the machine should provide the three translational motion (x, y, z) so that

$$(d_2 \sin\theta_2 \sin\theta_1, -d_2 \sin\theta_2 \cos\theta_1, z - d_2 + d_2 \cos\theta_2) = (x + P_x,\ y + P_y, P_z)$$

Therefore, the desired translational motions of the machine are

$$(x, y, z) = (-P_x + d_2 sin\theta_2 \sin\theta_1, -P_y - d_2 \sin\theta_2 \cos\theta_1, P_z + d_2 - d_2 \cos\theta_2) \tag{23}$$

The desired velocity components of the translational motion are

$$\dot{x} = -\dot{P}_x + d_2(\cos\theta_2 \sin\theta_1)\dot{\theta}_2 + d_2(\sin\theta_2 \cos\theta_1)\dot{\theta}_1 \equiv t_x V \tag{24}$$

$$\dot{y} = -\dot{P}_y - d_2(\cos\theta_2 \cos\theta_1)\dot{\theta}_2 + d_2(\sin\theta_2 \sin\theta_1)\dot{\theta}_1 \equiv t_y V \tag{25}$$

$$\dot{z} = \dot{P}_z + d_2 \sin\theta_2\, \dot{\theta}_2 \equiv t_z V \tag{26}$$

where

$$t_x \equiv -t_{px} + d_2(\cos\theta_2 \sin\theta_1)t_{\theta_2} + d_2(\sin\theta_2 \cos\theta_1)t_{\theta_1} \tag{27}$$

$$t_y \equiv -t_{px} - d_2(\cos\theta_2 \cos\theta_1)t_{\theta_2} + d_2(\sin\theta_2 \sin\theta_1)t_{\theta_1} \tag{28}$$

$$t_z \equiv t_{pz} + d_2 \sin\theta_2\, t_{\theta_2} \tag{29}$$

and t_x, t_y and t_z can be considered as the effective tangent of the motion.

The acceleration components of translational motion:

$$\begin{aligned}\ddot{x} = {} & -\ddot{P}_x + d_2(\sin\theta_2 \cos\theta_1)\ddot{\theta}_1 + d_2(\cos\theta_2 \sin\theta_1)\ddot{\theta}_2 - d_2(\sin\theta_2 \sin\theta_1)\dot{\theta}_1^2 - \\ & d_2(\sin\theta_2 \sin\theta_1)\dot{\theta}_2^2 + 2\, d_2(\cos\theta_2 \cos\theta_1)\dot{\theta}_1\dot{\theta}_2 \equiv t_x A + \kappa_x V^2\end{aligned} \tag{30}$$

$$\begin{aligned}\ddot{y} = {} & -\ddot{P}_y + d_2(\sin\theta_2 \sin\theta_1)\ddot{\theta}_1 - d_2(\cos\theta_2 \cos\theta_1)\ddot{\theta}_2 + d_2(\sin\theta_2 \cos\theta_1)\dot{\theta}_1^2 + \\ & d_2(\sin\theta_2 \cos\theta_1)\dot{\theta}_2^2 + 2\, d_2(\cos\theta_2 \sin\theta_1)\dot{\theta}_1\dot{\theta}_2 \equiv t_y A + \kappa_y V^2\end{aligned} \tag{31}$$

$$\ddot{z} = \ddot{P}_z + d_2 \sin\theta_2 \ddot{\theta}_2 + d_2 \cos\theta_2 \dot{\theta}_2^2 \equiv t_z A + \kappa_z V^2 \tag{32}$$

where

$$\begin{aligned}\kappa_x \equiv {} & -\kappa_{px} + d_2(\sin\theta_2 \cos\theta_1)\kappa_{\theta_1} + d_2(\cos\theta_2 \sin\theta_1)\kappa_{\theta_2} - d_2(\sin\theta_2 \\ & \sin\theta_1)t_{\theta_1}^2 - d_2(\sin\theta_2 \sin\theta_1)t_{\theta_2}^2 + 2\, d_2(\cos\theta_2 \cos\theta_1)t_{\theta_1}t_{\theta_2}\end{aligned} \tag{33}$$

$$\begin{aligned}\kappa_y \equiv {} & -\kappa_{py} + d_2(\sin\theta_2 \sin\theta_1)\kappa_{\theta_1} - d_2(\cos\theta_2 \cos\theta_1)\kappa_{\theta_2} + d_2(\sin\theta_2 \\ & \cos\theta_1)t_{\theta_1}^2 + d_2(\sin\theta_2 \cos\theta_1)t_{\theta_2}^2 + 2\, d_2(\cos\theta_2 \sin\theta_1)t_{\theta_1}t_{\theta_2}\end{aligned} \tag{34}$$

$$\kappa_z \equiv \kappa_{pz} + d_2 \sin\theta_2 \kappa_{\theta_2} + d_2 \cos\theta_2 t_{\theta_2}^2 \tag{35}$$

and κ_x, κ_y and κ_z can be considered as the effective curvature of the motion.

5. RELATIONSHIP BETWEEN TIME AND PARAMETRIC SPACE

For the control of a CNC machine tool, it is necessary to obtain the position or feedrate as a function of time. The time required to traverse the path from u_0 to u is

$$t = \int_{u_0}^{u} \frac{ds(u)}{V(u)} \tag{36}$$

5.1 Differential Arc Length of a Space Curve

The differential arc length of a space curve ds is

$$ds(u) = \overset{u}{s}(u)du = |\overset{u}{\underline{P}}(u)|\, du = \sqrt{\overset{u}{P}{}_x^{2}(u) + \overset{u}{P}{}_y^{2}(u) + \overset{u}{P}{}_z^{2}(u)}\, du \tag{37}$$

5.2 Feedrate at a Point of a Tool Path

Feed-acceleration $A(u)$ is the control variable of feedrate $V(u)$. Therefore, the specified feedrate can be obtained from a certain feed-acceleration by applying

$$V(u) = [V^2(u_0) + 2\int_{u_0}^{u} A(u)\sqrt{\overset{u}{P}{}_x^{2}(u) + \overset{u}{P}{}_y^{2}(u) + \overset{u}{P}{}_z^{2}(u)}du]^{1/2} \tag{38}$$

5.3 Relationship between Time and Parametric Space

The relationship between the parameter u and time t for a space curve can be obtained by substituting Eqs.(37) and (38) into Eq.(36):

$$t = \int_{u_0}^{u} \frac{\sqrt{\overset{u}{P}{}_x^{2}(u) + \overset{u}{P}{}_y^{2}(u) + \overset{u}{P}{}_z^{2}(u)}}{[V^2(u_0) + 2\int_{u_0}^{u} A(u')\sqrt{\overset{u'}{P}{}_x^{2}(u') + \overset{u'}{P}{}_y^{2}(u') + \overset{u'}{P}{}_z^{2}(u')}du']^{1/2}} du \tag{39}$$

6. THE COMMAND GENERATION OF FIVE-AXIS CNC MACHINES

Up to this point, a general theory for the command generation of coordinated motion of five-axis CNC/CMM machines has been derived and is summarized in Table 1:

7. RESULTANT JERK IN COORDINATED MOTION

In highly accurate machine control systems, big step changes in acceleration will cause machine structures vibration. The vibrations may lead to tracking inaccuracies and machine parts wear [15]. The rate change of acceleration is referred to as jerk. Control of the jerk can often be more cost-effective solution to these problems than redesigning mechanical structures. Therefore it is of practical interest to have the capability to evaluate the amount of jerk when a trajectory is assigned. For the fixed-bed type of machines with Euler-angle structure, the components of resultant jerk for the translational motion can be obtained from Eqs.(30), (31) and (32) as

$$J_x = t_x\frac{dA}{dt} + 3\kappa_x VA + \frac{d\kappa_x}{ds}V^3 \tag{40}$$

$$J_y = t_y\frac{dA}{dt} + 3\kappa_y VA + \frac{d\kappa_y}{ds}V^3 \tag{41}$$

$$J_z = t_z\frac{dA}{dt} + 3\kappa_z VA + \frac{d\kappa_z}{ds}V^3 \tag{42}$$

where $\frac{d\kappa_x}{ds}$, $\frac{d\kappa_y}{ds}$, and $\frac{d\kappa_y}{ds}$ are:

$$\begin{aligned}\frac{d\kappa_x}{ds} \equiv\ & -\frac{d\kappa_{px}}{ds} + d_2(\sin\theta_2\cos\theta_1)\frac{d\kappa_{\theta_1}}{ds} + d_2(\cos\theta_2\sin\theta_1)\frac{d\kappa_{\theta_2}}{ds} \\ & + 3d_2(\cos\theta_2\cos\theta_1)t_{\theta_2}\kappa_{\theta_1} - 3d_2(\sin\theta_2\sin\theta_1)t_{\theta_2}\kappa_{\theta_2} - 3d_2(\sin\theta_2\sin\theta_1)t_{\theta_1}\kappa_{\theta_1} \\ & + 3d_2(\cos\theta_2\cos\theta_1)t_{\theta_1}\kappa_{\theta_2} - 3d_2(\cos\theta_2\sin\theta_1)t_{\theta_2}t_{\theta_1}^2 - d_2(\cos\theta_2\sin\theta_1)t_{\theta_2}^3 \\ & - d_2(\sin\theta_2\cos\theta_1)t_{\theta_1}^3 - 3d_2(\sin\theta_2\cos\theta_1)t_{\theta_1}t_{\theta_2}^2\end{aligned} \tag{43}$$

$$\begin{aligned}\frac{d\kappa_y}{ds} \equiv \quad & -\frac{d\kappa_{py}}{ds} + d_2(\sin\theta_2 \sin\theta_1)\frac{d\kappa_{\theta_1}}{ds} - d_2(\cos\theta_2\cos\theta_1)\frac{d\kappa_{\theta_2}}{ds} \\ & + 3d_2(\cos\theta_2\sin\theta_1)t_{\theta_2}\kappa_{\theta_1} + 3d_2(\sin\theta_2\cos\theta_1)t_{\theta_2}\kappa_{\theta_2} + 3d_2(\sin\theta_2\cos\theta_1)t_{\theta_1}\kappa_{\theta_1} \\ & + 3d_2(\cos\theta_2\sin\theta_1)t_{\theta_1}\kappa_{\theta_2} + 3d_2(\cos\theta_2\cos\theta_1)t_{\theta_2}t_{\theta_1}^2 + d_2(\cos\theta_2\cos\theta_1)t_{\theta_2}^3 \\ & - d_2(\sin\theta_2\sin\theta_1)t_{\theta_1}^3 - 3d_2(\sin\theta_2\sin\theta_1)t_{\theta_1}t_{\theta_2}^2 \end{aligned} \tag{44}$$

$$\frac{d\kappa_z}{ds} \equiv \frac{d\kappa_{pz}}{ds} + d_2\sin\theta_2\frac{d\kappa_{\theta_2}}{ds} + 2d_2\cos\theta_2 t_{\theta_2}\kappa_{\theta_2} + d_2\cos\theta_2 t_{\theta_2}\kappa_{\theta_2} - d_2\sin\theta_2 t_{\theta_2}^3 \tag{45}$$

The equations of jerk from the rotational motion are derived from Eqs.(19) and (20) as

$$J_{\theta_1}(u) = t_{\theta_1}\frac{dA}{dt} + 3\kappa_{\theta_1}VA + \frac{d\kappa_{\theta_1}}{ds}V^3 \tag{46}$$

$$J_{\theta_2}(u) = t_{\theta_2}\frac{dA}{dt} + 3\kappa_{\theta_2}VA + \frac{d\kappa_{\theta_2}}{ds}V^3 \tag{47}$$

where $\frac{d\kappa_{\theta_1}}{ds}$ and $\frac{d\kappa_{\theta_2}}{ds}$ are as follows:

$$\begin{aligned}\frac{d\kappa_{\theta_1}}{ds} \equiv \quad & \overset{uuu}{\theta}_1(u)\frac{1}{[\sqrt{\overset{u2}{P_x} + \overset{u2}{P_y} + \overset{u2}{P_z}}]^3} - 3\,\overset{uu}{\theta}_1(u)\frac{(\overset{u}{P_x}\overset{uu}{P_x} + \overset{u}{P_y}\overset{uu}{P_y} + \overset{u}{P_z}\overset{uu}{P_z})}{[\sqrt{(\overset{u2}{P_x} + \overset{u2}{P_y} + \overset{u2}{P_z})}]^5} - \overset{u}{\theta}_1(u) \\ & \frac{(\overset{u}{P_x}\overset{uuu}{P_x} + \overset{u}{P_y}\overset{uuu}{P_y} + \overset{u}{P_z}\overset{uuu}{P_z} + \overset{uu2}{P_x} + \overset{uu2}{P_y} + \overset{uu2}{P_z})}{[\sqrt{(\overset{u2}{P_x} + \overset{u2}{P_y} + \overset{u2}{P_z})}]^5} + 4\,\overset{u}{\theta}_1(u)\frac{(\overset{u}{P_x}\overset{uu}{P_x} + \overset{u}{P_y}\overset{uu}{P_y} + \overset{u}{P_z}\overset{uu}{P_z})^2}{[\sqrt{(\overset{u2}{P_x} + \overset{u2}{P_y} + \overset{u2}{P_z})}]^7}\end{aligned} \tag{48}$$

$$\begin{aligned}\frac{d\kappa_{\theta_2}}{ds} \equiv \quad & \overset{uuu}{\theta}_2(u)\frac{1}{[\sqrt{\overset{u2}{P_x} + \overset{u2}{P_y} + \overset{u2}{P_z}}]^3} - 3\,\overset{uu}{\theta}_2(u)\frac{(\overset{u}{P_x}\overset{uu}{P_x} + \overset{u}{P_y}\overset{uu}{P_y} + \overset{u}{P_z}\overset{uu}{P_z})}{[\sqrt{(\overset{u2}{P_x} + \overset{u2}{P_y} + \overset{u2}{P_z})}]^5} - \overset{u}{\theta}_2(u) \\ & \frac{(\overset{u}{P_x}\overset{uuu}{P_x} + \overset{u}{P_y}\overset{uuu}{P_y} + \overset{u}{P_z}\overset{uuu}{P_z} + \overset{uu2}{P_x} + \overset{uu2}{P_y} + \overset{uu2}{P_z})}{[\sqrt{(\overset{u2}{P_x} + \overset{u2}{P_y} + \overset{u2}{P_z})}]^5} + 4\,\overset{u}{\theta}_2(u)\frac{(\overset{u}{P_x}\overset{uu}{P_x} + \overset{u}{P_y}\overset{uu}{P_y} + \overset{u}{P_z}\overset{uu}{P_z})^2}{[\sqrt{(\overset{u2}{P_x} + \overset{u2}{P_y} + \overset{u2}{P_z})}]^7}\end{aligned} \tag{49}$$

8. ILLUSTRATIVE EXAMPLE

An example is given here for the illustration of the theory developed in this investigation.

1. Description of the problem

A five-axis fixed-bed type CNC machine (shown in Fig. 2) is asked to cut a path defined by the following parametric equations:

$$\begin{aligned}P_x(u) &= 0.2881\,u^3 - 0.3841\,u^2 + 0.3841\,u + 0.1150\ m \\ P_y(u) &= -1.6935\,u^2 + 1.6935\,u + 0.0976\ m \\ P_z(u) &= -0.1481\,u^3 - 0.2722\,u^2 - 0.2772\,u - 0.1110\ m\end{aligned}$$

The path can be originally as this analytical form or as defined by a sequence of discrete positions and then parametrized by applying spline interpolation. For simplicity, the corresponding orientations are only defined at u=0, 0.5 and 1 as (−0.7276, −0.4851,0.4851), (0.2887, −0.4081, 0.8661) and (−0.6860, −0.5145, 0.5145) respectively. This cutter path is depicted in Fig. 3. Let the lengths of both links AB and BC be 0.2 m.

A software package is generated based on the formulas derived in this investigation. Giving below are the results from solving the problem at hand.

2. Desired displacement for coordinated motion

The desired displacement of each axis for coordinated motion is shown in Fig. 4.

3. **Desired velocity for coordinated motion**

   Let the feedrate be 0.03 m/sec, to exactly follow the path, each axis of the machine should move in the coordinated manner. The desired velocity of x, y, z axes and joints 1 and 2 for this coordinated motion are plotted in Fig. 5.

4. **Desired acceleration for coordinated motion**

   Coordinated velocity is originated from coordinated acceleration. Figure 6 shows the acceleration components of each axis for the coordinated motion

5. **Conversion between parameter and time**

   The relationship between parameter u and time t can be obtained by Eq.(39). Figure 7 shows this relationship.

6. **Command generation**

   The command generation for the position control of coordinated motion is shown in Fig. 8 of which horizontal axis represents time instead of parameter. Similarly, the command generations for the velocity and acceleration control are shown in Fig. 9 and Fig. 10, respectively.

7. **Resultant jerk in coordinated motion**

   The resultant jerks in coordinated motion are shown in Fig. 11.

9. CONCLUSION

This paper presents an analytical study on the command generation of five axis CNC/CMM machines to track paths which have both positions and approach orientations stored in CAD systems. The formulas derived relate the geometric properties of curves including, position, tangent and curvature to the kinematics properties of the coordinated motion including displacement, velocity, acceleration, and jerk. The result is a general one and is applicable to all types of space curves. We believe that this work should be useful in the integration of automatic CAD/CAM processes.

ACKNOWLEDGMENT

The financial support of IMAR/NSF is greatly appreciated. IMAR stands for the Institute for Manufacturing and Automation Research which is financially supported by some local companies in the great Los Angeles area and by the government of the state of California.

References

[1] Papaioannou, S. G., "Interpolation Algorithms for Numerical Control", Computers in Industry, Vol. 1, 1979, pp. 27-40.

[2] Koren, Y., "Interpolator for a Computer Numerical Control System", IEEE Trans. on Computers, Vol. C-25, No. 1, Jan. 1976, pp. 32-37.

[3] Koren, Y., "Design of Computer Control for Manufacturing Systems", Trans. of ASME, J. of Engineering for Industry, Vol. 101, Aug. 1979, pp. 326-332.

[4] Koren, Y., and O. Masory, "Reference-Pulse Circular Interpolators for CNC Systems", Trans. of ASME, J. of Engineering for Industry, Vol. 103, Feb. 1981, pp. 131-136.

[5] Masory, O., and Y. Koren, "Reference-Word Circular Interpolators for CNC Systems", Trans. of ASME, J. of Engineering for Industry, Vol. 104, Nov. 1982, pp. 400-405.

[6] Weck, M., Handbook of Machine Tools, Vol. 3, John Wiley & Sons Co., 1984.

[7] Lin C. S., P. R. Chang, and J. Y. S. Luh "Formulation and Optimization of Cubic Polynomial Joint Trajectories for Industrial Robots" IEEE Trans. on Automatic Control, Vol. AC-28, No. 12, Dec. 1983, pp. 1066-1074.

[8] Hollerbach, J. M., "Dynamic Scaling of Manipulator Trajectories" ASME Trans., Journal of Dynamic Systems, Measurement, and Control, Vol. 106, March 1984, pp. 102-106.

[9] Shin G. Kang, and Neil D. Mckay, "Minimum-time Control of Robotic Manipulators with Geometric Path Controls", IEEE Trans. on Automatic Control, Vol. AC-30, No. 6, June 1985, pp. 531-541.

[10] Bobrow J. E., S. Dubowsky, and J. S. Gibson "Time-optimal Control of Robotic Manipulators along Specified Paths" The International J. of Robotics Research, Vol. 4, No. 3, Fall 1985, pp.

[11] Chou, J. J., and D. C. H. Yang., "Command Generation For Three-Axis CNC Machining", in Proceedings of Advances in Manufacturing System Engineering, ASME Winter Annual Meeting, San Francisco, CA, pp. 29-37, Dec. 1989; to appear in the Trans. of ASME , Journal of Engineering for Industry.

[12] Millman S. Richard, and George D. Parker, Elements of Differential Geometry, Prentice-Hall Inc., 1977.

[13] Mortenson E. M, Geometric Modeling, John Wiley & Sons Inc., 1985.

[14] Chou, J. J., "Coordinated Motion and Optimal Speed of Multi-Axis CNC/CMM Machines", Ph.D Dissertation, University of California, Los Angeles, 1990.

[15] Bollinger, J. G., and N. A. Duffie, Computer Control of Machines and Processes, Addison-Wesley Publishing Co., 1988.

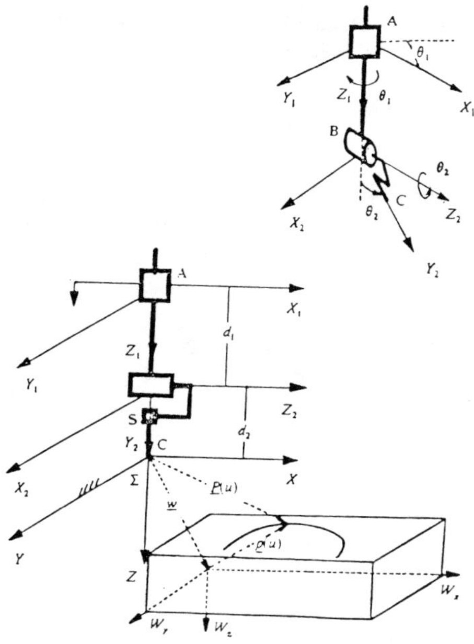

Fig.1: Euler-Angle Rotation Structure

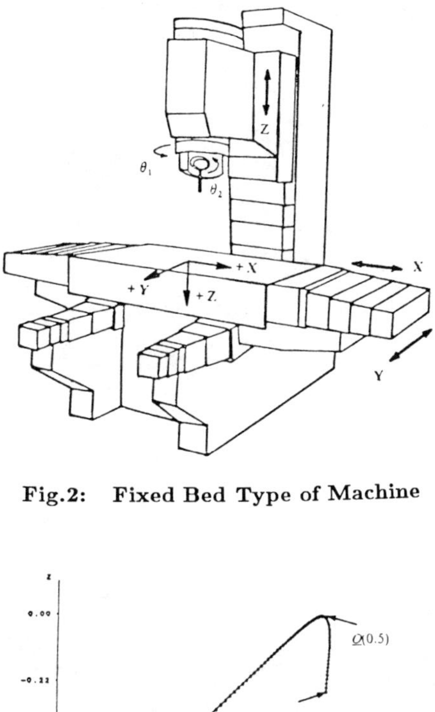

Fig.2: Fixed Bed Type of Machine

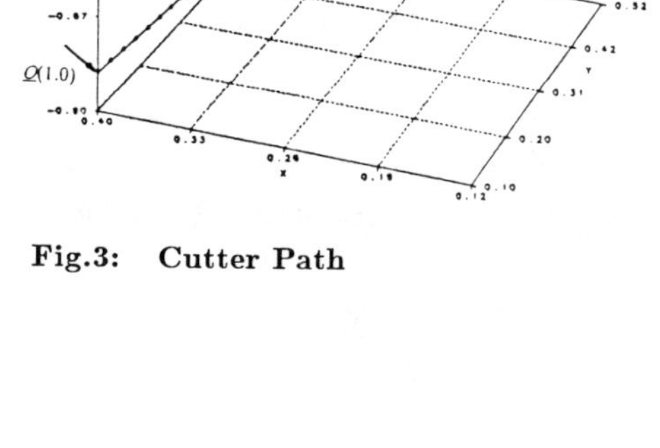

Fig.3: Cutter Path

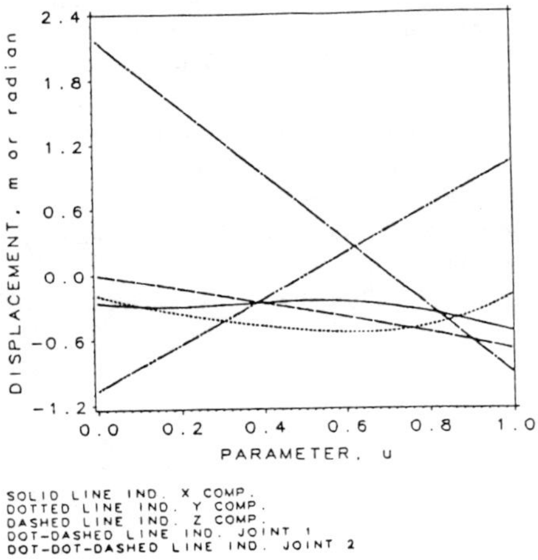

Fig.4: Desired Displacement for Coordinated Motion

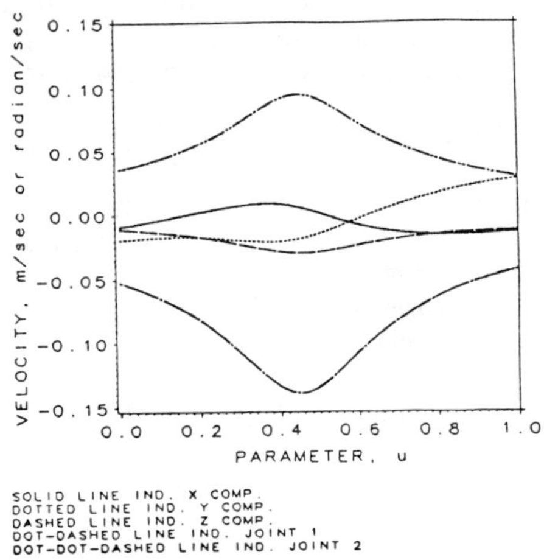

Fig.5: Desired Velocity for Coordinated Motion

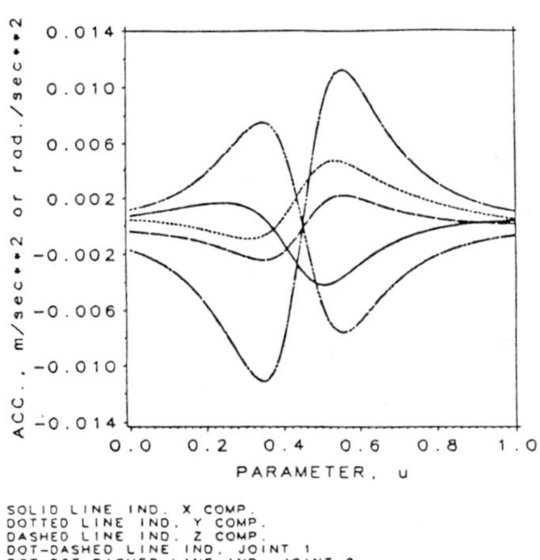

Fig.6: Desired Acceleration for Coordinated Motion

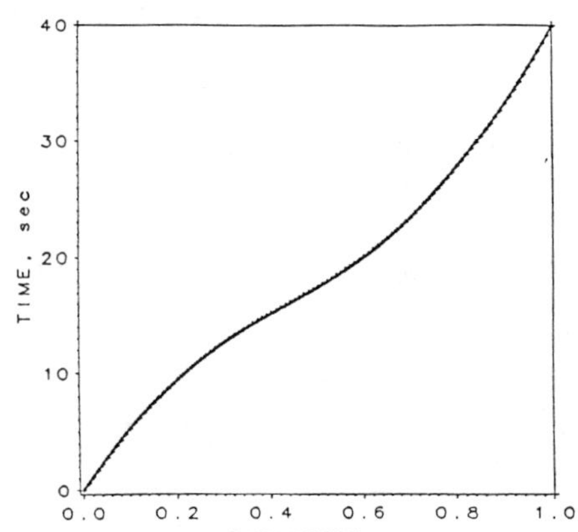

Fig.7: The Relationship Between Time and Parameter

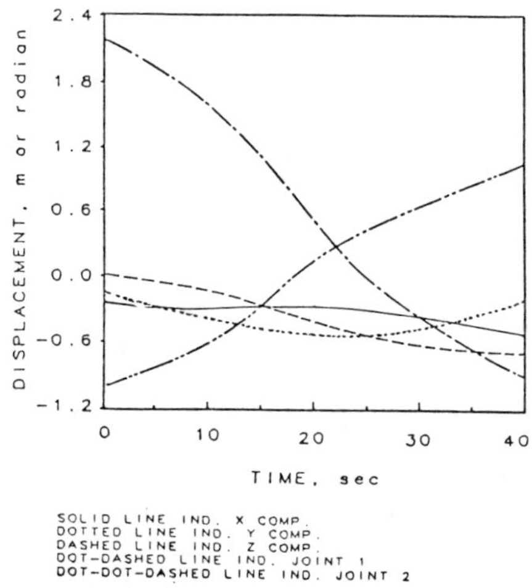

Fig.8: Displacement Curves for Command Generation

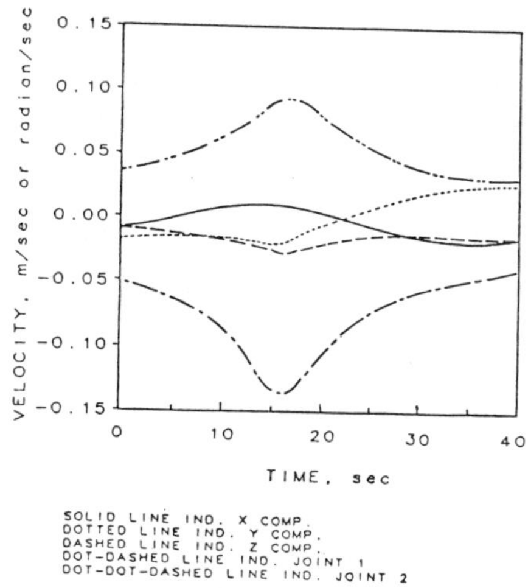

Fig.9: Velocity Curves for Command Generation

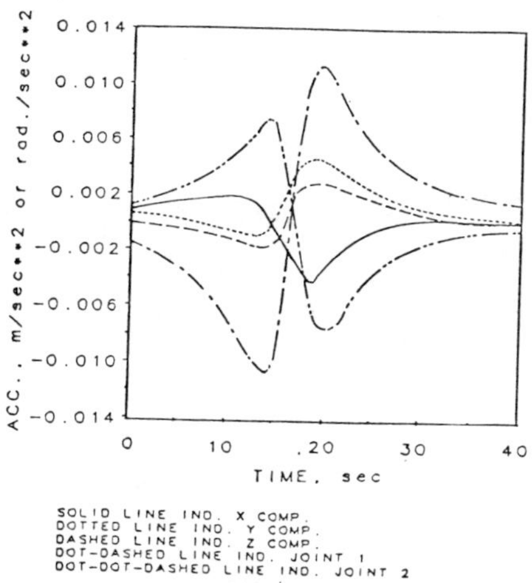

Fig.10: Acceleration Curves for Command Generation

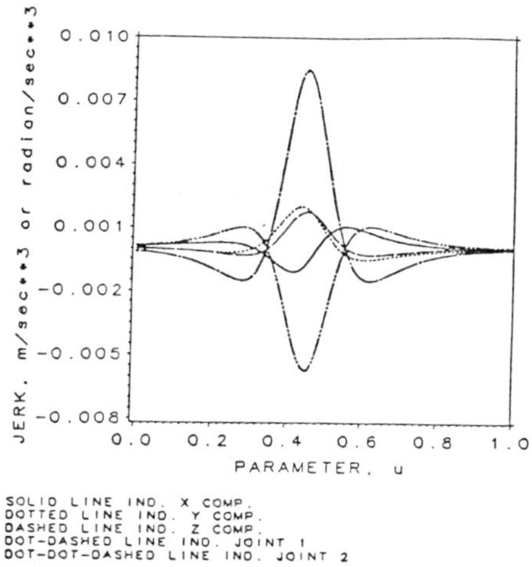

Fig.11: Resultant Jerk in Coordinated Motion

Table 1: **Summary of Command Generation for Coordinated Motion**

Summary of Command Generation for Coordinated Motion		
Command Generation	**Equations for Coord. Motion (Fct. of u)**	**Equation for Rel. bet. u and t**
For Position Control		
$[x,\, y,\, z] =$	$[Eq.(23)]$	
$[\theta_1,\, \theta_2] =$	$[Eq.(14)]$	
For Velocity Control		
$[\dot{x},\, \dot{y},\, \dot{z}] =$	$[Eq.(24),\ Eq.(25),\ Eq.(26)]$	$Eq.(39)$
$[\dot{\theta}_1,\, \dot{\theta}_2] =$	$[Eq.(17),\ Eq.(18)]$	
For Desired Acceleration		
$[\ddot{x},\, \ddot{y},\, \ddot{z}] =$	$[Eq.(30),\ Eq.(31),\ Eq.(32)]$	
$[\ddot{\theta}_1,\, \ddot{\theta}_2] =$	$[Eq.(19),\ Eq.(20)]$	

Sensors, Controls, and Quality Issues in Manufacturing
ASME 1991

LEARNING CONTROL BASED ON GENERALIZED SECANT METHOD AND OTHER NUMERICAL OPTIMIZATION METHODS

Homayoon S. M. Beigi
International Business Machines
Thomas J. Watson Research Center
Yorktown Heights, New York

C. James Li and R. W. Longman
Department of Mechanical Engineering
Columbia University
New York, New York

ABSTRACT

When control systems are given the same task to perform repeatedly, they will usually repeat the same errors in executing the command, except for certain random noise effects. The field of learning control has developed to produce controllers that can improve their performance at a given task with each repetition of the task. To date, learning control algorithms have concentrated on use of analogues of integral control to eliminate errors, and recently on conversions of adaptive control to the learning control problem. This paper studies an alternative of using numerical optimization techniques to determine ways to improve controller performance from one repetition to the next. Quasi-Newton methods, the direction set method, certain derivative free algorithms and the generalized secant method are studied. It is shown analytically that the generalized secant method converges to zero tracking error in fewer repetitions than other methods aimed at direct minimization of the Euclidean norm of the error. Simulations show good results when applied to the control of nonlinear systems that include a pendulum and a non-linear mass-spring-dashpot system.

I. INTRODUCTION

A large percentage of the practical applications of control theory involve systems which are repeatedly asked to perform the same task. Examples include tracking problems for robots on assembly lines, as well as a large number of manufacturing applications. Standard controller design methods produce systems that do an imperfect job of executing a command -- and when the command is repeated the system repeats these same errors everytime. It is perhaps a bit primitive to persist in repeating the same errors. And in the last few years a theory of learning control has been developed in the literature, generating controllers that can learn from previous experience at performing a specific task, see for example [1-7]. The fact that a specific task is involved repeatedly, distinguishes the learning control problem from the adaptive control problem.

To date, the learning control literature has been based on either analogues of integral control in the repetition domain [1-4], or more recently adaptive control ideas have been converted to apply to the learning control problem in [5-7]. This paper is devoted to a preliminary investigation of a third option -- the use of numerical optimization techniques as a method of accomplishing learning in repetitive tasks.

The approach bears a certain philosophical similarity to a few optimal control investigations that use the real world rather than a computer model to evaluate each iteration in the attempt to converge on the optimal control. Such an approach can avoid the sometimes lengthy process of obtaining a good model before the optimization problem can start. Reference [3] develops such an idea for minimizing the integral of the absolute value of the error in continuous systems.

Here, we consider the learning control problem for discrete systems, a system of linear algebraic equations relating the change in the state of the system to the change in the control action from one repetition of the task to another. From this set of linear equations which gives the transition between any two repetitions, we formulate an optimization function which is quadratic in the error. We study numerical optimization methods to determine how to adjust the control history from one repetition to the next in order to decrease this quadratic function of the error as the repetitions of the task proceed.

Furthermore, we utilize a generalized secant method to solve the system of linear algebraic equations directly for the appropriate control action and parameter estimates that will minimize the tracking error of the controlled dynamic system, without any prior knowledge of the system parameters except the order of the system. It is shown analytically that the generalized secant method requires the least number of repetitions to converge among all the methods given in this paper. Hence, simulations were conducted to study practical issues of this controller.

II. PROBLEM FORMULATION

Consider a general discrete-time linear time-varying or time-invariant system,

$$x^k(t+1) = A(t)\,x^k(t) + B(t)\,u^k(t) + w^k(t) \tag{1a}$$

$$y^k(t+1) = C(t+1)\,x^k(t+1) \qquad t = 0, 1, 2, \ldots, p-1 \qquad k = 0, 1, 2, \ldots \tag{1b}$$

where the state, x, is n-dimensional, the control, u, is m-dimensional, the output,y, is q-dimensional ($q \geq m$), t is the time step in the p-step repetitive operation, and k is the repetition number. For simplicity, the dimension n of the system is assumed known, but generalization to just knowing an upper bound on n is easily considered. Also, A, B, C, and w^k are assumed unknown -- otherwise one could determine in advance what control to use to minimize tracking error, and there would be no need for learning control. In the learning control problem (as contrasted with the repetitive control problem in [2]) the system is assumed to always start from the same initial state in each repetition of the task. Matrix A includes any state or output feedback control present in the system and the symbol u^k is reserved for the signal added to the control for learning purposes. A time varying model is considered because many applications, such as in robotics, involve nonlinear dynamic systems which when linearized produce linear models with coefficients that vary with time, and vary in the same manner each repetition. In such repetitive operations it is often the case that there will be disturbances $w^k(t)$ that repeat with each repetition of the task, and learning can be made to also correct for this source of errors in a natural way. We will limit ourselves to repetitive $w^k(t)$. Therefore the superscript k will be dropped. One of the purposes of the feedback control is to handle any non-repetitive disturbances, and these will be ignored for purposes of designing the learning control (some analysis of these random noise effects in learning control can be found in [4]).

The solution to (1) for the p-steps of the repetitive process can be written as

$$x^k(t+1) = \Big(\prod_{j=0}^{t} A(j)\Big)x^k(0) + \sum_{j=0}^{t}\Big(\Big(\prod_{r=j+1}^{t} A(r)\Big)(B(j)\,u^k(j) + w(j))\Big) \tag{2}$$

where the product symbol is taken to give the identity matrix if the lower limit is larger than the upper limit. Let p be the total number of time steps. Defining a difference operator $\delta_k z(t) = z^k(t) - z^0(t)$ for any variable z, and using the fact that $x^k(0)$ and $w(t)$ are repetitive, one can write,

$$y^k = y^0 + P\,\delta_k u \tag{3}$$

where,

$$y^k = [y^{kT}(1)\;\; y^{kT}(2)\;\; \ldots\;\; y^{kT}(p)]^T$$

$$u^k = [u^{kT}(0)\;\; u^{kT}(1)\;\; \ldots\;\; u^{kT}(p-1)]^T$$

$$P=\begin{bmatrix} C(1)B(0) & 0 & \cdots & 0 \\ C(2)A(1)B(0) & C(2)B(1) & & 0 \\ & & \cdots & \cdot \\ C(p)A(P-1)A(P-2)\ldots A(1)B(0) & \cdots & \cdots & C(p)B(p-1) \end{bmatrix}$$

The objective is to find a change in the control at each repetition k that will, as k progresses, minimize the quadratic error function

$$f(\delta_k u) = (y^k - y_D)^T (y^k - y_D) \tag{4}$$

where y_D is the desired output history for the p-step process. In previous work on learning control in discrete systems [4-7], as well as in the extensions of these papers for journal publication, attention had to be directed to the problem of ensuring that the special desired trajectory was in fact a feasible trajectory for the system to perform. For example, one method involved assuming knowledge of an upper bound on the system dimension n, and assuming the system is controllable. Then specifying the desired measured output variable history every n steps is guaranteed by modern control theory to be a feasible specification. In the present approach these conditions can be relaxed if one is satisfied with minimizing (4) rather than insisting that (4) be driven to zero.

Substitute (3) into (4),

$$\begin{aligned} f(\delta_k u) &= (y^0 + P\,\delta_k u - y_D)^T (y^0 + P\,\delta_k u - y_D) \\ &= \delta_k u^T (P^T P)\,\delta_k u + 2\,\delta_k u^T [P^T (y^0 - y_D)] + (y^0 - y_D)^T (y^0 - y_D) \end{aligned} \tag{5}$$

Note that the gradient of this objective function and the Hessian matrix of second partials are,

$$\frac{df}{d(\delta_k u)} = 2P^T P\,\delta_k u + 2P^T (y^0 - y_D) \tag{6}$$

$$\left[\frac{d^2 f}{d(\delta_k u)^2}\right] = 2P^T P \tag{7}$$

It is now possible to characterize the nonlinear optimization problem represented by our learning control objective. We wish to

1. minimize a function which is known to be quadratic in the control change variable $\delta_k u$,
2. but the first derivative of this function is not directly available, because we do not know the system matrices A, B, C, and hence do not know P,
3. and the second derivative is similarly unavailable.

Having characterized the problem, we present an overview and assessments of the possible approaches to the learning control problem using numerical optimizing method.

III. LEARNING CONTROL BY NUMERICAL OPTIMIZATION METHODS

3.1 Quasi-Newton Methods

Over the last couple of decades various finely tuned quasi-newton based methods have been generated for the minimization of unconstrained nonlinear functions. Among these methods are the rank 1 Broyden method, DFP, BFGS, and other members of the Broyden family of methods as well as SSVM methods. An important characteristic of these methods is the use of the iterative process to approximate the Hessian matrix so that no explicit expression for the second derivative is needed. If we could evaluate the first derivative $df/d(\delta_k u)$, then the BFGS method would converge to the optimal $\delta_k u$ in approximately mp repetitions. One might note that if one actually knows the $df/d(\delta_k u)$ of (6), then one could equate it to zero and solve for $\delta_k u$ and obtain the optimal $\delta_k u$ in one step -- something which could be classified as a Newton method in numerical optimization.

Since we do not know $df/d(\delta_k u)$, one can consider the use of Quasi-Newton methods with approximation of the gradient obtained by a finite difference method. In order to obtain these differences for all of the mp elements of the gradient vector it would require approximately mp repetitions of the task to make one evaluation of the gradient vector. The total number of repetitions required for convergence would exceed $(mp)^2$. This makes such methods a poor choice in learning control. Methods that do not require the use of a gradient in picking the search direction are called for. Note that the same difficulty eliminates the use of steepest descent and conjugate gradient methods.

3.2 A Direct Search Method

A method which does not require knowledge of gradients, is the direct search method of optimization as expressed by Wood [9] and later by Hooke and Jeeves [10]. The direct search method, which takes advantage of the quadratic nature of the objective function, is as follows:

1. Pick mp orthogonal directions in the $\delta_k u$ space. For each of these directions in succession, perform the line search as in the next steps.
2. Take a step along the chosen direction in $\delta_k u$ space, and apply the resulting control in the next repetition. Evaluate f from the data of this repetition.
3. If f decreased pick another step in the same direction, if it increased pick a step in the opposite direction, and apply to the system.
4. Since the quadratic objective function surface in the plane of the chosen direction is a parabola, the data from 2 and 3 determines this parabola, and can be used to find the minimizing $\delta_k u$ for this direction. This completes the line search.
5. Return to 2 with the next direction.

This algorithm will improve the tracking every three repetitions, provided there is no noise in the measurements. There is no guarantee for a finite convergence unless the directions of search are mutually conjugate about the Hessian matrix of the quadratic objective function. Care must be taken to avoid large disturbances to the repetitive process during the line search.

3.3 Conjugate Direction Methods

If the unidimensional search of the last section were used along mp mutually conjugate directions, then in the absence of noise, the system would converge after minimizing the quadratic objective function in all these directions. There are a few conjugate direction methods available in the literature [11-13]. These methods start with a set of mp orthonormal directions which could be coincident with the columns of an mpxmp identity matrix. Then, by doing unidimensional searches in these directions, a new set of directions is generated.

Rosenbrock's [11] method provides a direction coincident with one of the eigenvectors of the Hessian (H) matrix. Eigenvectors of H are mutually conjugate and point in the direction of the minimum of the quadratic objective function. One such eigenvector is reached after mp unidimensional searches along the orthonormal directions. The rest of the mp-1 directions are found by using the Gram-Schmidt [14] orthogonalization procedure. This cycle is then repeated to provide another eigenvector of H. There is no guarantee that the eigenvectors are not repeated. Therefore, there is no general finite convergence proof available.

Rosenbrock's method might converge to a non-minimum point in which case the method breaks down. To avoid or delay the failure of the minimization procedure, the method due to Davies, Swann, and Campey [12] (DSC) uses a renumbering system for the directions of search. Another method due to Powell [13] provides a direction set with one of the directions conjugate to the complementary mp-1 directions. This set is obtained after a cycle similar to that of Rosenbrock and DSC. However, Powell's method does not require any orthogonalization procedure after every cycle and it never breaks down as Rosenbrock and DSC methods do.

None of the above methods guarantees a finite convergence since the main direction after each cycle might be linearly dependent on directions obtained in previous cycles. However, if the set of directions are linearly independent, then the function minimization will converge within mp(2mp+1) repetitions. This applies to Powell's second method.

3.4 An Identification Test Method

It is always possible to identify a system from impulse response tests. In this section we study the use of such tests, and then in the next section we will study how to accomplish improvement with each repetition. If data is taken for a sequence of repetitions (j=0,1,2,...,N) then eq. (3) gives

$$\left[\delta_N y \; \delta_{N-1} y \ldots \delta_1 y\right] = P\left[\delta_N u \; \delta_{N-1} u \ldots \delta_1 u\right] \tag{8}$$

Once N is sufficiently large, this can be solved for P by use of an inverse or pseudo-inverse. If one makes each $\delta_j \mathbf{u}$ a different column of the identity matrix, which is the discrete equivalent of an impulse response test, then one can eliminate the need for inverting a matrix. In other cases, one can take advantage of the lower block triangular nature of P to efficiently compute the inverse. The noise level can be important in deciding the best approach. Once P is determined, then solving (6) gives

$$\delta_{j+1} u = P^{\dagger} (y^0 - y_D) \tag{9}$$

where the superscript † indicates the Moore-Penrose pseudoinverse.

In the case that the desired trajectory is feasible for system 1, and the desired trajectory exactly specifies all the freedom in the system, then the solution to (9) is the same as solving (3) for $\mathbf{v}_j$ to obtain $\mathbf{v}_j = P^{-1} (y^j - y_D)$. This approach requires mp+1 steps to optimize the tracking in the no noise case.

A Modification to Obtain Improvement Each Repetition: In practice one usually has a model of the system, call it P^m, when one starts the repetitive control, although one may not be very confident about the model. As the data from each repetition arrives one could modify the model to match the data, using the smallest possible change in the elements in the least squares sense. At step j, one would choose ΔP to satisfy

$$\left[\delta_j y \; \delta_{j-1} y \ldots \delta_1 y\right] = \left[P^m + \Delta P\right] \left[\delta_j u \; \delta_{j-1} u \ldots \delta_1 u\right]$$

or

$$\Delta P = \left\{ \left[\delta_j y \; \delta_{j-1} y \ldots \delta_1 y\right] - P^m \left[\delta_j u \; \delta_{j-1} u \ldots \delta_1 u\right] \right\} \left[\delta_j u \; \delta_{j-1} u \ldots \delta_1 u\right]^{\dagger} \tag{10}$$

where the superscript † indicates the Moore-Penrose pseudoinverse. One can show that this pseudoinverse will minimize $tr(\Delta P^T \Delta P)$, i.e. the sum of the squares of the changes of all elements of ΔP.

The method becomes:

1. Pick a direction for $\delta_1 \mathbf{u}$ and make the first repetition to obtain the $\delta_1 \mathbf{y}$, $\delta_1 \mathbf{u}$ pair.
2. Solve for ΔP from (10).
3. Obtain $\delta_{j+1} \mathbf{u}$ from (9) substituting $P^m + \Delta P$ for P. The singular value decomposition of $[\delta_{j+1} \mathbf{u} \; \ldots \; \delta_1 \mathbf{u}]$ involved in the pseudo-inverse will determine if $\delta_{j+1} \mathbf{u}$ is linearly independent. In the event that it is not, choose a linearly independent direction at this step.
4. Repeat steps 2 and 3 until convergence is obtained.

This approach takes full advantage of a priori information about the system, improves the control system behavior with every repetition, handles noise in a natural way that smooths its influence, handles in a logical manner specified desired trajectories that are not feasible for the system to perform, and in the no noise case produces convergence in mp+1 repetitions or less. The effectiveness of the method by comparison to the direction set method above is fundamentally related to the fact that the quadratic cost (5) is not a general quadratic function, but rather the linear and the quadratic matrix coefficients P^T and P^TP are related, and this extra information has been incorporated in the new algorithm. Rather than trying to identify the quadratic surface of f as in the direction set method, the linear relation (3) from which the quadratic is derived is identified, with a resulting savings in repetitions. A numerical optimization method with this same property is presented in the next section.

The above method involves identification of the system model in the repetition domain, which by analogy to learning control can be called learning identification. Various approaches to identification in the repetition domain are presented in [7,15,16], for both time-varying and time-invariant cases.

When the system is time-invariant one can accelerate the learning control convergence by using the identification of elements of P to compute the time domain description A, B, C, from which one can complete the elements of P before enough data has been taken to obtain all elements directly. The algorithm is as follows:

1. While applying the above algorithm, use the lower block triangular nature of P to determine successively CB, CAB, CA^2B, etc.

2. Form the Hankel matrix of these Markov parameters for the appropriate dimension, if one knows the order n of the system, for any repetition number large enough to have generated $CA^{2n-1}B$. Later repetitions allow more smoothing. If n is not known, monitor the rank of the Hankel matrix with repetitions.

3. Use the Eigensystem Realization Algorithm (see for example [17], or use the Ho Kalman algorithm) to obtain A, B, and C. Note that one does not need to identify w(t) of the time domain model.

4. Use A, B, and C to construct the full P matrix.

5. Find the optimizing control using this in (9).

Note that if the full state is measured, then the Eigensystem Realization Algorithm can be bypassed, by simply identifying A and B using a pseudoinverse solution of the sequence of eq. (1) for succeeding steps, after eliminating w(t) through differencing.

IV. THE GENERALIZED SECANT METHOD

Define the error, e^k, as the difference between the output y^k and the desired trajectory y_D,

$$e^k = y^k - y_D \tag{11}$$

Define the change of input between two consecutive repetitions as

$$v^k = u^{k+1} - u^k \tag{12}$$

Then from eq. (3), one could obtain

$$Pv^k = e^{k+1} - e^k \tag{13}$$

Since our goal is to reduce the tracking error, e_k, as the task is being repeated, we aim at finding a change in our input which would lead to zero tracking error at the next repetition, $e^{k+1}=0$. For that input change, eq. (13) becomes

$$P\tilde{v}^k = -e^k \tag{14}$$

Equation (14) suggests that the change of input, v^k, can be deduced from P and e^k. Since P is not available, an approximation to P at the kth repetition, P^k, has to be used. Hence

$$P^k v^k = -e^k \tag{15}$$

To obtain v^k, the Moore-Penrose pseudo-inverse can be used to minimize the following norm,

$$\| p^k v^k + e^k \|$$

namely,

$$v^k = -P^{k\dagger} e^k \tag{16}$$

Hence, the control action at the (k+1)st repetition, u^{k+1}, can be calculated according to eq. (12), $u^{k+1} = u^k + v^k$. In the following, the method for finding P^k will be described.

At repetition k, denote the difference between the actual system parameters P and the approximation P^k by the matrix D^k.

$$P = P^k + D^k \tag{17}$$

Substituting eq. (17) into (13) and rearranging,

$$D^k v^k = e^{k+1} - e^k - P^k v^k \tag{18}$$

Since D^k is a qp by mp matrix and the other terms in eq. (18) are vectors, one can not directly solve the complete D^k from eq. (18).

A "solution" that would solve D^k along a direction defined by z^k is,

$$D^k = \frac{(e^{k+1} - e^k - P^k v^k)\, z^{kT}}{z^{kT} v^k} \tag{19}$$

while z^k are chosen in the following way[18]:

- If $k \geq n-1$ then z^k is chosen orthogonal to the previous n-1 control steps $v^{k-n-1}, \ldots, v^{k-1}$.
- If $k < n-1$ then z^k is chosen orthogonal to the available k steps, $v^0, \ldots, v^{k-1}$, where $n \leq mp$ is the maximum number of linearly independent v's which could be found.

A few orthogonalization methods are available in the literature which may be used for the actual evaluation of the z vectors. These include the well-known Gram-Schmidt orthogonalization process[14] and a more advanced technique due to Fletcher [19] in which the number of vectors in the set may be changed.

Together with an orthogonalization method, eqs. (16), (17), and (19) form a learning control algorithm,

$$v^k = -P^{k\dagger} e^k \tag{20}$$

$$P^{k+1} = P^k + D^k$$

$$= P^k + \frac{(e^{k+1} - e^k - P^k v^k)\, z^{kT}}{z^{kT} v^k} \tag{21}$$

Proof of Convergence

The set of z^k picked by an orthogonalization method for eq. (21) has the property that $z^{kT} v^j = 0$ for $0 < k-j \leq n-1$. Consequently

$$D^k v^j = 0 \qquad 0 < k-j \leq n-1 \tag{22}$$

Therefore,

$$P^{k+1} v^j = [\,P^{j+1} + D^{j+1} + \ldots + D^k\,]\, v^j$$

$$= P^{j+1} v^j \qquad 0 < k-j \leq n-1 \tag{23}$$

Since $n \le mp$ is the maximum number of linearly independent v's, v^n has to be a linear combination of all previous n linearly independent control steps

$$v^n = \alpha_0 v^0 + \ldots + \alpha_{n-1} v^{n-1} = \sum_{j=0}^{n-1} \alpha_j v^j \tag{24}$$

Premultiply eq. (24) by P^n, one has

$$P^n v^n = \sum_{j=0}^{n-1} \alpha_j P^n v^j \tag{25}$$

Let k in (23) be n-1,

$$P^n v^j = P^{j+1} v^j = (P^j + D^j) v^j \qquad 0 \le j < n-1 \tag{26}$$

Substituting (18) into (26),

$$P^n v^j = e^{j+1} - e^j \tag{27}$$

According to (13), the right hand side of (27) must be equal to Pv^j. Hence

$$P^n v^j = P v^j \tag{28}$$

Substituting (28) into (25),

$$P^n v^n = P \sum_{j=0}^{n-1} \alpha_j v^j \tag{29}$$

According to (24), the summation in the right hand side of (29) is equal to v^n. Hence

$$P^n v^n = P v^n \tag{30}$$

Substituting (13) into (30),

$$\begin{aligned} P^n v^n &= P v^n \\ &= e^{n+1} - e^n \end{aligned} \tag{31}$$

If v^n satisfies (15), then,

$$e^{n+1} - e^n = -e^n \tag{32}$$

which means that e^{n+1} must be zero. However, if v^n only satisfies (15) in a least square sense, then,

$$e^{n+1} = P v^n + e^n \tag{33}$$

has minimum norm.

If n (the rank of P^0-P) is equal to mp, then after mp+1 repetitions of the task, a minimum norm tracking error will be achieved. (mp+1) is the maximum number before convergence. But if P^0 is a good guess, it can converge faster.

V. Simulation and Results

The learning control algorithm based on the generalized secant method was tested on a linear and two nonlinear dynamical systems. The first system is a linear mass-spring-dashpot system. The second system is a non-linear mass-spring-dashpot system and the third system is a pendulum with damping.

5.1. Governing Equations of Simulated Systems

System 1 is a linear Mass-Spring-Dashpot system. This system is described by the following differential equation

$$m\ddot{\alpha} + c\dot{\alpha} + k\alpha = T \tag{34}$$

where m, c, k are 1.0 kg, 1.0 Ns/m, 1.0 N/m.

System 2 is a nonlinear Mass-Spring-Dashpot system. The simulated system dynamics is given by the following differential equation:

$$m\ddot{\alpha} + c(1+\alpha^2)\dot{\alpha} + (k|\alpha|)\alpha = T \tag{35}$$

where m, c, k are the same as those of system 1.

System 3 is a one-degree-of-freedom pendulum of mass m, length l , angle from vertical α, and damping c. The dynamics of the simulated pendulum obeyed the following differential equation:

$$ml^2\ddot{\alpha} + c\dot{\alpha} + m g l \sin\alpha = T \tag{36}$$

where m, l, c are 1.0 kg, 0.1 m, 1.0 Ns/m.

The desired trajectory is given in Fig. 1. As shown in Fig. 1, the desired trajectory swing through two radians so that the system nonlinearity of system 3 can not be neglected.

5.3 Simulation Results

System 1 was first simulated to test the generalized secant controller. The results of this simulation are given in Fig. 2 which plots the sum of squares of the errors versus the repetition number. The first repetition was done using a self-tuning regulator with forgetting factor 1 (a complete theory of the self tuning regulator is given in [7]). The sum of squares of errors in this simulation converges to 1.08 and did not go to zero. The flat part of the curve is the result of control steps orthogonal to the direction of z_k, in which the P matrix is to be updated.

Barnes [18] proposes a practical stability criterion which translates to imposing a small positive lower limit to the inner product of the direction in which the P matrix is to be updated and the control step taken. Namely, a control step is checked before it is actually applied and if it does not satisfy the inequality (37) it will not be applied.

$$\frac{|z_k^T v_k|}{\|z_k\|_E \, \|v_k\|_E} > \rho \qquad 0 < \rho < 1 \tag{37}$$

This ensures the control step, v_k, is not going to be applied in a direction orthogonal to the direction of z_k, in which the P matrix is going to be updated. If a control step is rejected for not meeting (37), then a new step v_k should be found. The best choice for the new step is a step in the same direction as the P update direction, z_k, then the inequality of (37) is always satisfied.

Barnes [18] suggests taking the magnitude of the new step to be equal to that of the rejected step. This suggestion was shown to trigger instabilities in practice when applied to simulated nonlinear systems in this paper. This instability suggests that the size of the new steps ought to be small enough to keep the state of the controlled system from being driven too far away from the nominal desired trajectory. We find that a step size a little greater than the numerical accuracy of the computer usually works well.

The introduction of the checking of control steps worked well in reducing the tracking error to zero for system 1 (see Fig. 3). We then applied the controller to a set of more practical dynamic systems involving nonlinear dynamics such as those described in eqs. (35) and (36) which when linearized can be approximated by linear models with coefficients that vary with time step, and vary in the same manner each repetition. Figures 4 and 5 show the plots of the sums of squares of errors for system 2 and system 3. In both of these simulations the ρ in eq. (37) was set to 10^{-4}.

In application to the system 1 and system 2, zero tracking error was achieved after 8 repetitions which is less than the ceiling provided by the theory (10 repetitions). Similarly, the pendulum reached a zero trajectory error after only 9 repetitions which is again one repetition less than the ceiling provided by the theory.

The flat part of the curves in Figs. 3 through 5 are caused by rejecting of control steps. Since the amplitude of the new control steps replacing the rejected ones has to be kept very small, there is no significant change of state from one repetition to the next repetition. After a few repetitions involving the rejection of control steps, updates done to the P matrix bring it closer to the ultimate P and therefore a control action calculated from it usually leads to a significant improvement in tracking errors.

It should be noted that the theory in this paper was derived for control of a time-varying dynamic system while simulations were done on highly nonlinear systems. These simulations provide practical evidence on the robustness of this learning controller.

VI. CONCLUSION

This paper has given an overview of the possible numerical unconstrained optimization methods which might be used to generate learning control laws, and has identified the most applicable one among them. The lack of knowledge of the derivatives of the performance criterion constrains the range of the algorithms. It was shown that faster convergence can be obtained by identifying the system in the repetition domain and minimizing based on the current model, than can be obtained by a direct attempt to minimize the quadratic function of the error. The advantage comes from the fact that the quadratic function is not a general one, but one in which the coefficients are related. Knowledge of P is sufficient to determine all the coefficients of this function. The algorithms for minimizing the quadratic function do not take advantage of this special relationship whereas the generalized secant method constructs the coefficients from the knowledge of P. The fact that there are fewer unknown coefficients in P than there are in the quadratic cost function allows fast convergence of the generalized secant method.

Simulation results show that the learning controller based on secant method works well with some practical modifications such as the rejection of noninformative control steps.The generalized secant learning controller was shown to be very stable and robust in a practical sense when applied to the control of truly nonlinear systems.

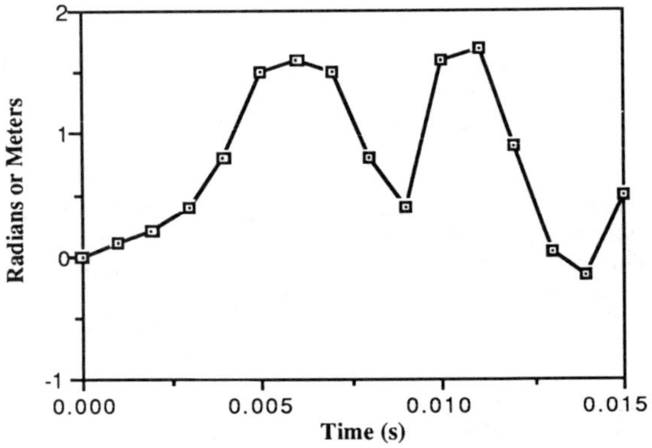

Figure 1 Desired Trajectory

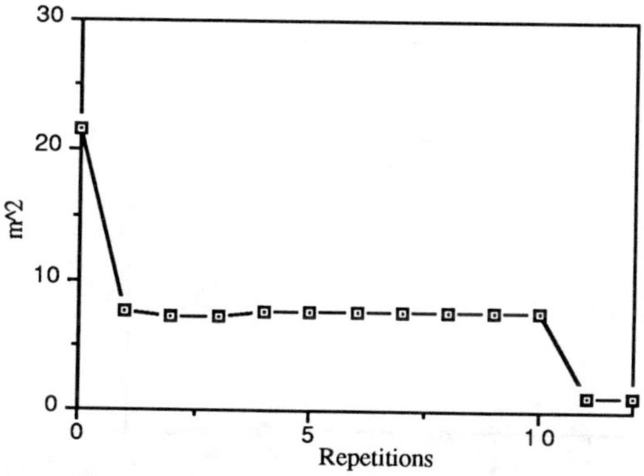

Figure 2 Squares of Errors for System 1
Using the Generalized secant Learning Controller without Rejections

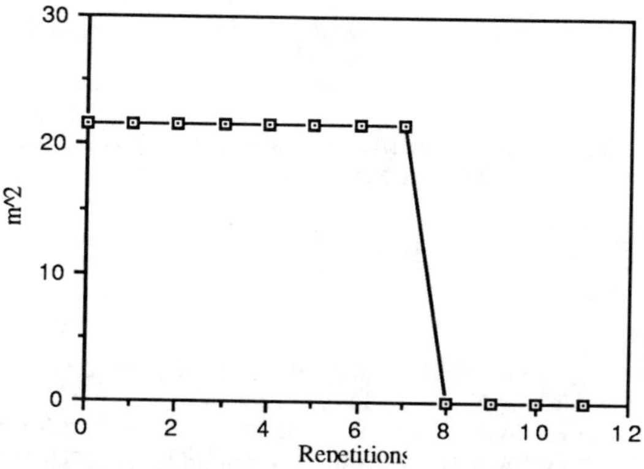

Figure 3 Squares of Errors for System 1
Using the Generalized secant Learning Controller with Rejections

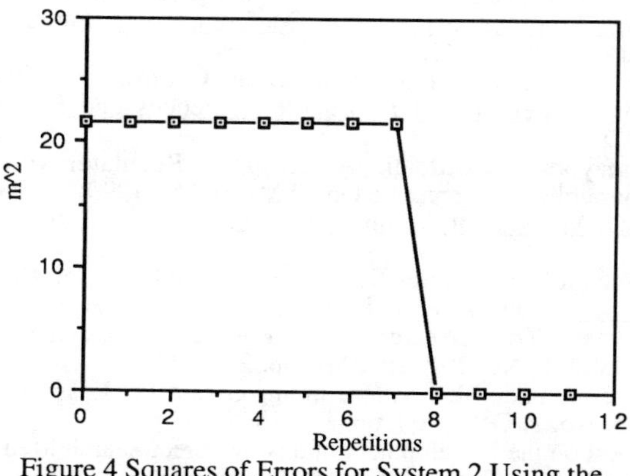

Figure 4 Squares of Errors for System 2 Using the
Generalized secant Learning Controller with Rejections)

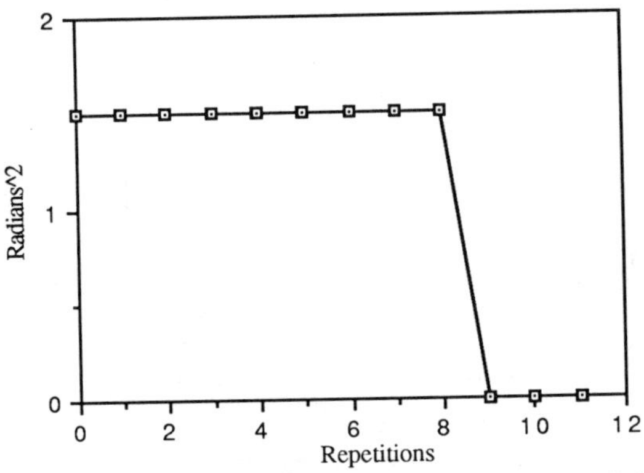

Figure 5 Squares of Errors for the Pendulum Using the Generalized secant Learning Controller with Rejections

Acknowledgment

The authors like to acknowledge the support from Department of Navy through Contract N00039-84-C-0165 P00019 and NASA through Grant NAG 1-649.

REFERENCES

1. S. Arimoto, S. Kawamura, and F. Miyazaki, "Bettering Operations of Robots by Learning," Journal of Robotic Systems, Vol. 1, No. 2, 1984, pp. 123-140.
2. R. H. Middleton, G. C. Goodwin, and R. W. Longman, "A Method for improving the Dynamic Accuracy of a Robot Performing a Repetitive Task," The International Journal of Robotics Research, Vol. 8, No. 5, October 1989, pp. 67-74.
3. E. G. Harokopos and R. W. Longman, "A Learning Controller for Tracking Arbitrary Inputs in Robot Motions," Proceeding of the Tenth IASTED International Symposium on Robotics and Automation, Lugano, Switzerland, June 29 - July 2, 1987.
4. M. Phan and R. W. Longman, "A Mathematical Theory of Learning Control for Linear Discrete Multivariable Systems," Proceedings of the AIAA/AAS Astrodynamics Specialist Conference, Minneapolis, Minnesota, August 1988.
5. M. Phan and R. W. Longman, "Liapunov Based Model Reference Learning Control," Proceedings of the Twenty-Sixth Annual Allerton Conference on Communication, Control, and Computing, Allerton House, Monticello, Illinois, 1988.
6. M. Phan and R. W. Longman, "Indirect Learning Control with Guaranteed Stability," Proceedings of the 1989 Conference on Information Sciences and Systems, Johns Hopkins University, Baltimore, MD, 1989.
7. C. James Li & Homayoon S. M. Beigi, "A Self-tuning Regulator with Learning Parameter Estimation", submitted for publication (presented in ASME WAM, 1990)
8. R. Fletcher, Practical Methods of Optimization, Second Edition, John Wiley & Sons, New York, 1987.
9. C. F. Wood, Application of "Direct Search" to the Solution of Engineering Problems, Westinghouse Res. Lab. Sci., Paper 6-41210-1-P1, 1960.
10. R. Hooke and T. Jeeves, "Direct Search Solution of Numerical and Statistical Problems," J. Assoc. Computer Mach., Vol. 8, No. 2, April 1961, pp. 212-229.
11. H. H. Rosenbrock, "Automatic Method for Finding Greatest or Least Value of Function," The Computer Journal, Vol. 3, No. 3, Oct. 1960, pp. 175-184.
12. W. H. Swann, "Report on the Development of a New Direct Search Method of Optimization," Imperial Chemical Industries, Ltd. Central Instr. Lab. Res. Note 6413, 1964.
13. M. J. D. Powell, "An Efficient Method for Finding the Minimum of a Function of Several Variables Without Calculating Derivatives," The Computer Journal, Vol. 7, 1964, pp. 155-162.
14. B. Noble and J. W. Daniel, Applied Linear Algebra, Second Edition, Prentice-Hall, 1977, pp. 138-143.

15. M. Phan and R. W. Longman, "Learning System Identification," Proceedings of the Modeling and Simulation Conference, Pittsburgh, PA, 1989, to appear.
16. J. N. Juang, L. Horta, and R. W. Longman, "ERA System Identification in the Repetition Domain,"Proceedings of the 1989 AIAA/AAS Astrodynamics Specialist Conference, 1989, to appear.
17. R. W. Longman and J. N. Juang, "A Variance Based Confidence Criterion for ERA Identified Modal Parameters," Astrodynamics, 1987, Advances in the Astronautical Sciences, Vol. 65, Part 1, 1988, pp. 581-601.
18. J. G. P. Barnes, "An Algorithm for Solving Nonlinear Equations Based on the Secant Method," Computer Journal, Vol. 8, 1965, pp. 66-72.
19. R. Fletcher, "A Technique for Orthogonalization," J. Inst. Maths Applics, Vol. 5, 1969, pp. 162-166.

Sensors, Controls, and Quality Issues
in Manufacturing
ASME 1991

IMPLEMENTATION OF THE FUZZY CONTROL STRATEGY TO THE MACHINING PROCESSES

M. K. Kim and K. Kim
Department of Mechanical Engineering
University of Illinois, Chicago
Chicago, Illinois

Abstract

An alternative approach is introduced for the investigation into the control strategies of complex and uncertain processes with a view to modelling these skills by a fuzzy algorithm. The basic idea of this approach is to incorporate the experience of a human operator in design of the controller. The control strategies are formulated as a number of rules which are simple to carry out manually but difficult to implement by using conventional algorithm. From a set of linguistic rules which describe the operator's control strategy, a control algorithm is constructed where the words are defined as fuzzy sets. In many industrial cases, the fuzzy algorithm can be very useful and successful in analyzing uncertain and complex systems.

1. Introduction

The application of adaptive control to manufacturing problems has been increased in recent years. This is due primarily to advances in adaptive control theory and microcomputer technology(1). The advanced theory has enabled the design of stable high-performance adaptive control systems, and advances in microcomputer technology have enabled the implementation of these sophisticated controller.

The adaptive control of metal-cutting processes is a logical extension of the Computerized Numerical Control(CNC) systems. The main idea in adaptive control is the improvement of production rate and of part quality in machining processes. Because of increased need for improvements in productivity, controllers for cutting process have introduced some new problems.

Basically, there are two different approaches in identifying the cutting state associated with the stability of the machining operation(2). One is a closed-loop analysis in which the structural and cutting dynamics are analyzed seperately and combined afterwards for stability analysis. It is suitable for better understanding of the cutting process and structural dynamics. Another is the analysis of a signal that contains the combined characteristics of the structural and cutting dynamics. This method is much more simple and useful in practical application.

The most commonly used constraints in adaptive control with constraints(ACC) systems are the cutting force, the machining power, and the cutting torque(3). The operating parameters are usually the feedrate and the spindle speed. The main cutting force is proportional to the depth of cut and feed. A conventional ACC system is a feedback control system(4).

One of the most earliest and intuitive approaches to adaptive control is gain scheduling(5,6). The idea is to find auxiliary process variable that correlate well with the changes in process dynamics. It, however, is an open-loop adaptation scheme without real learning or intelligence. Furthermore, the extent of design required for its implementation can be enormous.

Another way to adjust the parameters of the controller is the model reference adaptive system(5,8). The specifications are given in terms of a reference model, which tells how the process output ideally should respond to the command signal. The key problem is to determine the adjustment mechanism so that a stable system can be obtained. It is necessary to modify adjustment rules by using stability theory(5).

When a mathematical model is used in decision making processes, its validity must be questioned, especially if the actual model must be reduced to one that is deterministic even in a vague and fuzzy environment(7). Since decision processes have become increasingly very complicate, vague, and hard to analyze, the ability to arrive at decisions based on imprecise and qualitative data in contrast to formal mathematics and formal logic which demand the precise and quantitative data is necessary.

In order to control a machining process without knowing the process itself, one can change the input parameters only by identifying its effects upon the process and from previous experiences. The fuzzy set theory has proved to be very successful in analyzing uncertain and complex systems which can not be defined explicitly by mathematical tools because of their complicate properties. By using the fuzzy controller which is made by fuzzy set theory, the amount of computation task and time can be enormously reduced as compared to classical or modern control theory. So far, the fuzzy theory has been successfully applied to solve a wide variety of problems in many scientific fields.

In Section 2, the fuzzy control strategy is introduced by applying to the machining process. The effect of the new fuzzy control algorithm is evaluated through simulation study in Section 3.

2. Fuzzy Control Strategy

The idea of a fuzzy set allows imprecise and qualitative information to be expressed in an exact way and is a generalization of the ordinary notion of a set. A fuzzy set has a membership function which take all values between 0 and 1. The notion of a fuzzy set and the concept of a fuzzy conditional statement provide a basis for using fuzzy algorithms in systematic and effective ways. Thus, the fuzzy algorithms could become an important tool for an approximate analysis of systems and decision processes. A fuzzy instruction which is a part of a fuzzy algorithm can be assigned a precise meaning by making use of the concept of the membership function of a fuzzy set.

A schematic decision-making procedure of machining process for a regulation control task is shown in Fig.1. The activity is expressed as a fuzzy relationship or algorithm, relating significant observed variables to the control actions(10). The purpose of controllers is to compute values of action variables from observation of state variables of the process under control. The control strategy can be expressed linguistically as a set of imprecise conditional statements which form a set of decision rules. The control action is composed of several steps as follows(10):

(1) Calculate the present error and its rate of change.
(2) Convert the actual error values to the fuzzy variables.
(3) Examine the fuzzy control rules.
(4) Take fuzzy output variable from the decision table.
(5) Convert the fuzzy variable to the deterministic value.

The control rules are formulated in linguistic terms using fuzzy sets to describe the magnitudes of the appropriate control action. The elements of a universe of discourse and values are assigned to the grade of membership to give fuzzy set. The grade of membership values are assigned subjectively to define the meaning of the labels of the fuzzy sets. Fuzzy sets and decision tables are defined for the error, change of error, and control change respectively.

As far as control system is concerned, what we are interested in is the relation between inputs and outputs. A mapping from a fuzzy input variable to a fuzzy output variable in a system can be expressed by a fuzzy conditional statement(11). A control rule can be made by sequence of the fuzzy conditional statements. The major concern in the use of fuzzy controllers is the problem of

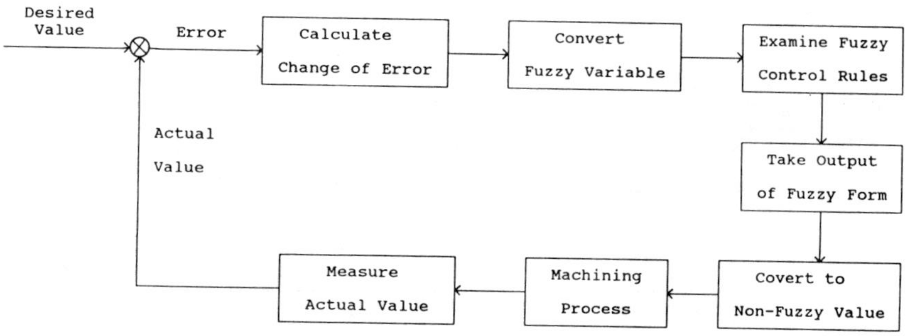

Fig.1 A Schematic Fuzzy Control System of Machining Process

finding an appropriate set of rules for designing the controller. The main advantage of this approach is to have the possibility of implementing experience, intuition, heuristics, and the fact that it does not need a model of the process(12).

For deciding the best control rules, there is no theory to analyze the relation between linguistic algorithm and system performance. Generally, there are several factors for deciding control rules as follows(13):

(1) An operator's experience and knowledge of a physical process
(2) A rule adaption mechanism which can be expanded and changed by the fuzzy controller on the basis of the controller's experience
(3) A mathematical model of the process concerned.

It is noted that the key point of solving a fuzzy control problem is how to find the fuzzy relation. To do so, we will need the determination of the primary fuzzy variables for the controller and the translation of fuzzy conditional statements.

3. Simulation Study based on Experimental Data

This section shows a simulation study to demonstrate the excellence of the proposed strategy. The objective of this simulation work is to manipulate the feedrate to maintain a constant cutting force in the end milling operation. An adaptive control algorithm was selected to make basic control rules of fuzzy controller for force control in the end milling operation(15).

3.1 Conditions for Simulation Work

The ACC systems which employ feedrate manipulation require models of machining process mechanics. A system identification of the cutting process parameters is also required to select an appropriate model structure for the machining process over the range of cutting conditions, which can be used to design a controller and to investigate how the model parameters change with cutting conditions. The data which was used for this study is experimental data(15). In order to achieve the research objective, the new approach was based on obtaining a control rule by using the measured force errors in machining process.

A series of slot-milling tests on the 1020 cold rolled steel is conducted using a four-tooth 12.7 mm High Speed Steel(HSS) helical cutter. The spindle rotation is set at 480 revolution per minute(RPM) and the corresponding sampling interval is 0.125 second. The desired constant force level is set at 350 Newtons for slot milling operation(15). A random sequence with a mean of 65.2 mm/min and a variance of 150 is used for the input feedrate of the stepping motor. The data which was used for simulation study is obtained at three different depths of cut which are 1.0 mm, 1.5 mm, and 2.0 mm. Fig. 2 shows the depth of cut variation. The sampling frequency of x and y force components is 32 Hz and the cutoff frequency of the low pass filters is set at 16 Hz. At each spindle revolution period, four x and y force components will be used to construct the quasi-mean

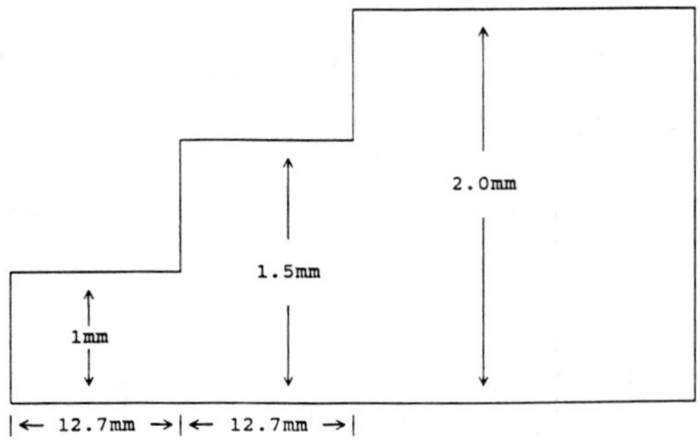

Fig.2 Variation of Depth of Cut

resulant force. That is, the sampling frequency of the quasi-mean resulant force is at the spindle frequency.

3.2 Input and output valuables.

Normally, the measured force value(F_N) of the current milling process is non-fuzzy. The reference value(F_R) is the desired force value for the operation. The deviation of the measured value from the desired value can be expressed by the relative error force(e). The change of error(c) is defined as difference of two adjacent error e_n and e_{n-1} follows:

$$e_n = F_R - F_N$$

$$c_n = F_n - F_{n-1}$$

where, n = n^{th} sample interval

For this research, F_R is set at 350 newtons for varying cutting conditions after considering stable cutting conditions. The measured force value F_N, which was obtained from the previous experiment(15) by using pole placement adaptive control algorithm, is used to establish the fuzzy control rules. Variation of the experimental force data is shown in Fig.3. Then, e_n and c_n are calculated based on the previous experimental data. As shown in Figs.4 and 5, most of e_n values

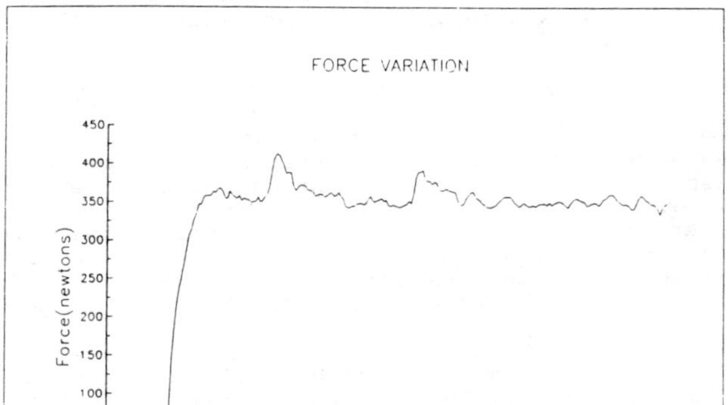

Fig.3 Variation of Previous Experimental Force Data

are scattered in between -20 newtons and 10 newtons, while most of c_n values are scattered in between -6 newtons and 6 newtons. According to the level of frequent occurrence for the values of e_n and c_n during a machining process, they are classified into 10 and 12 different categories, respectively.

3.3 D/F Converter

In this present research, the output from the comparator are non-fuzzy variables although fuzzy variables are necessary for the fuzzy controller. The D/F converter plays the role of converting between those variables. Now, fuzzy variables E and C are represented instead of non-fuzzy variables, respectively, as follows:

{E} = {NE,NH,NB,NM,NL,NS,NO,PO,PS,PB}

{C} = {NH,NB,NM,NL,NS,NO,PO,PS,PL,PM,PB,PH}

which are equivalent to some values in the non-fuzzy variables as shown in Table 1.

3.4 Fuzzy Controller

The fuzzy controller is composed of many control rules. The linguistic rules of a fuzzy controller consist of a series of conditional statements of the following form:

Fig.4 Scatter of Error Values

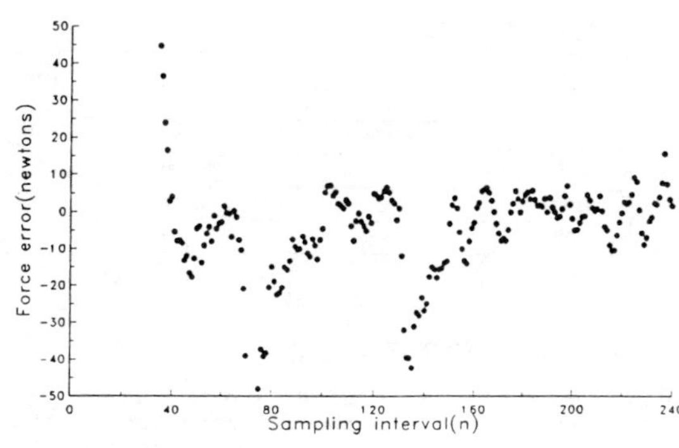

Fig.5 Scatter of Change of Error Values

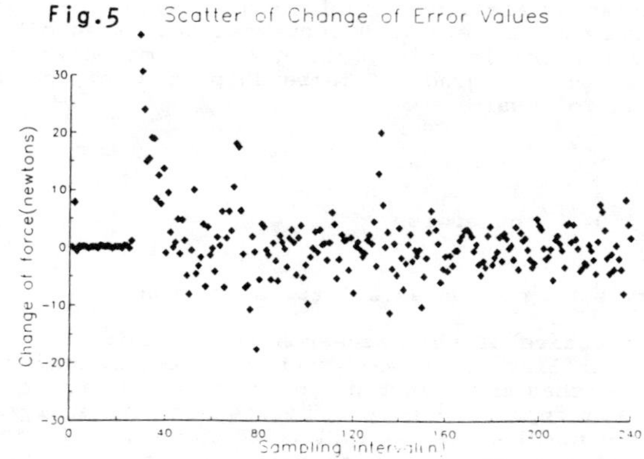

Table 1 Conversion from Non-fuzzy to Fuzzy Variables for Input

Error		
	Non-fuzzy variables(e)	Fuzzy Variables(E)
Category	Range	Name
1st	$e \le -30$	NE
2nd	$-30 < e \le -25$	NH
3rd	$-25 < e \le -20$	NB
4th	$-20 < e \le -15$	NM
5th	$-15 < e \le -10$	NL
6th	$-10 < e \le -5$	NS
7th	$-5 < e \le +0$	NO
8th	$+0 < e \le +5$	PO
9th	$+5 < e \le +10$	PS
10th	$+10 < e$	PB

Change of error		
	Non-fuzzy variables(c)	Fuzzy Variables(C)
Category	Range	Name
1st	$c \le -10$	NH
2nd	$-10 < c \le -8$	NB
3rd	$-8 < c \le -6$	NM
4th	$-6 < c \le -4$	NL
5th	$-4 < c \le -2$	NS
6th	$-2 < c \le 0$	NO
7th	$0 < c \le +2$	PO
8th	$+2 < c \le +4$	PS
9th	$+4 < c \le +6$	PL
10th	$+6 < c \le +8$	PM
11th	$+8 < c \le +10$	PB
12th	$+10 < c$	PH

"IF E is E_n THEN C is C_n THEN F is F_n"

In the above statement, E_n, C_n, and F_n are the linguistic values of fuzzy variables E,C, and F respectively, in the nth conditional statement. There are several ways for deciding control rules in the fuzzy control algorithm. Among them, the control rule based on mathematical model of the process concerned i.e. pole placement adaptive control logic(15) is selected for this research so as to explain the non-minimum phase sitiation in the end milling operations.

When the variables are converted into fuzzy variables, all elements of e_i or c_i have zero grade of membership except when the measured value of error(E) or change of error(C) are in exist, the grade of membership of that case is assigned as unity. Thus the grade of membership of e_i of the discrete universe E can be defined as follows:

$$\mu_E(e) = \begin{cases} 1 & \text{if } e = e_i \\ 0 & \text{otherwise} \end{cases}$$

This sort of fuzzy set is known as a fuzzy singleton.

Since the objective of this research is to manipulate the feedrate, the main goal of the controller is how well control rules, as main frame of the fuzzy controller is established and selected. In order to make the fuzzy control rule, the relation between input and output is checked and analyzed when the pole placement adaptive control system is applied to the end milling operations. After that, an outline of the control rules with 14 different categories of F is emerged and the decision rule is established as shown in Fig.6.

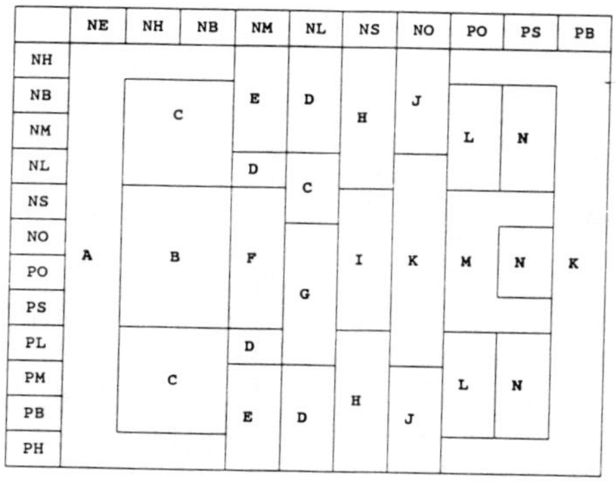

where,

Error(E)	Change of Error(C)
NE: E≤-30	NH: C≤-10
NH: -30<E≤-25	NB: -10<C≤ -8
NB: -25<E≤-20	NM: -8<C≤ -6
NM: -20<E≤-15	NL: -6<C≤ -4
NL: -15<E≤-10	NS: -4<C≤ -2
NS: -10<E≤ -5	NO: -2<C≤ 0
NO: -5<E≤ 0	PO: 0<C≤ 2
PO: 0<E≤ 5	PS: 2<C≤ 4
PS: 5<E≤ 10	PL: 4<C≤ 6
PB: 10<E	PM: 6<C≤ 8
	PB: 8<C≤ 10
	PH: 10<C

Fig.6 Decision-making Table of Control Rules

3.5 F/D Converter

The output from the fuzzy controller is a fuzzy variable, while the end milling process needs a deterministic value as its input. In order to change the feedrate of the milling machine, it is necessary to defuzzify the fuzzy variable into a non-fuzzy variable.

Like error(E) or change of error(C), feedrate change(F) which is output from the fuzzy controller was classified into 14 different categories. For convenience, the linguistic values of the fuzzy variable F are denoted from A to N as follows:

{F} = {A,B,C,D,E,F,G,H,I,J,K,L,M,N}

The output as the change of feedrate(f) of the F/D converter is ,then, the input to the machining process.

4. Conclusions

Cutting experiments were executed with the derived fuzzy rules and Fig.7 shows the variation of force as the fuzzy rules based on the pole placement adaptive control system are applied. When the process with derived fuzzy rules is compared to that of pole placement AC system, it is known that the calculation time can be enormously reduced by using the fuzzy lofic because computation tasks were done in off-line situation. That is, as the fuzzy system is applied to the on-line real time process, it is already not a kind of mathematical model but a decision-making rule. Even though the force values as controlled by the fuzzy rules are almost same as that of pole placement AC system, the fuzzy control logic can be modified to keep the cutting force as constant as possible through lots of experiments.

For good quality control of machining processes, fuzzy control algorithm is introduced. The advantage of a fuzzy controller is that it can simulate the human experiences without the need of the human operator. The fuzzy rules should be correctly formulated to control the processes effectively using heuristic rules based on fuzzy statements(16).

The purpose of this study is to implement a new single output fuzzy control algorithm as well as to examine the application of a fuzzy controller to the machining operations. The fuzzy controller is especially useful for time delay processes or non-linear processes and can simulate human experience and

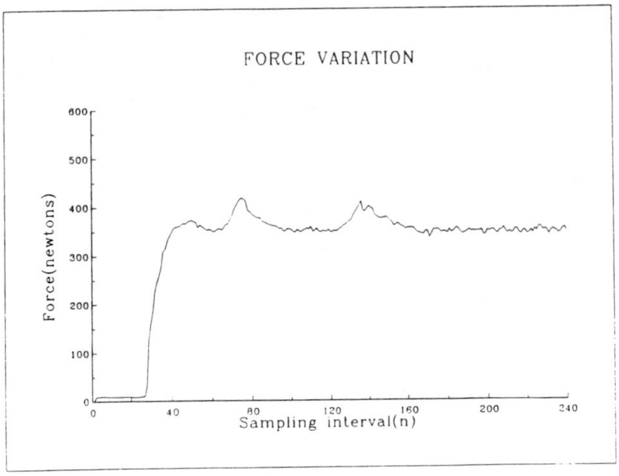

Fig. 7 Variations Based on the Fuzzy Rules

experimental information for controlling processes. In addition, it is easy to introduce new rules and to modify the existing rules. By using the fuzzy controller, computation time can be reduced as compared to conventional control theory. Thus, it is more applicable for on-line control system. For effective control, the fuzzy rules must be formulated correctly to take account of time delays when they occur. The multi-input and output variables for more complicate processes will be considered with more number of rules and will be applied for the real end milling process for future work.

References

1. Lauderbaugh, L.K. and Ulsoy, A.G., 1989, "Model Reference Adaptive Force Control in Milling," ASME Journal of Engineering for Industry, Vol.111, pp.13-21.
2. Szabo, B., Du, R., and Wu, S.M.,1989, "Tool Wear Monitoring for Milling Processes using DDS Methodology," 17th NAMRC, pp.160-165.
3. Koren, Y., 1984, Computer Control of Manufacturing Systems, McGraw-Hill, New York.
4. Ulsoy, A.G., Koren Y., and Rasmussen, F., 1983, "Principal Developments in the Adaptive Control of Machine Tools," ASME Journal of Dynamic Systems, Measurement, and Control, Vol.105, No.2, pp.107-112.
5. Astrom, K.J. and Wittenmark, B., 1984, "Computer Controlled Systems:Theory and Design," Prentice Hall, Eaglewood Cliffs, New Jersey.
6. Sastry, S. and Bodson, M., 1989, "Adaptive Control: Stability, Convergence, and Robustness," Prentice Hall, Eaglewood Cliffs, New Jersey.
7. Zimmermann, H.J., 1984, Fuzzy Set Theory and Its Applications, Kluwer-Nijhoff, Hingham.
8. Koren, Y., 1988, "Adaptive Control Systems for Machining," American Control Conference, pp.1161-1167.
9. Wick, C., 1977, "Automatic Adaptive Control of Machine Tools," Manufacturing Engineering, pp.38-45.
10. Umbers, I.G. and King, P.J., 1981, "An Analysis of Human Decision-Making in Cement Kiln Control and the Implications for Automation," Fuzzy Reasoning and Its Applications, pp.369-381.
11. Kikert, W.J.M. and Mamdani, E.H., 1978, "Analysis of a Fuzzy Logic Controller," International Journal of Fuzzy System.
12. Zadeh, L.A., 1973, "Outline of a New Approach to the Analysis of Complex Systems and Decision Processes," IEEE Transactions on Systems, Man, and Cybernetics, Vol.1.
13. Jain, R., 1976, "Outline of An Approach for the Analysis of Fuzzy Systems," International Journal of Control, Vol.23, 1976.
14. Zadeh, L.A., 1965, "Fuzzy Sets", Information and Control, Vol.8, pp.338-353.

15. Huang, S.D., 1990, Adaptive Force Control of End Milling Operations under Variable Cutting Conditions, Ph.D Dissertation, The University of Illinois at Chicago, IL.
16. Tong, R.M., 1977, "A Control Engineering Review of Fuzzy Systems," Automatica, Vol.13.
17. Mamdani, E.H. and Assilian S., 1975, "An Experiment in Linguistic Synthesis with a Fuzzy Logic Controller," International Journal of Man-Machine Studies, 7.
18. Zhu, J.Y., 1982, "Control of Machine Tools Using the Fuzzy Control Technique", Annals of the CIRP, Vol.31/1.
19. Mamdani, E.H. and Assilian, S., 1975, "A Fuzzy Logic Controller for a Dynamic Plant", International Journal of Man-Machine Studies, 7.

Sensors, Controls, and Quality Issues
in Manufacturing
ASME 1991

REAL-TIME CNC TOOL PATH GENERATION FOR MACHINING IGES SURFACES

Y. D. Chen, J. Ni, and S. M. Wu
University of Michigan
Ann Arbor, Michigan

ABSTRACT

A real-time CNC tool path generation algorithm has been developed for machining IGES surfaces. IGES-based CAD data files can be directly input to the CNC system and the tool paths generated in real time can be passed to motion controller via a Multibus II backplane structure during cutting. The development of such a real-time tool path generation method has eliminated the need for a tradeoff between the desired surface accuracy and the required memory size for storing off-line generated NC codes. The real-time NC path generation algorithm can deal properly with issues such as trimming lines, gouging detection, and adaptive tool step adjustment. The developed algorithm has been implemented on a multi-processor CNC system and verified through actual cutting tests. The test results show that no violating conditions occurred on machined part surfaces and the surface contour error of the cut part is less than the given tolerance, which was 0.02 mm in this particular test.

I. INTRODUCTION

Tool path planning determines methods of driving a cutting tool to generate a desired part surface. Cutter paths are usually generated in advance either manually for simple part geometries or automatically after a part is designed in a CAD station.

Many engineering parts such as car bodies and airplane wings have complex free-form surfaces. Since it is difficult to represent these surfaces by a single mathematic equation, they are in general represented by composite surfaces composed of many patches. Many kinds of surface patch equations have been developed and the most popular one is the bicubic parametric patch, entity type 114 defined in the Initial Graphics Exchange Specification (IGES).

For such kind of free-form surfaces, the current industrial practice uses an off-line tool path generation procedure by which the tool path is first generated from the CAD data files (surface equations) by a NC code generator, and then downloaded to a CNC controller. When a complex free-form surface with a tight tolerance is to be machined, the off-line NC path generators have to generate the cutter location in (X, Y, Z) format at very small steps in order to produce smooth surfaces within the given tolerances. This in turn requires a very large memory in the CNC controller to store such kind of NC tool path data. In many cases the memory of the available CNC controllers will not be able to accommodate this huge size of tool path data. For instance, to machine a die surface for an automobile quarter panel shown in Fig.1, which contains more than 600 surfaces, the tool path data could take as much as 20 Mbyte memory. As a result, the off-line generated tool path data have to be broken into many batches and downloaded one at a time after a previous batch is executed.

NC tool path generation has evolved over the last 30 years, influenced by available technologies, market needs and other forces outside of manufacturing. In 1960s, NC machine had to be programmed manually. The program process is very slow and error prone. The availiability of computer aided methods and the need for speed and responsiveness continues to reduce reliance on this method. Since 1970s, More research has been done in the computer aided programming for NC. In general, there are three types of techniques for the tool path generation .

- APT-based tool path generation method. Explicit tool guiding surfaces (drive surfaces) are introduced in this method. The basic idea is that the cutter is moved in one direction while maintaining contacts with both the part surface and drive surface [Faux and Pratt, 1981, Bobrow, 1985]. At each step numerical iteration searches are made to locate the cutter position within a specified tolerance limit. The disadvantage of this method is that the iterative computation is time consuming and there is no guarantee that the iterations will converge for irregularly curved sculptured surfaces .

- Cartesian machining method. Tool paths are planned on the XY-plane of Cartesian coordinates. The cutter path is the intersection of the part surface and a vertical plane perpendicular to the XY-plane [Choi and Lee etc. 1988 and 1989][Bobrow, 1985]. In this method, the cutter plane has to be a vertical plane. Because gouging points may be not in the vertical cutter plane, the search of gouging points must be performed on the entire part surface.

- Parametric machining method. Tool paths are planned on the parametric space. There are two ways to calculate the tool paths. (1) The tool is moved at equally spaced points on the u-line or v-line of the surface. Corresponding to each selected point on the u-v domain, a (x, y, z) point on the part surface can be calculated and used as tool paths. The drawback of this scheme is due to the fact that it is difficult to control the tolerance on the part surface only based on the u-v domain point interval. The equally spaced points in the parameter space may result in differently spaced points on the part surface [Bobrow, 1985]. (2) The points and normal vectors on a part surface are first evaluated at pre-defined small interval of both parameters u and v. The tool path is then generated by searching through this huge data array of part surface points on a given cutter plane. This method is widely used in the commercial NC software. The operations of filter are performed in order to generate smallest tool path point and reduce the tool path data size. The filler operations are also necessary to insert the points in the interval if it is not small enough. The operations of filler and filter are very time consuming.

All these off-line tool path generation methods have a common drawback. There is always a tradeoff between the level of path precision, in terms of the coarseness of tool path points on a sculptured surface, and the amount of memory requirements. The data format of the tool path has to be specified based on the NC machine.

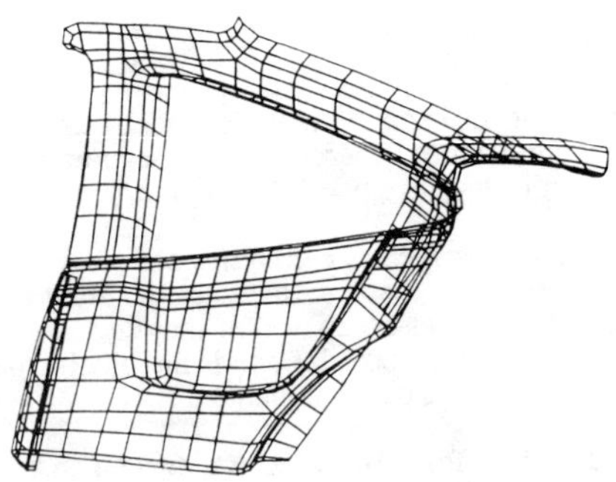

Fig.1 An automobile quarter panel surface represented in IGES format

This paper presents a new method for the real-time generation of NC tool paths to machine free-form sculptured surfaces. This new method is intended to eliminate the need for the tradeoff between the path precision and the memory requirements. The real-time path generation method can generate as many points as needed within every computational cycle time. The method is designed to generate NC tool paths directly from IGES data file since the IGES is the main standard electronic data translator recognized by the ANSI and ISO.

A new CNC controller is being developed based on the Multibus II structure and the iRMX-III operating system. The controller is a true multi-processor and multi-tasking system. One of the processors in this controller is used as a Free-Form Surface Milling Tool Path Processor, which can read IGES data directly from a CAD data file and generate the CNC tool path during machining operations. The tool path data is then passed to motion controller processors through the Multibus in real time.

The definitions of IGES surfaces and the derivation of their offset surfaces are given in Section 2. The proposed tool path generation algorithms for free-form sculptured surfaces are presented in Section 3. The issues of tool path planning, tool step calculation, gouging detection, tool path interval, and trimmed surfaces are discussed in details. These developed algorithms are verified through actual cutting tests and the results are presented in Section 4 and conclusions in Section 5.

II. IGES SURFACES AND THEIR OFFSET SURFACES

IGES is a widely used CAD data file structure. All major CAD software packages support the IGES input and output format. The IGES data format is used as the input data format of this new real-time tool path generation method.

The focus of this paper is on the machining of free-form parametric surfaces on a 3-axis milling machine. It is assumed that the surfaces can be reached by the milling cutter in the positive Z direction. The entity 114 in IGES can represent these free-form parametric spline surfaces. IGES gives all information used to define the free-form parametric spline surfaces. Detailed definition of IGES surfaces can be obtained from reference [1]. As discussed in appendix, the free-form parametric spline surface shown in Fig.2 can be expressed as:

$$\vec{S}(u,v) = x(u,v)\vec{i} + y(u,v)\vec{j} + z(u,v)\vec{k} \tag{1}$$

Let $\vec{S}(u,v)$ be a part surface defined by this specification parameterized and oriented by $\vec{N}(u,v)$, a differentiable normal vectors defined on the whole surface. An offset surface to S is a parameterized surface $\vec{O}(u,v)$ given by :

$$\vec{O}(u,v) = \vec{S}(u,v) + f(d)\cdot\vec{N}(u,v) \tag{2}$$

where

$$\vec{N}(u,v) = \frac{\dfrac{\partial\vec{S}(u,v)}{\partial u} \times \dfrac{\partial\vec{S}(u,v)}{\partial v}}{\left|\dfrac{\partial\vec{S}(u,v)}{\partial u} \times \dfrac{\partial\vec{S}(u,v)}{\partial v}\right|}$$

and f(d) is an offset distance function to be determined from cutter tip radius.

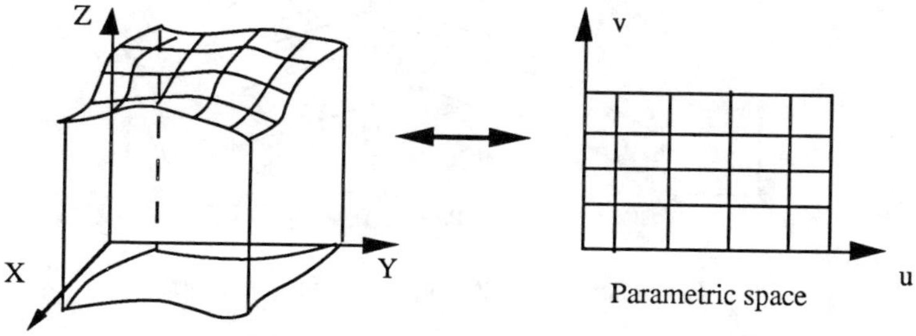

Fig. 2 Parametric Spline Surface (IGES Type 114)

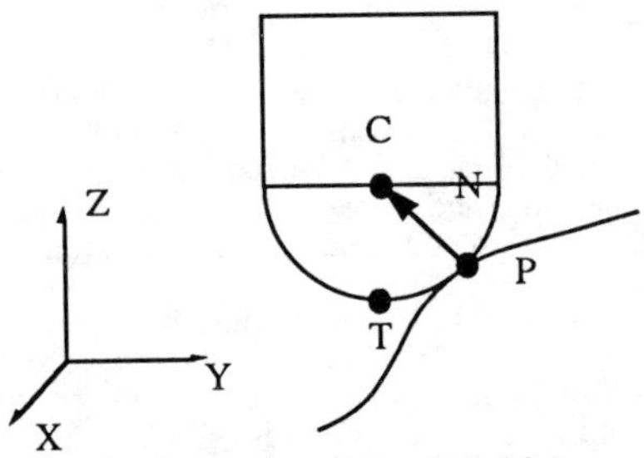

Fig. 3 Tool position definitions

Cutting tools are often not an idealized geometrical point and may have a finite radius ball tip as shown in Fig. 3, where point P is called the cutter contact (CC) position and point C is the cutter location (CL) and T stands for the tip position. The tool path is often referred to as the cutter location (CL) data, which are to be generated from designed part surface data and used by the CNC controller to drive the cutting tool.

III. PROPOSED TOOL PATH GENERATION ALGORITHM FOR FREE-FORM SURFACES

The basic requirements for the real-time tool path generation are that (1) the tool path on a cutter plane should have no gouging problem with points which are not in the cutter plane, (2) the length of tool steps should be adjustable based on the tolerance and condition of the surface to be machined, and (3) the algorithm should be able to handle the trimmed curves on the surfaces. In order to generate the CNC tool path for machining free-form surfaces, the following issues should be addressed:

- Tool path planning:
- Tool step length calculation;
- Gouging problem;
- Tool path interval; and
- Trimmed surface.

3.1 Tool Path Planning

The Cartesian and parametric machining methods mentioned above are combined in the development of the proposed new real-time path generation algorithm. It is important that the offset surface shown in Fig.4, instead of part surface, is intersected by a cutter plane in the new algorithm. This feature ensures that there will be no gouging points outside of the cutter plane. It is only necessary to check the gouging with points on the cutter plane.

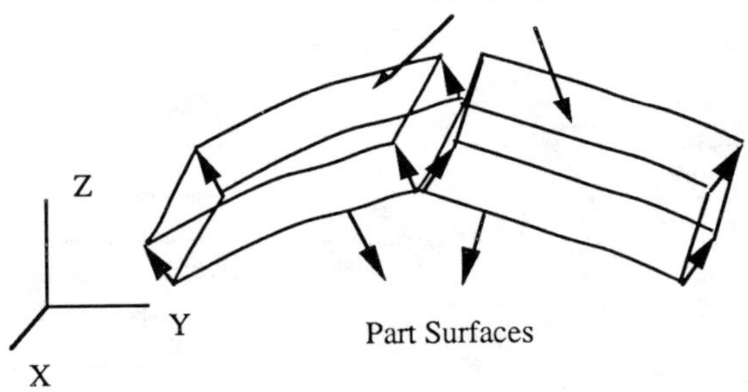

Fig. 4 Offset surfaces of IGES parametric spline surfaces

Unlike the Cartesian method, the cutter plane in this new algorithm will not necessarily be a vertical plane. It can be any general plane defined in the XYZ space:

$$AX+BY+CZ+D=0 \quad (3)$$

This cutter plane can be easily oriented and positioned in the XYZ space by varying the coefficients A, B, C, and D. The advantage of this selection of cutter planes compared with the Cartesian method lies in the fact that the new method can be better suited for the application to 5-axis contour cutting where it is desirable to keep the cutter plane always normal to the part surfaces.

The coefficients A, B, C, and D of the cutter plane can be calculated from the cutting direction (i,j,k) and starting point (X_0,Y_0,Z_0). The cutter path will be at the intersection line between the cutter plane and the offset surface. The equation for this curve is:

$$AX(u,v)+BY(u,v)+CZ(u,v)+D = 0 \tag{4}$$

where X(u,v), Y(u,v), and Z(u,v) are the equations for the offset surface and are given in the appendix. Since the X(u,v),Y(u,v), and Z(u,v) are nonlinear functions of parameters (u,v), above equation is a two-variable nonlinear equation.

The procedure to calculate the tool position is as follows. A cutter plane is first selected in the XYZ space and the intersection curve between the cutter plane and the offset surface will be found using Eq. (4). By mapping this intersection curve in the XYZ space into the u-v domain, a curve in the u-v domain can be obtained. For any given point in this u-v curve, a unique tool position can be determined in the intersection curve. Therefore, the task of tool path generation will become a task of finding this tool path in the u-v domain.

In order to solve for (u,v) in Eq.(4), i.e., mapping the intersection curve into a u-v curve, another equation or condition is required. The tool step length constrain is used to construct this equation, which can be represented as the following and will be discussed in detail in Section 3.2.

$$F(u,v) = 0 \tag{5}$$

The solutions of Eq.(4) and Eq.(5) (tool step length constrain) will give the (u,v) values corresponding to the intersection curve.

3.2 Tool step calculation:

The determination of the cutter step intervals depends on the accuracy requirements of part surfaces and on the interpolation scheme to be employed subsequently. In real-time cutter path generation, the step size can be controlled to small enough so that linear interpolation between different cutter positions will be sufficiently accurate. Thus, the intersection curve can be approximated by a sequence of linear segments.

The maximum deviation, δ , between a true intersection curve and the chord connecting two successive cutter positions, is measured normal to the chord as shown in Fig.5(a). For small steps, it is reasonable to approximate the curve by its osculating circle and the local curvature of the intersection curve may be used to determine the step length. From geometries, it can be shown that:

$$L^2 = 4\delta(2r-\delta) \tag{6}$$

where L is the step length, and r is the radius of the curvature at point P_j. When the maximum allowable deviation δ is given as an input parameter and the curvature of the surface is calculated from the equation of the surface, the step length at this position can be calculated based on the above equation.

Given the condition of a surface curvature and the permissible interpolation error, δ , a new algorithm is proposed to calculate the tool step interval recursively. First, the intersection points of the cutter plane and edges of the offset surface are calculated. The ending point is used as the reference point for this segment of the tool path as shown in Fig.5(b). At each tool position, P_i, the distance from this position to the reference point is denoted as D_i, and the radius of surface curvature at position P_i is r_i, which is approximated by the average of the two main components of the curvature radius at this position. L_i can be calculated using Eq.(6). Knowing D_i, r_i and L_i, the total number of tool steps can be calculated as:

$$N_i = \frac{\sin^{-1}\frac{D_i}{2r_i}}{\sin^{-1}\frac{L_i}{2r_i}} \quad (7)$$

where N_i represents the total number of linear segments required to approximate the intersection curve from position P_i to the ending point. This procedure is recursive in the sense that when the cutter proceeds to a new position, the parameters D_i, r_i and L_i will be updated to their new values at the new position and Eq. (7) will be applied to give a new total number of tool steps, N_i.

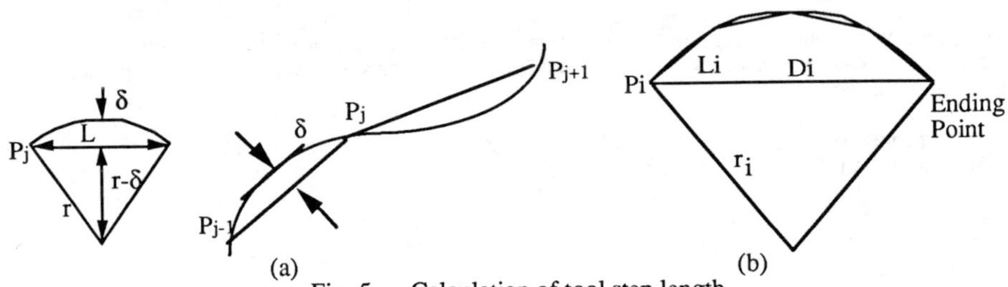

Fig. 5 Calculation of tool step length

After the total number of tool steps is determined, the tool path in the u-v domain will be divided into N_i segments. It is preferred to divide the tool path in the u-v domain along an axis which has a larger parameter range. For example, as shown in Fig. 6, the tool path from (u1, v1) to (u2, v2) has a relatively larger change along v axis. Therefore, the division of tool path will be made along v axis.

$$\Delta v_i = \frac{(v_i - v_2)}{N_i}$$

or

$$\Delta u_i = \frac{(u_i - u_2)}{N_i} \quad (8)$$

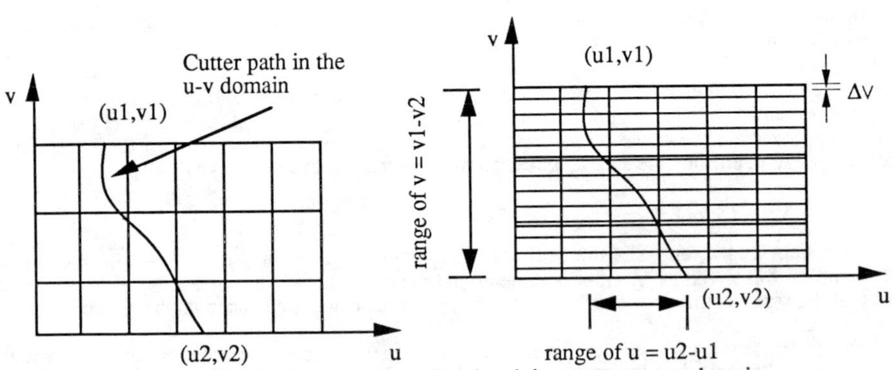

Fig. 6 Segmentation of tool path in parameter u-v domain

This method is fast and easy to implement. It can adjust the parameter interval based on the curvature of the surface.

3.3 Gouging Problem

After calculating the intersection between the offset surface and the cutter planes, the tool path will become a series of curves which are in the cutter planes. These curves will discontinue at the joint section of the surfaces. Algorithms must be developed to deal with these discontinuities.

Curve sorting:

The order of the surfaces given in IGES format is not arranged by the geometric structure and the information about how to link these curves is not provided by IGES. A curve sorting algorithm is developed to find the connection between different intersection curves. The objective of the curve sorting algorithm is to determine which two curves ought to be connected and how the ends of these two curves are connected.

Two additional orientation index points are added to each intersection curve in this sorting process. These index points are the intersection points of the cutter plane and the part surface boundary to represent starting and ending points of a part surface boundary. The ending point of one surface will be the starting point of another surface if they are connected to each other.

In order to determine whether the index points are put in the proper positions or not, the triple product of vectors **n1**, **n1'** and **n** is used to check the relative positions of the starting and ending points of both part surface and offset surface, as shown in Fig. 7. Vector **n** is the surface normal of the cutter plane (which is the paper plane in the drawing), and vectors **n1** and **n1'** are the vectors connecting the starting and ending index points, respectively, to the edges of the offset surface. If $\mathbf{n} \cdot (\mathbf{n1} \times \mathbf{n1'}) < 0$, the index points on a part surface is properly connected to the starting and ending points on its offset surface. If $\mathbf{n} \cdot (\mathbf{n1} \times \mathbf{n1'}) > 0$, these points are not properly connected.

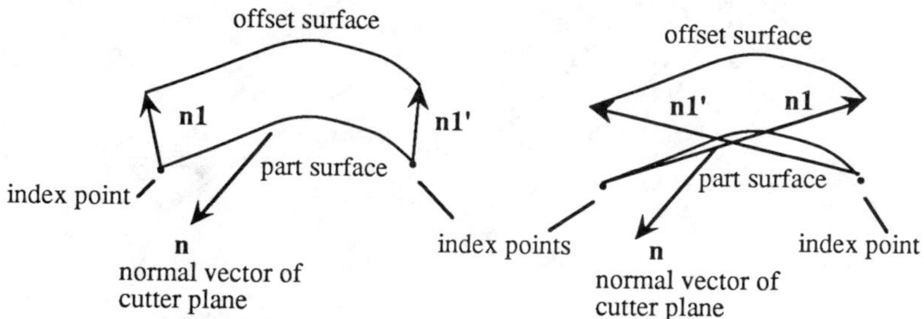

Fig. 7 Triple product of vectors for curve sorting

After the index points associated with each tool path segment are properly labeled with a starting or ending mark, they are used to identify the connection between different tool path segments. If the ending index point of a tool path segment is determined to be equal to the starting index point of another segment, these two segments should be connected together.

Another faster method to sort the order of the curves is to sort them based on the distances from their starting and ending points to a given fixed reference point. These distances are calculated using Eq.(9). For instance, let us assume that the fixed reference point be on the far left side of all curves and tool path segments start from left to right when properly oriented. Then the distance from the starting point of the tool path segment to the reference point should be smaller than that from the ending point. Otherwise, the tool path segment is placed in a wrong orientation. By checking the distances from both the starting and ending points of each tool path segment, its orientation can be determined.

$$DIS_s(i) = \sqrt{(X_s(i) - X_r)^2 + (Y_s(i) - Y_r)^2 + (Z_s(i) - Z_r)^2}$$
$$DIS_e(i) = \sqrt{(X_e(i) - X_r)^2 + (Y_e(i) - Y_r)^2 + (Z_e(i) - Z_r)^2} \quad (9)$$

where subscript s stands for starting point, subscript e stands for ending point and subscript r stands for the reference point.

After sorting the curves, all calculated tool path curves can be linked from the ending index point of one curve to the starting index point of the next curve as shown in Fig. 8.

Fig. 8 Connection of tool paths in the cutter plane.

Connecting surfaces:

It can be seen from Fig.8 that for two surfaces with a convex joint there is a gap between their offset surfaces and for surfaces with a concave joint there is overlap between their offset surfaces. In order to connect these tool paths, corner fillets and/or edge removals are necessary. To check if the tool path segment need to be removed or filleted, another triple product is used. If $\mathbf{n} \cdot (\mathbf{n}1\times\mathbf{n}2) < 0$, an interpolation or fillet is needed between these two tool path segments as shown in Fig. 9(a). If $\mathbf{n} \cdot (\mathbf{n}1\times\mathbf{n}2) > 0$, these two tool path segments are gouged and a removal operation is needed as shown in Fig. 9(b).

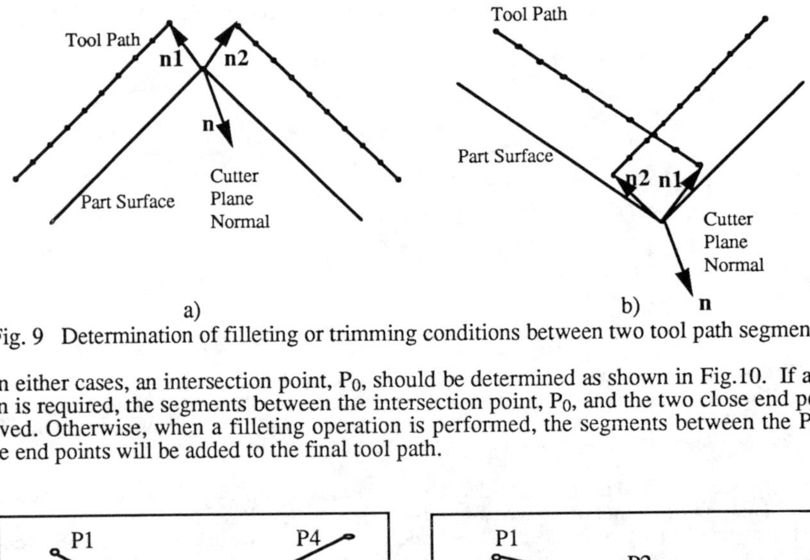

Fig. 9 Determination of filleting or trimming conditions between two tool path segments

In either cases, an intersection point, P_0, should be determined as shown in Fig.10. If a removal operation is required, the segments between the intersection point, P_0, and the two close end points will be removed. Otherwise, when a filleting operation is performed, the segments between the P_0 and the two close end points will be added to the final tool path.

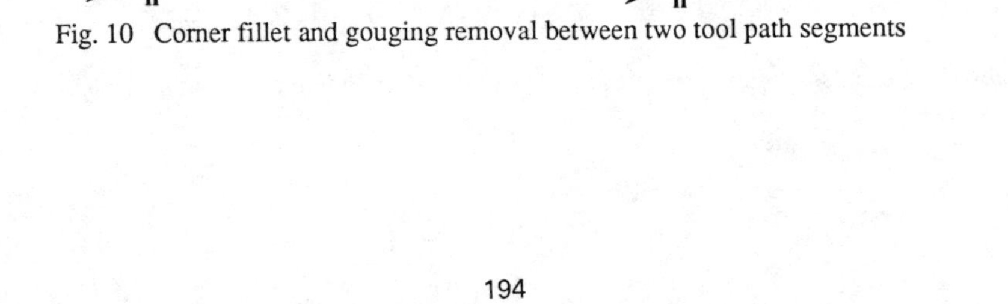

Fig. 10 Corner fillet and gouging removal between two tool path segments

3.4 Tool pass interval

Tool pass interval is the interval between two subsequent passes of a cutter on the surface. It is the distance between the two cutter planes. The cutter plane is moved parallelly by a distance D, which can be calculated from a specified cusp error, as shown in Fig.11.

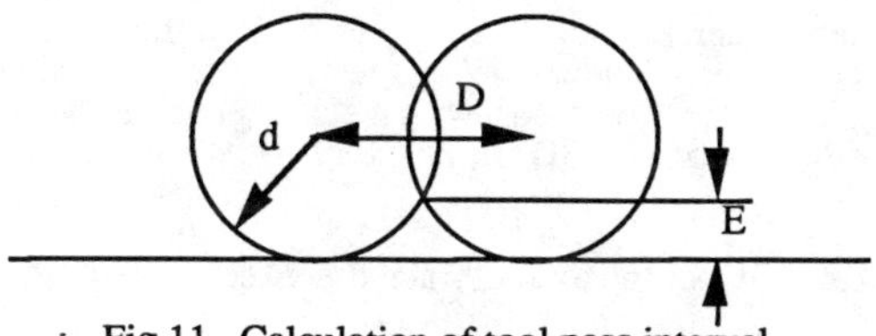

Fig.11 Calculation of tool pass interval

If E is the specified cusp error and d is the diameter of a ball-end-miller, the distance between the two cutter planes can be calculated from:

$$D_i=2\sqrt{2dE-E^2} \tag{10}$$

3.5 Trimmed Surface

Above discussion on corner fillet and surface trimming is mainly for the conditions resulting from offsetting a free-form surface. In IGES data format, there are specific entities such as entities 142, 144, and 106 used to specify the trim curves in the given parametric surfaces.

For instance, IGES entity 106 specifies a trim line by a series of (u,v) data in u-v domain. These data can be mapped to the XYZ space using Eq. (4) to form a trim curve as shown in Fig.12.

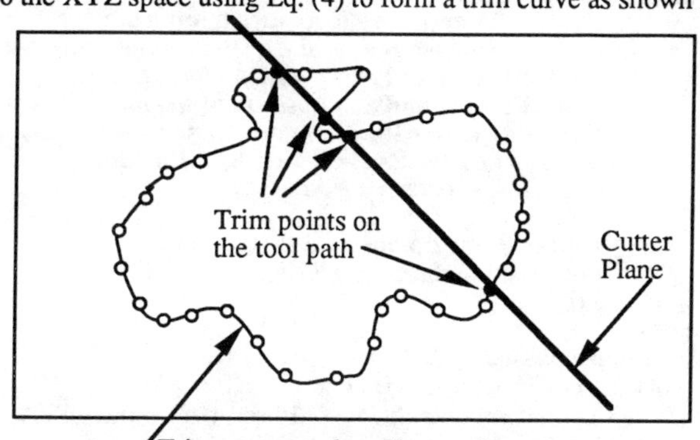

Fig.12 A trim curve specified using entity 106

The mapped points on the trim curve will be used to find the trim points in the cutter plane. For each given mapped trim point, its distance to the cutter plane can be treated as an index and calculated using:

$$\text{Distance (from trim point i to cutterplane)} = \frac{AX(i) + BY(i) + CZ(i) + D}{\sqrt{A^2 + B^2 + C^2}} \tag{11}$$

for i=1, ...,number of trim line points.

This calculated distance index has different signs for points at different side of the cutter plane. If the points are in the positive side of the cutter plane, the distances will be positive number, otherwise they will have negative sign. The zero-cross points are detected by checking the signs of the distances. The intersection point between a line linking two adjacent zero-cross points and the cutter plane is found to be the trim point in the cutter plane.

The tool path calculated in section 3.2 is first trimmed by these trim points and is then used to analyze the connection condition by algorithms discussed in section 3.3.

IV. EXPERIMENTAL RESULTS

The real-time tool path generation algorithms developed in this paper have been implemented on a Multibus II based CNC controller. The algorithms reside on a 80386 single board computer, called Free-Form Surface Milling Tool Path Processor, with a clock rate of 25 Mhz. CAD data files in IGES format can be read by the processor and 3D milling tool paths can be generated in real time during cutting.

A procedure is sketched below to illustrate the execution of this real-time NC tool path generator:

```
Tool_Path_Generation
Input Parameter:
        IGES File Name
        Cutter Size
        Cutter Direction (i,j,k)
        Tolerance for Chord Error
        Tolerance for Cusp Error
Main {
        Load IGES CAD data file /* read IGES to memory */
        Calculate boundaries for all surfaces;
        Do {  Put Cutter Plane at a start position;
              For (Each Surfaces of IGES 114) {
                    If (The Cutter Plane  pass:
                                    - Trim Curve if surface trimmed;
                                    - Surface Boundary if surface not trimmed) {
                          Find the start and end position of part surface;
                          Find the start and end position of normal offset surface;
                          Calculate cutter location data on normal offset surface;
                          Trim the cutter location data by trim curve;
                          not_pass_surface == FALSE;
                    } /* end of if */
                    else {  Cutter Plane do not pass this surface
                          not_pass_surface == TRUE;
                    }  /* end of else */
              } /* end of for */
              For ( All surface passed by the cutter plane) {
                    Sort the cutter location data of all surface;
                    If (i^th segment index point == i+1^st segment index point) {
                          If ( i^th segment gouged with i+1^st segment){
                                    Find Intersect Point;
                                    Trim cutter location data by Intersect Point;}
                          else { /*  i^th segment not gouged with i+1^st segment*/
                                    Find Interpolation Point between these segments}
                    else { /*i^th segment index point ≠ i+1^st segment index point*/
                          Move tool to safe zone between these two segments;
              } /* end of for */
              Pass Cutter Location data to Motion Processor through Multibus;
              Move cutter plane to another position based on cusp error tolerance;
        } while ( not_pass_all_surface == FALSE )  /* end of do */
} /* end of main */
```

A fender of truck is used to generate the NC tool path based on the algorithm proposed in this paper. Figure 13(a) shows the IGES definition data. Figure 13(b) shows the surface representation and its normal offset surface of this fender. The NC tool path for rough cutting is calculated as shown in Fig.

13(c).

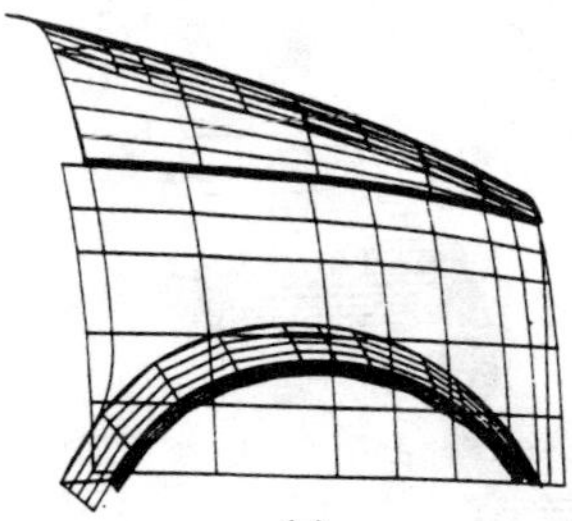

(a)
A truck fender in IGES representation

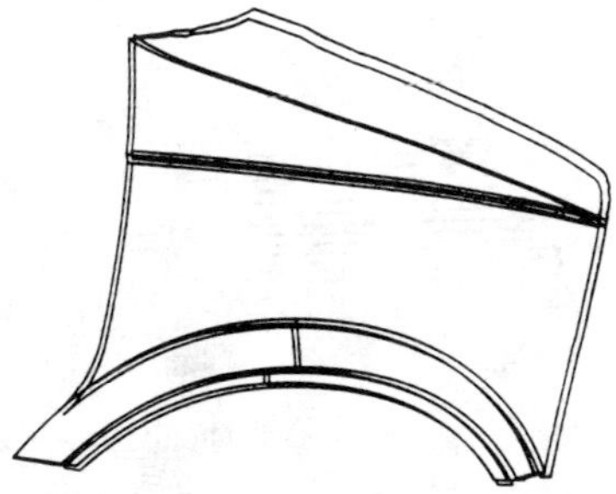

(b)
Fender surface and its normal offset surface

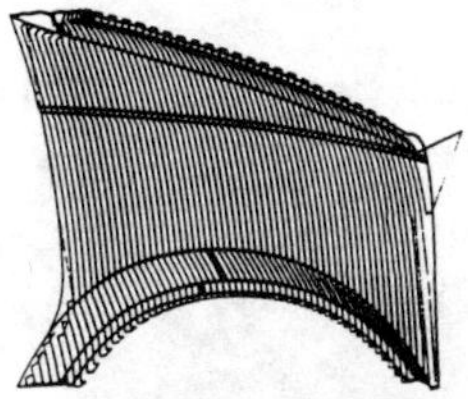

(c) Tool path for fender rough cutting

Fig. 13 Surface of truck fender and NC tool path

Another part shown in Fig.14 is used in a real cutting experiment since this part contains more gougings which the real-time algorithm has to process correctly. Using above tool path generation algorithm, the tool paths have been generated in real time and are stored for illustration purpose. In order to display properly, only rough cutting tool path has been shown in Fig.15.

Actual machining has been conducted based on the following experimental conditions: (1) The radius of the ball-end milling cutter is 3.175 mm. (2) The rough cutting is conducted with 2 mm liftoff, 0.02 mm chord error and 0.5 mm cusp error (3) Tolerable cusp and chord errors for finishing cutting are specified to be 0.02 mm. (3) Spindle speed is 2000 rpm and feedrate is 150 ipm. (4) Workpiece material is wax. It takes less than 1 second to generate one tool path across the entire work piece surface, which contains on average more than 100 cutter location points.

Figure 16 shows the part machined using the real-time tool path generation algorithm. It can be concluded from the machined part that the developed algorithm can properly handle problems such as trimming lines, gouging detection, and control of tool step sizes in real time. The surface contouring accuracy of the machined part is also checked on a coordinate measuring machine. It is found that the surface contour error of the cut part is less than 0.02 mm, which is within the specified tolerance.

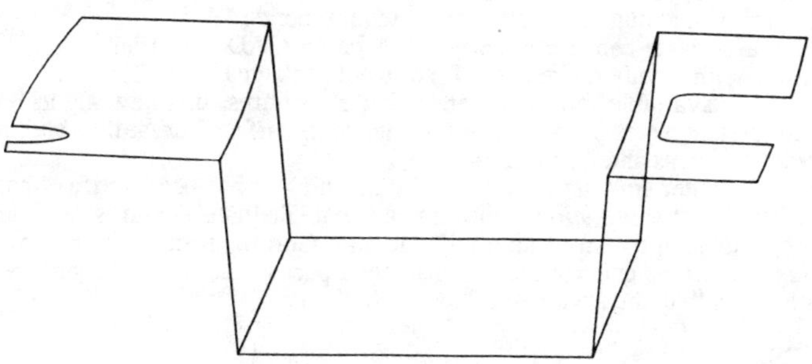

Fig. 14 Computer representation of the test part

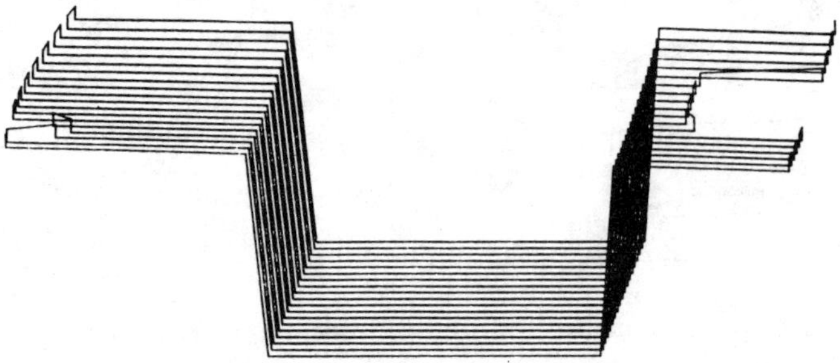

Fig. 15 Tool paths for the test part generated by the developed algorithm

Fig. 16 The photograph of the machined test part

V. CONCLUSION

A novel real-time NC tool path generation algorithm has been developed. Features of this new algorithm include:

(1) Three dimensional NC tool paths can be generated in real time during cutting, which has eliminated the need of compromising between the desired surface contouring tolerance and the required memory sizes for storing off-line generated NC tool path data. This provides a capability of greatly increasing the surface contouring accuracy without the lengthy NC code download time and huge memory storage requirements.

(2) This new algorithm can use directly IGES based CAD data files as its input, which makes it compatible with a wide range of CAD software packages.

(3) Unlike many available tool path generation algorithms, this new algorithm can automatically adjust its tool step size according to the change in surface curvatures, and, thus, can yield better surface smoothness and uniformity.

(4) Through computer simulation and actual machining, it has been verified that the developed real-time NC tool path generation algorithm can deal adequately with issues such as trimming lines, gouging, and adaptive tool step adjustment. Cutting results have shown that there are no violating conditions occurred on the machined part surfaces and the surface contour error of the cut part is less than the specified tolerance, 0.02 mm.

REFERENCES

1. Anderson, R. O., " Detecting and Eliminating Collisions in NC Machining", Computer Aided

Design, 1978, Vol 10, pp231-238.
2. Armstrong, G.T., Garey, G. G. and Penningtion, A. De,"Numerical Code generation from A Geometric modeling System", Solid Modeling By Computers, from Theory to Applications, Pickett and J. W. Boyse eds., Plenum Press, New York, 1984, pp139-154.
3. Barnhill, R.E.," Surface in Computer Aided geometric Design: A Survey with New Results", Computer Aided Geometric Design, Vol. 2, No. 1, 1985,pp1-17
4. Bobrow, J. E., "NC Machine Tool Path Generation From CSG Part Representations", Computer-Aided Design, Vol.17, No.2, pp69-76, 1985
5. Boehm, W. and Farin, G and Kahmann J., " A Survey of Curve and Surface Methods in CAGD",Computer Aided Geometric Design, Vol. 1, No. 1, 1984, pp1-60.
6. Choi, B K, Barash, M M and Anderson, D C, "Automatic Recognition of Machined Surface from 3D Solid Model", Computer Aided Design, Vol. 16,No. 2,1984,pp81-86.
7. Choi, B K, Jun, C. S., "Ball-end Cutter Interference Avoidance in NC machining of Sculptured Surfaces", Computer Aided Design, Vol. 21,No. 6,1989,pp371-378.
8. Choi, B K, Lee, C. S., Hwang, J.S. and Jun, C. S., "Compound Surface Modeling and Machining", Computer Aided Design, Vol. 20,No. 3,1988,pp127-136.
9. Culbreth, D. N., "Manufacturing Model: an Integrated Approach to Plnning, Design and Managing Industrial Facilities",Computer Aided Design, Vol. 21,No.1, 1989, pp49-53.
10. Drysdale, R. L and Jerard, R.B.,"discret Simulation of NC Machining", Proc. ACM Symp. Computational Geometry, June 1987.
11. Duncan, J. P. and Mair, S. G., "Sculptured Surfaces in Engineering and Medicine", Cambridge University Press, 1983.
12. Eversheim, W. and Holz, B.,"Computer Aided Programming of NC-Machine Tools by Using the System AUTAP-NC", annal of the CRIP, 31,1982, p323.
13. Farouki,R.T, " The Approximation of non-degenerate Offsert Surfaces", Computer Aidded Design, Vol.4, 1987, pp3-16
14. Faux, I. D. and Pratt, M. J., "Computational Geometry for Design and Manufacture", John Wiley, 1981.
15. Ferstenberg, R., Wang, K. K. and Muchstadt, J., "Automatic Generation of Optimized 3-Axis NC Programs Using Boundary Files", IEEE, 1986.
16. Gregory, C. L. and Ozsoy, T. M., " NC Machining of Free Form Surfaces", Computer Aided Design, Vol. 19, No. 2,1987,pp84-89.
17. "Initial Graphics Exchange Specification (IGES) Version 4.0, NBSIR 88-3813, National Bureau of Standards, U. S. Department of Commerce, 1988.
18. Jayaram, S., and Myklebust,A. "Automatic Generation of geometry Interfaces between Applications Programs and CADCAM Systems.", Computer Aided Design, Vol. 22, No. 1, 1990, pp50-56.
19. Jerard, R. B., Drysdale, R. L., Hauck, K., Schaudt, B. and Madgewick, J., "Methods for detecting Errors in Numerically Controlled Machining of Sculptured Surfaces", IEEE Comput. Graph. Appl. Vol. 9, No. 1, 1989.
20. Kakazu,Y., Okino,N. and Hoshi, K., "The Penalty Method of Determining Automatically a NC Cutter Path", CIRP Conference, 1974.
21. Mills, R. B., " Computer Graphics for NC Programming", CAE, Nov. 1984, pp32-42.
22. Okawa, Y., Miyazaki,M., "Automatic Generation of NC Tapes for 3-D Milling Machine"
23. Oliver,J. H., Goodman,E. D.,"Direct Dimensional NC Verification",Computer Aided Design, Vol. 22,No. 1,1990,pp3-10.
24. Persson, H, "NC Machining of Arbitrarily Shaped Pockets", CAD, Vol. 10, No. 3, May 1978.
25. Press, K., Kaplansky,E., " Automated Part Programming for CNC Milling By Artificial Intelligence Techniques:, Journal of Manufacturing System, Vol. 4, No. 1, pp51-63.
26. Shpitaini,M. and Lemaistre,G. "The Problem of NC Program Verification- Analysis and Approaches".
27. Suh, Y. S. and Lee,K., "NC Milling tool Path Generation for Arbitrary Pockets Defined by Sculptured Surfaces"", Computer Aided Design, Vol. 22 No. 5,1990,pp274-284.
28. Thomas, S. W., "Scanline Rendering for 3-Axis NC Toolpath Generation, Simulation, and Verification", CSE-TR-43-90, The University of Michigan, 1990
29. Wang, K.K. and Wang, W. P., "A PADL Based Numerical Control Machining Data Generation Program", Annual of the CRIP, Vol. 30, 1981, pp 359-362.
30. Wang, W. P. amd Wang, K. K.," Real-Time Verfication of Multiaxis NC Programs with Raster Graphics.", In IEEE Proc. Int'l Conf Robotics and Automation, Los Alamitos, CA, CS Press, 1986, p166-171.

Sensors, Controls, and Quality Issues
in Manufacturing
ASME 1991

IMPROVEMENT OF SERVOMOTOR CONTROL USING TWO-STEP REAL TIME COMPENSATION

Kazuo Yamazaki
Department of Mechanical,
Aeronautical, and Materials Engineering
Intelligent Manufacturing System
and Mechatronics Laboratory
University of California, Davis
Davis, California

Yiqi Zhou
Department of Mechanical Engineering
Shandong University of Technology
Shandong, People's Republic of China

Jae Hun Chung
Department of Mechanical, Aeronautical and Materials Engineering
Intelligent Manufacturing System and Mechatronics Laboratory
University of California, Davis
Davis, California

Abstract

Conventional servomotor systems have controllers with fixed gains which are designed from knowledge of servomotor dynamics or by manual tuning experiments. For the purpose of high speed and high precision response performance a feedforward two-step compensation algorithm composed of step reference command compensator and self-orgnized , self-adjusted real time gain tuning network can be used. With the help of flexible servo processor technology, the algorithm can be implemented easily. This paper describes the principle of two step compensation, the structure of real time gain tuner, and the mechanism of the self-adjustment of the tuning network. The practical applications of the method on AC, DC, and BLDC servo motor are also discussed.

1. Introduction

In recent development of manufacturing process high speed motion with high precision is an important feature which heavily depends on the characteristics of servomotor driving system. Also, the speed and torque control loop and the response must be quick, stable with no shock even in the case where the control target is subjected to the disturbances. The typical algorithm for the speed and torque control is fixed gain control algorithm aiming at zero steady state error (for example, the PI --- Proportional and Integral control). This kind of control algorithm, however, cannot provide the highest response with minimum possible settling time and zero overshoot that are the most desirable characteristics for driving control in a mechatronics system.

There are a variety of works that have been worked out by many researchers for improving the control performances and one of the effective methods is adaptive control such that the superior performance can be obtained compared to fixed gain control technique [1],[2],[3]. Those past researcher's works have used sophisticated control algorithm based on the modern control theory. Since the algorithms developed are complicated it is often difficult to implement in a practical controller that should be a cost effective.

One method that can be considered appropriately for cost effective solution would be a variable P.I. control with self-tuning and self-learning method. This paper deals with the new method for improving dynamic control characteristics in servomotor control for mechatronics system. The method proposed is so-called two step compensation with self gain tuning based on the application of neural network. The proposed method consists of two functions associated with reference command compensation and self gain tuning. The function of compensation is to change step reference command to sequenced two-step command to make the overshoot of the transient process smaller so that the faster system dynamic response and shorter settling time can be obtained. In order to get the best compensation result in any case of variation of servo system properties, the transient process is monitored and the gains are tuned as soon as the change of the system characteristics is detected. The principle of gain tuning is that the gain tuning network recognizes gain tuning pattern by matching memorized pattern with sampled data on transient response and then tunes the gains according to the tuning pattern. The gain tuning network is also under the self adjusting process including the generation and the refreshment of memorized pattern and the adjustment of gain tuning network are all automatically executed by the trained knowledge based on neural network.

The principle of two-step compensation is described in section 2, the mechanism of self gain tuning in section 3, the self adjustment of the neural network in section 4 and the experiment system survey and experiment results are discussed in section 5.

2. Two-step compensation

In the transient process of servomotor system, settling time Ts and overshoot ϕp are two major factors to be evaluated(see Fig. 1a). They can be adjusted by tuning the gains of the control system. However , there is a contradictory changing tendency between Ts and ϕp when gains are tuned in regular P.I. controller , in which the performance improvement on Ts (becoming shorter) is almost always companied by unexpected increment of ϕp, and vice versa.

In order to get minimum possible settling time Ts without increasing ϕp a certain method which is different from direct PI gain tuning must be used. Since the overshoot ϕp can not be limited in the range of [-d,d] when the shortest settling time Ts is achieved by tuning the gains of PI controller, the proposed method is based on a pre-adjustment on the step reference command R to meet the real output of the servo system depending upon known information on the transient response characteristics. The thought is that a given reference command R at first is changed into R' so that the maximum value of the response coresponding to modified input R' is in the range of [R-d, R+d], and then at a suitable moment t_p the reference command is changed back to R to get the correct stable response (see Fig. 1b). So the problem can be defined as how to find a suitable R' (or the difference R-R') and a suitable changing back time t_p.

The motor can be represented as a second-order motor model. The indicial response to the step reference command A is expressed by Equation (1) as follows :

$$y(t) = A\cdot [1 - \frac{e^{-\omega_n \zeta t}}{\sqrt{1-\zeta^2}}\cdot \sin (\omega_n \sqrt{1-\zeta^2} + \theta)] , \qquad (1)$$

where $y(t)$: system response to the step reference command,

ζ : damping coefficient, $\geqq 0$,

ω_n : natural frequency (1/T),

and θ : is equal to $\tan^{-1}(1-\sqrt{1-\zeta^2}/\zeta)$.

When

$$t = t_0 = \frac{\pi-\theta}{\omega_n \sqrt{1-\zeta^2}},$$

then

$$y(t_0) = A.$$

When

$$t = t_p = \frac{\pi}{\omega_n \sqrt{1-\zeta^2}},$$

then

$$y(t_p) = A\cdot (1 + \frac{e^{\frac{\zeta\pi}{\sqrt{1-\zeta^2}}}}{\sqrt{1-\zeta^2}}\cdot \sin \theta) \qquad (2)$$

$y(t_p)$ is the maximum value of the response to the reference command A and $y(t_p) - y(t_o)$ is the overshoot to A. If at the instance $t=t_p$ a second reference command A' is given to the servo motor system, the response will be the sum of two processes. One is started at t=0 with reference amplitude A and the other is started at $t=t_p$ with reference amplitude A'. (See eq.(3))

$$y(t) = A + A' - \frac{e^{-\omega_n \zeta t}}{\sqrt{1-\zeta^2}} * \sin(\omega_n \sqrt{1-\zeta^2} + \theta) * [A + A' * e^{-\frac{\zeta\pi}{\sqrt{1-\zeta^2}}}] \qquad (t \geqq t_p) \qquad (3)$$

By substituting the following A'

$$A' = A * e^{-\frac{\zeta\pi}{\sqrt{1-\zeta^2}}} \qquad (4)$$

The equation (3) will become

$$y(t) = A + A * e^{\frac{\zeta\pi}{\sqrt{1-\zeta^2}}} \qquad (t \geqq t_p) \qquad (5)$$

The equation (5) is a constant and independent with time t ($t \geqq t_p$), which means that the response curve will be a horizontal line after $t=t_p$. Therefore if the parameter A is set such that

$$R = A + A * e^{\frac{\zeta\pi}{\sqrt{1-\zeta^2}}} \qquad (6)$$

then it can be expected that the system dynamic response will be much quicker and with no overshoot.

The above analysis is based on the ideal second-order motor open loop control model, and the analysis result can not directly be applied to closed loop controller. But the modelling error can be eliminated to a minimum extent by tuning the gains of P.I. controller and the compensation parameters during the control process because the two step compensation algorithm is composed of two main actions: two step compensation and intelligent gain regulation.

Structure of compensator

For simplifying control model, two other parameters S and T are adopted instead of using A and t_p. S is the pre-adjustment value of the reference command in equation (7):

$$S = R - A \qquad (7)$$

and T is the time delay after the instance $t=t_o$ so that system response reachs level A (see Fig 1b). The structure of the compensator is described in Fig. 2.

The main operation in compensator is as follows:

1. When the compensator receives a new reference command R, it checks the amplitude of incoming command to evaluate if a two step compensation process is needed at first.
2. If the compensation is needed the compensator gives feedforward compensation S, so that the real input command to the P.I. control loop is the algebraic combination R and S.
3. When the compensation process is started, the compensator simultaneously begins to observe the servo system output y(t) in order to decide the starting point of delay counting. When y(t) reaches the point $y(t_o)$, the delay counting will be started.
4. After the delay counter reaches T, the compensation process is finished and the real input command to the control loop is changed back to R.

3. Mechanism of self gain tuning

In practical situation, it is difficult to deduce the exact values of parameters S and T when PI controller is designed because of modelling error and the co-effection from compensator to the feedback system. The two step compensation algorithm adopts an additional gain regulator to tune the parameters of compensator and the gains of P.I. controller (kp and ki) such that the characteristics of the servo system from both sides of compensator and controller is optimized. The function of the regulator is mainly divided into two part : the one is the gain tuning mechanism and the other is the self learning mechanism. The gain tuning mechanism is the execution part of the regulator and is discussed in this section. The self learning mechanism is discussed in the next section.

The structure of the gain tuning mechanism is illustrated in Fig. 3. It consists of a pattern recognizer , a tuning command generator and a gain tuner. The pattern recognizer is a comparison network and the sampled data of the system response are compared with the data of memorized tuning pattern. By using the network, matching score between the system response and every memorized tuning pattern is calculated. After maximization only the pattern having maximum matching score is selected and the output related to selected pattern is set to "1". Also, the outputs of other patterns are cleared to "0". The tuning command generator is a decision network that connects the tuning pattern and some gain tuning rule (for example to increase or decrease some gains). Also, the decision network gets inputs from matching network and generates tuning command (the address of the tuning rule in real algorithm). The gain tuner changes the gains according to the tuning command. The matching network is constructed by a set of so called Fuzzy-And element. In Fig. 4a, one Fuzzy-And element is illustrated. It has three nodes, two inputs and one output . The two inputs are sampled data and tuning pattern data corresponding to a particular sampling point. The set of Fuzzy-And element generates the relative relation during a term of sampling period between the system response and tuning pattern (see Fig. 4b). If the sampled response data fall into the matching area, they are considered being matched with the pattern. The matching relation in one Fuzzy-And element is described in eq.(8)

$$y_o^i = f[x_a+x_b-2] \qquad (8)$$

where

$x_a = f[-(x_s^i-x_p^i-\theta_s)]$, the output of node A

$x_b = f[-(x_p^i-x_s^i-\theta_p)]$, the output of node B

x_s^i: sampled data

x_p^i: pattern data

y_o^i: matching score at a point

$f[\alpha]$: hard limit function, $f[\alpha] = 1$ when $\alpha \geqq 0$; $f[\alpha] = 0$,else.

The matching relation between the system response signal and memorized tuning pattern is a fuzzy area instead of fixed line that gives a robustness to the gain tuning process. If every tuning pattern is matched with sampled response data then the tuning pattern output is achieved. By adjusting the thresholds θs and θp the matching area can be adjusted easily. In the experimental system the threshold θs and θp are fixed at 0 and 5 rpm. Though in the matching network the connections are fixed, it still gives good performance for robust control. The decision network is a one-layered neural network. Its input is the output of the matching network and the every output node is connected with a tuning command address. Between the input nodes and output nodes there are connection weights, and the gain tuning relation is adjusted by changing the weights. If the value of the output is "1" , then the related tuning command vector is sent to the gain tuner and the gains can be adjusted.

4. The mechanism of self adjustment

Self learning and self adjusting is the most important part of the algorithm and at the beginning stage of running the algorithm there is no memorized tuning pattern in the memory and no recognized tuning relation in the decision network. All of these are in a random condition . When the two step compensation process is started, the tuning patterns are identified, refreshed and stored into memory, and the gain tuning relation is also fixed and adjusted. Those learning and adjusting processes are carried on during every compensation process. The mechanism of self adjustment is mainly based on some direct optimizing search rules which are listed briefly as follows:

1. If the last tuning is successful, then the tuning relation is accepted.

2. If the last tuning is unsuccessful, then the changed gains are replaced by the latest successful gain data and the last tuning relation in the decision network will be readjusted in the next compensation cycle.
3. If the overshoot is in the range of [-d,d], then the new tuning relation in the decision network will not be accepted until the system transient response has the minimum overshoot among all the overshoot records.
4. The adjustment of tuning relations is only carried on those in which the tuning relation is not fixed.
5. The tuning relation of adjusting method is a co-ordination turn around direct search algorithm and the relation under the adjusted tuning command vectors will be tried on continuously until the tuning result (response characteristics) is successful.
6. After one successful gain tuning, the same tuning command vector will be kept using until one unsuccessful tuning is met and the gains and the sampled data with respect to the latest successful tuning result will be written into memory.

The terms used in the above rules are defined as follows:

1. Criteria: two criteria are used to evaluate the tuning result

a. Emax: $\text{Emax} = \max | R - X^i_s(t) |$ (i=0,n)

b. Earea: $\text{Earea} = \Sigma | R - X^i_s(t) |$ (i=0,n)

where R is reference speed command

X^i_s is the ith sampled system response data

2. Successful tuning: when the last tuning process is P(m) and the tuning process before last one is P(m-1), the last tuning process is called successful as long as one of Emax and Earea in P(m) is less than the one in P(m-1).
3. Best tuning : the criterion Emax in P(m) is the smallest one among all the tuning processes that have been done.
4. Unsuccessful tuning: both of the criteria Emax and Earea in P(m) are larger than ones in P(m-1).
5. Accepting tuning relation: to set the weight between the selected tuning pattern and the used tuning command vector to "1", the weights of the selected tuning pattern and the other tuning command vectors (not used in the present tuning process) are set to "0", and the sampled data are written into the tuning pattern data memory.
6. Unfixed relation: the tuning relation is known as incorrect and under searching.
7. Oldest relation: the tuning relation has the longest record that has not been used.

5. Implementation and Experiment

FSP-2(Flexible Servo Control Peripheral Version 2) designed by the Mechatronics-CIM Laboratory of Toyohashi University of Technology and fabricated by Fujitsu Corporation has been employed in the servomotor control system for the experiment. The FSP-2 has been developed by using the ASIC(Application Specific Integrated Circuit) Technology and can do the current loop servo control calculation including the feedforward control within 60 micro seconds. The main structure of the experiment system is illustrated in Fig. 5. The system structure is divided into two parts by a dotted line in the figure. Under the dotted line is the ordinary multiple loop controller in which an enhancement of responce and damping performance can be achieved by having the minor feedback loops of velocity and current. In the speed and current controller, a proportional-integral(PI) control algorithm has been employed to improve the steady state characteristics. The upper part is the additional two -step compensation mechanism including compensator and gain regulator. The system output is sent to the compensator as well as fedback to the speed control loop. After a certain transient process, the sampled data are stored in a temporary memory to be used for pattern recognition and the self adjustment of gain tuning network. During the same time of gain tuning operation, the controller also works for steady state response.

The experiment results on DC, BLDC, and AC motor have shown the effectiveness and feasibility of this method. The settling time of the transient responses of all the three kinds of motors are shortened about 10 - 15 percent, and the overshoots are limited within 0.001% of the target value. The gain tuning process is completely automatic and started as soon as the variation of the system characteristics is detected. The compared experiment results under the same condition are illustrated in Fig. 6, 7, 8.

6. Conclusion

This research has been carried out aiming at improving the performance of the servomotor control system. The results of the research are as follows:

(1) Two-step compensation method has been employed to eliminate the contradictory relation between settling time and overshoot in regular P.I. controller that the performance on settling time is almost always companied by unexpected increment of overshoot, and vice versa.

(2) In order to implement two-step compansation results in case of variation of servo system characteristics self gain tuning method using neural network has been adopted, so that the transient process is monitored and gains are tuned as soon as the change of the system characteristics is detected based on the trained knowledge.

(3) In order to verify the feasibility of the proposed compensation scheme for the servomotor control system the experiment has been performed by employing FSP-2 (Flexible Servo Control Peripheral Version 2) developed by using ASIC(Application Specific Integrated Circuit).

(4) The experiment results have shown that two step compensation method can be used as an adaptive regulator to the ordinary P.I.DSP(digital signal processor) controller to improve the transient response performance without special limitation. The algorithm is simple, flexible, and easily applied to the controller that has been used with digital processor. The experiment algorithm is realized by software, so that the calculation is still carried in sequence. However, the algorithm itself is totally parallel, which can be considered as the foundation of the adaptive regulator of some general purpose hardware.

Reference

[1].N.K.Sinha(ed);'Microprocessor-based control systems'.D.Reidel Publishing Company,(1986)

[2] George N.Saridis;'Self-organizing control of stochastic systems'.Marcel Dekker, inc

[3] Graham C.Goodwin and Kwai Sang Sin ;'Adaptive filting prediction and control'.Prentice-Hall,inc.

[4] Frank de Scheper;'A study on flexible and Intelligent servo controller for actuator in Mechatronics systems'.Ph.D theme Toyohashi University of technology (1989)

[5] Marvin Minsky and Seymour Papert;'Perceptrons: an introduction to computational geometry'.The MIT Press,(1988)

[6] Bernard Widrow, Rodney G. Winter, and Robert A. Baxter;'Layered neural nets for pattern recognition'.IEEE Trans on ASSP Vol.36, No. 7,pp1109-1117,July 1988

[7] Richard P. Lippmann;'An introduction to computing with neural nets'.IEEE ASSP Magazine, pp4-22, April 1987

[8] Nils J. Nilsson;'Principles of artificial intelligence'.Springer-Verlag , (1982)

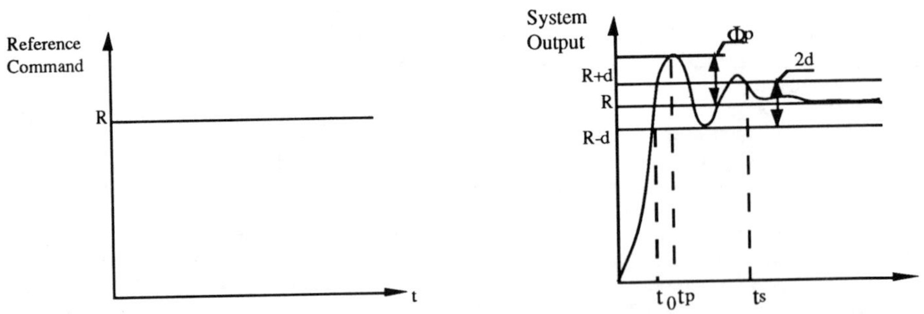

Fig.1a Transient response without compensation

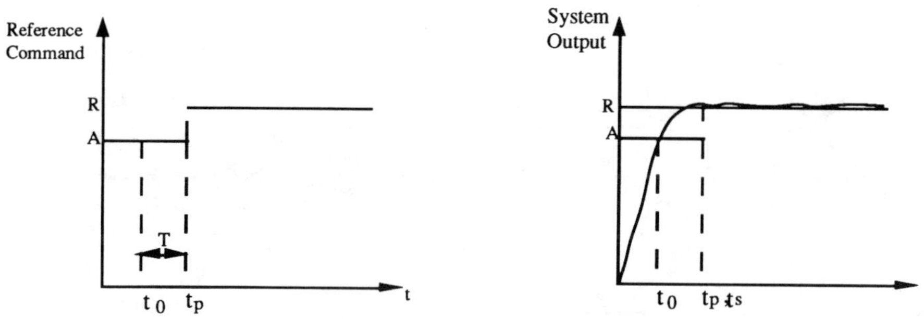

Fig. 1b Transient response with two-step compensation

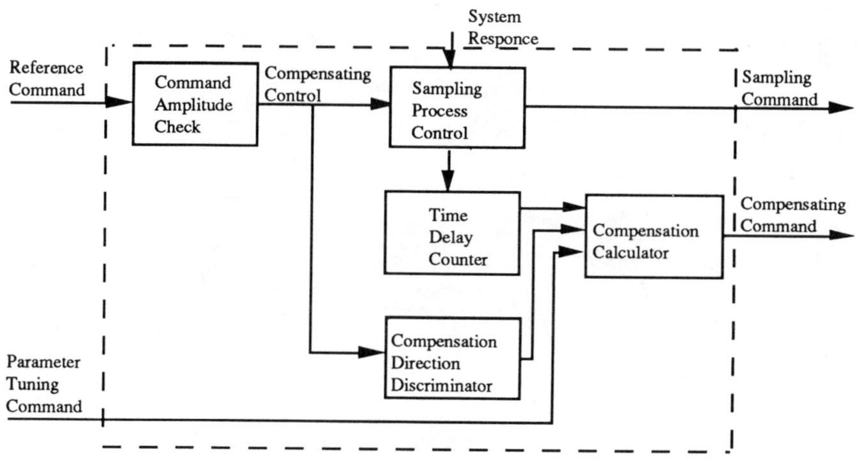

Fig.2 Compensator structure

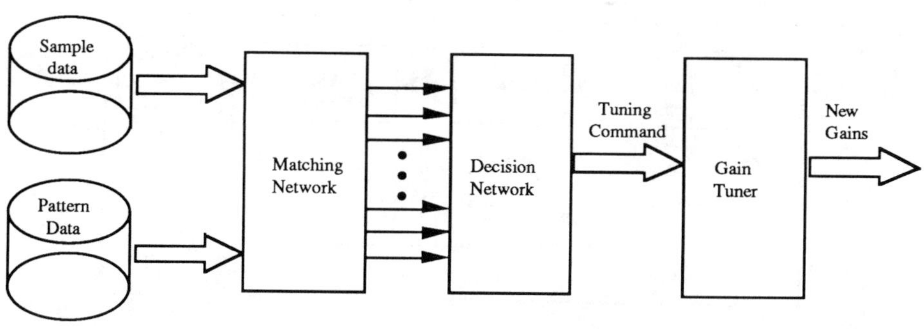

Fig. 3 Gain tuning mechanism

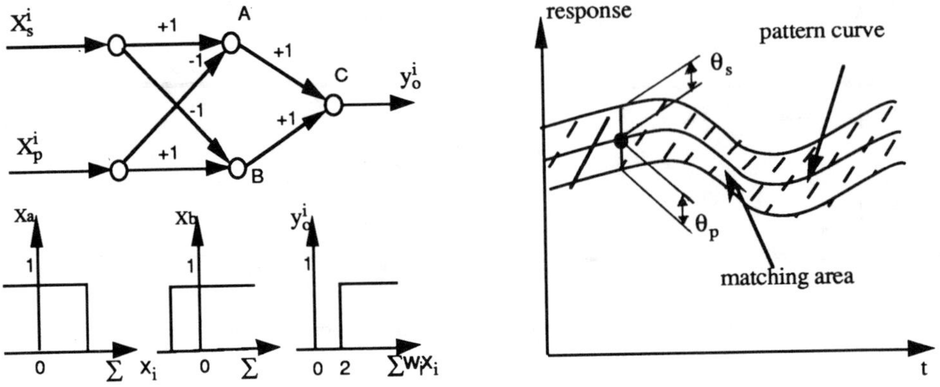

a. One fuzzy-and unit

b. Matching relation

Fig.4 principle of matching

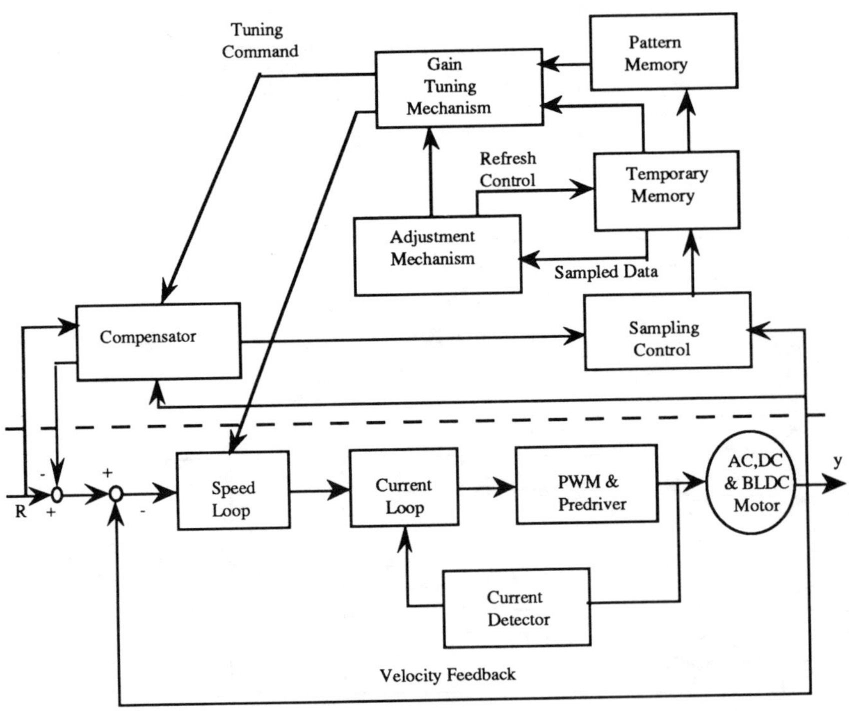

Fig.5 System configuration

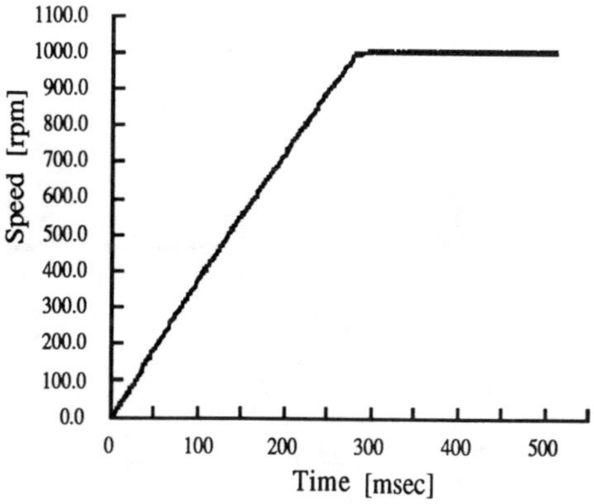

Fig.6.a DC motor with two step compensation

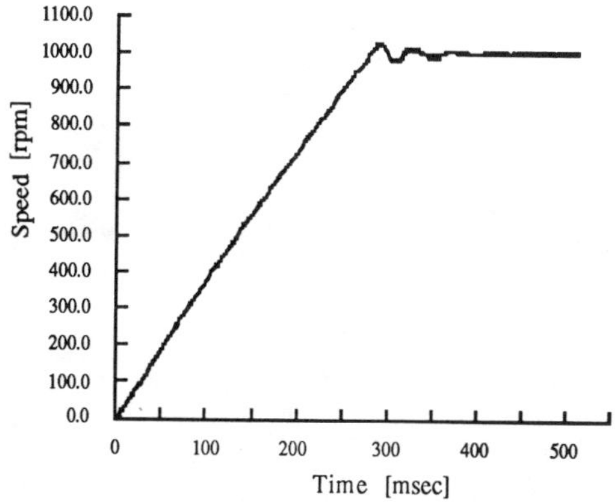

Fig.6.b DC motor without compensation

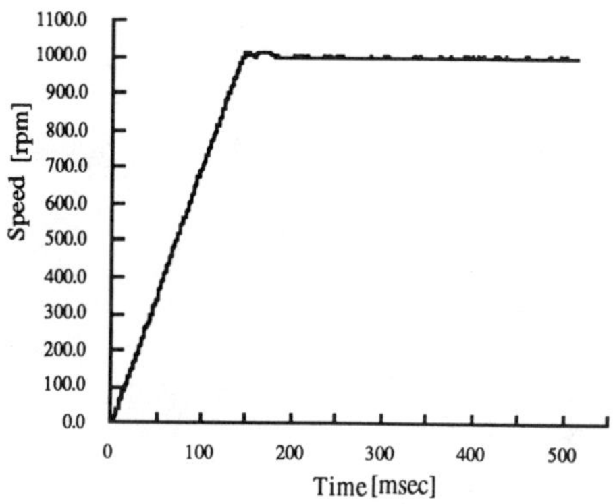

Fig.7.a BLDC motor with two step compensation

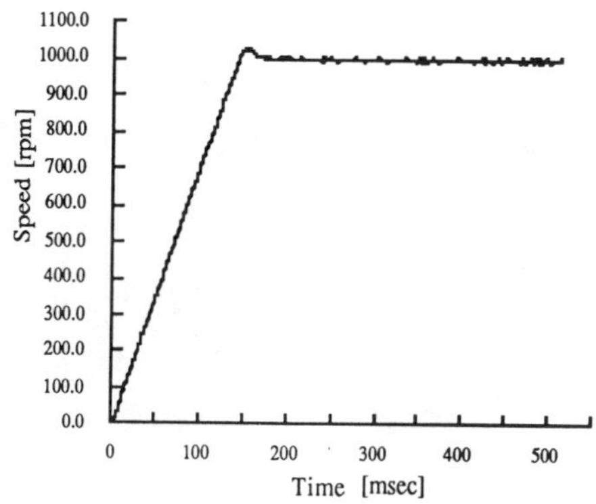

Fig.7.b BLDC motor without compensation

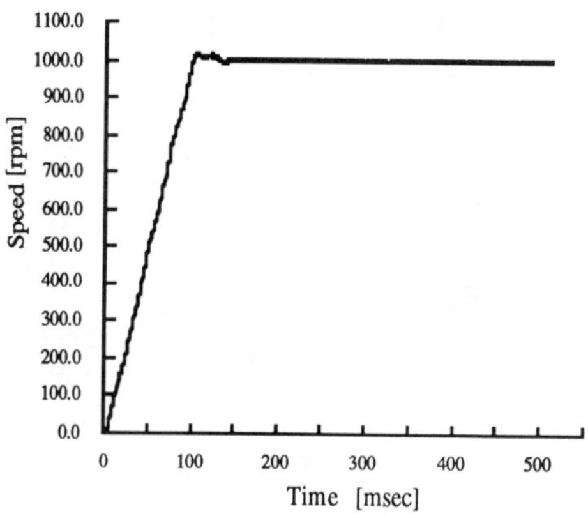

Fig.8.a AC motor with two step compensation

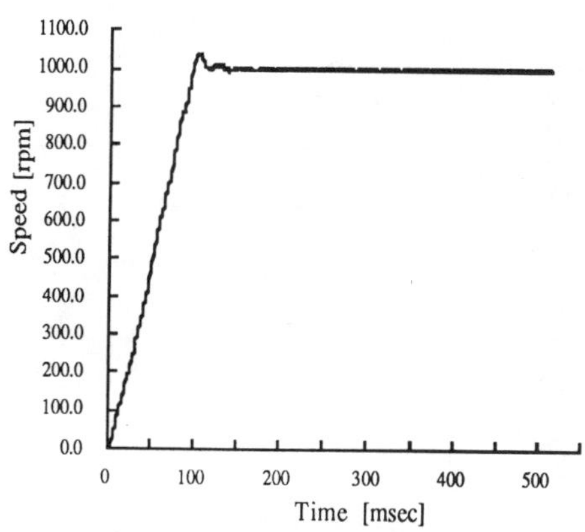

Fig.8.b AC motor without compensation

Sensors, Controls, and Quality Issues
in Manufacturing
ASME 1991

AN INTEGRATED APPROACH TO CALIBRATE AN UNTENDED MACHINING SYSTEM

J. E. Parker, Guangming Zhang, J. A. Kirk, and D. K. Anand
Department of Mechanical Engineering and Systems Research Center
University of Maryland
College Park, Maryland

ABSTRACT This paper presents a new methodology for the calibration of an untended machining system. The methodology requires the calibration conducted under both static and dynamic loading conditions. The transformation matrix method was used to establish the mapping function between the performance measure of interest (input) and the control signal (output). A prototype system, building on a CNC machining center equipped with a magnetic spindle, was developed to demonstrate a strategy for monitoring the tool wear progress through the cutting force prediction. The effect of phase distortion due to the presence of time lag on the output prediction was studied. A performance index to characterize the phase distortion error was suggested for further improvement of untended machining.

1. INTRODUCTION

Today's technological innovations in manufacturing are driven by demands to maintain a consistent, high level of product quality in the modern manufacturing environment. At the same time, the ever increasing competition in the world market pushes the manufacturing industry to reduce production costs and to increase productivity. These challenges are forcing the manufacturing industry to consider untended machining as one of the major technical innovations to succeed in the modern manufacturing environment.

The effectiveness of utilizing untended machining has attracted great attention from the manufacturing industry. Its impact on the improvement of production automation and machining accuracy are evident. Under untended machining, the operator is not in a position to sense the operation. To replace the operator, an on-line monitoring system is used to ensure the product quality. It is expected that the on-line system, as an integral part, is built in the machine tool. Consequently, the success of an untended machining system is entirely dependent on the methods to perform sensing and control during machining. During the past 30 decades, various sensors, based on force, torque, power, vibration, acoustic emission, and vision, have been developed. Many of them have been successfully applied in monitoring the machining process. The availability of sophisticated computers further enables us to incorporate signature analysis techniques into untended machining for the control of machining processes. The adaptive control system reported in (Tlusty et al., 1978) was used in a CNC milling machine. A dynamometer was attached to the spindle. Its on-line monitoring of the milling operation was through the detection of cutting force. A control algorithm was developed to adjust the table movements for keeping the dynamic variation of the cutting force within a preset level. It was reported in (Lan and Naerbeim, 1985) that chipping of the cutting tool during machining was detected through an on-line monitoring of cutting force. The detected cutting force at each sampling instant was used as an input to a discreet autoregressive model that estimated the status of tool wear during machining. A common characteristic in the previous approaches is that the sensors were attached to the machine tool and were not built in the machine tool. Consequently, these researchers performed the calibration process, which is a complex undertaking, on their own. It is felt that the complexity involved in the calibration process and in the determination of monitoring strategy has slowed down the wide acceptance of untended machining significantly.

There has been a deterministic trend that a new generation of machine tools should be built to carry out untended machining. In these machine tools, sensors are built to function as an integral part for the on-line monitoring of various aspects of machining such as tool wear, tool breakage, material handling, and in-process inspection. It can be expected that the success of building these machine tools is heavily dependent on the availability and capability of the built-in sensors, or the built-in monitoring system. One of the major issues involved in designing the monitoring system is its calibration because the calibration data provides a basis for diagnostics of the machining process.

The work reported in this paper is the calibration of an on-line monitoring system built in a Matsuura CNC machining center. The on-line monitoring system mainly consists of a S2M magnetic bearing spindle with an Intel based data acquisition system and a control system (Zivi, 1989, Zivi et al., 1990). During machining, the magnetic bearing spindle is subjected to the cutting force. The S2M control system reacts to the cutting force by varying the currents in the built-in electromagnetic coils to maintain the spindle in the center of its air gaps, both horizontally and vertically. To carry out a specific monitoring target such as tool wear, an indirect model-based method is used (Koren et al., 1989, Danai and Ulsoy, 1987). Through the detection of the cutting force variation, the tool wear progress during machining is retraced. we need to know the dynamic variation of the cutting force present during machining. Therefore, the calibration between the cutting force and the coil currents, or the air gaps, or both forms a basis for the establishment of a quantitative mapping function between them.

In section 2 of this paper, calibration under static loading is presented. The transformation matrix method used to establish the transfer function between the coil currents and the cutting force is discussed. Section 3 presents calibration under dynamic loading to characterize the dynamics of the magnetic bearing spindle in relation to the coil current variation. The integration of the two calibration processes forms a systematic approach to perform the calibration of an untended machining system.

2. CALIBRATION UNDER STATIC LOADING

2.1 Description of the Equipment for Untended Machining

In order to carry out an intended machining process, a Matsuura 500 Vertical Machining Center was retrofitted with an S2M magnetic bearing spindle in place of the conventional spindle (Zivi, 1989). The basic structure of the S2M spindle is shown in Fig. 1. There are two sets of magnetic bearing coils located in two horizontal planes and one set in the vertical axis. When electrical currents flow through these coils, the generated magnetic forces maintain a uniform air gap around the spindle. During machining, the magnetic spindle is subjected to the generated cutting force. As a result, the air gap becomes uneven. At the same time, the position sensors, which are shown in Fig. 1, detect such changes and send feedback signals to the controller. The controller adjusts the coil currents to recover the uniform air gap. This monitoring cycle provides the capability of performing an untended machining process.

2.2 Cutting Force Measurement

During an untended machining process, there is always a need to replace the cutting tool when it becomes worn. It has been known that direct measurements of tool wear seem difficult (Koren et al., 1989). On the other hand, indirect and model-based measurement methods to retrace tool wear through the cutting force measurement seem promising (Danai and Ulsoy, 1987, Chryssolouris et al., 1987, Park and Ulsoy, 1989). It has been reported that the thrust force acting on the drill during machining is a good indication for drill replacement

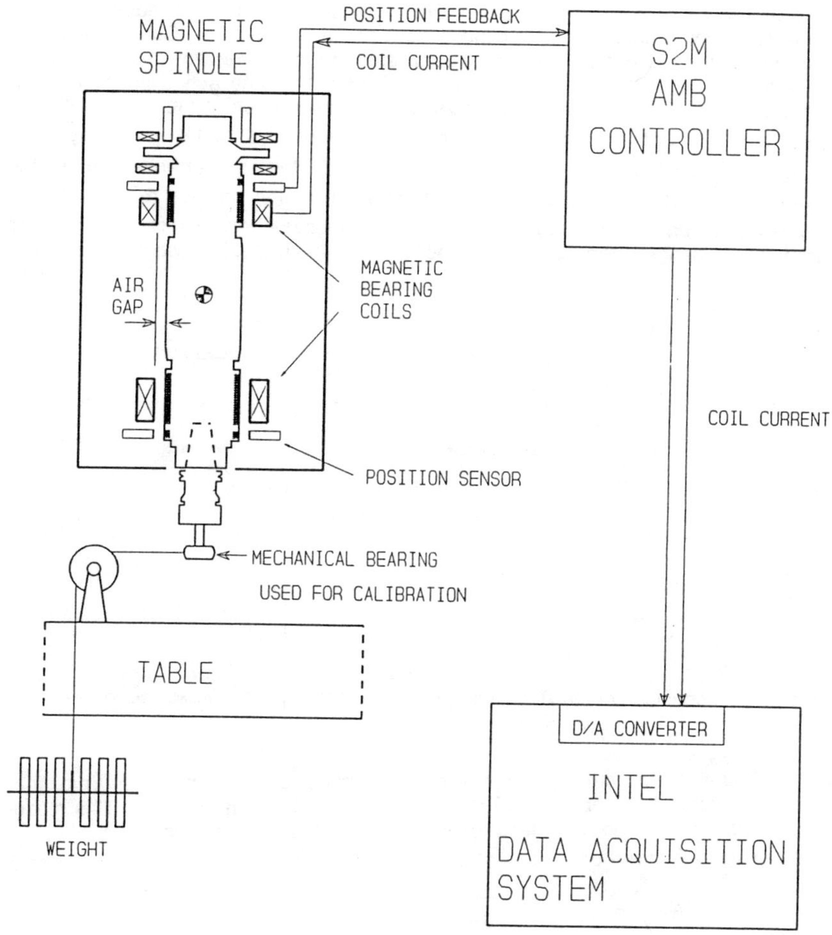

Figure 1 Matsuura 500 Machining Center equipped with the S2M Magnetic Bearing Spindle and the Experimental Setup for Static Loading Calibration.

(Bandyopadhyay and Wu, 1985). Therefore, it is imperative to study the relation between the cutting force, such as the thrust force, and the coil currents in order to carry out the untended drilling process on the retrofitted CNC machining center. The necessity to calibrate this relation can also be felt when machining composite/steel laminate materials. An on-line adjustment for feedrate and spindle speed during machining is needed when the drill reaches the boundary between the composite and steel materials. A reliable signal to actuate this adjustment would be again the thrust force.

2.3 Calibration under Static Loading

Since the magnetic bearing spindle functions as a sensing system, it is a common practice to apply a known and static force at the end of the spindle during the calibration process. When the force is applied, the corresponding changes in the coil currents are recorded. By increasing, or decreasing, the applied force, the relationship between the applied force and the coil currents, or the transfer function of the sensing system, can be established (Zhang and Kapoor, 1986).

It should be pointed out that the cutting force generated during machining is a vector in the three-dimensional space. The cutting force acts on the spindle. Consequently, there exist reaction forces at the bearing locations. As indicated in Fig. 2, the two reaction forces at the two radial bearings balance the two components of the cutting force acting in the horizontal plane (x-y plane), i.e., F_x and F_y. From the modeling viewpoint, each of the two reaction forces can be treated as a sum of two components, namely, the pair of F_{1x} and F_{1y}, or the pair of F_{2x} and F_{2y}. In the vertical plane, the reaction force at the thrust bearing balances the vertical component of the cutting force, F_z. Therefore, the calibration process requires analysis of the force transmission and establishment of the transfer functions between the applied force and the coil currents. In general, the calibration process consists of the following four steps.

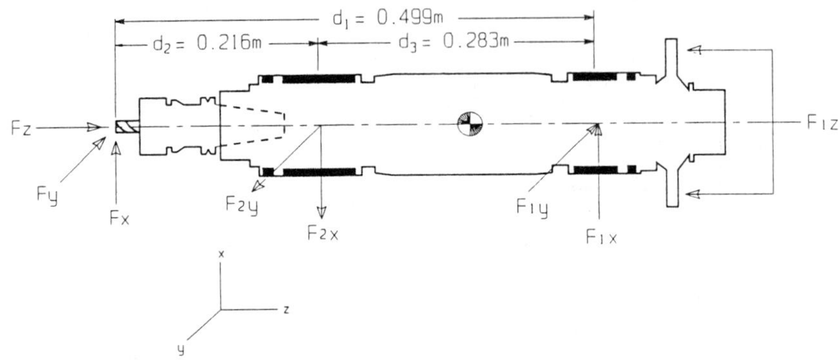

Figure 2 Decomposition of the Reaction Force Components when Subjected to the Applied Loads

Step 1: Transfer the cutting force, or the applied calibration force, into the five (5) components of the reaction forces at the three bearing locations. This means that the applied force is first decomposed into its three components, i.e., F_x, F_y, and F_z. Then, the three components are transferred into F_{1x}, F_{1y}, F_{2x}, F_{2y}, and F_z.

$$\begin{bmatrix} F_{1x} \\ F_{1y} \\ F_{2x} \\ F_{2y} \\ F_z \end{bmatrix} = \frac{1}{d_3} \begin{bmatrix} (d_2) & 0 & 0 \\ 0 & (d_2) & 0 \\ -(d_1) & 0 & 0 \\ 0 & -(d_1) & 0 \\ 0 & 0 & d_3 \end{bmatrix} \begin{bmatrix} F_x \\ F_y \\ F_z \end{bmatrix} \qquad (1)$$

where d_1 and d_2 are distances between the applied force and the two radial bearings, and d_3 is the distance between the two radial bearings. They are shown in Fig. 2.

Step 2: Identify the transfer function between the corresponding coil current and the reaction force component for each of the five pairs, i.e.,

$$\begin{aligned} TF_{1x} &= \frac{(\text{Coil Current})_{1x}}{(\text{Applied Force})_{1x}} = \frac{I_{1x}}{F_{1x}} \\ TF_{1y} &= \frac{(\text{Coil Current})_{1y}}{(\text{Applied Force})_{1y}} = \frac{I_{1y}}{F_{1y}} \\ TF_{2x} &= \frac{(\text{Coil Current})_{2x}}{(\text{Applied Force})_{2x}} = \frac{I_{2x}}{F_{2x}} \\ TF_{2y} &= \frac{(\text{Coil Current})_{2y}}{(\text{Applied Force})_{2y}} = \frac{I_{2y}}{F_{2y}} \end{aligned} \qquad (2)$$

$$TF_z = \frac{(\text{Coil Current})_z}{(\text{Applied Force})_z} = \frac{I_z}{F_z}$$

Step 3: Assemble the five identified transfer functions into a transformation matrix. The transformation matrix is a diagonal matrix, assuming that there does not exist any cross-talking among the coil currents when subjected to the cutting force.

$$[\,TF\,] = \begin{bmatrix} TF_{1x} & 0 & 0 & 0 & 0 \\ 0 & TF_{1y} & 0 & 0 & 0 \\ 0 & 0 & TF_{2x} & 0 & 0 \\ 0 & 0 & 0 & TF_{2y} & 0 \\ 0 & 0 & 0 & 0 & TF_z \end{bmatrix} \qquad (3)$$

Step 4: Combine Eqs. (1), (2), and (3) to derive the mapping function between the cutting force components and the coil currents. The mapping function in a matrix form is given below.

$$\begin{bmatrix} I_{1x} \\ I_{1y} \\ I_{2x} \\ I_{2y} \\ I_z. \end{bmatrix} = \begin{bmatrix} TF_{1x} & 0 & 0 & 0 & 0 \\ 0 & TF_{1y} & 0 & 0 & 0 \\ 0 & 0 & TF_{2x} & 0 & 0 \\ 0 & 0 & 0 & TF_{2y} & 0 \\ 0 & 0 & 0 & 0 & TF_z \end{bmatrix} \begin{bmatrix} \frac{d_2}{d_3} & 0 & 0 \\ 0 & \frac{d_2}{d_3} & 0 \\ \frac{-d_1}{d_3} & 0 & 0 \\ 0 & \frac{-d_1}{d_3} & 0 \\ 0 & 0 & 1 \end{bmatrix} \begin{bmatrix} F_x \\ F_y \\ F_z \end{bmatrix} \qquad (4)$$

In the present work, the calibration process was carried out in three separate directions. For example, when the applied force is oriented in the x direction, the force vector is equivalent to $[\,F_x \;\; 0 \;\; 0\,]^T$. This enables us to estimate TF_{1x} and TF_{2x}. Using the relation indicated in Eq. (1), we identify the two reaction forces in the x-direction, i.e., F_{1x} and F_{2x}. It has been observed that, among the five magnetic coils, only two magnetic coils respond to the applied force acting in the x-direction. Through the recording of the two coil currents, i.e., I_{1x} and I_{2x}, the two transfer functions can be established by the two ratios, $TF_{1x} = \frac{I_{1x}}{F_{1x}}$ and $TF_{2x} = \frac{I_{2x}}{F_{2x}}$ where I_{1x} and I_{2x} are the recorded coil currents during the calibration. Figure 1 illustrates the experimental setup of static loading to identify the five diagonal elements in the transformation matrix. During the calibration process, known weights were gradually added, from 10 Newtons to 510 Newtons with an interval of 50 Newtons in the x direction. The current changes in the five coils were recorded. Linear relations between the applied force and the coil currents were observed. The solid line shown in Fig. 3 is the plot of the applied force in the x direction F_x .vs. the coil current I_{1x}. Consequently, the slope of the line through regression analysis characterizes the transfer function $TF_{1x} = \frac{\Delta I_{1x}}{\Delta F_{1x}}$. In a similar manner, we estimate TF_{1y} and TF_{2y} when applying the static loading in the y direction, and estimate TF_z when applying the static loading in the z direction.

During machining, the detected signals from the on-line monitoring are the five coil currents. The corresponding three cutting force components can be known from the following matrix multiplication.

$$\begin{bmatrix} F_x \\ F_y \\ F_z \end{bmatrix} = \begin{bmatrix} 1 & 0 & -1 & 0 & 0 \\ 0 & 1 & 0 & -1 & 0 \\ 0 & 0 & 0 & 0 & 1 \end{bmatrix} \begin{bmatrix} \frac{1}{TF_{1x}} & 0 & 0 & 0 & 0 \\ 0 & \frac{1}{TF_{1y}} & 0 & 0 & 0 \\ 0 & 0 & \frac{1}{TF_{2x}} & 0 & 0 \\ 0 & 0 & 0 & \frac{1}{TF_{2y}} & 0 \\ 0 & 0 & 0 & 0 & \frac{1}{TF_z} \end{bmatrix} \begin{bmatrix} I_{1x} \\ I_{1y} \\ I_{2x} \\ I_{2y} \\ I_z \end{bmatrix}. \qquad (5)$$

It is worth noting that special cares have been taken during the calibration under static loading. First, a mechanical ball bearing was attached to the end of the magnetic spindle, as shown in Fig. 1. The calibration force was directly applied to the mechanical ball bearing, then transmitted to the spindle. This attachment allowed the calibration process to be carried out while the spindle was rotating in order to more closely mimic the machining environment. Second, the calibration process was duplicated at five different spindle speeds to study possible effects of the spindle speed on the transformation matrix. Among the five calibration data sets, three were plotted in Fig. 3. They represent the three calibration processes where the spindle speeds were set at 1000 rpm, 3000 rpm, and 5000 rpm, respectively. Examining the data shown in Fig. 3, there appears to be only a single straight line due to the fact that the estimated slope and intercept of each of the three calibration lines are so close to each other. The difference among them can be hardly displayed by the scale used in Fig. 3. Although the three lines are not exactly collinear, using the calibration line at spindle speed setting 1000 rpm for other spindle speed settings can be justified from the statistical point of view. Consequently, this observation could justify using one transformation matrix when the spindle is under operation from 1000 rpm to 5000 rpm.

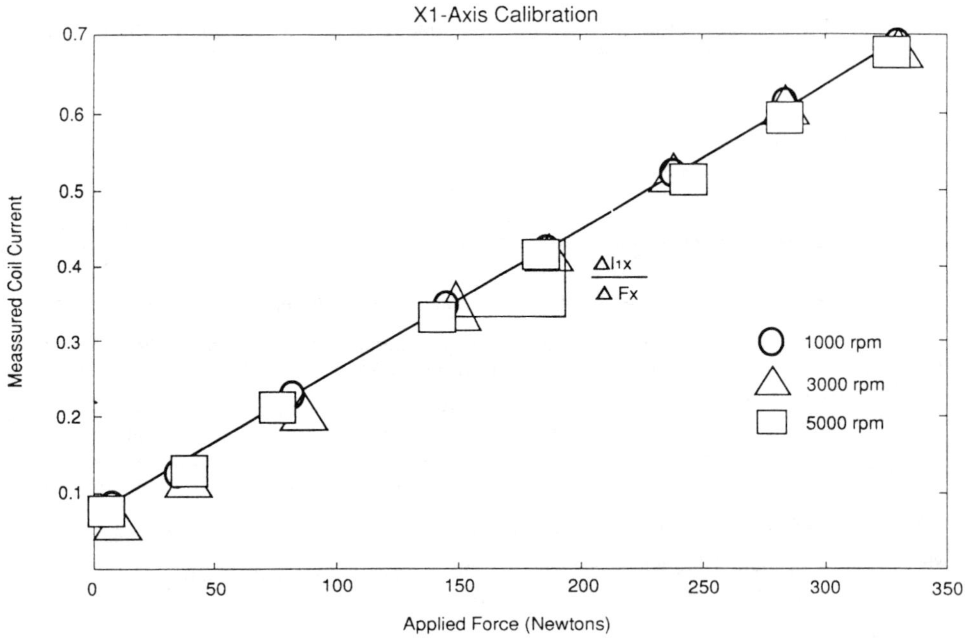

Figure 3 Calibration Data Plot to Illustrate the Determination of Transfer Function $\frac{I_{1x}}{F_{1x}}$.

3. CALIBRATION UNDER DYNAMIC LOADING

The need to perform a calibration under dynamic loading comes from the need to identify patterns of the dynamic variation of the cutting force. It has been reported that methods of using pattern recognition in the frequency domain are effective in detecting the tool wear progress during machining (Bandyopadhyay and Wu,1985). In order to perform the dynamic analysis of the cutting force variation in the frequency domain, and detect the dynamic variation of the cutting force during machining , the calibration process was also carried out under dynamic loading. The concern for calibrating the magnetic bearing spindle dynamically arises from the observation from previous research (Park and Ulsoy, 1989) that tool wear can be detected by monitoring the frequency components of the cutting force.

The calibration under dynamic loading consists of three parts, namely, the impulse response test, the frequency response test, and the phase distortion test. Each test is designed to perform specific tasks, such as identification of natural frequency, damping coefficient, gain factor, and phase shift. An integration of these test results presents a comprehensive description of the system dynamics of the magnetic bearing spindle in the frequency domain.

The impulse response test was used to determine the natural frequency and the damping ratio of the magnetic bearings. Figure 4 is the plot of the response of the magnetic bearing spindle to impulsive loading. The response pattern shown in Fig. 4 strongly suggests that the dynamic characteristics of the magnetic bearing spindle can be modeled as a second-order system. The recorded time interval for a single cycle is approximately equal to 0.008 second. The natural frequency of the magnetic bearing spindle is estimated to be 125 Hz. The damping coefficient, equal to 0.13, is also estimated from the decay envelope.

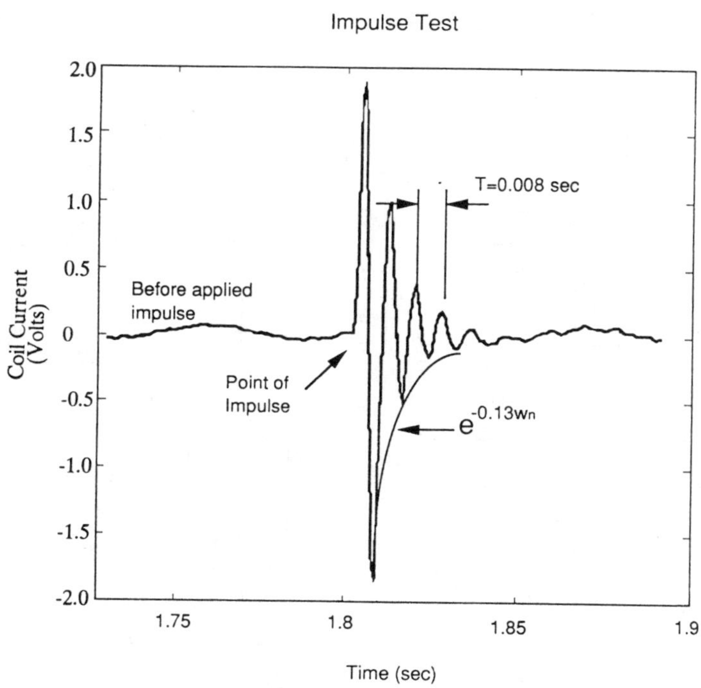

Figure 4 Experimental Data Recorded during the Impulse Test

To perform the frequency response test, a HP Digital signal analyzer and a shaker amplifier were used. The experimental setup is illustrated in Fig. 5. The signal analyzer injected white noise into the shaker amplifier to generate an excitation characterized with a flat frequency spectrum. Figure 6 presents the Bode plot from the signal analyzer. The response pattern is typical of a second-order system when subjected to white noise excitation. At the low frequency range from 0 Hz to 80 Hz, the gain factor is almost kept at a unity (db) level. At the high frequency range, the gain factor decreases as the excitation frequency increases. At the frequency range between 80 Hz and 200 Hz, the gain factor changes dramatically, indicating that the excitation frequency is closed to the natural frequency of the magnetic bearing spindle.

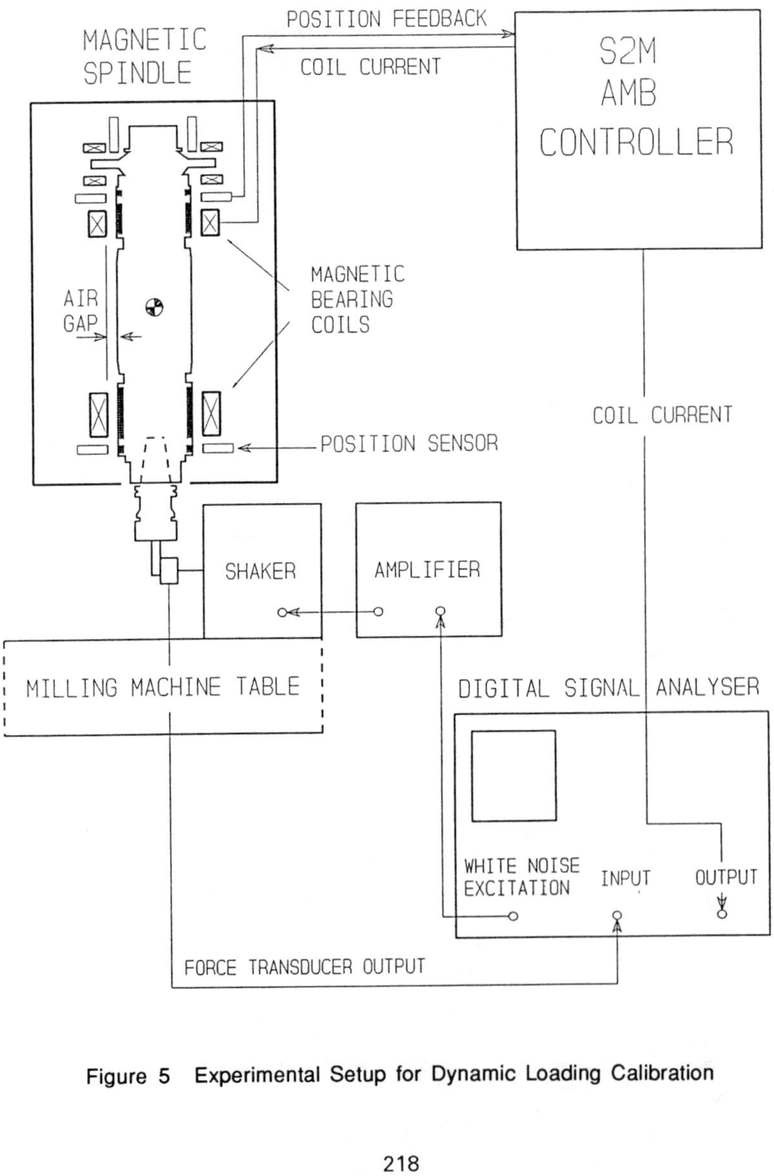

Figure 5 Experimental Setup for Dynamic Loading Calibration

As shown in Fig. 6a, the resonance frequency is about 125 Hz, which matches the experimental results obtained during the impulse response test. These observations confirm the validity of using a second order model to predict the response of the magnetic bearing spindle in the frequency domain. Examining the Bode plot carefully, the gain factor remains close to unity between 0 and 80 Hz. The gain factor increases significantly when the excitation frequency approaches 125 Hz. Figure 6b is the phase shift plot. It indicates that the phase shift is relatively constant between 0 and 80 Hz. The phase shift angle is about 12.5^{o} . This shift angle indicates that the response of the magnetic bearing spindle, or the dynamic variations of the coil currents, lag behind the dynamic variation of the cutting force during machining. The time lag is given by the phase shift angle divided by the angular frequency $2\pi f$ (Rao, 1990). It is evident that the time lag in the frequency range from 0 to 80 Hz varies and decreases as the excitation frequency increases. Consequently, the dynamic variation of the coil currents may not give an accurate picture of the dynamic variation of the cutting force during machining. This indicates that the distortion in the wave form of the applied force with respect to the recorded wave form of the coil currents is likely to occur.

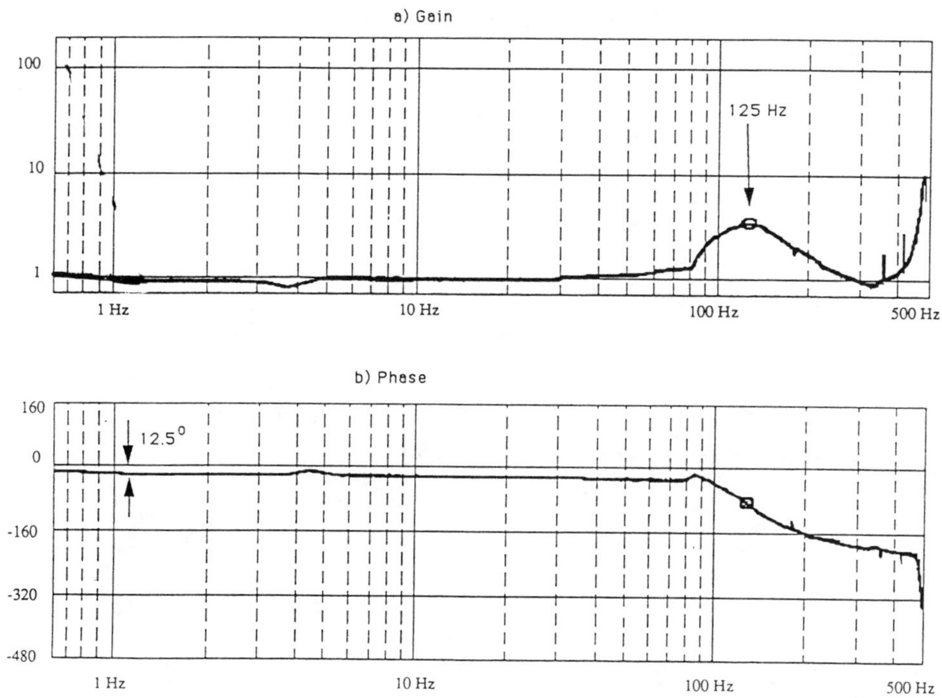

Figure 6 Plots of Amplitude and Phase Angle of the Coil Currents vs Excitation Frequency

In order to verify just how faithfully the magnetic bearing spindle sensing system will be recording the dynamic cutting force during machining, the phase distortion test was designed to quantify the time lag under different excitation frequencies. When the time lag is known, the difference between the recorded coil currents and the actual cutting forces can be evaluated. The actual cutting force can be recovered from the recorded coil currents. Figure 7 shows the experimental set up used. Two function generators were connected in parallel to the shaker amplifier. They excited the shaker with sinusoidal waves of two specified and different frequencies. As a result, the excitation force generated by the shaker amplifier consisted of two frequency components. It was expected that the recorded coil currents also contained the same frequency components. During the test, the excitation force acting on the spindle and the resulting coil current signals were recorded by an Haitachi 20 MHz digital storage oscilloscope, and down loaded into an IBM/AT. The input and output wave forms were then compared in the time domain. Figure 8 shows the input wave as a dotted curve and the output wave as a solid curve. The excitation frequencies were 10 Hz and 30 Hz. The time lag between the input and output was measured through a shifting process to mate the two patterns. As illustrated in Fig. 8a, the time lag is 10.5 ms. After shifting the output curve to the left for mating the input pattern, the areas between the two curves shown in Fig. 8b reflects the distortion between the input and output.

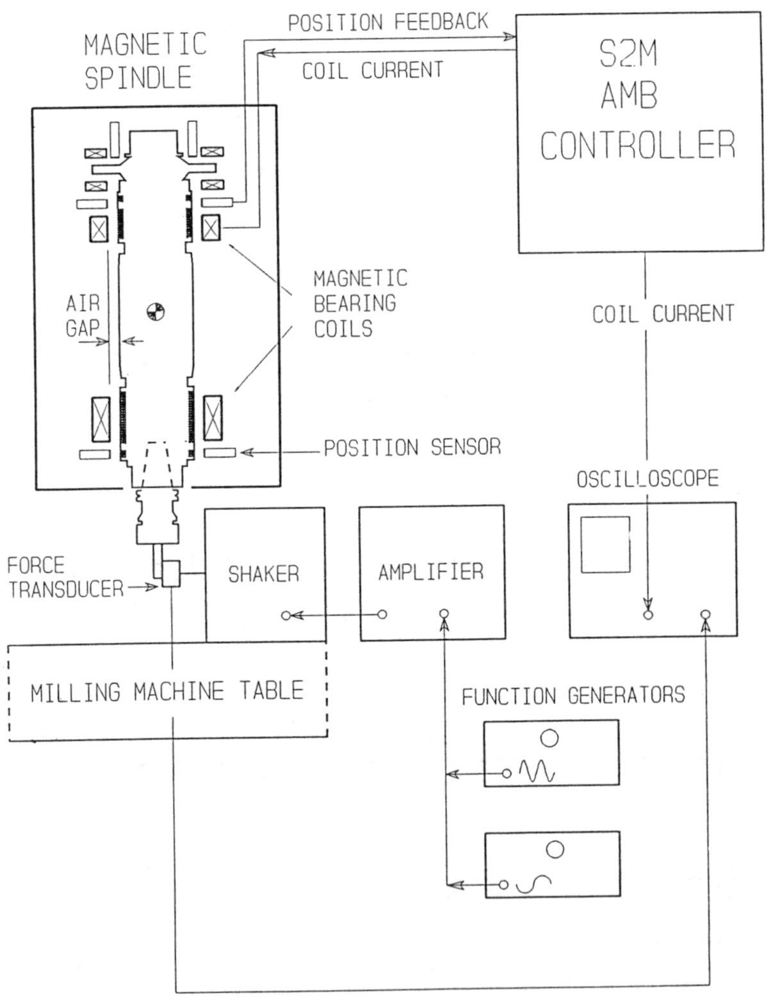

Figure 7 Experimental Setup for Testing the Phase Distortion

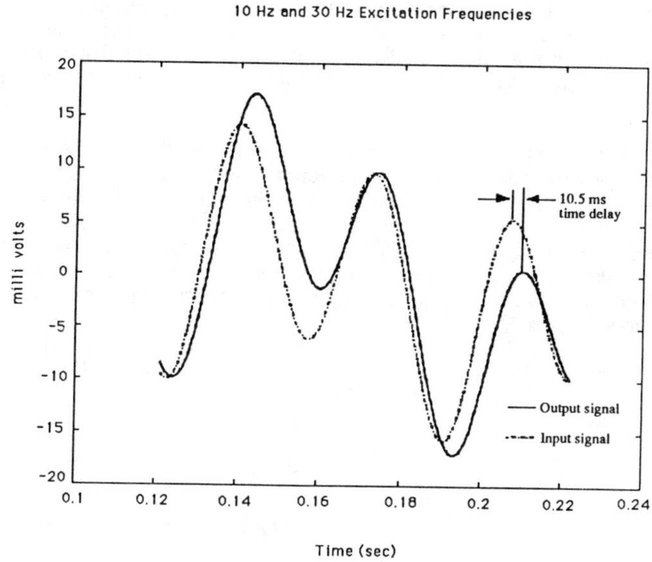

(a) Determination of Shift Time

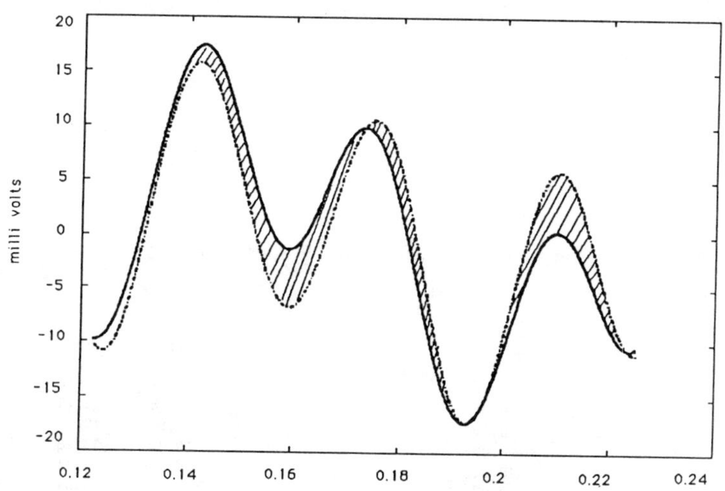

b) Error Due to Phase Distortion

Figure 8 Shift of the Coil Current Response with Respect to the Excitation Signal Consisting Two Frequencies (10 Hz and 30 Hz)

4. DISCUSSION OF RESULTS

4.1 A Case Study: On-Line Monitoring of Tool Wear

In order to demonstrate the on-line monitoring of tool wear during machining, a prototype system was developed using an indirect model-based method. We assumed that the magnitude of the cutting force would increase as tool wear progressed. In addition, a sudden drop in the magnitude of the cutting force would also indicate the occurrence of tool breakage. Therefore, as long as the magnitude of the cutting force varied within two preset limits during the monitoring, the machining process should continue on. Otherwise, human intervention should be called in.

Because of the availability of the transformation matrix through the calibration process, the monitoring of tool wear can be performed through the on-line detection of the coil currents. The machining process was to mill slots in a composite material. The end mill diameter was 12.7 mm. The cutting parameters were set at spindle speed = 1000 rpm, feedrate = 60 mm/min, and depth of cut = 4 mm. Table 1 is a list of the measured coil currents in a consecutive time order. The three predicted cutting force components are calculated using the calibrated transformation matrix. As an example, the three cutting force components listed in row 1 were calculated using Eq. (4).

$$\begin{bmatrix} 3.04 \\ 74.58 \\ 92.4 \end{bmatrix} = (4.45) \begin{bmatrix} 1 & 0 & -1 & 0 & 0 \\ 0 & 1 & 0 & -1 & 0 \\ 0 & 0 & 0 & 0 & 1 \end{bmatrix} \begin{bmatrix} 92.4 & 0 & 0 & 0 & 0 \\ 0 & 99.66 & 0 & 0 & 0 \\ 0 & 0 & 158 & 0 & 0 \\ 0 & 0 & 0 & 151 & 0 \\ 0 & 0 & 0 & 0 & 86.5 \end{bmatrix} \begin{bmatrix} .11 \\ -.15 \\ .06 \\ -.21 \\ .24 \end{bmatrix}$$

Table 1 Measured Coil Currents and Predicted Cutting Forces

Current Measered at Bearings (Volts)					Calculated Cutting Forces (Newtons)		
Ix1	Iy1	Ix2	Iy2	Fz1	Fx	Fy	Fz
0.11	-0.15	0.06	-0.21	0.24	-0.48	-51.45	-90.44
-0.03	-0.03	0.00	0.04	-0.13	11.52	-10.72	48.61
0.00	-0.09	0.04	0.09	0.31	-19.25	-13.26	-118.63
-0.04	0.02	0.07	-0.02	-0.02	-19.63	5.71	9.15
0.05	0.01	-0.01	0.02	0.10	-10.17	-11.34	-39.70
0.08	-0.01	-0.05	0.07	0.04	3.17	-30.78	-17.15
-0.05	0.01	-0.07	0.01	-0.28	51.18	-7.49	103.75
0.04	-0.09	0.05	0.23	0.34	-35.45	-81.43	-129.90
-0.06	0.03	-0.02	0.03	-0.30	28.32	-25.96	116.26
-0.06	-0.05	0.12	0.05	0.14	-40.66	-7.71	-52.86
0.01	0.02	-0.01	0.05	-0.58	2.19	-29.13	223.37
0.00	0.02	-0.04	-0.02	0.31	18.58	1.86	-118.63
0.07	-0.05	-0.04	0.06	1.37	-0.27	-13.89	-528.27
-0.01	-0.04	-0.02	0.06	0.26	15.55	-13.09	-101.71
-0.05	-0.16	0.06	0.17	1.05	-13.36	-29.17	-404.25

Figure 9 was the control chart designed for monitoring the thrust force acting on the end mill. The center line of the control chart was determined by the mean value, -65.12 Newtons, calculated from the 15 detected thrust forces (note: the time period should be longer for calculating this mean value in a practical application). The upper and lower control limits were established using the ± 3s principle where parameter s stands for the standard deviation of the thrust force about its mean level and is equal to 229 Newtons calculated from the 15 detected thrust forces. The decision-making policy was that the milling should be stopped when the on-line detected thrust force(s) beyond the upper limit, or below the lower limit.

4.2 Evaluation of Phase Distortion Effect

For an accurate monitoring of the cutting force during machining, it would be desirable to quantitatively evaluate the effect of the phase distortion on the cutting force prediction, and, if possible, to compensate the error for the recovery of the true cutting force signal (Rao, 1990). A performance index to characterize the phase distortion error was developed in this research. As shown in Fig. 8a, the displayed input and output signals were recorded directly from the oscilloscope during the phase distortion test. In Fig. 8b, the output signal was shifted to the left to match the input signal pattern. The shifted distance represents the time lag, which indicates that the dynamic variation of the measured coil currents lags behind the excitation force. By integrating the two curves to get the two covered areas and subtracting the two areas, a quantitative measure of the phase distortion error can be obtained. For comparison, the evaluated difference was normalized with respect to the time period used during the integration process. The normalized difference was used as the performance index to characterize the phase shift effect. The larger the difference, the greater the phase shift error. Several calculated normalized difference is listed below.

Combination of Frequencies	5 & 6	10 & 30	50 & 55
Normalized Difference	0.1	28.5	0.2

Examining the listed values, it is evident that the normalized difference increases as the difference between the two combined frequencies. The research is going on at the University of Maryland, is to compensate the phase distortion error to recover the true cutting force signal for further improvement of on-line monitoring.

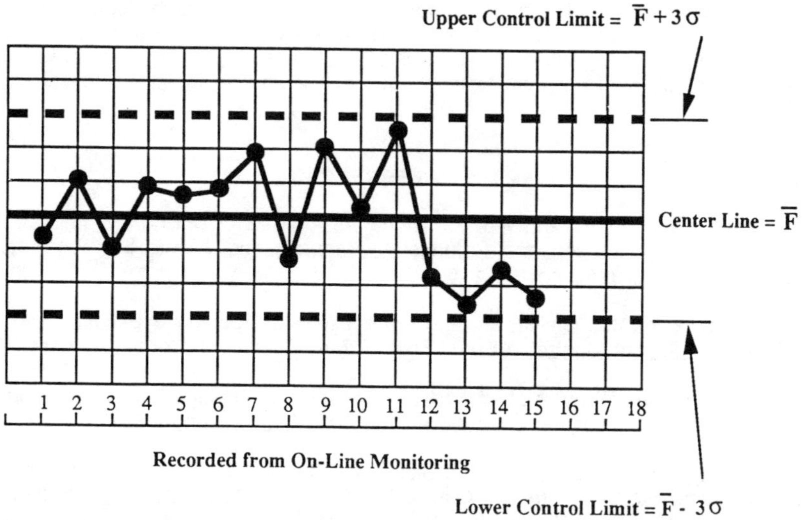

Figure 9 Control Chart Designed for Monitoring the Thrust Force during the Machining of a Composite Material

V. CONCLUSIONS

1. A CNC machining center equipped with a magnetic bearing spindle was used as a test bed to perform untended machining. In order to monitor the dynamic variation of the cutting force during machining, a strategy of measuring the coil currents in the magnetic bearings was used. The dynamic variation of the cutting force was being indirectly monitored through the detection of the coil current variation. A calibration procedure was developed to establish the relationship between the three components of the cutting force and the coil currents for the purpose of performing untended machining.

2. Calibration under static loading was carried out to identify the transfer function between the statically applied force and the coil currents. The matrix transformation method was used. The force decomposition matrix establishes the relationship between the applied force and the reaction force on each of the magnetic bearing locations. The mapping function matrix establishes the relationship between the reaction forces and the coil currents. Combination of these two matrices leads to the identification of the system transfer function under static loading.

3. Calibration under dynamic loading identified the dynamic characteristics of the coil current variation in relation to the dynamic characteristics of the magnetic bearing spindle. It has been found that system dynamics of the magnetic bearing spindle can be represented by a second order system with a natural frequency equal to 125 Hz and a damping coefficient equal to 0.13. The most significant finding is the distortion in the wave form of the applied force with respect to the recorded wave form of the coil currents. Time lags were experimentally determined to quantify the effect of phase distortion on the prediction accuracy of the cutting force through the measurements of the coil currents.

4. As a demonstration example, a prototype system to perform an on-line monitoring of the tool wear progress was presented. Using the indirect model-based method, the coil currents were detected during machining to predict the cutting force, which retraced the tool wear status during machining. A system performance index, called normalized phase distortion error, was developed to characterize the phase distortion effect on the prediction accuracy, showing good promise for further improvement of monitoring sensitivity.

ACKNOWLEDGEMENTS

The authors acknowledge the support of the Systems Research Center at the University of Maryland at College Park under Engineering Research Centers Program: NSFD CDF 8803012. They also express their gratitude to the College of Engineering for the financial support through Minta Martin funding. They wish to thank Dr. M. Abdulhamid for his technical support in this research.

REFERENCES

Bandyopadhyay, P., and Wu, S., "Signature Analysis of Drilling Dynamics for On-Line Drill Life Monitoring", Sensors and Controls for Manufacturing, ASME, San Fransico, CA, November 17-22, 1985, pp.101-110.

Chryssolouris, G., Guillot, M., and Domroese, M., "Tool Wear Estimation for Intelligent Machining", Intelligent Control, ASME, Boston, MA, December 13-18, 1987. pp.35-43.

Danai, K., and Ulsoy, A., "An Adaptive Observer for On-Line Tool Wear Estimation in Turning, Part I and II," Mechanical System and Signal Processing, 1987, pp. 211-240.

Koren, Y., Ko, T., Danai, K., and Ulsoy, A., "Methods for Tool Wear Estimation from Force Measurements Under Varying Cutting Conditions", Control Issues in Manufacturing Processes, ASME, Miami Beach, FL, December 10-15, 1989, pp.45-53.

Lan, M. and Naerheim, Y., "In-Process Detection of Tool Breakage in Milling", Sensors and Controls for Manufacturing, ASME, San Fransico, CA, November 17-22, 1985, pp. 49-56.

Park, J., and Ulsoy, A., "On-Line Tool Wear Estimation Using Force Measurement and a Nonlinear Observer", Control Issues in Manufacturing Processes, ASME, Miami Beach, FL, December 10-15, 1989, pp.55-63.

Rao, S., Mechanical Vibrations, second edition, Addison Wesley, 1990, pp. 501-502.

Tlusty, J., Cowley, A., and Elbestawi, M., "A Study of an Adaptive Control System for Milling with Force Constraint",Sixth NAMRC Proceedings, Society of Manufacturing Engineers, 1978, pp. 364-371.

Zhang, G., and Kapoor, S., "Development of An Instrumented Boring Bar Transducer," The 14th North American Manufacturing Research Conference Proceedings, pp. 194-200, May 1986.

Zivi, E., "Robust Control of a Magnetic Bearing Spindle for Increased Milling Tool Accuracy," Ph. D. Thesis, University of Maryland, 1989

Zivi, E., Anand, D., Anjanappa, M., and Kirk, J., "Magnetic Bearing Spindle Control for Accuracy Enhancement in Machining," 1990 ASME Winter Annual Meeting, Dallas, Texas, November 1990.

Sensors, Controls, and Quality Issues
in Manufacturing
ASME 1991

A GAME THEORY APPROACH TO THE OPERATION OF MACHINING PROCESSES

George Chryssolouris, Velusamy Subramaniam, and Michael Domroese
Laboratory for Manufacturing and Productivity
Massachusetts Institute of Technology
Cambridge, Massachusetts

ABSTRACT

This paper presents an approach to the operation of manufacturing processes which is based on game theory. The input settings of the process are selected by evaluating, on-line, a set of feasible alternatives with respect to several criteria. Relevant performance measures such as process cost and production rate can be directly influenced in this approach by establishing appropriate definitions for the decision criteria. In this paper this PROcess DEcision MAking (PRODEMA) approach is compared with an adaptive control (AC) scheme through simulations of a turning process. Results show that PRODEMA provides the ability to operate a process in accordance with overall performance measures.

1. INTRODUCTION

Efforts to control manufacturing processes have primarily focussed on the development of adaptive control (AC) schemes in which the input variables are determined on-line based on measurements of a process parameter and a model of the process. Among manufacturing processes, machining processes have been researched extensively and automatic control techniques have been suggested as a method of automatically adjusting the cutting parameters of a machining process. The application of such techniques requires a mathematical description of the process which may be very complex because a number of physical phenomena interact during the actual process [10,21]. This problem has been addressed by means of adaptation (Figure 1) whereby efforts are made to adapt the model parameters in such a manner that the model behavior closely reflects the behavior of the actual process.

The design and implementation of a model reference adaptive control (MRAC) scheme has been considered for force control in milling and turning [10,15,20]. In these control schemes the speed is held constant while the feedrate is adjusted in an attempt to maintain a constant cutting force. The MRAC approach consists of two control loops. The first loop is the ordinary feedback loop consisting of a controller and the process. The parameters of the controller are adjusted in the second loop so that the error between the outputs of the process and a reference model is minimized [1]. The small gain theorem has been used to design an adaptive controller for a turning operation in which the feedrate was used as the control input [3]. A self tuning regulator (STR) has also been used to control a turning process [9]. The STR also consists of two control loops, where the first is the ordinary feedback loop consisting of the process and a controller. In the second control loop, a recursive estimation of the parameters in the process model is performed based on the input and output of the process. The estimated parameters are then used to adjust the parameters of the controller [1]. An adaptive control optimization (ACO) approach was investigated in an attempt to optimize a milling process [21]; in this approach many parameters in several complex relationships

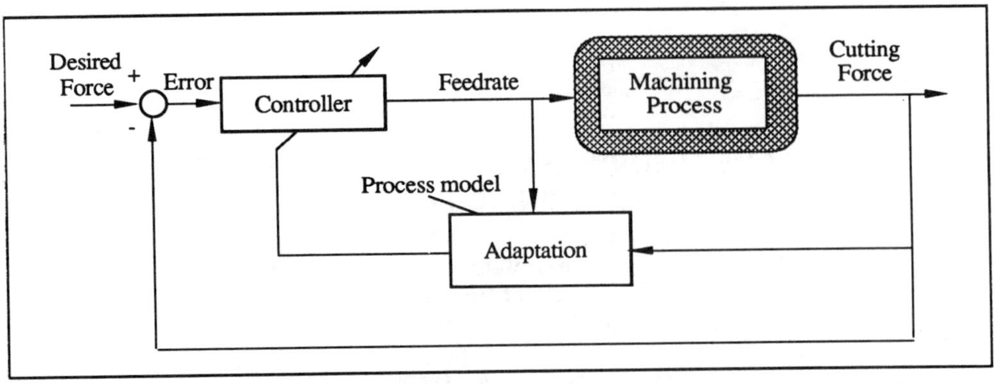

Figure 1. Adaptive Control Scheme for Controlling Force in Machining

can be selected off-line in order to obtain a model which reflects the complexity of the process. The optimal locus approach was used as the basis for an optimization procedure for the control of a grinding process [13]; in this approach the optimal input parameters are selected in-process based on the value of a measured constraint variable.

Adaptive control (AC) schemes attempt to maintain a measured variable at a reference value. Therefore, implementation of these approaches first requires the choice of an appropriate, measurable variable, such as force or temperature, to be controlled, and the selection of a reference value for this variable. With a few exceptions [4,5,18], adaptive control approaches for machining reported in the literature rely on a single sensing variable, which in many cases is force. These schemes typically attempt to maintain the cutting force at a preselected value because it is believed that such a strategy leads to a "better" machining process. However, it is not evident that maintaining the cutting force at a constant value will necessarily optimize a machining process with respect to any criteria such as cost or production rate. In the attempt to maintain the measured variable at a reference value, the process operation may become unstable. Furthermore, the issue of selecting an appropriate reference cutting force which will lead to adequate process operation has not been sufficiently considered.

As an alternative to an AC scheme this paper discusses a PROcess DEcision MAking (PRODEMA) approach which is based on Game Theory. Unlike the AC schemes, PRODEMA does not select the process input variables based on the magnitude of the error between a reference value and a measured signal, but rather selects on-line optimal, or at least "good" input variables based on relevant decision-making criteria. Since the inputs of the PRODEMA approach are selected from a set of feasible alternative inputs that will not lead to undesirable process behavior, the PRODEMA approach is not susceptible to instability in the sense of a typical closed-loop control scheme.

2. METHOD OVERVIEW

A typical machining process requires several operating parameters; for example, in a machining/turning process, the operating parameters could include the feed and cutting speed. For generality, let us assume that there are k such operating parameters, i.e., (OP_1, OP_2,......., OP_k). An operating point of the process is described by the specific values of each of these operating parameters. If these parameters are continuous in nature, they can be discretized so that the interval of separation determines the number of operating points. Each of these operating points is an alternative for the process. If the discretization is fine, the number of alternatives is large and if the discretization is coarse, the number of alternatives is small. Due to process constraints, some of the operating parameters may not be feasible and would have to be removed from the set of alternatives. The remaining elements of the set form a feasible solution space of operating points.

These operating points form the options of a "decision maker;" the criteria that are used to evaluate and select an alternative are considered by Game Theory to be the "states of nature/process." Thus the operation of a manufacturing process can be formulated as a two-person non-zero sum game [11,16,19] where the decision maker is Player 1 and the process is Player 2. A consequence or outcome can be calculated for each alternative criterion pair. A typical game for a machining/turning process (Figure 2) can be conceived having as alternatives combinations of feed, f_r, cutting speed, v, production rate and workpiece quality as criteria.

Alternatives	Criteria		
	Production rate	Workpiece quality	...
$(f_{r\,1}, v_1)$	a_{12}	a_{13}	...
$(f_{r\,2}, v_2)$	a_{22}	a_{23}	...
$(f_{r\,3}, v_3)$	a_{32}	a_{33}	...
⋮	⋮	⋮	⋮
⋮	⋮	⋮	⋮
$(f_{r\,n}, v_n)$	a_{n2}	a_{n3}	...

Figure 2. A Game Matrix for Operating a Turning Process
$(f_{r\,i}, v_i)$: alternative pair operating point,
a_{ij} : value of alternative i with respect to criterion j

A matrix entry, a_{ij}, for an alternative i and a criterion j is calculated based on the definition of the criterion j and the particular value(s) of the alternative i. In order to make the comparison among the different alternatives independent of the different units, a normalization procedure can be applied to normalize the matrix entries so that their values do not have units and are bound between 0 and 1 [16]. Once the matrix has been normalized, a number of different methods can be applied [11,16,19] in order to select the best alternative.

Considering game theory as the basis for the operation of a manufacturing process (Figure 3) provides a number of advantages:

- For many manufacturing processes it is desirable to optimize the process with respect to several different criteria. A game theory approach can easily consider multiple criteria for the selection of the optimal process input variables (Figure 2), and thus account for various "global" performance measures.

- A game theory approach can easily consider multiple sensors and process models [4,5] so that multiple-sensor information can be synthesized or fused.

- The operation of manufacturing processes and their control is associated with a degree of uncertainty, primarily due to changes in material properties and process modelling inaccuracies. These uncertainties in the process parameters can be addressed by utilizing concepts developed in game theory for making decisions based on uncertain information. Work has been done concerning the application of these concepts within the PRODEMA approach [6].

3. IMPLEMENTATION

In order to demonstrate the differences between AC and PRODEMA, the turning process was used as a "test bed" for comparison purposes. The AC approach considered in this paper attempts to maintain a constant cutting force [20] by controlling the feed, f_r, whereas PRODEMA chooses the cutting speed, v, and feed, f_r, based on the evaluation of several decision-making criteria.

The Turning Process

The comparison of the two approaches was made with the help of the turning process, which was simulated based on a model with the feedrate command U_c as input (Figure 4) and the cutting force F_m as output [20]. This model, in its original form [20] accounted for:

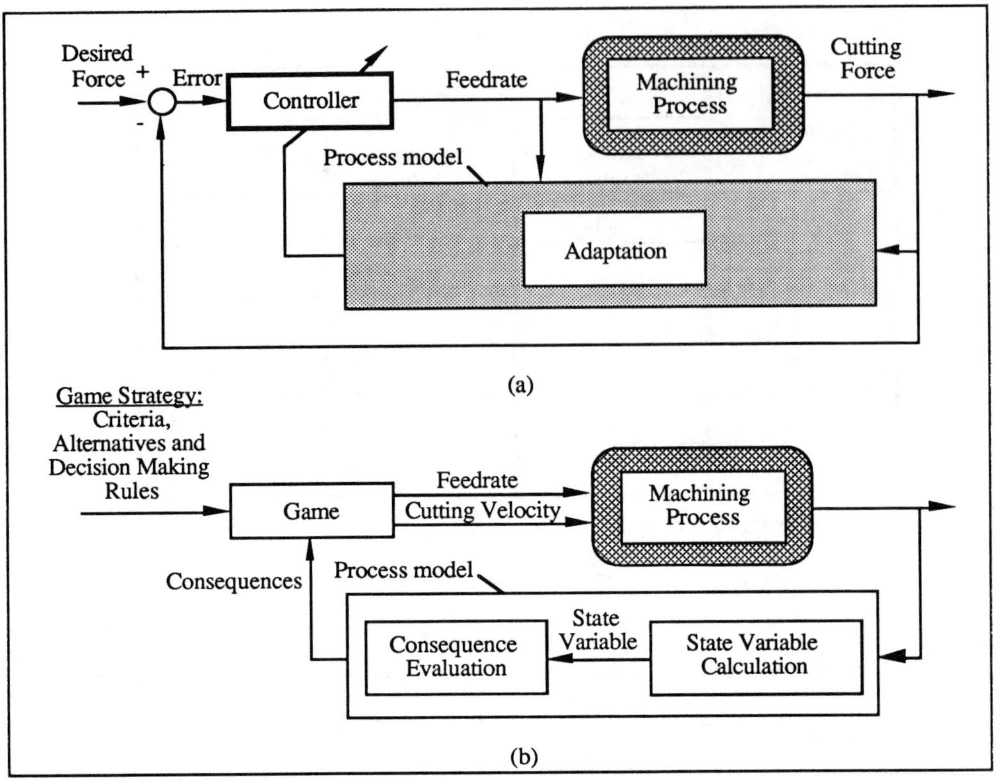

Figure 3. Adaptive Control (a) vs. Game Theory (b) Approach for Machining

a) Carriage motion dynamics

$$\frac{U_a}{U_c} = \frac{0.552z + 0.4529z}{z^2 + 0.215z + 0.2466} \tag{1}$$

where

U_c : command feedrate (volts)
U_a : actual feedrate (cm/sec)

b) Feedrate/chip thickness dynamics

$$\frac{f_r}{U_a} = \frac{B\,(z^4 + z^3 + z^2 + z + 1)}{z^5} \tag{2}$$

where B = 0.01 for the uncut chip thickness/feed, f_r, and actual feedrate, U_a, in cm and cm/sec, respectively. This transfer function [20] depends on the sampling time and the spindle speed.

c) Cutting process

$$F = K\, d\, f_r^p \tag{3}$$

where

F : cutting force
K : constant
d : depth of cut
p : constant exponent
f_r : feed

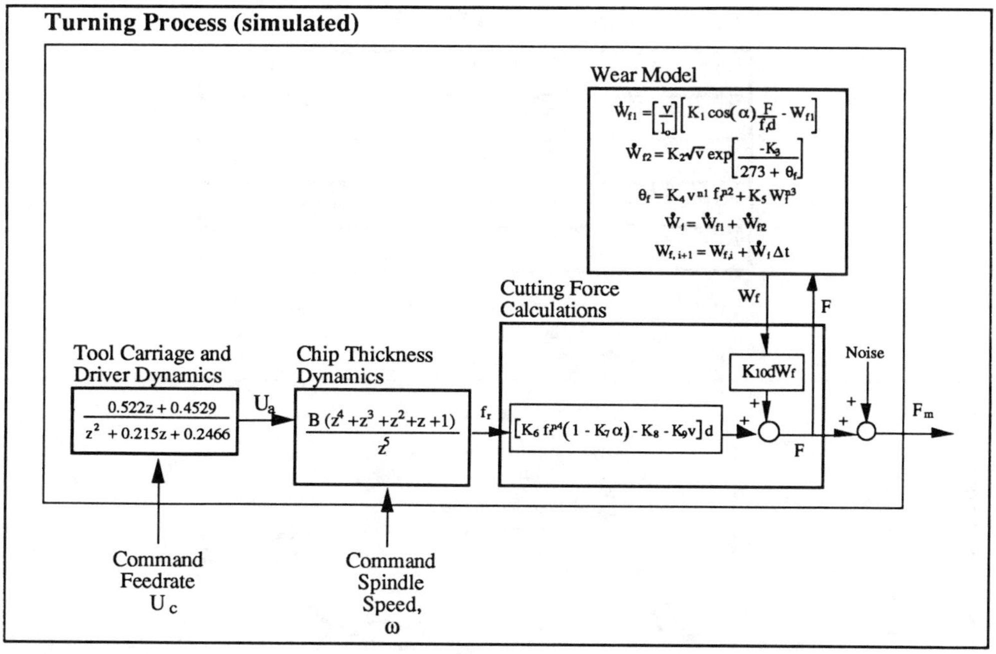

Figure 4. Model of a Turning Process

This initial model was augmented by considering the effect of tool wear on the cutting force [17] so that instead of Equation (3) the following relationship was used.

$$F = \left[K_6 f_r^{n4}(1 - K_7\alpha) - K_8 - K_9 v\right] d + K_{10} d\, W_f \tag{4}$$

where

K_6, K_7, K_8, K_9, K_{10} : constants
n_4 : constant exponent
α : rake angle
v : cutting speed
W_f : flank wear

The flank wear, W_f, was determined assuming [17] that abrasion and diffusion were the dominant wear mechanisms for the carbide tools considered in this model. The flank wear rate, $\dot{W}_f$, is separated into a component caused by abrasion, $\dot{W}_{f1}$, and a component caused by diffusion, $\dot{W}_{f2}$.

$$\dot{W}_{f1} = \left[\frac{v}{l_o}\right]\left[K_1 \cos(\alpha)\frac{F}{f_r\, d} - W_{f1}\right] \tag{5}$$

$$\dot{W}_{f2} = K_2 \sqrt{v} \exp\left[\frac{-K_3}{273 + \theta_f}\right] \tag{6}$$

where

l_o : constant
K_1, K_2, K_3 : constants

where the flank temperature, θ_f, in Equation (6) is estimated using the following relationship.

$K_1 = 5.2 \times 10^5$	$K_7 = 0.57$	$n_3 = 1.45$
$K_2 = 15$	$K_8 = 86$	$n_4 = 0.76$
$K_3 = 8000$	$K_9 = 0.1$	$l_o = 300$
$K_4 = 72$	$K_{10} = 500$	$\alpha = 5$ degrees
$K_5 = 2500$	$n_1 = 0.4$	$d = 2.0$ mm
$K_6 = 2160$	$n_2 = 0.48$	

Table 1. Constants and Cutting Conditions Assumed for the Turning Process

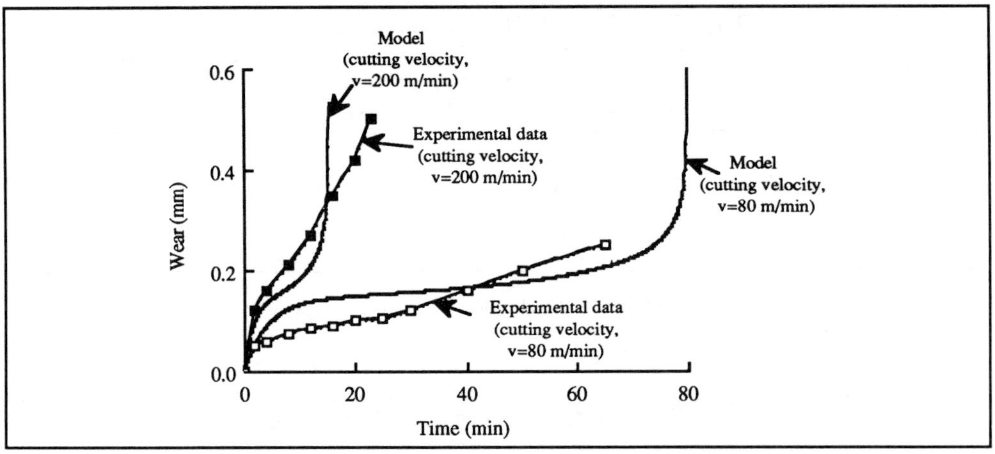

Figure 5. Experimental and Calculated Tool Wear (feed = 0.25 mm/rev, depth of cut = 2.0 mm)

$$\theta_f = K_4 \, v^{n1} \, f_r^{n2} + K_5 \, W_f^{n3} \tag{7}$$

where

K_4, K_5 : constants
n_1, n_2, n_3 : constant exponents

The total wear rate, $\dot{W}_f$, is then

$$\dot{W}_f = \dot{W}_{f1} + \dot{W}_{f2} \tag{8}$$

In the wear model presented above, the cutting conditions and constants K_i ,n_i and l_o (Table 1) were assumed to have values partially reported in [17], and partially to reflect experimental data reported in [14] (Figure 5).

In an actual turning operation, the cutting force signal provided by a force sensor may contain a noise component, the magnitude of which may vary as a function of conditions such as the tool and workpiece materials or the machine on which the turning is performed. In order to account for this noise, the turning process model used in this investigation included a noise component for the cutting force. For a force, F, calculated by the deterministic expressions of the process model (Figure 4), a "measured" force F_m was estimated using the following expression

$$F_m = F\,[1 + N_F \times N_R] \tag{9}$$

where

F_m : "measured" force value assumed to be provided by a force sensor
F : force value calculated using the model of Figure 4
N_F : noise factor (assumed to be 0.1 for this investigation)
N_R : A normally distributed random number with mean zero and variance of one

In the simulations for both the PRODEMA and the AC approach, the measured force, F_m, was filtered using the lowpass filter used in [20].

$$\frac{F_f}{F_m} = \frac{0.454}{z - 0.5464} \tag{10}$$

where

F_f : filtered force signal

Performance Measures

Production rate, process cost and part quality are typical measures used to evaluate the performance of a manufacturing process. Therefore the comparison of the PRODEMA approach with the AC scheme is based on the following performance measures.

a) **Process cost per part, C**
This performance measure is the ratio of the total cost, including machining cost ($M\,t_m$), part changing cost ($N_p\,M\,t_L$), and tool replacement cost ($N_t\,C_c + N_t\,M\,t_d$), to the number of parts machined (N_p):

$$C = \frac{M\,t_m + N_t\,C_c + N_t\,M\,t_d + N_p\,M\,t_L}{N_p} \tag{11}$$

where

t_m : total time the tool is in contact with the workpiece
M : machine operating rate including machinist's wages, machine depreciation and expendables
N_t : number of tools used
C_c : cost of a new tool
t_d : time required for tool change
N_p : total number of parts completed
t_L : time required for a part change

The number of parts completed, N_p, was determined by dividing the total distance traveled by the tool when in contact with the workpiece divided by the length of one part.

b) **Production rate, P**
This performance measure is the ratio of the number of parts completed (N_p), to the length of time during which the parts were machined ($t_m + N_t\,t_d + N_p\,t_L$):

$$P = \frac{N_p}{t_m + N_t\,t_d + N_p\,t_L} \tag{12}$$

c) **Surface roughness, Q**
This performance measure was assumed to be related to the feed, f_r, with which the tool was advancing on the workpiece surface. A lower value for the time average of the feed indicates that the surface roughness of the completed parts is reduced.

$$Q = \frac{\Sigma\left(f_r \times \Delta t\right)}{t_m} \tag{13}$$

where Δt is the sampling time for the AC approach and the decision interval for the PRODEMA approach.

$M = \$50.00$	$C_c = \$10.00$	$t_d = 1$ minute	$t_L = 1$ minute

Table 2. Assumed Values for the Estimation of Performance Measures

Using the performance measures defined above and the values in Table 2, one can compare the effectiveness of a turning process controlled by an AC scheme with one controlled by the PRODEMA approach. For this purpose, the turning process described above was "active" for one hour and its performance measures were calculated when the process was controlled by an AC and by PRODEMA.

PRODEMA Implementation

In the PRODEMA approach, a game is formulated as a matrix, and is periodically played during the process; the evaluation of the game matrix leads to the selection of an operating point during a decision interval. In the application of PRODEMA to a turning process, the operating point may correspond to a pair of settings for the cutting speed, v, and feed, f_r. The performance of the PRODEMA approach is influenced by the number of alternatives considered in the game and by the frequency of the game matrix evaluation; both of these factors are limited by the available computation power. The criteria in the game matrix can be defined to reflect relevant performance measures. However, there are different definitions for criteria and performance measures. The relationship for a particular criterion is used to evaluate, on-line, the consequence of each individual alternative with respect to this criterion in order for the decision making process to select the best alternative. Performance measures, on the other hand, are "globally" estimated, over a relatively long period of time. Thus the major differences between criteria and performance measures are that the former must be estimated on line and for the decision interval Δt, while the latter are estimated over a longer period of time, e.g. a shift, and off-line at the end of the selected time period. The point here is that if the performance metrics according to which the overall process is to be measured are known, then one can establish properly defined criteria which can be used for on-line decision making. The criteria implemented in the PRODEMA approach for the purpose of this discussion were production rate, p, and surface quality, q, and are defined below.

a) **Production rate criterion, p.** The use of this criterion will tend to maximize the production rate and minimize the cost per part. The definition of this criterion includes in its numerator the fraction of the part(s) completed during the decision interval, Δt, which is the time between two subsequent decisions. The denominator must include not only the decision interval, Δt, but also a portion of the tool change time, t_d, prorated by the fraction of the tool life used during the decision interval, and a portion of the part change time, t_L, prorated by the fraction of a part(s) completed during the decision interval. Thus the production rate decision criterion p can be expressed as:

$$p = \frac{\Delta N_p}{\Delta t + \frac{\dot{W}_f}{W_{max}} \Delta t\, t_d + \Delta N_p\, t_L} \qquad (14)$$

where the fraction of a part completed during the interval Δt is determined as:

$$\Delta N_p = \frac{\omega\, f_r\, \Delta t}{L} \qquad (15)$$

The variables in the above expressions are defined as:

t_L : time required for part change (Table 2)

$\dot{W}_f$: flank wear rate
W_{max} : tool life criterion
t_d : time required for tool change (Table 2)

ω : spindle speed
f_r : feed
L : distance the tool must travel to complete one part as measured axially along the part

b) **Surface quality criterion, q.** The surface roughness produced by the turning process has a direct correlation with the feed. As the feed is decreased, the surface roughness of the machined part tends to also decrease. The surface quality criterion is defined as the feed used during the decision interval, Δt. Use of this criterion will tend to reduce the surface roughness of the completed parts.

$$q = f_r \tag{16}$$

After the criteria values have been calculated, the game matrix can be evaluated using several different methods [11,16,19]. Since this investigation included a random signal to simulate sensor noise, the outcome of each alternative was somewhat uncertain. In this implementation of PRODEMA this uncertainty was not taken into account in the decision-making process. Use of a method for handling uncertainty may have improved the performance of PRODEMA, and such an evaluation will be considered in future work.

In the implementation of PRODEMA reported in this paper the game matrix was evaluated by first normalizing the criteria in unitless values between 0 and 1 according to the following expressions.

For a "benefit" criterion where it is desirable to select an alternative which maximizes the criterion:

$$a_{i,j}^{norm} = \frac{a_{i,j} - a_{min\ i,j}}{a_{max\ i,j} - a_{min\ i,j}} \tag{17}$$

For a "cost" criterion where it is desirable to select an alternative which minimizes the criterion:

$$a_{i,j}^{norm} = 1 - \frac{a_{i,j} - a_{min\ i,j}}{a_{max\ i,j} - a_{min\ i,j}} \tag{18}$$

where

$a_{i,j}^{norm}$: normalized criterion value
$a_{i,j}$: original criterion value
$a_{min\ i,j}$: minimum over all i alternatives for criterion j
$a_{max\ i,j}$: maximum over all i alternatives for criterion j

The relative importance of each criterion depends on the overall objectives for the manufacturing system. For example, in a finishing operation it is generally most important to reduce the surface roughness of the parts, whereas the production rate is more significant in a roughing operation. Thus in the utility function used in PRODEMA each decision criterion is assigned a weight which reflects its importance.

$$u_i = w_1 a_{i,1}^{norm} + w_1 a_{i,2}^{norm} + \ldots + w_n a_{i,n}^{norm} \tag{19}$$

where

u_i : utility for alternative i
w_j : weight for criterion j

The alternative with the highest utility, u_i, is selected.

The production rate criterion in Equation (14) contains a wear rate term, $\dot{W}_f$, and therefore the tool wear rate must be estimated for each alternative in order for this criterion to be evaluated. PRODEMA can use a variety of sensor inputs to estimate the process state variables [7]. However, for the work presented in this paper, PRODEMA was provided with only a measurement of the cutting force, F_f, for the purpose of estimating the tool wear rate. Equations (5-8) were used to estimate the wear rate, $\dot{W}_f$, and the measured cutting force F_f. A detailed illustration of the PRODEMA approach as applied to the turning process is shown in Figure 6.

The minimum decision interval that can be utilized by the PRODEMA approach is limited by the speed of the computing hardware. Longer decision intervals may lead to suboptimal performance of PRODEMA with respect to the performance measures; however, for the reasons discussed in the introduction of the paper, longer decision intervals will not cause the PRODEMA approach to become unstable. For the results presented in this paper, a decision interval of five seconds was used in the simulations of PRODEMA. Sixty-five alternatives were evaluated at each decision point, which

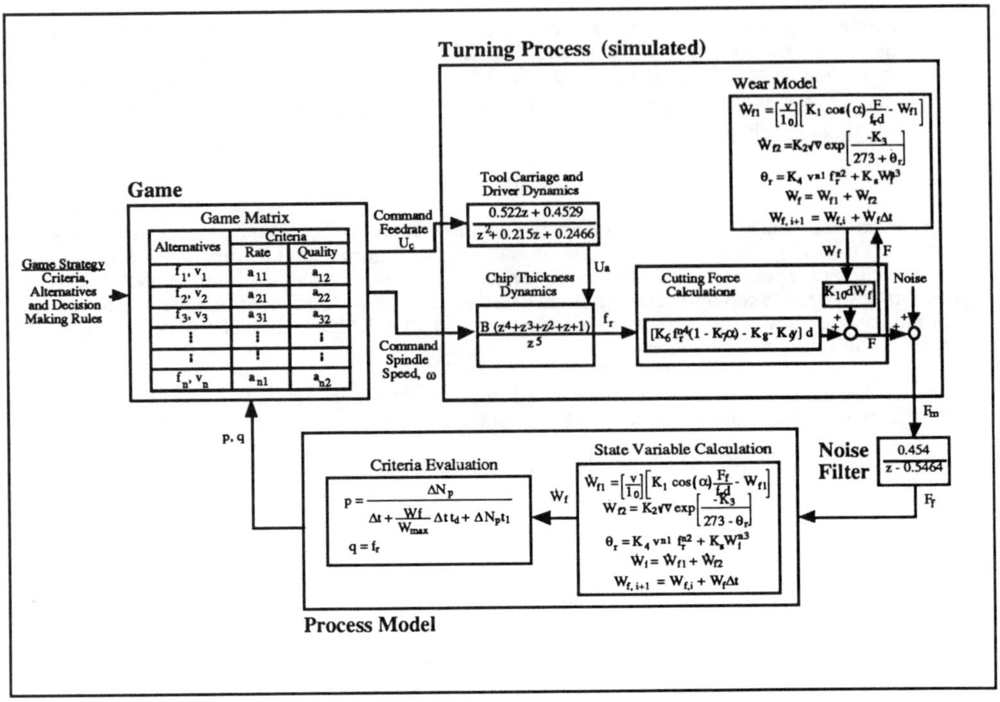

Figure 6. PRODEMA Approach Applied to a Turning Process

consisted of all possible combinations of five different cutting speeds, v, (150, 175, 200, 225 and 250 m/min) and 13 different feeds, f_r, (ranging from 0.100 to 0.400 in increments of 0.025 mm/rev). The tool life criterion used was 0.3 mm; the tool was replaced when the flank wear reached 0.3 mm.

Adaptive Control Implementation

The adaptive control approach implemented is a model reference adaptive control (MRAC) scheme in which a proportional-integral (PI) controller with an adjustable gain is employed [20]. The objective of this control approach is to maintain the cutting force, F_f, at a desired value, F_d, by making appropriate adjustments to the command feedrate, U_c while the cutting speed remains constant. The controller consists of two control loops (Figure 7). The first loop is a standard feedback loop utilizing a PI controller; the second loop tunes the gain, $\widehat{K}_p$, so that the difference between the cutting force estimated by the model, $\widehat{F}_f$, and the measured cutting force, F_f, is minimized. The adaptation model contains the Equations for the tool carriage driver dynamics (Equation 1), the chip thickness dynamics (Equation 2), and the lowpass filter for the measured force (Equation 10).

In the simulations of the adaptive control approach, the cutting speed remained constant at v = 200 m/min. Tools were replaced when the flank wear land, W_f, reached 0.3 mm. K_{po} was set to K_6Bd, where K_6 is a proportionality constant relating the cutting conditions with the cutting force (Equation 4), B depends on the spindle speed and sampling rate (Equation 2), and d is the depth of cut. The controller gains k_p and k_i were selected to be $0.0294/K_{po}$ and $0.0235/K_{po}$, respectively [20]. $\widehat{K}_p$ was initialized to K_{po} at the beginning of each turning process. The forgetting factor, λ, was set at 0.95.

Short sampling times are preferred in an AC approach [1,2]. The ideal case is to have sampling intervals that approach a continuous system, but this is not possible because of computational limitations. The general tendency is to have about two to four samples per rise time. The simulations of the adaptive control approach were based on a 10 ms sampling interval.

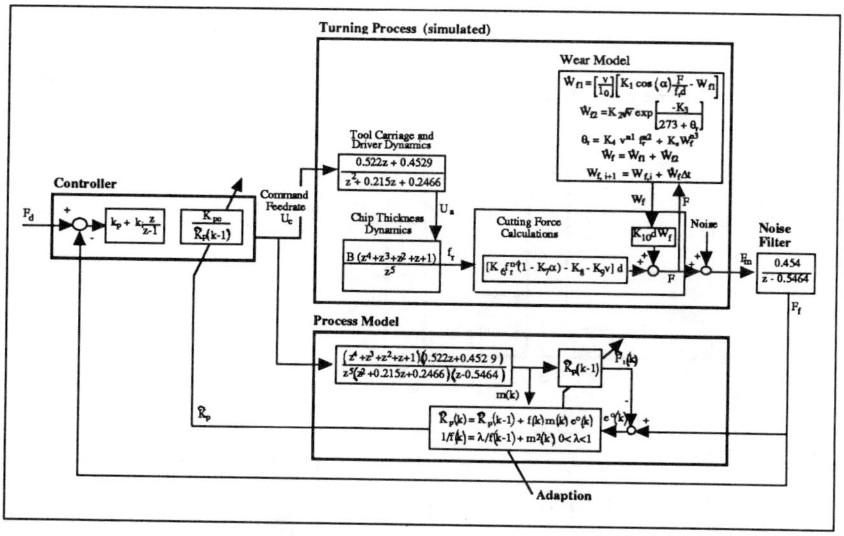

Figure 7. Adaptive Control Scheme Applied to Simulated Turning Process

5. RESULTS AND DISCUSSION

The AC approach was simulated for a 1000 N value of the reference force, F_d (Figure 8). In the simulations of the PRODEMA approach, increasing the weight on the production rate criterion, p, (cases 2 and 5 in Figure 8) tended to increase the production rate, P, reduce the cost per part, C, and increase the surface roughness, Q. Conversely, increasing the weight on the surface quality criterion, q, (cases 3 and 6 in Figure 8) tended to increase the performance measure of cost per part, C, decrease the production rate, P, and decrease the surface roughness, Q.

The decision criteria used in PRODEMA are directly related to the performance measures used to evaluate the methods of operating the process over the period of one hour. Thus the weight selected for a criterion directly influences the corresponding performance measure. For example, weighting only the surface roughness criterion, q, minimized the surface roughness performance measure, Q, (Case 3 in Figure 8). This clear relationship does not exist between the performance measures and the reference value selected for the force in an adaptive control approach. Therefore the PRODEMA control approach can be more easily interfaced with optimization and scheduling decisions made at higher levels in the manufacturing system hierarchy. In addition, the inherent tradeoffs between the values of the different performance measures, i.e. production rate and surface roughness, are clearly evident in the PRODEMA approach during the selection of the relative weights on the decision criteria. Such performance tradeoffs are less easily identifiable in any AC approach.

The speeds and feeds commanded by the AC approach and by PRODEMA during a turning process (*Figure 9*) reveals that the range of feeds utilized by PRODEMA is greater than the range utilized by the AC approach. The AC approach attempted to maintain the cutting force at 1000 N and as the tool wear increased it was necessary to decrease the feed somewhat in order to maintain this force. For the simulation case shown in Figure 9, PRODEMA attempted to optimize the process with respect to the production rate and the surface quality. An excessively high tool wear rate requires frequent replacement of the tool which will tend to reduce the production rate. Consequently, in the stages of the process when the tool wear is high such as at the beginning and end of the tool life (*Figure 5*) PRODEMA selected lower feeds than in the middle stage of the tool life where the wear rate is lower. If the surface quality criterion had not been weighted PRODEMA would have selected higher feeds than those shown in *Figure 9*. The AC approach controls only the feed; the speed of 200 m/min was selected a priori and must remain constant during the process. PRODEMA controls the feed and the speed. For the simulation case shown in *Figure 9*, PRODEMA selected a speed of 150 m/min at each decision point during the process.

It is conceivable that a different value for the reference force, F_d, in the AC approach would influence the performance measures. However, this is a rather indirect approach; if one is interested in the

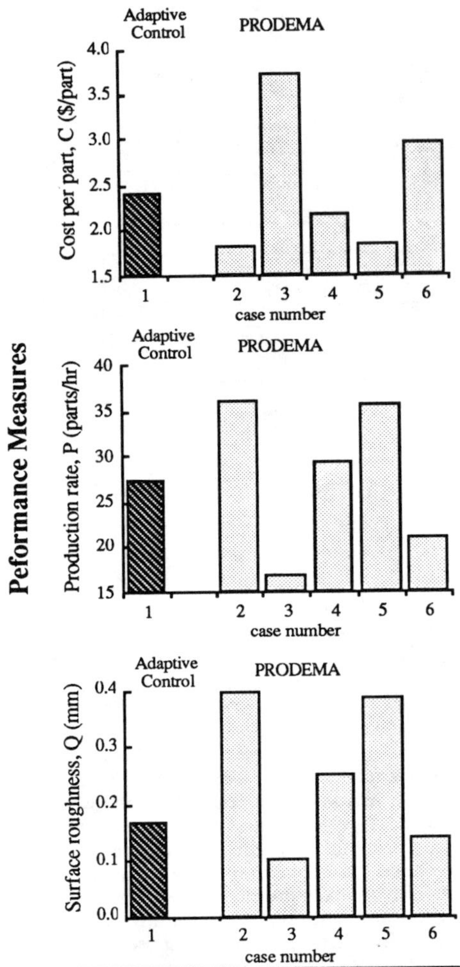

Case No.	Case details	Weight for Production rate criterion, p	Weight for Surface quality criterion, q
1.	AC at 1000 N	-	-
2.	PRODEMA	1.00	0.00
3.	PRODEMA	0.00	1.00
4.	PRODEMA	0.50	0.50
5.	PRODEMA	0.75	0.25
6.	PRODEMA	0.25	0.75

Figure 8. Comparative Performance of the AC and PRODEMA Approaches

performance of a turning process with respect to measures such as cost per part and production rate one might not be able to establish the impact of the reference force on the performance measures.

6. CONCLUSIONS

As an alternative to adaptive control approaches to manufacturing process control, this paper has presented the PROcess DEcision MAking (PRODEMA) approach, which is based on game theory. In PRODEMA the optimal process input settings are selected by evaluating a set of feasible input settings with respect to one or more decision criteria. The decision criteria can be defined in a manner which will tend to optimize the process with respect to relevant performance measures.

In this paper the operation of PRODEMA was compared with an adaptive control approach using the simulation of a turning process. The values of cost per part, production rate and surface roughness performance measures provided by PRODEMA were directly influenced by selecting appropriate values of the relative weights on each of the corresponding decision criteria.

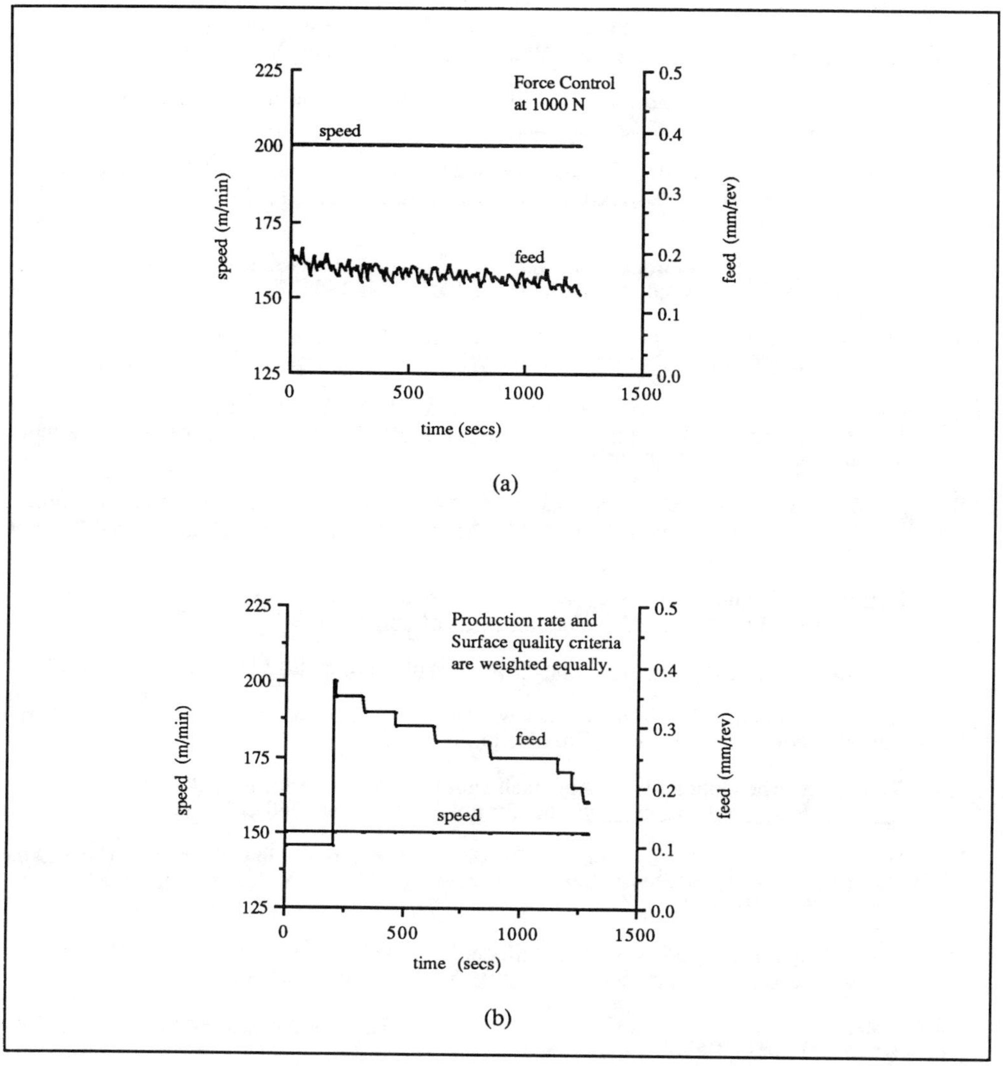

Figure 9. Speed and Feed vs. Time for the Turning Process
a) Adaptive Control (corresponds to case 1 in *Figure 8*)
b) PRODEMA (corresponds to case 4 in *Figure 8*)

Acknowledgements

This work was partially funded by NSF's Strategic Manufacturing Research Initiative project "Decision Making in Manufacturing Systems" (N.S.F. Grant DMM-8914181) and the Leaders for Manufacturing Program at M.I.T. The authors would also like to thank Professors Karl Ulrich and Steve Graves of the Sloan School of Management at M.I.T. and Mr. Rajan Ramaswamy, graduate student at the Artificial Intelligence Laboratory in M.I.T.

REFERENCES

[1] Astrom, K.J. and Wittenmark, B., Adaptive Control, Addision Wesley Publishing Company, 1989.

[2] Astrom, K.J. and Wittenmark, B., Computer Controlled Systems: Theory and Design, Prentice Hall International, 1984.

[3] Chen, B.S. and Chang, Y.F., "Constant Turning Force Adaptive Controller Design," ASME, Journal of Engineering for Industry, Vol. 111, pp. 125-132,1989.

[4] Chryssolouris, G., Domroese, M. and Zoldos, L., "A Decision Making Strategy for Machining Control," Annals of the CIRP, Vol. 39, No. 1, pp. 501-504, 1990.

[5] Chryssolouris, G., Guillot, M. and Domroese, M., "A Decision Making Approach to Machining Control," ASME, Journal of Engineering for Industry, Vol. 110, No. 4, pp. 397-398, Nov. 1988.

[6] Chryssolouris, G., Domroese, M., and Beaulieu, P., "A Statistical Approach to Sensor Synthesis," to be presented at NAMRC-XIX, SME, May 1991.

[7] Chryssolouris, G., Domroese, M., and Beaulieu, P., "Sensor Synthesis for Control of Manufacturing Processes," presented at the Winter Annual Meeting of ASME, 1990.

[8] Danai, K. and Ulsoy, A. G., "A Dynamic State Model for On-Line Wear Estimation in Turning," Sensors and Controls for Manufacturing, PED-Vol. 18, ASME Winter Annual Meeting, pp. 137-148, Nov. 1985.

[9] Fassois, S.D., Eman K.F., and Wu, S.M., "A Fast Algorithm for On-Line Machining Process Modelling and Adaptive Control," ASME, Journal of Engineering for Industry, Vol. 111, pp. 133-139, 1989.

[10] Fussell, B.K. and Srinivasan, K., "On Line Identification of End Milling Process Parameters," ASME, Journal of Engineering for Industry, Vol. 111, pp. 322-330, 1989.

[11] Jones, A.J., Game Theory: Mathematical Models of Conflict, Ellis Horwood Limited, 1980.

[12] Koren, Y. and Lenz, E., "Mathematical Model for the Flank Wear While Turning Steel with Carbide Tools," CIRP Seminar, Trondheim, June 1970.

[13] Koren, Y., "The Optimal locus Approach with Machining Applications," ASME, Journal of Dynamic Systems, Measurements and Control, Vol. 111, pp. 260-267, 1989.

[14] Langhammer, K., "Cutting Forces as Parameters for Determining Wear on Carbide Lathe Tools and as Machinability Criterion for Steel," Carbide J., Vol. 8, No. 3, pp. 5-13, May-June 1976.

[15] Lauderbaugh, L.K. and Ulsoy, A.G., "Model Reference Adaptive Force Control in Milling," ASME, Journal of Engineering for Industry, Vol. 111, pp. 13-21, 1989.

[16] Luce, R.D., and Raiffa, H., Games and Decisions: Introduction and Critical Survey, John Wiley and Sons, 1957.

[17] Park, J.J., and Ulsoy, A.G., "On line Flank Wear Estimation Using Adaptive Observers," Automation of Manufacturing Processes, Winter Annual Meeting of ASME, Dallas, Texas, pp. 11-20, Nov. 25-30, 1990.

[18] Rangwala, S., and Dornfeld, D., "Integration of Sensors via Neural Networks for Detection of Tool Wear States," Intelligent and Integrated Manufacturing Analysis and Synthesis, ASME Winter Annual Meeting, Boston, MA, Dec 12-13, 1987.

[19] Thomas, L.C., Games Theory and Applications, Ellis Horwood Limited, 1984.

[20] Tomizuka, M. and Zhang, S., "Modelling and Conventional Adaptive PI Control of a Lathe Cutting Process," ASME, Journal of Dynamic Systems, Measurements and Control, Vol. 110, pp. 350-354, 1988.

[21] Watanabe, T., "A model Based Approach to Adaptive Control Optimization in Milling," ASME, Journal of Dynamic Systems, Measurements and Control, Vol. 108, pp. 56-54, 1986.

Sensors, Controls, and Quality Issues
in Manufacturing
ASME 1991

REAL-TIME COMPENSATION OF TIME-VARIANT VOLUMETRIC ERROR ON A MACHINING CENTER

J. S. Chen, J. X. Yuan, J. Ni, and S. M. Wu
Department of Mechanical Engineering and Applied Mechanics
University of Michigan
Ann Arbor, Michigan

ABSTRACT

An error compensation scheme was developed to enhance the time-variant volumetric accuracy of 3-D machines by correcting existing machine errors through off-line calibration, on-line sensing and real-time computer control techniques. A general methodology was developed to synthesize both geometric and thermal errors as a time-variant volumetric error field. A compensation controller based on an IBM/PC was developed and linked with a DynaPath Delta20 CNC controller to compensate for volumetric error in real time. The scheme implemented on a horizontal machining center was proven to improve machine accuracy by an order of magnitude. Inspected by a coordinate measuring machine, dimension errors of a cut part were reduced from 92.4 μm to 18.9 μm in a dimension of 404x310 mm^2 and the depth difference of milled surfaces was reduced from 196 μm to 8 μm.

NOMENCLATURE

$\vec{P}_A(t)$, $\vec{P}_B(t)$: coordinates of point P based on frames A and B

$T_A^B(t)$: homogeneous coordinate transformation matrix from frame A to frame B

$\delta x(t)$, $\delta y(t)$, $\delta z(t)$: translational errors of a slide deviated from its nominal movement

$\alpha(t)$, $\beta(t)$, $\gamma(t)$: rotational errors of a slide in the x, y and z axes

O_x, O_y, O_z : offset distances between the origins of frames B and A when x is zero

$\delta O_x(t)$, $\delta O_y(t)$, $\delta O_y(t)$: drifts at the origin of frame B.

$\vec{T}$, $\vec{W}$: position vectors of the tool and the workpiece

$\delta_{xx}(t)$, $\delta_{yx}(t)$, $\delta_{zx}(t)$, $\delta_{xy}(t)$, $\delta_{yy}(t)$, $\delta_{zy}(t)$, $\delta_{xz}(t)$, $\delta_{yz}(t)$, $\delta_{zz}(t)$: translational errors of slides (The first subscript represents the direction of errors and the second subscript represents the axes of slides.)

$\varepsilon_{\alpha x}(t)$, $\varepsilon_{\beta x}(t)$, $\varepsilon_{\gamma x}(t)$, $\varepsilon_{\alpha y}(t)$, $\varepsilon_{\beta y}(t)$, $\varepsilon_{\gamma y}(t)$, $\varepsilon_{\alpha z}(t)$, $\varepsilon_{\beta z}(t)$, $\varepsilon_{\gamma z}(t)$: angular errors of slides

$\eta_{xy}(t)$, $\eta_{yz}(t)$, $\eta_{xz}(t)$: squareness errors between machine axes

T_x, T_y, T_z : coordinates of the tool tip based on frame D

$\delta T_x(t)$, $\delta T_y(t)$, $\delta T_z(t)$: variations of the tool tip due to thermal effects

W_x, W_y, W_z : coordinates of workpiece based on frame C

$\delta O_{xy}(t)$, $\delta O_{yy}(t)$, $\delta O_{zy}(t)$, $\delta O_{xz}(t)$, $\delta O_{yz}(t)$, $\delta O_{zz}(t)$: thermal drifts at the origins of y and z axes

$\varepsilon_{\alpha s}(t)$, $\varepsilon_{\beta s}(t)$: inclinations of the spindle axis due to thermal effects

L : tool offset distance when the tool length is changed.

$\Delta V_x(t)$, $\Delta V_y(t)$, $\Delta V_z(t)$: components of the volumetric error in the x, y and z directions

$\{A_1\}$, $[A_2]$, . . . : coefficients of regression model

$\{\Delta T\}$: a vector of temperature field variations of machine .

δ, δ_G , δ_T : the summed, geometric and thermal errors

b_0, b_1, . . : coefficients of stationary error profile model

$K_1(t)$, $K_2(t)$: empirical models for slide-position dependent error

1. INTRODUCTION

Performance of machine tools in terms of accuracy is defined by the relative error between movements of the cutting tool and workpiece. For a 3-axis machine, this relative error is called volumetric error which is spatial-variant across the machine working zone due to angular errors of machine linkages. Besides the spatial-variant characteristic, volumetric error is also time-variant due to the dynamic effect of machine's thermal errors. A typical magnitude of spindle growth could be more than one hundred micrometers from machine cold-start to warm-up. By studying a large amount of data, Peklenik [3] found that thermal errors could comprise 40-70% of the workpiece error in precision machining. Therefore, the problem is not only to make higher accuracy machines but also to maintain the machine accuracy under varying environmental and machine temperatures.

Making higher accuracy and thermally stable machine could be achieved through improving machine designs. However, error reduction by design can be expensive and is difficult to apply to existing machine tools. In addition to design modifications, machine tool researchers have utilized error compensation method to enhance machine accuracy. In many cases, error compensation method which identifies and corrects existing machine errors is cheaper and more efficient than design modification method. Some CMM manufacturers have used this technique to maintain the volumetric accuracy of their products by correcting repeatable errors of CMMs. Usually, a 30 to 60% error reduction can be gained after compensation.

Although error compensation is very effective on CMMs, few of developed schemes are actually implemented on machine tools. The difficulties in handling time-variant thermal errors and the limitation of the capability of CNC controllers are the main reasons. The issue of thermal errors is much more complicated in machine tools than in CMMs. Major thermal errors of CMMs are from the variation of environmental temperature which is usually corrected as uniform temperature effect. However, major thermal errors of machine tools are created by internal heat sources within machines such as motors, bearings and hydraulic systems. These internal heat sources can create local temperature gradients scattered across some parts of machine structures with much higher magnitude and frequency than the environmental effect. The dynamic and local temperature gradients scattered throughout the machine structure make the thermal effect of machine tools a very complicated issue.

Implementation of an error compensation scheme is also more difficult on machine tools than on CMMs. For CMMs, implementation can be achieved by simply editing inspection results reported to users without interfering machine servo-control loops. Since machine tools must precisely cut off excessive materials of parts, the relative error between the cutter and workpiece must be corrected in real time by machine servo-control system. To date, the capability of error compensation of most commercial CNC controllers is restricted to off-line stored 1-D errors such as backlash or leadscrew error.

Recently, compensation of 2-D planer error for a turning machine or 3-D volumetric error for a milling machine has gained much attention in response to the need of higher machining accuracy [1, 5, 7, 10]. One example is the work of "Quality in Automation" at the National Institute of Standards and Technology [8]. This work will try to compensate for 21 geometric errors, thermal growth of the spindle, thermal expansions of leadscrews and thermal deformations of machine structures.

The objective of this paper is to develop an error compensation scheme which compensates for time-variant volumetric error of a machining center in real time. A general synthesis model of volumetric errors was developed to combine both geometric and thermal errors. Instead of the well-known 21 errors, 32 machine linkage errors were synthesized as a time-variant volumetric error field by this general model. A compensation controller based on an IBM/PC was developed and linked with a DynaPath Delta20 CNC controller to compensate for the time-variant volumetric errors in real time. This scheme was implemented on a 4-axis horizontal machining center and a series of tests were conducted to verify the effectiveness of this compensation scheme.

2. FORMULATION OF TIME-VARIANT VOLUMETRIC ERROR

Under the rigid body kinematic assumption, 21 error components of machine linkages are found enough to represent geometric errors of a 3-D machine. The rigid body kinematic assumption implies that linkage errors of each slide only depend on its own coordinate and are not influenced by movements of the other slides. That is, there are no coupling effects of linkage errors between machine slides. For each slide there are six errors (three translational and three rotational errors). In addition, machine axes may not be orthogonal each other for which three out of squareness errors may be observed.

By examining the machine structure, it was found that the well-known 21 errors are not sufficient to represent all thermal effects of a machining center. Therefore, a new volumetric error model with 32 error components was developed. 11 additional errors were introduced because of some thermal effects of the machining center.

The first group of additional thermal errors is from thermal drifts at the tool tip as shown in Fig. 1. For a machining center, three translational thermal drifts at the spindle head during machining could affect the workpiece precision. Besides, two inclinational errors of the spindle axis must be concerned because the tool length of a machining center can vary widely for different machining processes and workpieces.

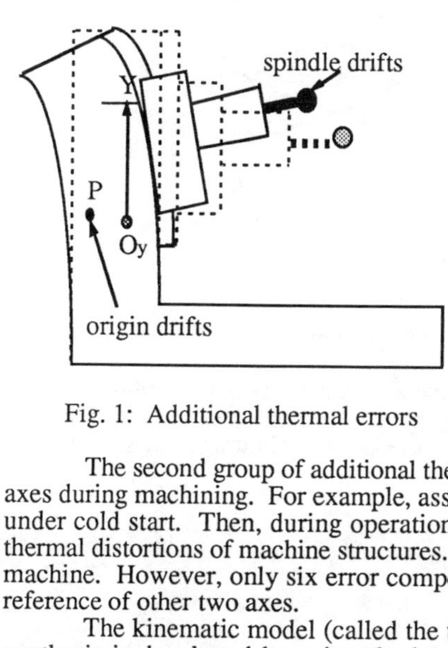

Fig. 1: Additional thermal errors

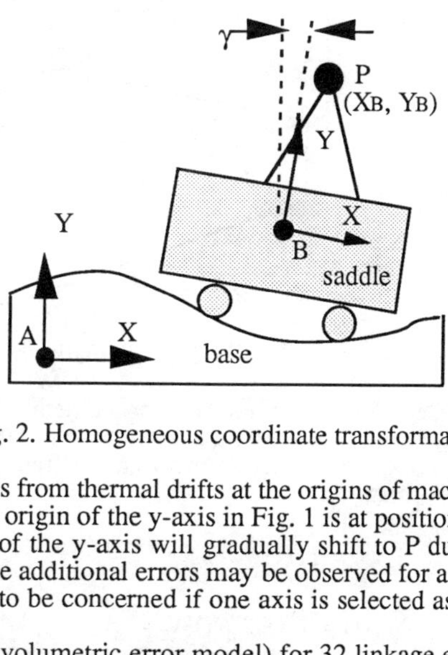

Fig. 2. Homogeneous coordinate transformation

The second group of additional thermal errors is from thermal drifts at the origins of machine axes during machining. For example, assume that the origin of the y-axis in Fig. 1 is at position O_y under cold start. Then, during operation, the origin of the y-axis will gradually shift to P due to thermal distortions of machine structures. Totally, nine additional errors may be observed for a 3-D machine. However, only six error components need to be concerned if one axis is selected as the reference of other two axes.

The kinematic model (called the time-variant volumetric error model) for 32 linkage error synthesis is developed by using the homogeneous coordinate transformation method [6] which transforms the position and orientation of a rigid body in one coordinate frame to another coordinate frame. Assume that a movable saddle with a coordinate frame B is mounted on a base with a coordinate frame A (see Fig. 2). Then, under small angular error assumption, a point P on the movable saddle can be transformed into the coordinate frame A as

$$\vec{p}_A(t) = T_A^B(t)\vec{p}_B \tag{1}$$

and

$$\mathcal{T}_A^B(t) = \begin{bmatrix} 1 & -\gamma(t) & \beta(t) & O_x + \delta O_x(t) + x + \delta x(t) \\ \gamma(t) & 1 & -\alpha(t) & O_y + \delta O_y(t) + \delta y(t) \\ -\beta(t) & \alpha(t) & 1 & O_z + \delta O_z(t) + \delta z(t) \\ 0 & 0 & 0 & 1 \end{bmatrix}. \qquad (2)$$

An important feature of the homogeneous coordinate transformation method is that the coordinates of one object with respect to several different coordinate frames can be found by a series of multiplications as

$$\vec{P}_i = \mathcal{T}_i^{i+1} \bullet \mathcal{T}_{i+1}^{i+2} \bullet \ . \ . \ . \ \bullet \mathcal{T}_{j-1}^{j} \vec{P}_j \ . \qquad (3)$$

The machining center studied has a mechanical structure shown in Fig. 3. The y-axis spindle carrier mounted on the machine column provides a vertical movement for the tool. Movement of the workpiece is achieved by two movable and stacked slides mounted on the machine base which consist of a horizontal xz plane movement .

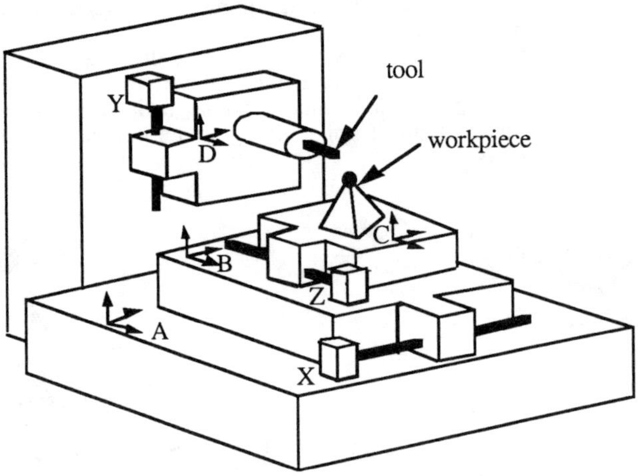

Fig. 3: The kinematic chain of the machining center

Now, consider the three slides as a kinematic chain of linkages and assume that coordinate frames A, B, C and D are assigned to the base and three movable slides. Then position vectors of the tool and workpiece can be transformed to frame A by using corresponding homogeneous coordinate transformation matrices

$$\vec{T}_A(t) = \mathcal{T}_A^D(t)\vec{T}_D(t) \qquad (4)$$

$$\vec{W}_A(t) = \mathcal{T}_A^B(t)\mathcal{T}_B^C(t)\vec{W}_C \qquad (5)$$

and

$$\vec{T}_D(t) = \begin{Bmatrix} T_x + \delta T_x(t) + \varepsilon_{\beta s}(t)L \\ T_y + \delta T_y(t) - \varepsilon_{\alpha s}(t)L \\ T_z + L + \delta T_z(t) \\ 1 \end{Bmatrix} \qquad (6)$$

$$
\mathbf{T}_A^D(t) = \begin{bmatrix} 1 & -\varepsilon_{\gamma y}(t) & \varepsilon_{\beta y}(t) & O_{xy} + \delta O_{xy}(t) + \delta_{xy}(t) - \eta_{xy}(t)y \\ \varepsilon_{\gamma y}(t) & 1 & -\varepsilon_{\alpha y}(t) & O_{yy} + \delta O_{yy}(t) + y + \delta_{yy}(t) \\ -\varepsilon_{\beta y}(t) & \varepsilon_{\alpha y}(t) & 1 & O_{zy} + \delta O_{zy}(t) + \delta_{zy}(t) \\ 0 & 0 & 0 & 1 \end{bmatrix} \tag{7}
$$

$$
\mathbf{T}_B^C(t) = \begin{bmatrix} 1 & -\varepsilon_{\gamma z}(t) & \varepsilon_{\beta z}(t) & O_{xz} + \delta O_{xz}(t) + \delta_{xz}(t) - \eta_{xz}(t)z \\ \varepsilon_{\gamma z}(t) & 1 & -\varepsilon_{\alpha z}(t) & O_{yz} + \delta O_{yz}(t) + \delta_{yz}(t) - \eta_{yz}(t)z \\ -\varepsilon_{\beta z}(t) & \varepsilon_{\alpha z}(t) & 1 & O_{zz} + \delta O_{zz}(t) + \delta_{zz}(t) + z \\ 0 & 0 & 0 & 1 \end{bmatrix} \tag{8}
$$

$$
\mathbf{T}_A^B(t) = \begin{bmatrix} 1 & -\varepsilon_{\gamma x}(t) & \varepsilon_{\beta x}(t) & x + \delta_{xx}(t) \\ \varepsilon_{\gamma x}(t) & 1 & -\varepsilon_{\alpha x}(t) & \delta_{yx}(t) \\ -\varepsilon_{\beta x}(t) & \varepsilon_{\alpha x}(t) & 1 & \delta_{zx}(t) \\ 0 & 0 & 0 & 1 \end{bmatrix} \tag{9}
$$

$$
\vec{W}_C = \begin{Bmatrix} W_x \\ W_y \\ W_z \\ 1 \end{Bmatrix} = \begin{Bmatrix} O_{xy} - x - O_{xz} + T_x \\ O_{yy} + y - O_{yz} + T_y \\ O_{zy} - O_{zz} - z + T_z + L \\ 1 \end{Bmatrix} \tag{10}
$$

Note that the origin of frame A is assigned at the starting position of the x-axis and also as the reference of the origin drifts of the y and z axes. For squareness errors between slides, the x-axis is the reference for the xy and the xz squareness errors. The y-axis is the reference for the yz squareness error. The offset distance L is used to accommodate a change in the tool length when different tools are used.

If there are no motion errors of machine linkages, then $\vec{T}_A(t) = \vec{W}_A(t)$. However, a relative error between the tool and workpiece is created by the mentioned 32 linkage errors as

$$
\Delta\vec{V}(t) = \vec{T}_A(t) - \vec{W}_A(t) = \mathbf{T}_A^D(t)\vec{T}_D(t) - \mathbf{T}_A^B(t)\mathbf{T}_B^C(t)\vec{W}_C \ . \tag{11}
$$

Ignoring the higher order error terms, Eq. (11) can be expanded as

$$
\begin{aligned}
\Delta V_x(t) = {} & -\varepsilon_{\gamma y}(t)T_y + \varepsilon_{\beta y}(t)(T_z + L) + \delta_{xy}(t) - \eta_{xy}(t)Y + (\varepsilon_{\gamma x}(t) + \varepsilon_{\gamma z}(t))W_y - (\varepsilon_{\beta x}(t) + \varepsilon_{\beta z}(t))W_z \\
& - \delta_{xz}(t) - \varepsilon_{\beta x}(t)(O_{zz} + Z) + \eta_{zx}(t)Z + \varepsilon_{\gamma x}(t)O_{yz} - \delta_{xx}(t) + \delta T_x(t) + \varepsilon_{\beta s}(t)L \\
& - \delta O_{xz}(t) + \delta O_{xy}(t) \\
\Delta V_y(t) = {} & \varepsilon_{\gamma y}(t)T_x - \varepsilon_{\alpha y}(t)(T_z + L) + \delta_{yy}(t) - (\varepsilon_{\gamma x}(t) + \varepsilon_{\gamma z}(t))W_x + (\varepsilon_{\alpha x}(t) + \varepsilon_{\alpha z}(t))W_z - \varepsilon_{\gamma x}(t)O_{xz} \\
& - \delta_{yz}(t) + \eta_{yz}(t)Z + \varepsilon_{\alpha x}(t)(O_{zz} + Z) - \delta_{yx}(t) + \delta T_y(t) - \varepsilon_{\alpha s}(t)L - \delta O_{yz}(t) + \delta O_{yy}(t) \\
\Delta V_z(t) = {} & -\varepsilon_{\beta y}(t)T_x + \varepsilon_{\alpha y}(t)T_y + \delta_{zy}(t) + (\varepsilon_{\beta x}(t) + \varepsilon_{\beta z}(t))W_x - (\varepsilon_{\alpha x}(t) + \varepsilon_{\alpha z}(t))W_y + \varepsilon_{\beta x}(t)O_{xz} \\
& - \varepsilon_{\alpha x}(t)O_{yz} - \delta_{zz}(t) - \delta_{zx}(t) + \delta T_z(t) + \delta O_{zy}(t) - \delta O_{zz}(t)
\end{aligned} \quad . \tag{12}
$$

Note that Eq. (12) is a general purpose model for a machining center. In practice, it is not necessary to identify all 32 errors in Eq. (12) because for every specific machine only a few of the 32 errors may have accountable effects on machining accuracy.

3. THERMAL ERROR PREDICTION

Prediction of thermal errors is achieved by the empirical model approach which correlates thermal errors to temperature rises of the machine through experiment. The 11 additional thermal errors and squareness errors have empirically multiple regression models as

$$\delta = \{A_1\}'\{\Delta T\} + \{\Delta T\}'[A_2]\{\Delta T\} \quad (13)$$

Note that higher order terms more than order two can also be included in Eq. (13).

In Eq. (13), thermal errors are predicted by a temperature field but not by the temperature rise information at a single position of the machine. This is a necessary step because the thermal time constants of machine components vary greatly with their heat capacities. Due to the difference in time constants, thermal drifts at the spindle are sensitive to the variation of the spindle speed (Fig. 4) and the warm-up or cool-down period (Fig. 5). Therefore a multiple-variable model is necessary to include the thermal effects from the spindle, column and even environment.

As found by Attia [2], thermal errors of machine tools exhibit highly nonlinear interaction between machine components. The so-called stepwise regression analysis was used to select the higher order and cross-product terms in Eq. (13). It was also used to screen out unrelated and highly intercorrelated terms which may cause serious round off problem on regression analysis. For more information about the stepwise regression search procedure, see Neter [9].

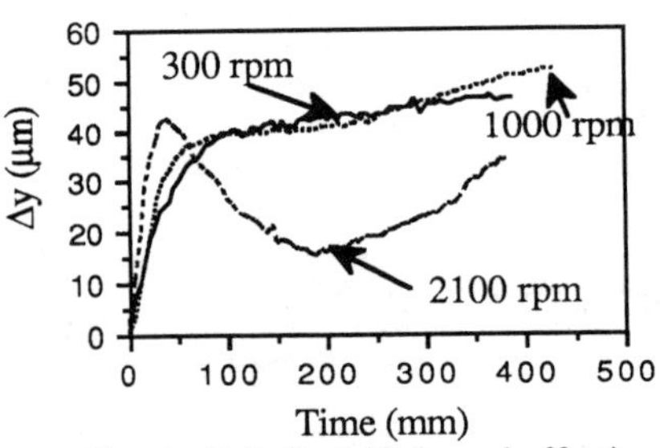

Fig. 4: Spindle drift (speed effect)

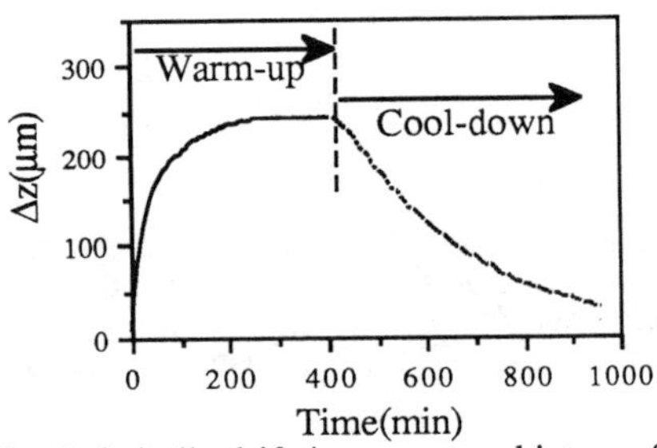

Fig. 5: Spindle drift (temperature history effect)

The other 18 errors of three movable slides depend on not only machine temperatures but also positions of machine slides. Separation of geometric and thermal errors is done by estimating the slope variation of error profiles. Fig. 6 shows distance errors in the y-axis measured periodically during a machine operation period. Although distance errors varied with temperature, their basic profiles did not change drastically and only the slopes differed. This phenomenon also applies to thermal bending errors such as angular errors of the column.

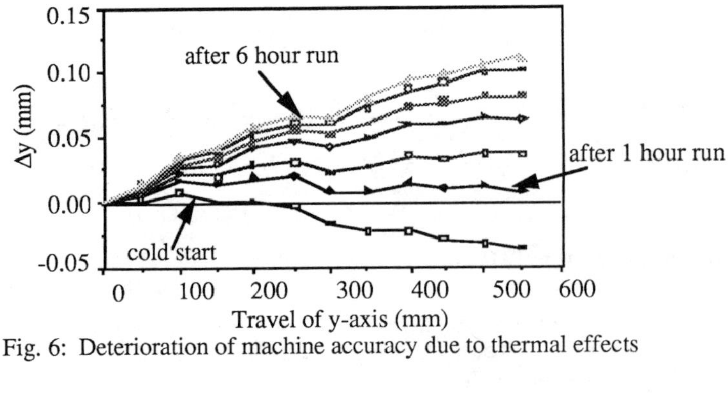

Fig. 6: Deterioration of machine accuracy due to thermal effects

These 18 errors are then seen as the summation of stationary error profiles and time-variant slopes. Under cold start, only stationary error profiles (called geometric-induced errors) exist in a machine. During operations, time-variant slopes (called thermal-induced errors) are added to geometric-induced errors. Empirical models for position-dependent errors are then as:

$$\delta = \delta_G + \delta_T(t)$$
$$\delta_G = b_0 + b_1X + b_2X^2 + \ldots$$
$$\delta_T(t) = K_1(t)X + K_2(t)X^2 + \ldots \qquad (14)$$

where K1(t) and K2(t) are empirical models which need to be established.

Note that K1(t) and K2(t) have mathematical form as Eq. (13). The major advantage of this approach is that it reduces the number of empirical models needed to represent the slide-position errors. For example, in this research, most slide-position dependent errors can be approximated accurately with only the first order term. For more information about the modeling of thermal errors, see [4].

4. COMPENSATION OF NON-RIGID BODY KINEMATIC EFFECT

Since a heavy z-axis slide is stacked on the x-axis slide (Fig. 3), the pitch errors of the x-axis were affected by the movement of the z-axis slide (Fig. 7). The result implies that the rigidities of the foundation, the machine structures and the guideway systems are not sufficient to support the heavy deadweight of the working table and to keep deformations within permissible limits.

When the accuracy requirement becomes stricter, volumetric error model derived from the kinematic rigid body assumption may not be precise enough. This problem was corrected by updating the pitch errors in the x-axis according to the position of the z-axis slide in five intervals. That is, z = 0, 100, 200, 300, 400, 500 mm. Within each interval, the coupling effect was corrected by a linear interpolation.

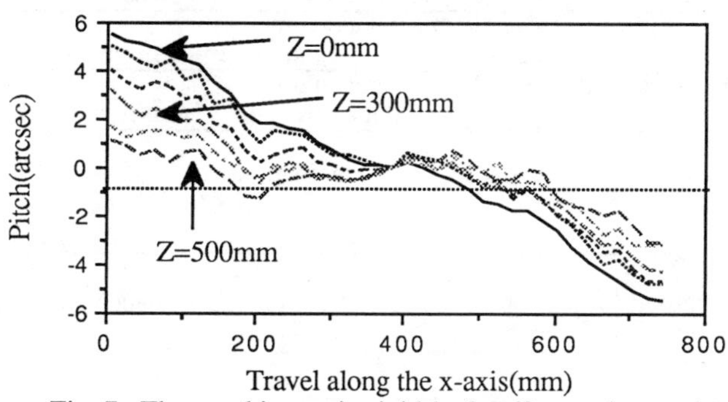

Fig. 7: The non-kinematic-rigid-body effect at the x-axis

5. IMPLEMENTATION OF ERROR COMPENSATION SCHEME

The block diagram of the compensation scheme is depicted in Fig. 8. There is a data bank of the off-line identified geometric errors. Although these errors may be quite stationary over a period of time, they need to be periodically updated on a yearly basis to accommodate the long term stability of the material and foundation.

Machine temperatures are monitored on-line by a temperature measurement system made by Omega Co. Based on measured temperatures, the pre-established empirical models estimate the thermal error components in real time. During the preliminary experiments, 40 critical points at the machine structure were measured. Twenty three thermal couples have been permanently installed on the machine structure. Real-time estimated thermal error components include thermal expansion of leadscrews, thermal drifts at spindle head, thermal variation of squareness errors, thermal drift of axis references and thermal bending (angular errors) of the machine structure and spindle. Table 1 lists the relationships between thermal couples and thermal errors. Positions of three movable slides are captured by a Quadrature Encoder Input /Digital Output board in real time.

Based on the determined volumetric error, an interface module which consists of three digital I/O boards sends out the compensation signals to the digital I/O ports of the Delta20 controller. The Delta20 controller will import and compensate for the compensation signals in real time. This was done by simply editing the software program of the programmable logic controller (PLC) of the Delta20 without any hardware change on the servo-drive loops. According to the execution speed of the PLC, the Delta20 can correct the volumetric error by shifting references of three slides within 10ms. Therefore, this scheme can do contour compensation in real time. For more information about the implementation, see [4].

	Thermal error components	Monitored temperature
1	distance error (x)	bearing and nut on the x-axis ballscrew
2	distance error (z)	bearing and nut on the z-axis ballscrew
3	reference drift (x-axis)	bearing and nut on the x-axis ballscrew
4	reference drift (z-axis)	bearing and nut on the z-axis ballscrew
5	distance error (y)	bearing and nut on the y-axis ballscrew machine column, spindle
6	pitch angular error (y)	front and rear sides of machine column
7	yaw angular error (y)	left and right sides of machine column
8	squareness error (xy)	left and right sides of machine column
9	squareness error (yz)	front and rear sides of machine column
10	spindle thermal drift (x)	spindle, machine column bearing and nut on the y-axis leadscrew
11	spindle thermal drift (y)	spindle, machine column bearing and nut on the y-axis leadscrew
12	spindle thermal drift (z)	spindle, machine column bearing and nut on the y-axis leadscrew
13	spindle angular drift (tilt-up)	spindle, machine column bearing and nut on the y-axis leadscrew
14	spindle angular drift (lateral)	spindle, machine column bearing and nut on the y-axis leadscrew

Table 1 : Monitored thermal errors and temperatures

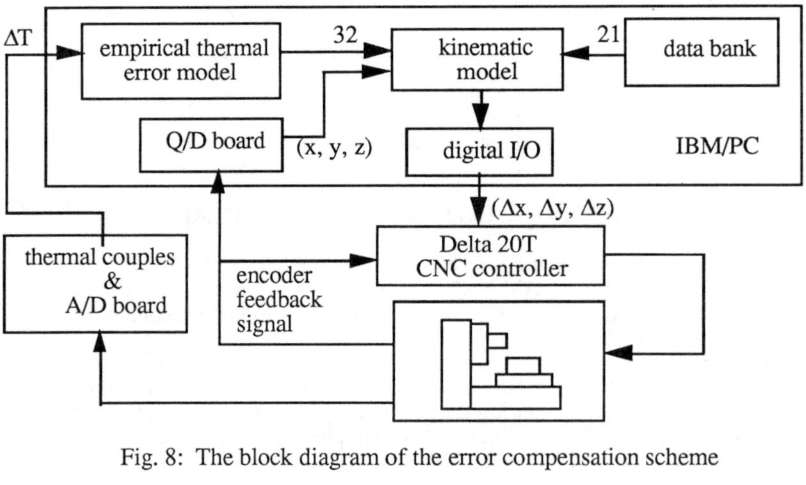

Fig. 8: The block diagram of the error compensation scheme

6. EXPERIMENTAL RESULTS

6.1 Machine Error Source Identifications

Four kind tests for error source identification were conducted. Geometric error identification under machine cold start conditions was to identify machine errors created by manufacturing imperfections, misalignments and mounting effects. The identification of thermal effects recognized thermal distortions of machine structures during operation. The identification of the non-rigid body kinematic effect characterized coupling effects among the machine slides, i.e. the error variations of one linkage due to the movements of two other linkages. All of the tests mentioned above were conducted under no cutting or air cutting conditions. Consequently, no cutting forces were exerted on the machine structure. Cutting force effect test will complement the tests mentioned above.

The largest geometric errors came from distance errors of the three axes (90 μm of the x-axis, 30 μm of the y-axis and 150 μm of the z-axis). The Abbe errors created by pitch errors and large offset distances are mainly responsible for these large distance errors.

For the thermal errors, thermal expansions of leadscrews with magnitudes of 25 μm, 120 μm and 40 μm were found in the x, y and z-axes respectively after a six hour run. Thermal expansion of the z-axis also caused an 82 μm drift at the reference of the z-axis. Translational thermal drifts of the spindle in all three directions were significant after a seven hour run with 1000 rpm spindle speed (100 μm, 60 μm and 200 μm in the x, y and z directions). Thermal inclinations of the spindle axis were about 30 arcsec and 20 arcsec in the tilt-up and lateral directions. The column was also subject to significant thermal bending which caused a 12arcsec thermal variation in the pitch error of y-axis and a 15arcsec thermal variation in the y-z squareness. A 12 arcsec thermal variation of the yaw error in the y-axis was also found despite the fact that the machine structure is symmetric. Examining the temperature field of the machine structure, the column had a 4°C temperature gradient from the left hand side to the right hand side. The reason is that some heat sources of the machining center such as the hydraulic system and the power amplifiers of motors are clustered in the left hand side of the machine column which caused unsymmetrical thermal design.

Major non-rigid body kinematic effects were found at the distance (15 μm) and pitch (8 arcsec) errors of the x-axis due to the movement of the z-axis. This happens because the z-axis slide is stacked above the x-axis slide. The non-rigid body kinematic effect between y and z axes or y and x axes was also checked, but no significant coupling effect was found. Different machining processes (including drilling, milling and boring) at an aluminum part were studied. It was found that the cutting force effect of the machine structure was less than 5 μm in finish cutting.

In summary, for this machining center, thermal errors are the largest error sources to the volumetric error, followed by geometric errors, non-rigid body kinematic effect and finally the cutting force effect.

Volumetric error under 1000 rpm spindle speed synthesized from the 32 error components is shown in Fig. 9. Maps for the bottom and top planes of the machine working zone are plotted to show the spatial-variant characteristic of the volumetric error. Maps under cold start and after an 8 hour run are also plotted to show the time-variant characteristic of the volumetric error. For a simple workpiece requiring a relatively short machining time, the spatial variation of the volumetric error would be the major error source of the workpiece. However, when machining a very complicated workpiece or a batch jobs, which may take several hours of the machining time, both the spatial-variant error and the thermal drift of the machine reference would be important.

Under cold start, the volumetric error had a 320 μm error in terms of spatial variation (from a minimum value of zero at the bottom plane to a maximum value of 320 μm at the top plane). After an 8 hour run, the spatial-variant error increased to a value of 400 μm (from a minimum value of 250 μm at the bottom plane to a maximum value of 650 μm at the top plane). Also, when compared to the maps under cold start, a 250 μm thermal drift of the machine reference was observed after the 8 hour run.

6.2 Performance Evaluation of Compensation Scheme

The effectiveness of the error compensation scheme has been verified by a HP laser interferometer system and by a cutting test. Instrumental evaluation has the advantage in that the machine accuracy can be evaluated covering a wide range of the machine working zone. On the other hand, the cutting test can provide the information much closer to the real machining process.

The first test is for spatial-variant error compensation. In this test, the distance errors of the z-axis at two XZ planes within the machine working zone were measured without and with compensation. The first xz plane is at y = 65 mm and the second xz plane is at y = 385 mm. For each plane, 6 lines distributing equally across the plane were tested. Notice that this test covers the positional variations in all three axes. Let's single out one line from each plane for discussion. The maximum errors without compensation were 160μm at line 1 and 250μm at line 2 (Fig. 10). With compensation they were reduced to within a 10μm range. Actually, the errors along all 12 lines were all within ±10μm after compensation compared to 160μm ~ 250μm before compensation.

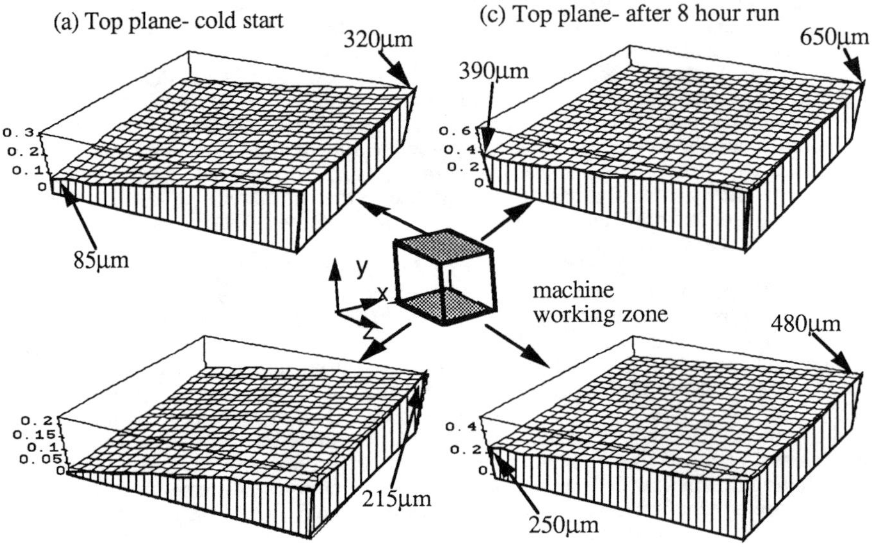

Fig. 9: The spatial-variant and time-variant volumetric error (under 1000rpm).

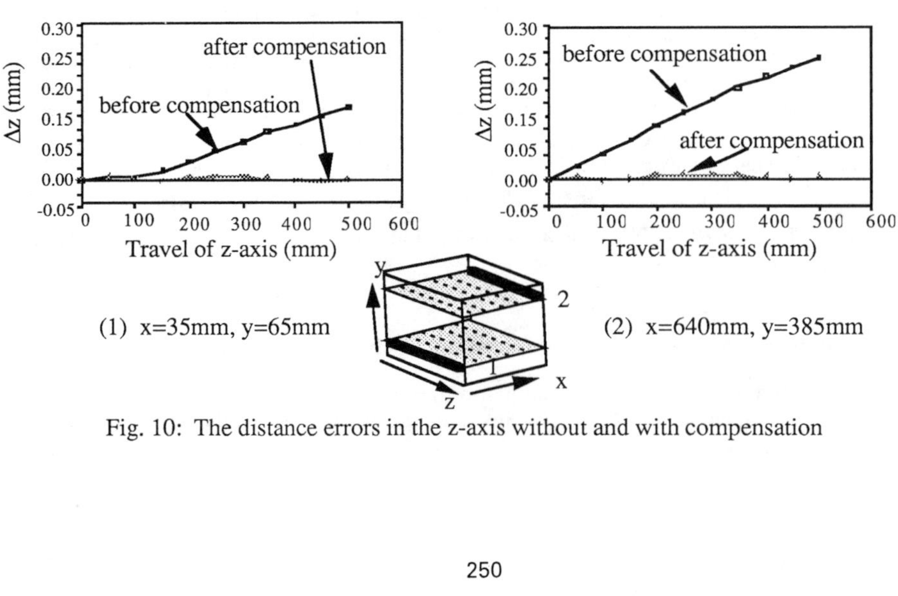

Fig. 10: The distance errors in the z-axis without and with compensation

For the evaluation of the time-variant error compensation, the machine was run under a typical cutting cycle for 6 hours. The distance errors of the y-axis were measured periodically with and without compensation. Fig. 6 shows that without compensation the distance errors along the y-axis varied from -35 µm at cold-start to 110 µm after a six hour run due to the time-variant thermal effect. With compensation, machine errors have all been kept within ± 10 µm for the entire 6 hour run (Fig. 11).

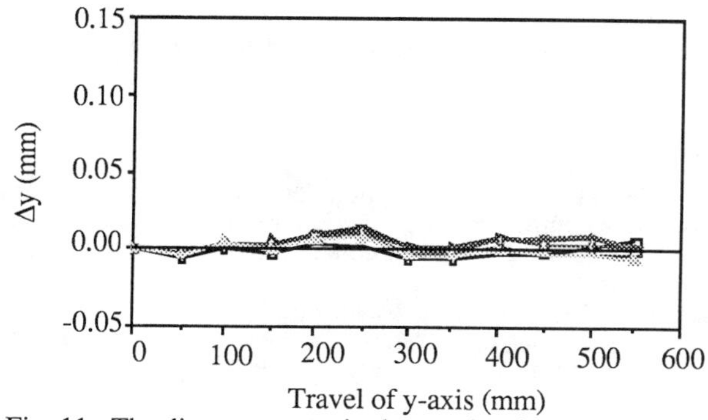

Fig. 11: The distance errors in the y-axis with compensation

Two kinds of machining jobs were evaluated on an aluminum workpiece. First, several holes were made under cold start and after an 8 hour run. This is to evaluate the effectiveness of the compensation in the x-axis and the y-axis. Second, several strip-like surfaces were milled under cold start and after an 8 hour run. The depth difference of the milled surface was used to evaluate the compensation in the spindle growth (in the z-axis of the machine).

The cut workpiece was inspected using a Sheffield Cordax CMM with a volumetric accuracy about 14 µm. The results shows that the maximum positional errors of these holes was reduced from 92.4µm to 18.9µm (Fig. 12) in a dimension of 404x310 mm^2. Also notice that the squareness errors of the workpiece were improved from 8.7arcsec to 2.3arcsec. The difference of these milled surfaces was reduced from 196µm to 8µm (Fig. 13).

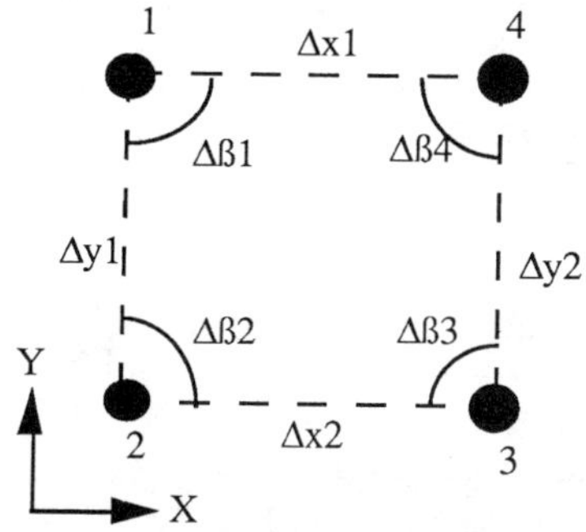

	uncompensated	compensated
Δx1	28.9µm	-18.9µm
Δx2	30.9µm	-16.8µm
Δy1	92.3µm	0.2µm
Δy2	92.4µm	7.2mm
Δß1	7.4arcsec	2.2arcsec
Δß2	-7.4arcsec	-1.4arcsec
Δß3	8.7arcsec	2.3arcsec
Δß4	-8.7arcsec	-1.5arcsec

x1=x2=410mm, y1=y2=304mm, β1=β2=β3=β4=90

Fig. 12: The dimensional errors without and with compensation (after 8 hour run)

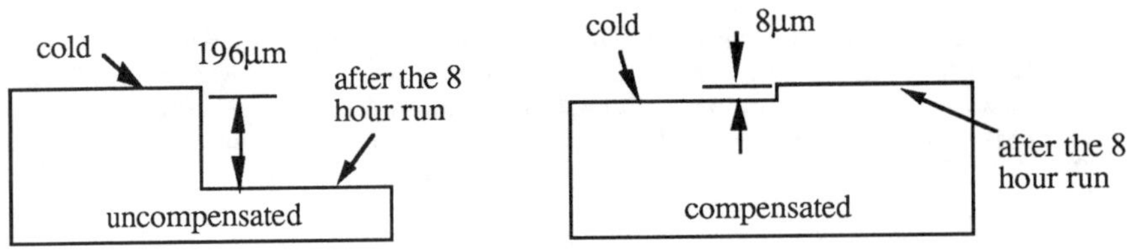

Fig. 13: The surface jumps without and with compensation

7. CONCLUSIONS

A volumetric error model was developed to synthesize machine error sources and predict the result volumetric error between the cutting tool and workpiece at any location within the machine working zone at any time. This model can accommodate the effects of varying tool length and spindle position of a machining center.

Thermal errors were predicted in real time by the empirical models with multiple variables through on-line monitoring of the machine temperatures. The test results show that both models are well able to predict the new observations of thermal errors under the variation of spindle speed and temperature history.

A non-rigid body kinematic effect was found to cause the pitch error of the x-axis slide varying with the movement of the z-axis slide. A correction was made to enhance the precision of the volumetric error model derived from the kinematic rigid body assumption. The pitch errors in the x-axis were updated according to the z-axis slide position in five intervals (z=0, 100, 200, 300, 400, 500mm). Within each interval, the coupling effect was corrected by the linear interpolation.

A compensation controller based on an IBM/PC was developed to upgrade current CNC controllers for the volumetric error compensation. The compensation controller employs a software approach to work with the CNC controller without any hardware modification of the machine servo-drive loops. It can do not only a point to point compensation but also a contour compensation. Furthermore, all interface boards required for this compensation controller are mass produced at commercial markets. These features make this compensation scheme very cost competitive.

The developed error compensation scheme was implemented on a 4-axis horizontal machining center equipped with a DynaPath Delta20T CNC controller. It was proven that this error compensation scheme is able to improve the machine accuracy by one order of magnitude. A cut workpiece was inspected using a coordinate measuring machine (CMM) to confirm that the dimensional errors had been reduced from 92.4 μm to 18.9 μm with a nominal dimension of 404x310 mm^2 and that the depth difference of milled surfaces was reduced from 196 μm to 8 μm.

REFERENCES

[1]. Anjanappa, M., et al., "Error Correction Methodologies and Control Strategies for Numerical Control Machine", Control Methods for Manufacturing Process, DSC-Vol. 7, 1988

[2]. Attia, M. H. and Kops, L.,"Nonlinear Thermoelastic Behavior of Structural Joints-Solution to a Missing Link for Predication of Thermal Deformation of Machine Tools", Trans. of ASME, Journal of Engineering for Industry, Aug. 1979, Vol. 101

[3]. Bryan, J.B., "International status of thermal error research," Annals of CIRP, Vol.16(1), pp:203, 1968

[4]. Chen, J. S., "Real-time Compensation of Time-variant Volumetric Error on a Machining Center", Ph. D. Dissertation, University of Michigan, Ann Arbor, 1991

[5]. Donmez, M. A. et al., "A general methodology for machine tool accuracy enhancement by error compensation", Precision engineering, Oct. 1986,Vol. 8, No. 4

[6]. Koren, Y., "Robotics for Engineers", McGraw-Hill Book Company, 1987

[7]. Kurtoglu, "The Accuracy Improvement of Machine Tools", Annals of the CIRP, Vol. 39/1/1990

[8]. Lovett, D. C. edited, “Progress Report of the Quality in Automations”, NISTIR 89-4045

[9]. Neter, J., Wasserman, W., and Kuther, M. H., "Applied Linear Statistical Models", 1985, published by Richard D. Irwin, Inc.

[10]. Woytowitz, M.A. et al., "Tool Path Error Analysis for High Precision Milling with a Magnetic Bearing Spindle", ASME 1989, pp:129-142

Sensors, Controls, and Quality Issues
in Manufacturing
ASME 1991

A MULTI-DEGREE-OF-FREEDOM MEASURING SYSTEM FOR CMM GEOMETRIC ERRORS

J. Ni, P. S. Huang, and S. M. Wu
Department of Mechanical Engineering and Applied Mechanics
University of Michigan
Ann Arbor, Michigan

ABSTRACT

A precision multi-degree-of-freedom measuring (MDFM) system has been developed and implemented for the simultaneous measurement of straightness, pitch, yaw, and roll errors of the moving axes of a CMM. The system is based on the principles of laser alignment and autocollimator. Its measurement principles and the influence of laser beam drifts on its measurement quality have been investigated and some improvement schemes have been implemented. Through the measurements of actual as well as artificially created geometric errors of the CMM, it has been found that the system's accuracy of measuring straightness error components is better than 1 μm and its accuracy for angular error measurements is better than 0.5 arcsec.

I. INTRODUCTION

The use of error compensation techniques has been recognized as an effective way in the improvement of the performance of coordinate measuring machines (CMMs). An essential part of the error compensation technique is the ability to measure the geometric errors of the CMMs. The measurement of these geometric errors has always been a tedious and time-consuming task.

Conventional methods of measuring the geometric errors of a CMM range from the use of laser interferometers, electronic levels, mechanical squares, straight edges, and other artifacts [1,2]. Laser interferometers have the features of high accuracy, high complexity and high cost. Commercial systems such as the Hewlett-Packard 5528A systems are available for the measurement of displacement, straightness, pitch and yaw errors. For the measurement of roll errors, differential electronic levels are frequently used and a surface plate combined with two offset electronic indicators is developed to measure the roll of a vertical axis [3]. Dial indicators and other electronic comparators are also used together with mechanical squares and straight edges to measure squareness, straightness, parallelism, and some angular errors [4,5]. An optical method using a single laser beam with a single four-quadrant photodiode was developed for measuring radial run out, tilt of a rotation axis, straightness and parallelism [6]. All these methods are suited for off-line applications. That is, only one or two error components can be measured at one particular setup of these measurement devices.

To increase the number of error components measured at a single setup, an experimental laser optical system has been developed on a milling machine to demonstrate the simultaneous measurement of 5 geometric errors (two straightness components, plus pitch, yaw and roll errors)

[7]. Most recently, a similar laser optical system has also been developed in Japan to measure two straightness components and two angular errors (pitch and yaw) of a mechanical slide [8]. However, both these systems are still not suited for the implementation on CMMs because of limited measurement range and accuracy, and their vulnerability to environmental influences.

The applications of a system to CMMs impose different requirements on the system as opposed to the applications to metal cutting machines. The former involves no cutting and no coolant, but requires a high level of accuracy and stability of the system. Due to these stringent requirements, substantial improvement have been made in a new MDFM system based on the work in reference [7].

This paper presents this new Multi-Degree-of-Freedom Measurement (MDFM) system for simultaneously measuring two straightness, pitch, yaw and roll errors of each moving axis of a CMM. Section 2 presents the overall system layout of this new MDFM system along with the principle of measurements. Section 3 presents some significant improvement schemes used in this MDFM system. The system performance is tested and evaluated in Section 4. The implementation of a prototype MDFM system on a CMM is discussed in Section 5 and final experimental evaluation of the system is also given by comparing the measurements with a commercially certified HP 5528A system.

II. PRINCIPLE OF THE MDFM SYSTEM

2.1 Overall system layout

Figure 1 shows the optical layout of the MDFM system. This system consists of two straightness measurement units which determine two straightness errors and roll, and a rotational error measurement unit based on an auto-collimator which identifies the pitch and yaw.

A linearly polarized He-Ne laser tube is used as the light source. The quality of its output beam is improved by the use of a beam expander as will be explained later. The direction of polarization of the laser beam lies in a plane neither parallel nor perpendicular to the incident plane of polarizing beam splitter PBS1 and thus results in that one portion of the beam (namely, P-polarized beam) passing through PBS1 while the rest (namely, S-polarized beam) being reflected.

The beam passing through PBS1 is reflected from a retroreflector RR1 and is projected by

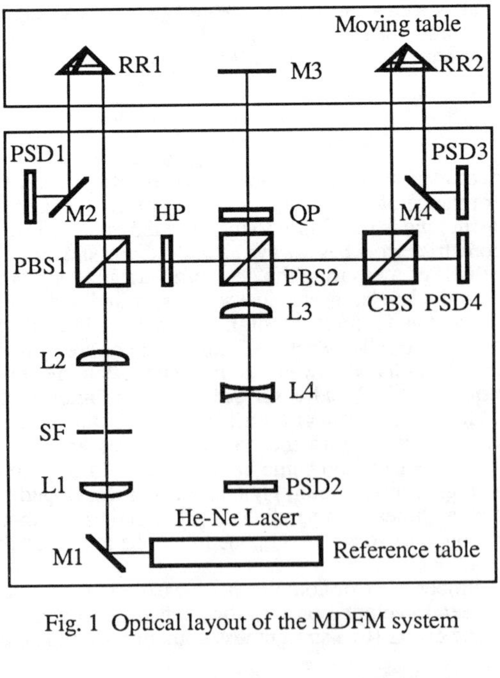

Fig. 1 Optical layout of the MDFM system

mirror M2 onto a 2-D position sensitive detector PSD1 to measure both lateral and vertical straightness errors at the position. The reflected beam from PBS1 passes through a half wave plate, HP, which alters the direction of polarization of the beam from S polarization, normal to the incident plane, to a skewed plane. This rotation of the direction of polarization is required in order to further split the beam at PBS2 where part of the beam is reflected and propagates toward the reflection mirror M3 for pitch and yaw measurement, and part of the beam passes through to the next straightness measurement unit and a beam drift compensation unit.

The beam passing through PBS2 is split into two by cube beam splitter CBS. This CBS is such a beam splitter that it splits beam into two with approximately same intensity and is insensitive to the direction of polarization of the incident beam. The beam reflected at CBS traverses to the second straightness measurement unit consisting of RR2, M4 and PSD3. Sensor PSD4 is fixed relative to the laser source and monitors the drift of the laser beam itself. Therefore it can be used for on-line beam drift compensation as is needed in a high accuracy system.

There are four measurement beams in this system. As they are sensed by the same types of detectors and electronic circuits, it is desirable that the intensities of all these four beams be the same. In the designed system, this intensity homogeneity can be achieved by rotating the laser tube and the half wave plate HP. Specifically, rotate the laser tube first and let the beams passing through and reflected from PBS1 have the intensity ratio of about 1 to 3. Then rotate HP so that the beam reflected from PBS2 has the intensity of half of that of the beam passing through PBS2. Since for each measurement unit, the number of optical components used is different, the loss of beam power due to unexpected reflection in each unit is different. Therefore fine adjustment is needed for both the laser source and the half wave plate HP.

2.2 Measurement of two straightness errors

The measurement of straightness errors of a table movement is accomplished by the straightness measurement units, RR1 and RR2. These two identical units are specifically designed for the measurement of the rolling error of a moving table, but can provide a redundant measurement of straightness errors of the table movement. Thus, averaging the outputs from these two units can achieve better measurement results.

A retroreflector, the RR1 or RR2, which is also called a corner cube, is attached rigidly to a moving table and used to emulate the motion of the table. When the table has a lateral displacement, D, in a plane perpendicular to the travelling direction, the laser beam reflected by the retroreflector will shifted on the photosensor plane by a distance of 2D, as illustrated in Fig.2 That is, there is an amplification factor of 2 built in this measurement scheme and it results in an increase of the resolution of the straightness error measurement by a factor of 2.

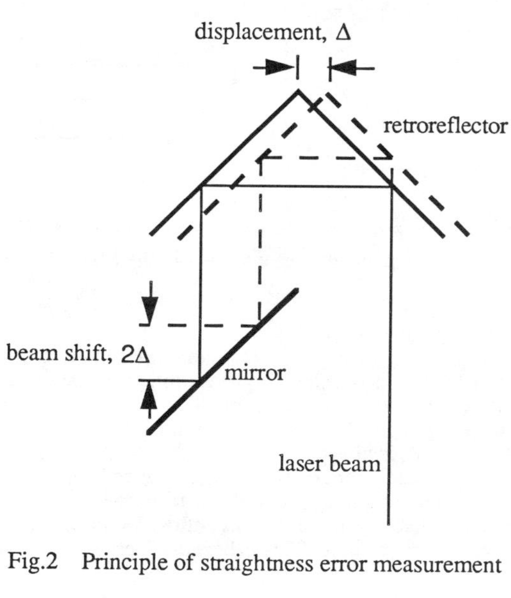

Fig.2 Principle of straightness error measurement

Since there are two straightness measurement units available, averaging the readings from these two units will give the measurements of straightness errors at the midpoint between these two units. If the rotational errors of the table are also measured, the straightness errors of any point on the table can be calculated from the straightness errors at the midpoint and the angular rotations of the table, based on the kinematics of rigid body. Algebraically, the straightness errors at the midpoint are calculated as:

$$S_{lateral} = \frac{x_1 + x_3}{4} \tag{1}$$

$$S_{vertical} = \frac{y_1 + y_3}{4} \tag{2}$$

where (x_1,y_1) and (x_3, y_3) are the outputs of PSD1 and PSD3, respectively.

2.3 Measurement of roll error

The roll of a moving table is measured indirectly by the two straightness measurement units as illustrated in Fig.3. The straightness errors at points A and B are measured in both lateral (left and right in the paper plane) and vertical (in and out perpendicular to the paper plane in the top view of Fig.3) directions. The roll is calculated from the difference of the straightness errors in the vertical direction as follows:

$$\text{roll} = \frac{\text{straightness at A - straightness at B}}{\text{separation between A and B}} = \frac{\frac{y_1}{2} - \frac{y_3}{2}}{D} = \frac{y_1 - y_3}{2D} \tag{3}$$

where D is the separation distance between the two straightness measurement units. It can be seen that the accuracy of the roll measurement depends on that of the straightness measurements. In order to improve the resolution of roll measurement, it is desirable to place the two straightness measurement units apart as far as physically possible in the actual setup.

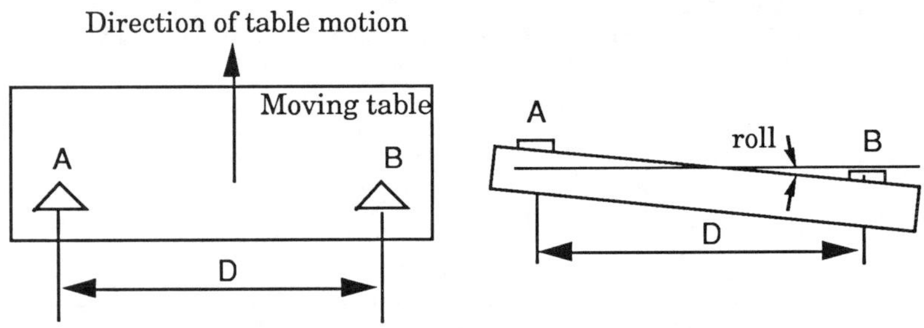

Fig.3 Principle of roll measurement in the MDFM system

2.4 Measurements of pitch and yaw errors

For the measurements of pitch and yaw errors, a modified auto-collimator is used instead of a conventional type auto-collimator to achieve high resolution of measurement. Figure 4 shows a conventional type of auto-collimator in which a positive or a convex lens is used to focus the incident beam. The sensor PSD is fixed at the focal point of the collimator lens, L. Therefore the beam reflected from the mirror, M, is always focused on the surface of the sensor. This ensures that the lateral displacement of the focused spot on the sensor surface is only related to the orientation of the incident beam and independent on the position of the mirror M.

However, there is a problem with this conventional type of auto-collimator when high resolution of angle measurement is required. This is because the resolution of this auto-collimator is

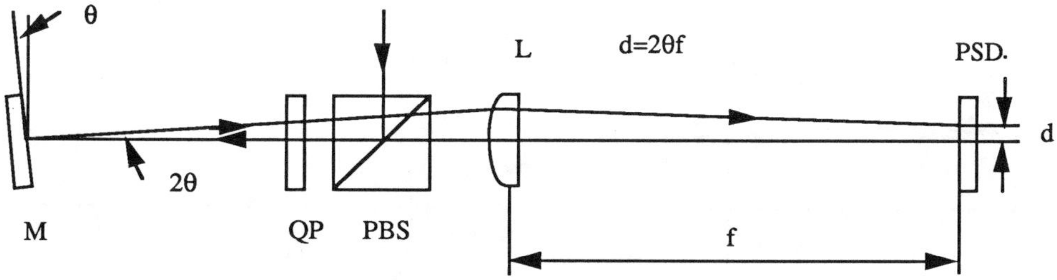

Fig.4 A conventional autocollimator for angular error measurement

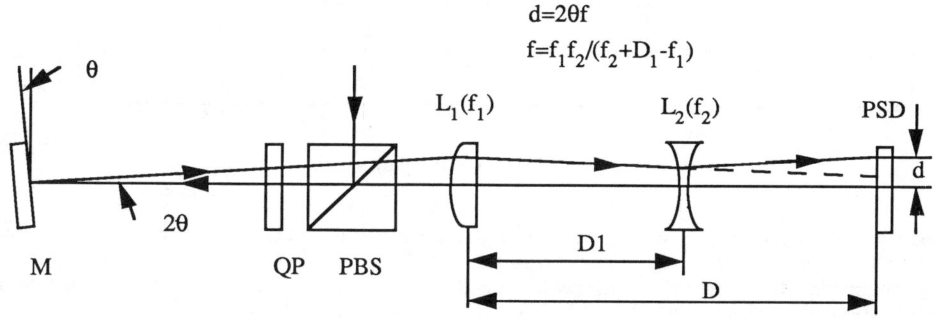

Fig.5 The new high resolution autocollimator in the MDFM system

directly proportional to the focal length, f, of the collimator lens. If high resolution is needed, then a collimator lens with a very long focal length should be chosen. This focal length, however, is limited in the physical implementation by the overall configuration size of the system and apparently cannot be designed very long. Therefore there will be a tradeoff between the size of the system and the resolution of the angle measurement.

A high resolution and compact auto-collimator is designed to meet the need of high angular measurement resolution in the MDFM system. Figure 5 shows the principle of this new auto-collimator. Instead of a single positive lens a combination of a positive and a negative lenses is used to increase the equivalent focal length while keeping the distance between the positive lens and the sensor PSD short. For instance, in the present system, f_1, f_2 are chosen to be 80mm and -10mm respectively. With this combination an equivalent focal length of 650mm is obtained while the distance D is only about 143mm.

Following is the derivation of determining the optimal focal lengths of the positive and negative lenses L1, L2 and their distance. The objective is to obtain a largest equivalent focal length with a shortest distance from Lens L1 to sensor PSD. In Fig.6, the parameters used in this derivation are defined. From the theory of geometric optics [9], the following equations are derived:

$$f = \frac{f_1 f_2}{D_1 - f_1 + f_2} \tag{4}$$

$$D = D_1 + \frac{f_2(f_1 - D_1)}{D_1 - f_1 + f_2} \tag{5}$$

$$\frac{D}{f} = 1 - \frac{D(f_1 - D_1)}{f_1 f_2} \tag{6}$$

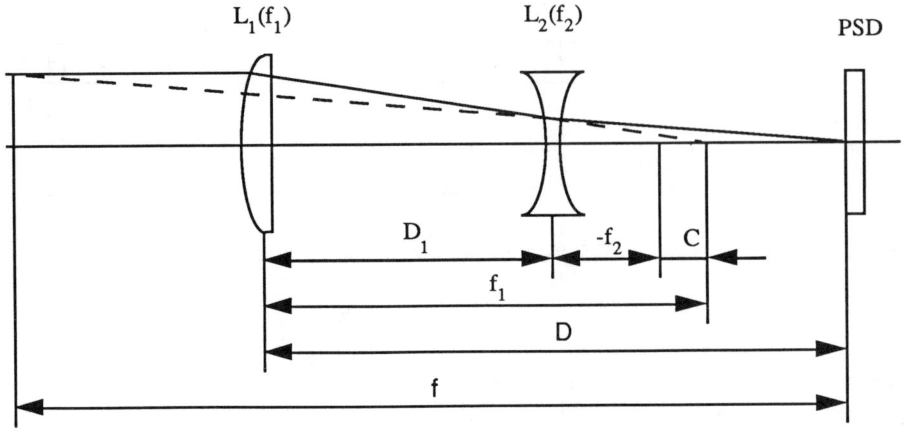

Fig.6 Definition of the parameters

where f is the equivalent focal length of the combination lenses and D is the distance between the Lens L1 to sensor PSD. Since $(D_1-f_1+f_2)=-C$, from Eq.(4), $f=-f_1f_2/C$. Therefore, to acquire a larger equivalent focal length, the smaller C the better. Now assuming that C has been selected for a small value, we need to adjust D_1 and f_2 so that D/f is the smallest. Rewriting Eq.(6) as a function of f_1, f_2 and C, we obtain

$$\frac{D}{f}=\frac{(f_2-C)^2+Cf_1}{f_1f_2} \tag{7}$$

Suppose f_1 has been selected. By setting $d(D/f)/df_2=0$ and solving for f_2, we obtain the optimal f_2 and therefore the optimal D_1 as following:

$$f_2=\sqrt{C(f_1+C)} \tag{8}$$

$$D_1=f_1+C-\sqrt{C(f_1+C)} \tag{9}$$

In the present system, f_1 and C are selected to be 80mm and 1.23mm respectively. Calculating from Eqs.(4), (8) and (9), f=650.41mm, f_2=10mm, D_1=71.23mm. From Eq.(6), it can be found that D/f=21.9% in this case. This means that by using the high resolution auto-collimator, the resolution is increased by approximately 4.6 times compared with that of the conventional type of auto-collimator while maintaining the same physical size.

The pitch and yaw errors are measured by using this new high resolution autocollimator. Algebraically, the following equations are used to calculate the pitch and yaw errors.

$$\text{pitch}=\frac{y_2}{2f} \tag{10}$$

$$\text{yaw}=\frac{x_2}{2f} \tag{11}$$

III. IMPROVEMENT OF THE MDFM SYSTEM PERFORMANCE

The principle of the MDFM system is based on the use of fixed laser beams as measurement references and, thus, the performance of the system will be affected by the quality of the reference laser beams. In this section, various efforts of improving the reference laser beam quality are discussed.

One of the biggest concerns about the laser optical measurement systems is the quality of the laser beam in terms of beam drifts, and beam noise. A beam expander is designed into the MDFM system to improve the quality of the reference laser beams.

A beam expander is basically a telescope used in reverse. Traditionally there are two different types of beam expanders, namely Keplerian and Galilean. Since a Keplerian type of beam expander can accommodate a spatial filter for beam noise reduction, it is more suitable for the MDFM system. A Keplerian beam expander usually consists of a focus lens, a spatial filter and a collimator lens as illustrated in Fig. 7.

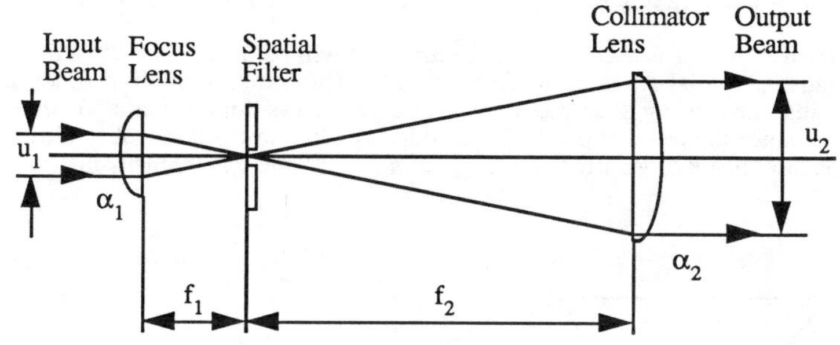

Fig. 7 Keplerian beam expander

The advantages of integrating a beam expander into the MDFM system are as following:

1) Beam divergence reduction

One nature of a parallel laser beam is that the smaller the beam diameter is, the larger the beam divergence is. Therefore reducing beam divergence can be accomplished by simply enlarging the beam diameter. Following are the equations relating the divergence and diameter of the output beam to that of the input beam.

$$\alpha_2 = \alpha_1 \frac{f_1}{f_2} \qquad (12)$$

$$u_2 = u_1 \frac{f_2}{f_1} \qquad (13)$$

where α_1, α_2 : Divergences of input and output beams respectively.
u_1, u_2 : Diameters of input and output beams respectively.
f_1, f_2 : Focal lengths of the focus lens and collimator lens respectively.

This means that if f_2 is chosen larger than f_1, the divergence of the resultant beam will be decreased by a factor of f_2/f_1 while the diameter be increased by the same ratio. In the MDFM system, f_1=14.8mm, f_2=50.2mm, therefore f_2/f_1=3.39.

2) Beam drift reduction

The drift or pointing instability of a laser beam can be divided into two parts, namely the angular drift and the lateral drift. For measurements over a long distance, obviously the error contributed by the angular drift will be the dominant factor. With the use of a beam expander, this angular drift can be reduced by a ratio of f_2/f_1 or 3.39 in the MDFM system.

In contrast to its effect on angular drift, however, the beam expander will enlarge the lateral drift at the same time by the same ratio. Although this increasing of lateral drift is undesirable, it will not be a problem in the MDFM system since the lateral drift can be compensated by the use of PSD4.

The PSD4 is especially designed to compensate for the lateral drift of the laser beam from a laser source. As can be seen in Fig. 1, the beam received by PSD4 is directly from the laser source and is not affected by any moving elements. Therefore, any positional change of the beam centroid on the PSD4 can be considered as the effect of the laser beam pointing instability.

3) Beam noise reduction

When dust or other particles exist alone the path of a laser beam, they will scatter the laser beam and therefore add noise to it. As shown in Fig.8, the result of this process can be depicted as a high frequency noise superimposed to a smooth Gaussian distribution of intensity. This noise will not affect the output of sensors as long as it stands still, since the sensors used in this system are supposed to detect only the intensity centroid of the laser beam. However, in reality the noise usually has a time-variant distribution which makes the intensity centroid of the laser beam wandering. This wandering of the intensity centroid will be perceived by PSD as the pointing error of the laser beam.

If a beam expander is used, then the beam noise will be obstructed when the focused beam passes through the spatial filter of a 25μm pinhole. This noise reduction function of a beam expander is illustrated in Fig.8. In the MDFM system, this is not the main advantage of using a beam expander since the beam expander is placed very close to the laser tube. However, it is still effective if the laser beam originally has noise and/or a non-Gaussian intensity distribution.

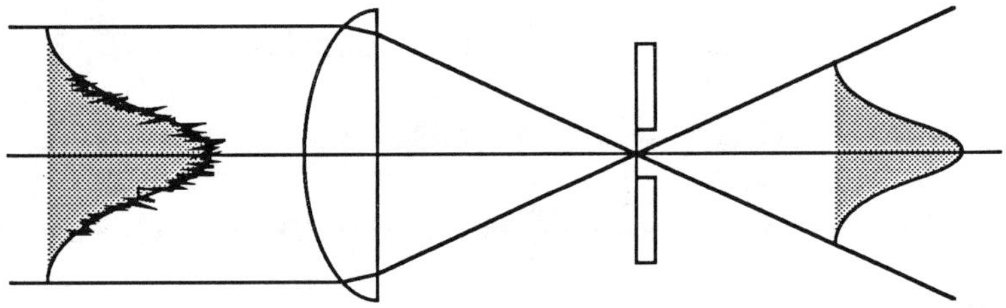

Fig.8 Noise reduction effect of a spatial filter

Table 1 summarizes the effect of a beam expander on the quality of laser beams.

Table 1 Effect of a beam expander on a laser beam

	Before beam expander	After beam expander
Beam diameter	0.63mm	2.14mm
Beam divergence	1.3mrad	0.38mrad
Diameter after 1m travel	3.23mm	2.90mm
Angular drift	<0.01mrad (2.08arcsec)	<0.003mrad (0.61arcsec)
Lateral drift	<1.5μm	<5.09μm

IV. TEST OF THE STABILITY OF THE MDFM SYSTEM

The constructed MDFM system was aligned on an optical table for stability evaluation where the effects of beam shielding and beam drift compensation were investigated. The shielding is introduced to reduce the air turbulence. Figure 9 is the result of a 2 hour test of the outputs from PSD3, without shielding and drift compensation, which shows a maximum drift of 7μm. This drift is mainly due to the drift of the laser source. The amount is approximately the same as specified by the laser tube manufacturer. In addition to this long term drift, there is a high frequency noise of about 2μm superimposed to the drift curve. This noise is postulated to be the result of air turbulence from the experimental environment.

The beam drift from the laser source can be compensated for by subtracting the outputs of PSD4 from the corresponding components of PSD3. As shown in Fig.10, with compensation, the drift of the output from PSD3 is reduced from 7 μm to 2 μm in range. It is also clear that the noise still maintains the same level of magnitude. From this test evaluation, one can conclude that the compensation technique for laser beam drift is effective.

To eliminate the air turbulence, paper pipes are introduced to restrict the air flow by enclosing the laser beam passage. A test is designed based on this concept and the result of the

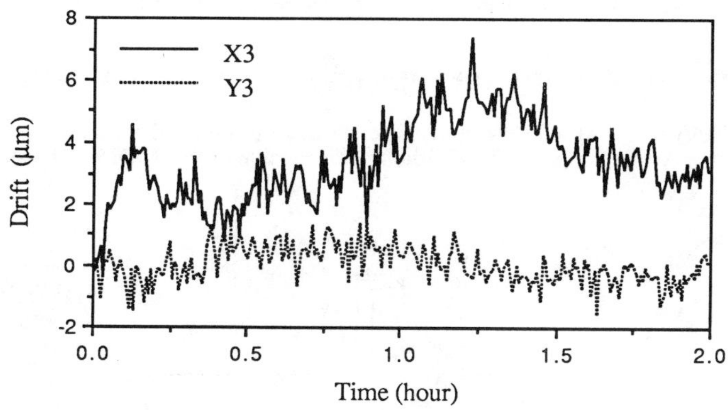

Fig.9 Unshielded, uncompensated 2 hour drift

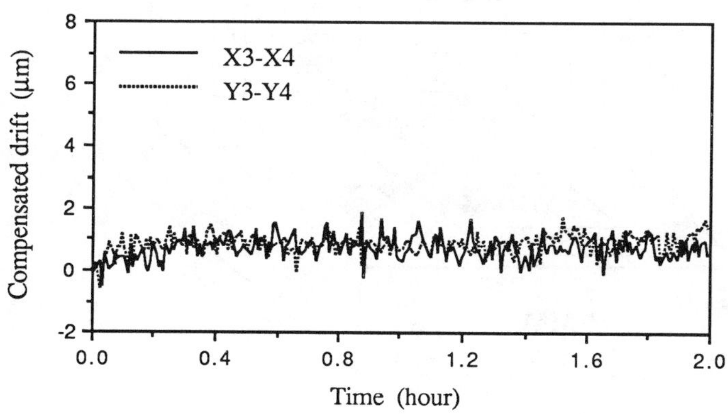

Fig.10 Unshielded, compensated 2 hour drift

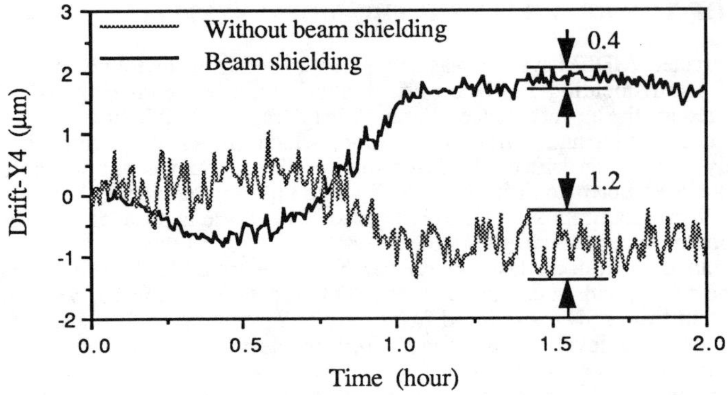

Fig.11 Effect of beam shielding

shielded, uncompensated 2 hours drift test is shown in Fig.11. It can be seen that with the use of shielding pipe the noise level of sensor output has been reduced from 1.2 μm to 0.4 μm.

V. THE IMPLEMENTATION OF THE MDFM SYSTEM ON A CMM

A bridge type CMM with a volumetric error of 12 μm is used for the purpose of testing the MDFM system. As illustrated in Fig.12, the reference table of the MDFM system is fixed on the

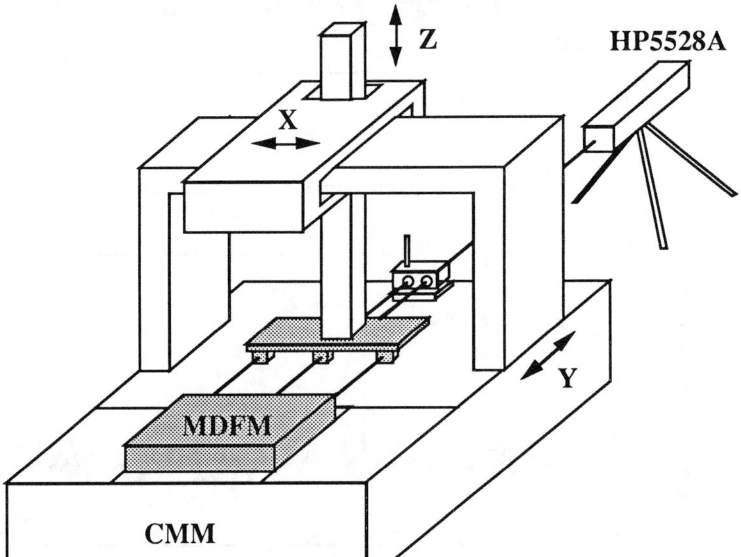

Fig.12 Test of the MDFM system on a CMM

rear part of the CMM table whereas the moving table of the MDFM system is attached to the Z-axis column. The task of aligning the MDFM system is to adjust the three measurement beams so that they are parallel in both X and Z direction. A HP interferometer measurement system is used as the base reference to compare with the new MDFM system.

5.1 The calibration of the MDFM system

The calibration of the MDFM system consists of two parts, the calibration of the straightness error measurement and that of angular error measurement. In both calibration processes, the HP interferometer measurement system is used as the reference. Also in order to check if the calibration curve is independent on the position of the moving table, the calibrations of angular error measurement are performed in two different positions, the near and far positions. The near and far positions correspond to the position where the measuring head is 100mm and 750mm, respectively, away from the reference table of the MDFM system. Also, some of the calibrations are repeated twice to check the repeatability of the calibration results.

For the calibration of the straightness error measurement, a retroreflector used for the MDFM system and a straightness retroreflector for HP system are attached to the Z-axis column as shown in Fig.13. At each position, the CMM is driven to move at a 10 μm increment along X-axis and Z-axis respectively. The output of the MDFM system is then compared with that of the HP system. Figures 14 and 16 show the results of these calibrations in both horizontal and vertical directions. Figures 15 and 17 are the residuals of the calibration curves when they are fitted by a liner line. It can be seen that the nonlinearity error plus noise is about 0.7% in X-axis or horizontal direction and about 0.8% in Z-axis or vertical direction.

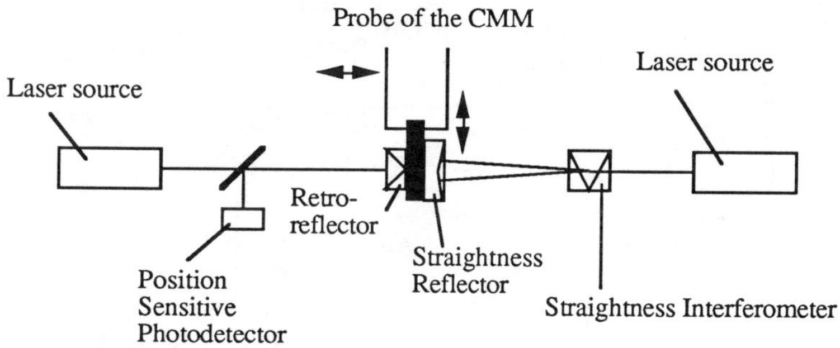

Fig.13 Setup for the calibration of straightness measurement

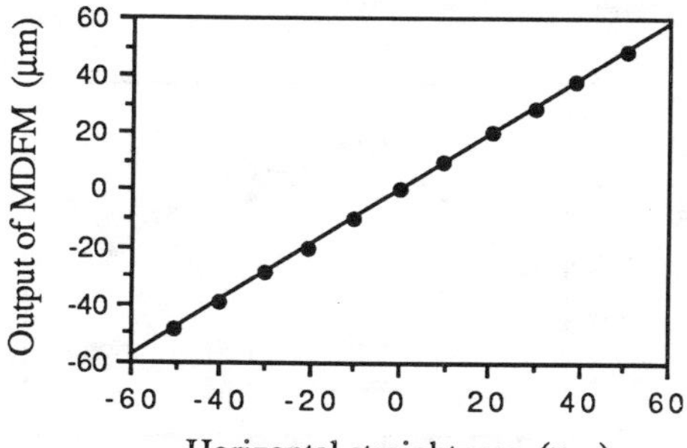

Fig.14 Horizontal straightness calibration

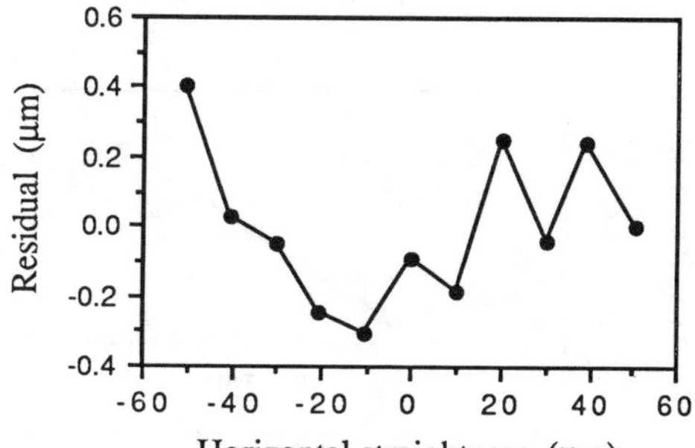

Fig.15 Residual of horizontal straightness calibration

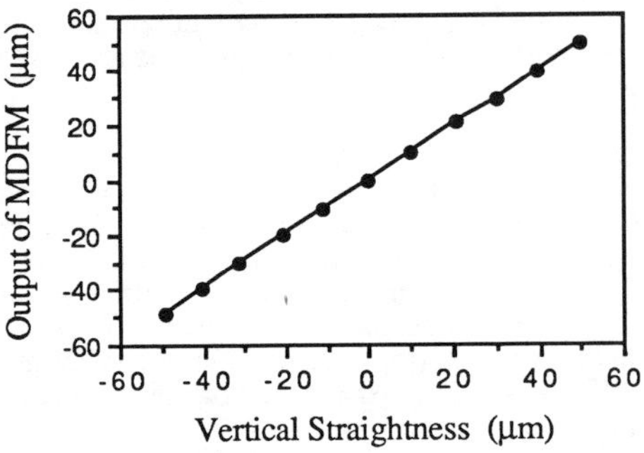

Fig.16 Vertical straightness calibration

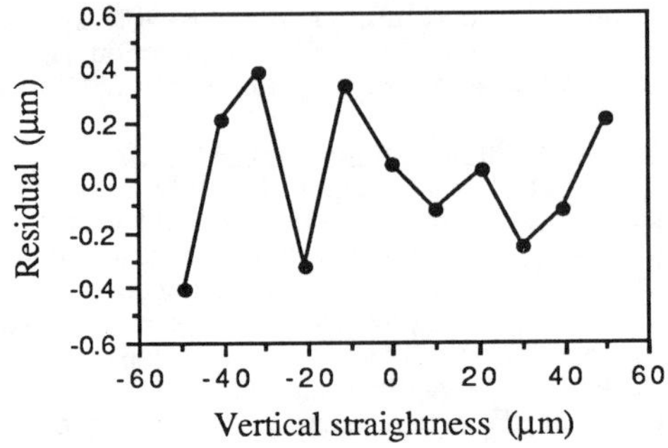

Fig.17 Residual of vertical straightness calibration

For the calibration of the angular measurement, a rotational stage is attached to the z-axis column and is used as an angular displacement driving device. On the rotational stage a reflection mirror of the MDFM system and a reflector of the HP system are fixed as shown in Fig.18. Calibration is done by rotating the rotational stage at a step of approximately 10 arcsec and recording both the readings of the MDFM system and the HP system. Figures 19 and 21 are the

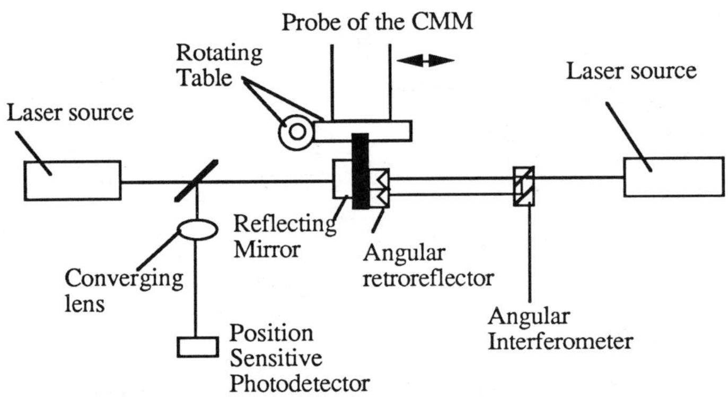

Fig.18 Setup for the calibration of angular measurements

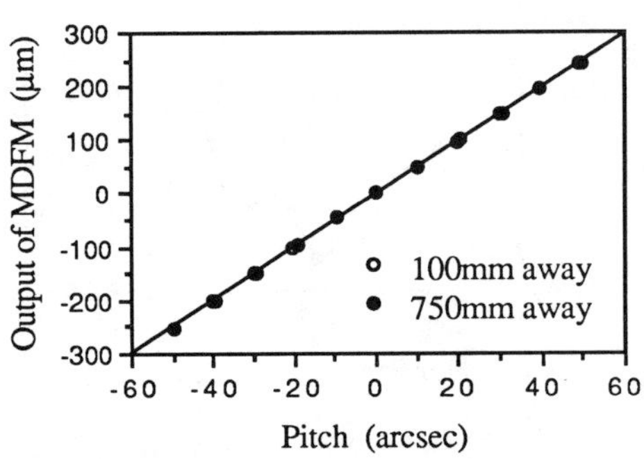

Fig.19 Pitch calibration

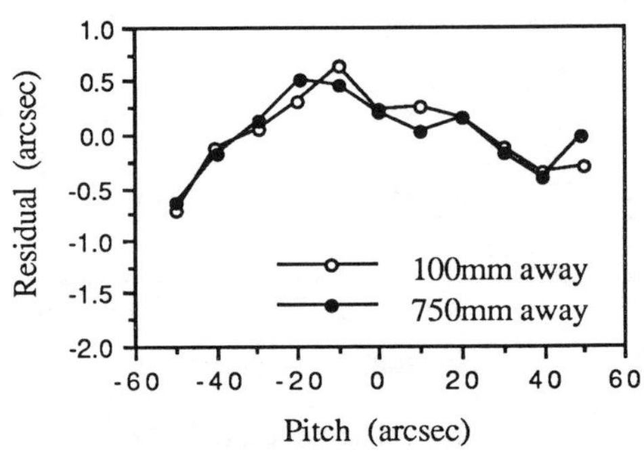

Fig.20 Residual of pitch calibration

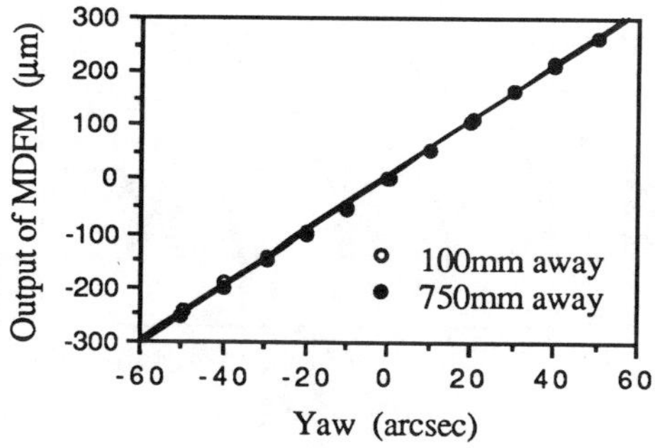

Fig.21 Yaw calibration

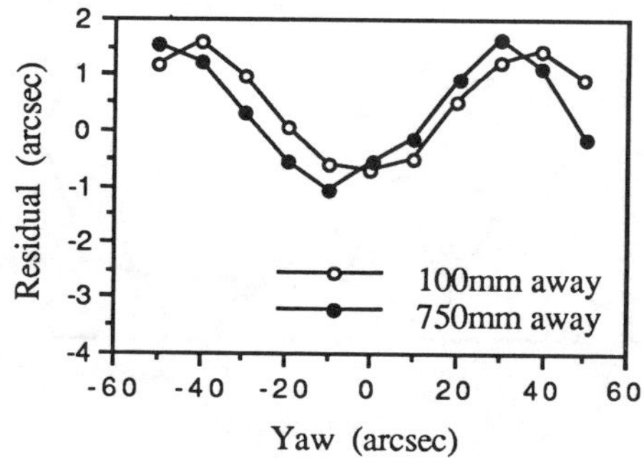

Fig.22 Residual of yaw calibration

results of pitch and yaw calibrations. Figures 20 and 22 are the residuals from linear fitting. The nonlinearity errors of the angular measurements are less than 1.5% for pitch error measurement and less than 2.8% for yaw measurement. Note that the repeatabilities of this calibration at two different positions are reasonably good. This indicates that the software compensation of these nonlinearity errors for higher accuracy is possible.

5.2 Measurements of the geometric errors of the CMM

After the implementation and calibration of the MDFM system on the CMM, it is used to measure the geometric errors of the CMM and the measurement results are compared with those obtained by a HP 5528A system. Two kinds of measurement experiments have been done on the CMM. In the first test, the real geometric errors (two straightness errors plus pitch and yaw errors) of the CMM along Y-axis are measured. In the second test the CMM is artificially driven to simulate the geometric errors. To simulate the geometric errors, as the CMM is moving farther, the geometric errors are induced by driving the CMM along either X or Z axis for straightness measurement, or rotating the rotational stage attached to the moving table for pitch and yaw measurement. The reason for the second experiment is that the errors of the CMM used are so small that the performance of the developed MDFM system may not be effectively evaluated.

Figures 23 to 26 show the results of straightness measurement, both the real error measurement and the measurement of artificially created geometric errors on a CMM. Compared with the results obtained from the HP system, one can see that the differences between these two

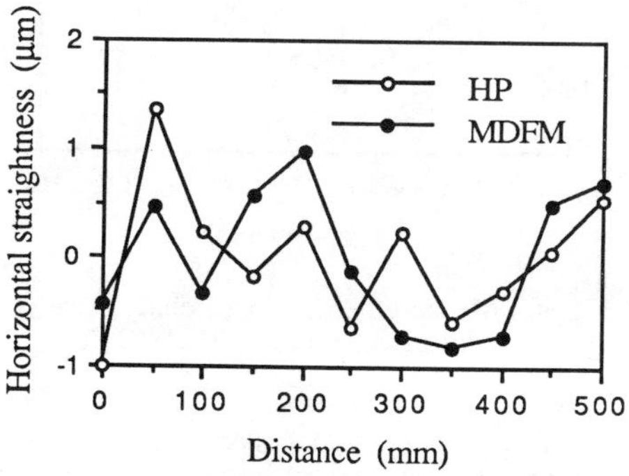

Fig.23 Measurements of horizontal straightness of the CMM

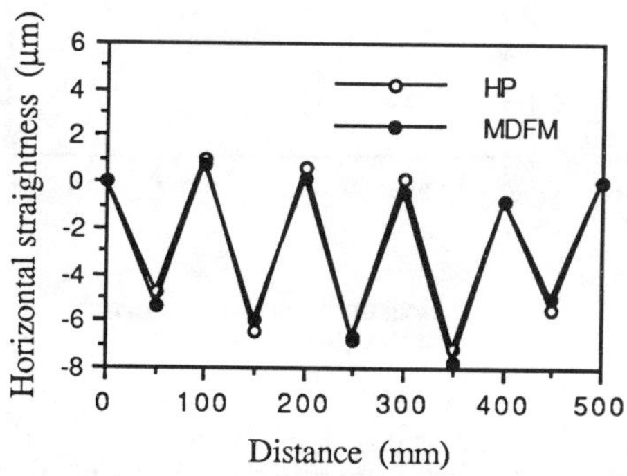

Fig.24 Measurements of artificially created horizontal straightness on the CMM

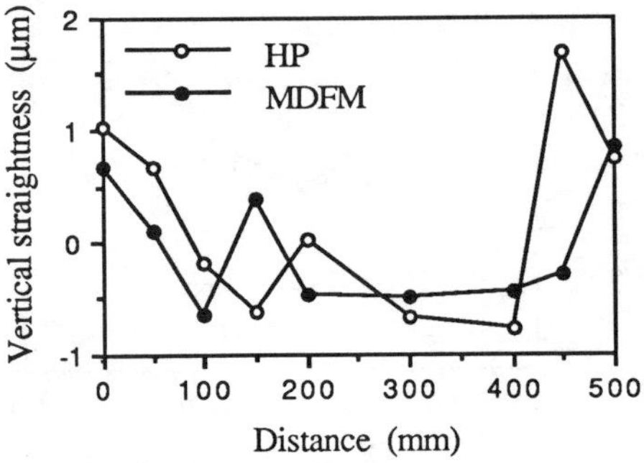

Fig.25 Measurements of vertical straightness of the CMM

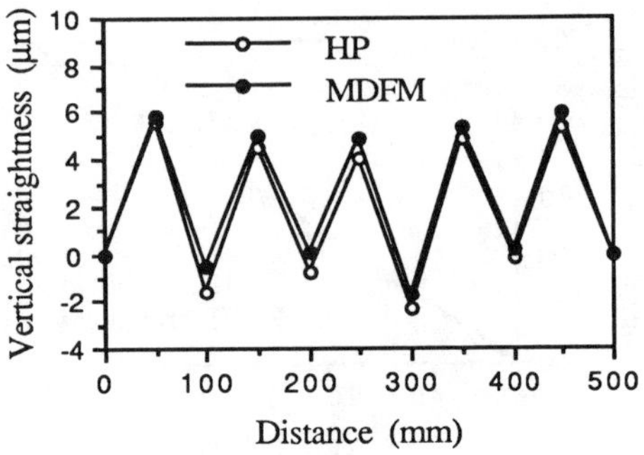

Fig.26 Measurements of artificially created vertical straightness on the CMM

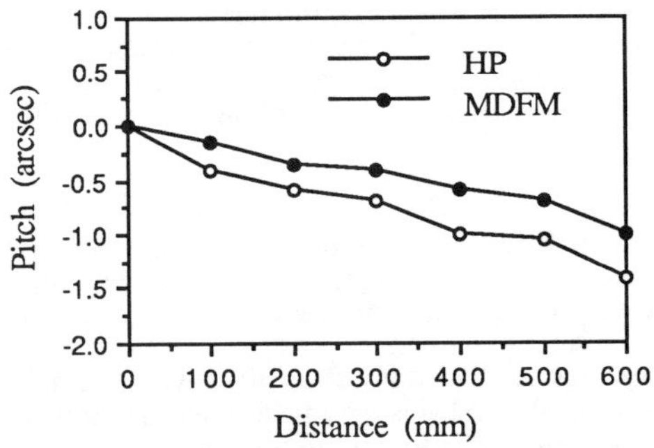

Fig.27 Measurements of pitch errors of the CMM

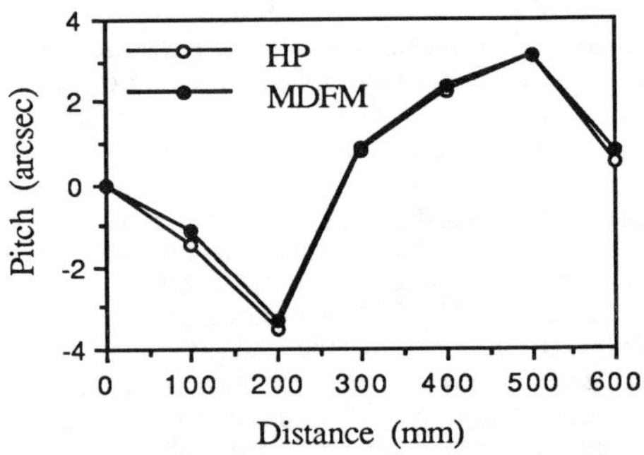

Fig.28 Measurements of artificially created pitch errors on the CMM

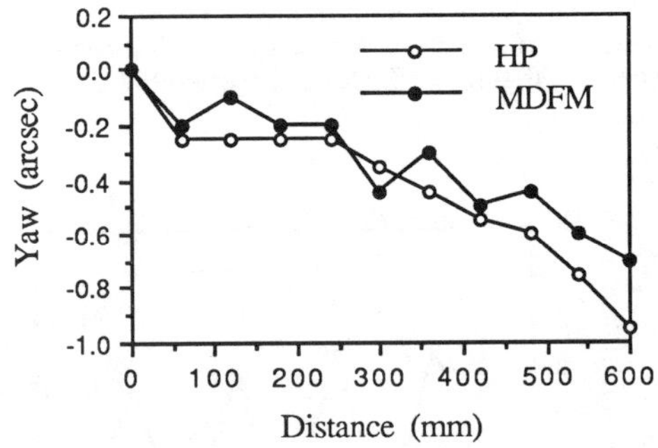

Fig.29 Measurements of yaw errors of the CMM

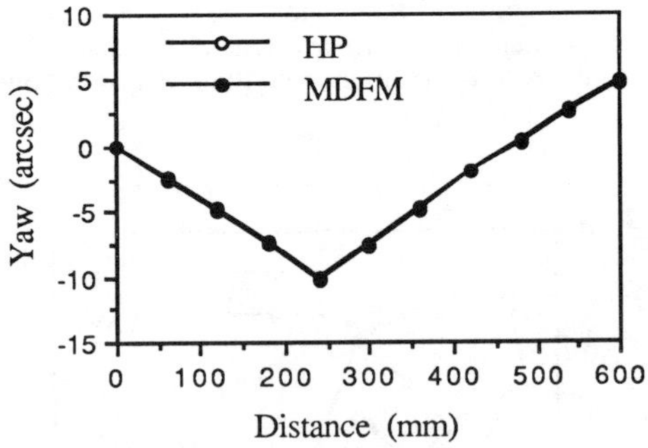

Fig.30 Measurements of artificially created yaw errors on the CMM

systems are less than 1 μm except at one point where the reading of the HP system is apparently unreliable. This 1 μm difference is mainly due to the noise level and errors introduced by the air turbulence and vibration of the measurement environment.

Figures 27 to 30 are the results of pitch and yaw measurements. In this case the maximum difference between the two measurement systems is less than 0.5arcsec. While comparing the pitch and yaw measurements individually, it is found that the yaw measurement, with 0.25 arcsec maximum difference, is better than the pitch measurement. The measurement of roll is not compared here since it cannot be measured by the HP system.

In summary, the developed MDFM system has measurement accuracy of better than 1 μm for straightness and 0.5arcsec for pitch and yaw.

VI. CONCLUSION

The following conclusions can be made based on the implementation and experimentation of the MDFM system:

(1) A MDFM system has been designed and implemented on a CMM to perform the on-line simultaneous measurements of 5 geometric error components, including two straightness errors, pitch, yaw, and roll errors, for each moving axis.
(2) A high resolution and compact autocollimator has been developed in this MDFM system for the precision measurement of pitch and yaw errors.
(3) Various schemes for improving the robustness of the MDFM system have been implemented, including a Keplerian beam expander for reducing angular drift of a laser beam, a PSD4 for the compensation of lateral drift of the beam, and shieldings for reducing the effect of air turbulences. Both short and long term stabilities of the MDFM system have been tested under experimental environments.
(4) Through the measurements of actual as well as artificially created geometric errors of the CMM, it has been found that the system's accuracy of measuring straightness error components is better than 1 μm and its accuracy for angular error measurements is better than 0.5 arcsec.

ACKNOWLEDGEMENT

The authors would like to express their thanks to the support from the University Research Program Fund, Ford Motor Company.

REFERENCES

1. Hocken, R., Simpson, J.A., Borchardt, B., Lazar, J., Reeve, C., and Stein, P., "Three Dimensional Metrology", Annals of the CIRP, Vol. 26/2, 1977
2. Busch, K., Kunzmann H., Waldele F.: "Calibration of Coordinate Measuring Machines",Precision Engineering, Vol.7 No.3, 1985.
3. Zhang, G., Veale, R., Charlton, T., Borchardt, B., and Hocken, R.,"Error Compensation of Coordinate Measuring Machines", The Annals of CIRP, Vol.34/1, 1985
4. Weck, M "Report of the Machine Tool Task Force", Vol.5: Machine Tool Accuracy", UCRL-52906, 1980
5. Tlusty, J., "Testing of Accuracy of NC Machine Tools", Supplement of the Report of Machine Tool Task Force, 1980
6. Weck, M. and Schmidt, M.,"A new method for determining geometric accuracy in the axis of movement of machine tools", Precision Engineering, Vol.8, No.2, April 1986
7. Ni,J. and Wu, S.M., "A New On-Line Measurement System for Machine Tool Geometric Errors", NAMRC XV, Vol.2, 1987
8. CHUO Precision Industrial Co. Ltd., "Straightor TC-2" Product Manual, 1990
9. Hecht, E. and Zajac, A.,"Optics", Addison-Wesley, 1979

Sensors, Controls, and Quality Issues
in Manufacturing
ASME 1991

A MODEL FOR THE PREDICTION OF BORE CYLINDRICITY DURING MACHINING

G. Subramani
Cummins Engine Company
Columbus, Indiana

S. G. Kapoor and R. E. DeVor
Department of Mechanical and Industrial Engineering
University of Illinois at Urbana–Champaign
Urbana, Illinois

ABSTRACT

In this paper, a model is developed for computation of bore cylindricity of machined cylinder bores. The model takes into account the effect of cutting process variables and the bore design on bore cylindricity. Surface error, which is a measure of the lack of bore cylindricity, is caused by both the elastic deflection of the bore wall due to cutting forces and the thermal expansion of the bore during machining. The effect of both the cutting forces and the bore temperature is included in the model. Experiments have been conducted to measure the bore cylindricity using an Incometer device. Comparisons between the shape and magnitude of predicted and measured surface error show a good agreement.

1. INTRODUCTION

The cylinder boring process is a machining process commonly used in the automotive industry, and by other engine manufacturers, for the production of cylinder bores in cast engine blocks. The process is used for both roughing and finishing operations to produce cylinder bores that meet both the dimensional tolerance and surface finish requirement specifications. Although the purpose of the cylinder boring process is to produce cylindrical surfaces, the machined bore surfaces are very often not cylindrical. The lack of cylindricity of machined bores often leads to poor engine performance and excessive wear of cylinders and pistons. Consequently, in order to obtain cylindrical bore surfaces, it is not uncommon for the boring operation to be performed in three or more passes (stages), followed by a few passes of the micro-sizing operation to obtain the desired cylindricity.

For many machined surfaces, three components may be identified, namely, surface error, surface waviness and surface finish. Surface error is usually of the order of 25 μm and is caused by the inherent flexibilities of the system. Surface waviness is typically of the order of 2.5 μm and is the result of the tooth marks left on the surface by the geometry of the cutting tool. The surface finish is associated with the microstructure variations of the material being machined. Thus, the surface error is a measure of the deviation from the nominal dimensions and is the only component of a machined surface addressed in this paper. There is much reported work in the literature addressing surface waviness and surface finish [3,4].

Kline et al. [5,6] have addressed the issue of predicting surface error when end milling thin walled components. The concept of flexibility influence coefficient was introduced by Kline and is used in this paper. However, in Kline's work, the effect of thermal expansion on surface accuracy has been ignored. On the other hand, McClure and Watts [7] and Ichimiya and Kawahara [8] have attempted to determine the effect of thermal expansion on surface accuracy in turning. McClure and Watts assumed that the tool-work interface could be modeled as a linearly traversing ring source of heat and consequently the model prediction does not correlate well with experimental data. Ichimiya and Kawaharas modeled the tool-work interface in turning as a helically moving heat source, and assumed a simplified equation for computing surface error.

The cylinder boring process affects the cylindricity of machined bores through the cutting forces and the temperature field generated during the operation. Surface error, which is a measure of the lack of bore cylindricity, is caused by both the elastic deflection of the bore wall due to cutting forces and the thermal expansion of the cylinder bore during machining. The surface error due to elastic deflection is dependent upon the magnitude of cutting forces and the flexibility of the bore structure, while the surface error due to thermal expansion is a function of the temperature distribution in the bore during the cut. Both the magnitude of the cutting forces and the temperature rise in the bore are dependent upon the boring process variables, such as cutting conditions and tool geometry. Figure 1 shows an overall scheme to predict the surface error produced in cylinder boring. Models that characterize the effect of process variables on cutting forces and bore temperature have been presented elsewhere [1,2].

In this paper, a model for the prediction of surface error of machined cylinder bores is presented. The model takes into account the effect of both cutting forces and bore temperature on surface error. The reminder of this paper evolves in four sections. The next two sections presented the model for surface error caused by elastic deflections due to cutting forces and bore thermal expansion, respectively. This is followed by experimental verification of the developed model. Finally, some conclusions are drawn from this work.

2. MODEL FOR SURFACE ERROR RESULTING FROM CUTTING FORCES

The cutting forces generated during the cylinder boring operation cause both the bore wall and the cutter to deflect in the radial direction. Consequently, at the point of surface generation, the true radial depth of cut is less than the pre-specified nominal depth of cut. After the cutter passes, the material that was left uncut due to the elastic deflection manifests itself as surface error of the bore. A cross-sectional view of a cylinder bore showing machined surface error along with the polar representation of surface error is shown in Figure 2. Surface error for a cylinder bore is thus defined as the difference between the machined inner surface of the bore and the perfectly cylindrical inner surface that would result if the bore were "perfectly rigid". Hence, surface error at the point of surface generation (on the inner surface of a machined bore) is a function of both the angular and axial location of the point. In this paper, only the deflection of the bore wall is included in the computation of surface error; the deflection of the cutter is neglected since the boring tool used in cylinder boring operations is much more stiffer than the bore wall.

The elastic deflection of the bore wall is computed through a finite element analysis of the cylinder bore. The finite element model of a cylinder bore workpiece used in this analysis is shown in Figure 3. One approach to the computation of surface error is to link together the force model and the finite element model such that a finite element analysis is performed at every incremental step as the boring process is simulated. This would, however, entail an enormous computational effort. This problem is circumvented through the development of a "flexibility influence coefficient" database which may be used for predicting the surface error of the machined bore.

The flexibility influence coefficient of any point on the bore surface is defined as the static radial deflection of the point due to a unit force in any of the three directions (i.e tangential, radial, or axial) applied at that point or any other point in the bore. Since the deflections are computed through a finite element analysis, the influence coefficients are defined only at the nodal locations. A point force acting at any point in a structure may be applied only as equivalent nodal loads in a finite element analysis. Consider an n-node three-dimensional finite element defined by the local coordinate system, ξ, η, and ζ as shown in Figure 4. Let a point force F act at a point (ξ_F, η_F, ζ_F) within the element. Then the nodal loads equivalent to the point force may be calculated through the following considerations:

(i) The nodal loads must be equivalent to the point force in a static sense, i.e.,

$$\sum_{\alpha=1}^{n} F_\alpha = F \tag{1}$$

where F_α refers to the nodal force at node α, and,

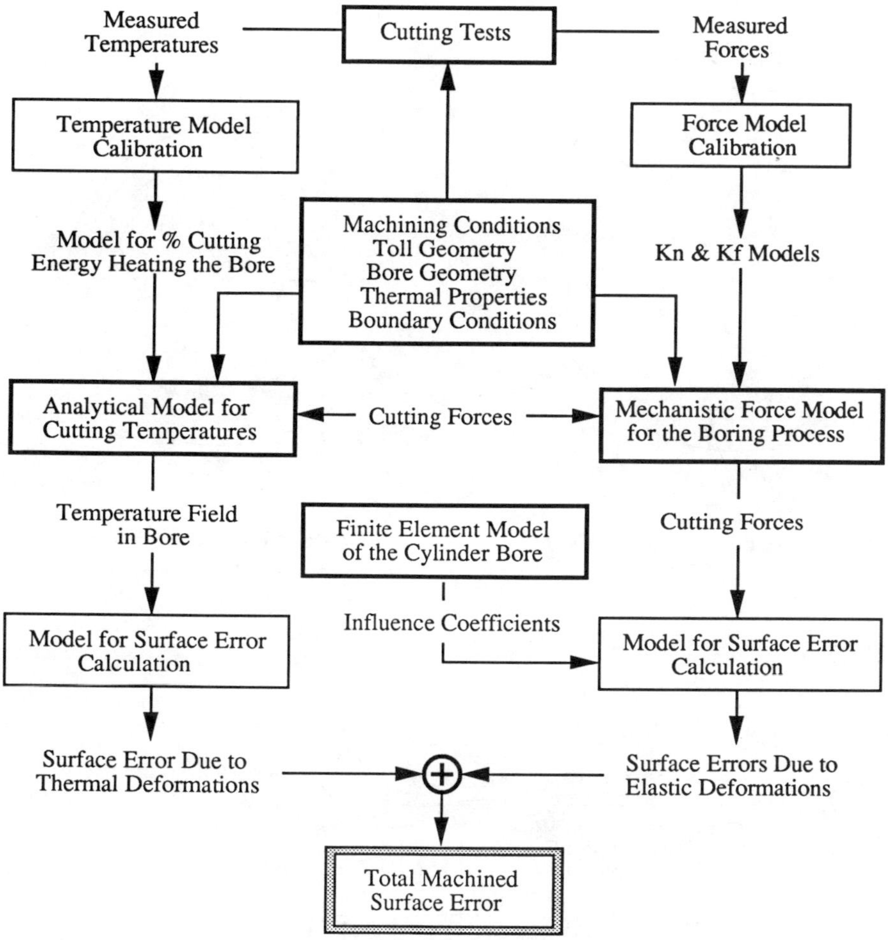

Fig. 1 Flow Chart for the Prediction of Surface Error

(ii) The work done by the point force must be equivalent to the work done by the nodal loads, i.e.,

$$\sum_{\alpha=1}^{n} F_\alpha u_\alpha = F\, u(\xi_F,\eta_F,\zeta_F) \tag{2}$$

where u_α refers to the nodal displacement at node α and $u(\xi_F,\eta_F,\zeta_F)$ is the displacement at the point (ξ_F,η_F,ζ_F). But $u(\xi_F,\eta_F,\zeta_F)$ may be written in terms of u_α and the element shape functions $h_\alpha(\xi,\eta,\zeta)$ as follows:

$$u(\xi_F,\eta_F,\zeta_F) = \sum_{\alpha=1}^{n} h_\alpha(\xi_F,\eta_F,\zeta_F)\, u_\alpha \tag{3}$$

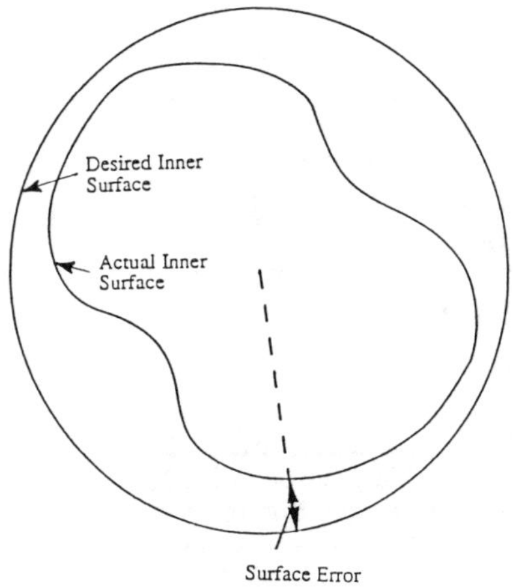

Figure 2a Cross-Sectional View of a Cylinder Bore Showing Machined Surface Error

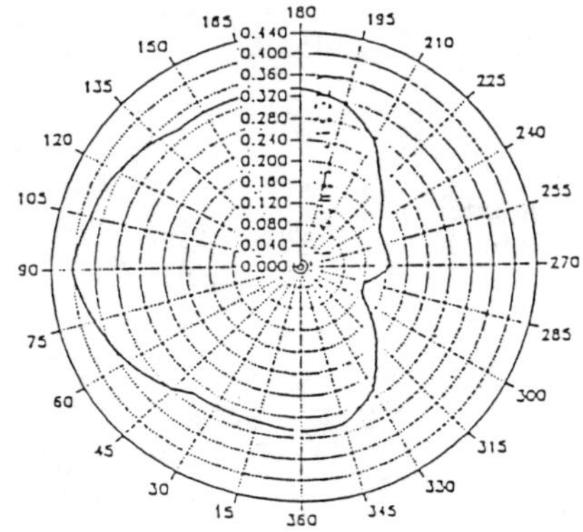

Figure 2b Polar Representation of a Machined Surface Error

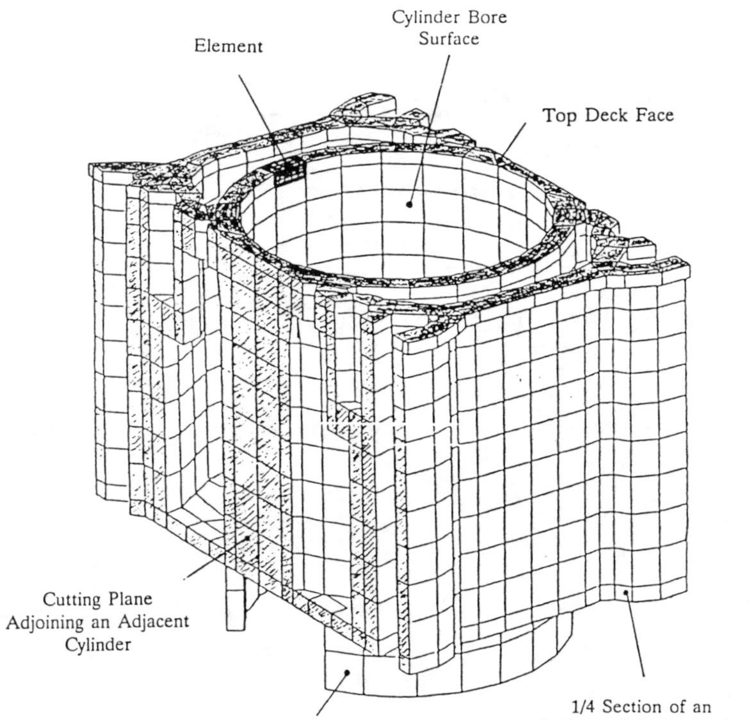

Figure 3 Finite Element Model of the Quarter Block

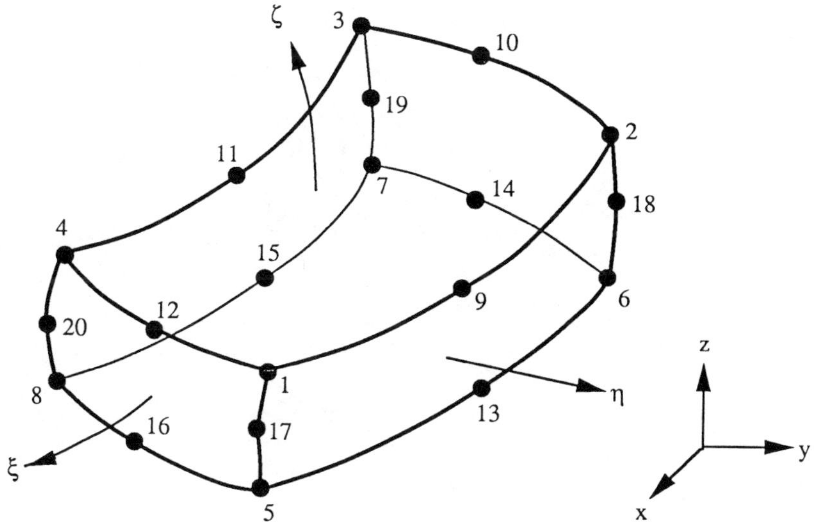

Figure 4 Eight to Twenty Variable-Number-Nodes Three-Dimensional Finite Element

Substituting Eq.(3) into Eq.(2), and noting that each u_α is an independent degree of freedom, we get,

$$F_\alpha = F\, h_\alpha(\xi_F,\eta_F,\zeta_F) \tag{4}$$

Note that the equivalent nodal loads satisfy Eq.(1) identically since

$$\sum_{\alpha=1}^{n} h_\alpha(\xi,\eta,\zeta) = 1 \tag{5}$$

at any point within the element. Thus, the nodal forces equivalent to the cutting forces may be computed from Eq.(4) and used with the influence coefficients of the node points to compute the nodal deflections.

The cross-section of the cylinder bore workpiece used in this study at four axial locations is shown in Figure 5. The influence coefficient of the node points located 12.5 mm below the top deck face due to a unit 100 lbf (444.8 N) radial force and a unit 100 lbf (444.8 N) tangential force are shown in Figures 6a and 6b, respectively. The double-sinusoidal shape of these profiles may be understood with reference to Figure 5. The cross-section over most of the length of the bore resembles that denoted by 2 in Figure 5. As can be seen, the thickness of the bore wall changes around the bore due to the presence of the water jacket and the Siamese nature of the bore configurations for the engine block used in these experiments.

Using the following equations (6-8) for the tangential, radial and axial force components acting on the ith insert as developed in Reference [1], and the flexibility influence coefficient data, the surface error arising from elastic deflection can be obtained.

$$F_t^i = K_n \cdot CA \cdot \{1+K_f \cos(\gamma_l) \tan(\gamma_a)+K_f \sin(\gamma_l)\tan(\gamma_r)/ \cos(\gamma_a)\} \tag{6}$$

$$F_r^i = K_n \cdot CA \cdot \{\tan(\gamma_r)-K_f \sin(\gamma_l) / \cos(\gamma_a)+ K_f \cos(\gamma_l)\tan(\gamma_a)\tan(\gamma_r)\} \tag{7}$$

$$F_a^i = K_n \cdot CA \cdot \{(-\tan(\gamma_a)+ K_f \cos(\gamma_l))/ \cos(\gamma_r)\} \tag{8}$$

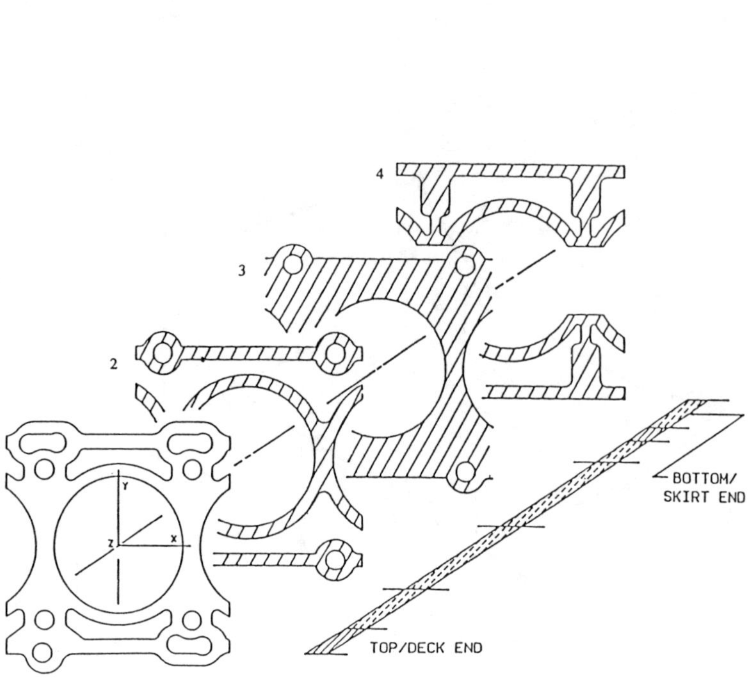

Figure 5 Cross-sections of the Cylinder Bore Workpiece

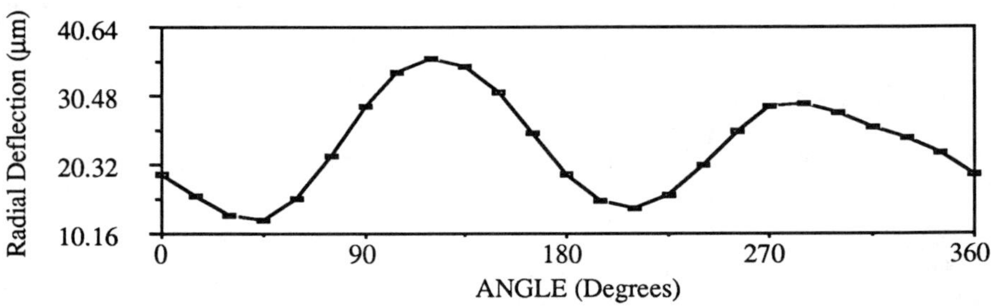

Figure 6a Radial Deflection of the Bore Wall Due to a 100 lbf (444.8 N) Radial Force

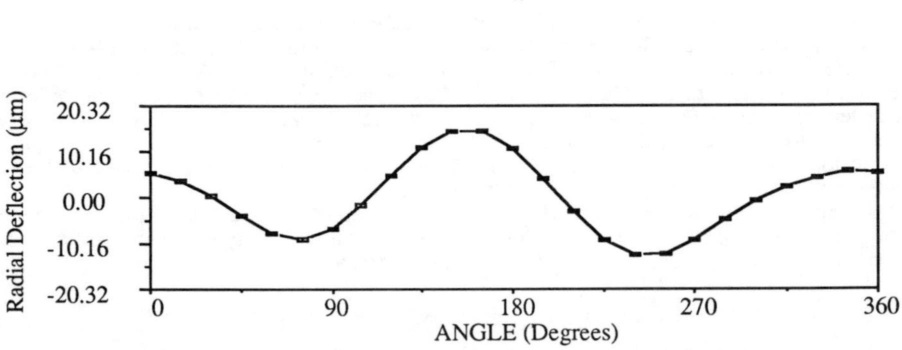

Figure 6b Radial Deflection of the Bore Wall Due to a 100 lbf (444.8 N) Tangential Force

where Kn, Kf are the cutting force coefficients, also known as the normal cutting pressure and the friction pressure, CA refers to the chip cross-sectional area or chip load and is a function of feed per insert, depth of cut and radial & axial throws, γ_l , γ_a and γ_r are the effective lead, axial and rake angles respectively. More detail discussion on the chip area calculation and effective lead angle can be found in reference [1].

Mathematically, the surface error due to elastic deflection, $SE_E(\varphi_s(t),z_s(t))$ can be expressed as:

$$SE_E\,(\phi_s(t),z_s(t)) = -\sum_{i=1}^{NT}[F_r^{bi}\,WC_{r;r}^{i} + F_t^{bi}\,WC_{r;t}^{i} + F_a^{bi}\,WC_{r;a}^{i}] \tag{9}$$

where (ϕ_s,z_s) represents the point of surface generation at time t. Note that the instantaneous location of an insert is related to the angular velocity of the cutter, ω and the feedrate, F such that

$$\phi_s = \omega\, t + \phi_0 \tag{10}$$
$$z_s = F\, t + z_0 \tag{11}$$

where ϕ_0 and z_0 are the initial angular and axial locations of the cutter. 'i' denotes the insert number, F_j^{bi} represents the force on the bore wall in the j-th direction at insert i, $WC_{r;j}^{i}$ is the influence coefficient at the point of surface generation due to a unit force in the j-th direction at the location of insert i. The direction j refers to either radial, tangential or axial direction. The negative sign in Equation (9) indicates that when the forces are positive (in the conventional sense), then the diameter of the bore is smaller than the nominal diameter due to surface error.

3. MODEL FOR SURFACE ERROR RESULTING FROM BORE TEMPERATURE FIELD

In reference [2] , a model has been presented for the computation of the temperature field in a hollow cylindrical workpiece during machining. The model may be used to calculate the thermal strains in the body, and hence the magnitude of thermal expansion of the bore during machining. The chip removal process sets up both mechanical and thermal stresses and strains in the workpiece. In this paper, the effects of the cutting forces and the temperature field are treated as uncoupled. The model developed for computation of temperature field in the cylinder bore assumes azimuthal symmetry in geometry. This assumption is valid since for the cylinder boring process, the temperature gradient along the radial direction is much larger than that along the axial or azimuthal directions. Consequently, an analysis at a cross-section of the cylinder bore may be performed for computation of bore thermal expansion. This assumption implies that the displacement along the ϕ-direction and the shear stresses and strains are all zero.

The equilibrium equation satisfied by the stress field in the hollow cylindrical workpiece is [9]:

$$\frac{\partial \sigma_{rr}}{\partial r} + \frac{\sigma_{rr} - \sigma_{\phi\phi}}{r} = 0 \tag{12}$$

where σ_{rr}, and $\sigma_{\phi\phi}$ denote the radial, and hoop stress components, respectively. The body force terms are assumed to be zero. The stress strain relationships are given by:

$$\begin{aligned} \varepsilon_{rr} - \beta\theta &= \frac{1}{E}[\sigma_{rr} - \nu\,(\sigma_{\phi\phi} + \sigma_{zz})] \\ \varepsilon_{\phi\phi} - \beta\theta &= \frac{1}{E}[\sigma_{\phi\phi} - \nu\,(\sigma_{rr} + \sigma_{zz})] \\ \varepsilon_{zz} - \beta\theta &= \frac{1}{E}[\sigma_{zz} - \nu\,(\sigma_{rr} + \sigma_{\phi\phi})] \end{aligned} \tag{13}$$

where ε_{rr}, $\varepsilon_{\phi\phi}$, and ε_{zz} denote the radial, hoop, and axial strain components, respectively, σ_{zz} denotes the axial stress component, θ denotes temperature differential $(T(r,\phi,z,t)-T_\infty)$, and β is the coefficient of linear thermal expansion. Since we are dealing with the plane strain problem, $\varepsilon_{zz} = 0$. The strain-displacement relations take the form:

$$\varepsilon_{rr} = \frac{\partial u}{\partial r} \quad \text{and} \quad \varepsilon_{\phi\phi} = \frac{u}{r} \tag{14}$$

where u represents the radial component of the displacement. The radial component of displacement causes the depth of cut to change, and hence represents the surface error caused by thermal expansion of the bore. By rearranging Eqs.(13), the stress components may be expressed in terms of strain components. The substitution of these equations, along with Equation (14), into Eq.(12) transforms the equilibrium equation into one expressed in terms of the displacement component [10]. The resulting differential equation for the radial component of displacement is:

$$\frac{d^2u}{dr^2} + \frac{1}{r}\frac{du}{dr} - \frac{u}{r^2} = (1+\nu)\,\beta\,\frac{d\theta}{dr} \tag{15}$$

where ν is the Poisson's ratio. Eq.(15) is integrated and the constants of integration determined from the conditions that the radial stresses are zero at both the inner and outer surfaces of the bore. Then using Eqs.(13) and (14), the three components of stress are then found to be:

$$\sigma_{rr} = \frac{\beta E}{1-\nu}\frac{1}{r^2}\left\{\frac{r^2 - a^2}{b^2 - a^2}\int_a^b \theta\, r\, dr - \int_a^r \theta\, r\, dr\right\}$$

$$\sigma_{\phi\phi} = \frac{\beta E}{1-\nu}\frac{1}{r^2}\left\{\frac{r^2 + a^2}{b^2 - a^2}\int_a^b \theta\, r\, dr - \int_a^r \theta\, r\, dr - \theta\, r^2\right\} \tag{16}$$

$$\sigma_{zz} = \frac{\beta E}{1-\nu}\left\{\frac{2}{b^2 - a^2}\int_a^b \theta\, r\, dr - \theta\right\}$$

where E is the Young's Modulus, and a and b denote the inner and outer radii of the hollow cylinder, respectively. The substitution of Eq.(16) into the equation for hoop strain, which is the second of Eq.(13), and using the second of Eq.(14), the radial displacement on the inner surface may be computed. The negative of this radial displacement is the surface error due to thermal expansion at the point of surface generation, (ϕ_s, z_s), i.e.,

$$SE_T\,(\phi_s(t),z_s(t)) = -\,\beta\, a\left\{\frac{2}{(b^2-a^2)}\int_a^b \theta\,(r,\phi_s(t),z_s(t))\; r\, dr\right\} \tag{17}$$

The mathematical expression for $\theta\,(r,\phi_s(t),z_s(t))$ has been developed in [2] and is given as:

$$\theta\,(r,\phi_s(t),z_s(t)) = \sum_{n=0}^{\infty}\sum_{m=1}^{\infty}\sum_{l=1}^{\infty}\left[K_l(r)\,G_m(z)\,\bar{\theta}_{n;m;l}\,(\phi,t)\,/\,N_n M_m L_l\right] \tag{18}$$

$G_m(z)$ and $K_l(r)$ are the Kernels (eigen functions) of the integral transforms in the z- and r-coordinates, respectively, and N_n, M_m, and L_l are Normalization Constants. $\bar{\theta}_{n;m;l}$ is the solution of the transformed Ordinary Differential Equation (ODE) characterizing the temperature field in the hollow cylindrical workpiece. More details about Equation (18) can be found in [2]. It must be noted that the effect of thermal expansion of the workpiece during machining is to produce a bore whose diameter is smaller than the nominal and changes along the axis of the bore.

4. VALIDATION OF THE MODELS FOR PREDICTION OF SURFACE ERROR

A set of cutting tests was designed to validate the above developed models for surface error prediction. The test conditions are listed in Table 1 along with the cutting force components predicted using Equations (6-8). The models for the cutting force coefficients (K_n & K_f) used for force prediction were obtained through a different set of tests. Note that the cutting tests were conducted using tool cartridges with positive rake angles (0^0 radial rake and 5^0 axial rake) and negative rake angles (-5^0 radial rake and -6^0 axial rake). The models for the cutting force coefficients are:

Positive Cartridge

$$K_n = 19092\,(CT)^{-0.528}$$
$$K_f = 0.0489\,(d)^{-0.716}\,(f)^{-0.111} \tag{19}$$

Negative Cartridge

$$K_n = 28668\,(CT)^{-0.482}$$
$$K_f = 0.015\,(d)^{-1.073}\,(f)^{-0.056} \tag{20}$$

where CT refers to average chip thickness, d and f are the radial depth of cut and feed per insert, respectively. The predicted cutting force components are used in conjunction with the flexibility influence coefficients to compute the radial elastic deflections due to cutting forces from Eq. (9).

Similarly, for the test conditions listed in Table 1, Eq. (18) was used to compute the temperature differential in the cylinder bore at any given point of surface generation (ϕ_s,z_s). Since Eq. (17) for surface error prediction due to thermal expansion requires an integration of the thermal field in the radial direction for r=a to r=b, the bore wall thickness was divided into equi-distant intervals and Eq. (18) was used to predict the temperature at these points. The integration in Eq. (17) was then performed numerically using Simpson's rule. It must be noted that although Eq. (18) includes triple infinite summations, convergence for temperature is achieved quite rapidly at the cutting conditions studied in this paper.

Table 1 Machining test conditions and Cutting Force Predictions

No	FPT (mmpr)	DOC (mm)	Nose radius (mm)	Lead angle (deg)	Tool geom.	Ft (N)	Fr (N)	Fa (N)
1	.254	0.764	1.192	5	pos	- 439.1	346.9	-209.2
2	.254	1.528	1.192	15	pos	- 756.6	291.7	-288.3
3	.508	0.764	0.794	15	pos	- 585.3	376.7	-324.2
4	.508	1.528	0.794	5	pos	- 976.3	261.1	-337.6
5	.508	0.764	1.192	15	neg	- 655.6	549.3	-213.8
6	.254	0.764	0.794	5	neg	- 379.2	309.6	-262.6
7	.508	1.528	1.192	5	neg	-1127.	398.5	-455.9
8	.254	1.528	0.794	15	neg	- 408.3	319.8	-342.7

For the purpose of validating the surface error prediction models developed above, comparisons were made between predicted surface error and the surface error of machined cylinder bores measured using a special purpose measurement system. Discussions on the prediction of surface error due to the additive effects of cutting force and bore temperature, the experimental test facility, the surface error measurement system, and the comparison between measured and predicted surface errors are presented in this section.

4.1 Prediction of Surface Error: Additive Effect of Error Due to Cutting Force and Bore Temperature. Figures 7 and 8 show the predicted surface error due to cutting force, temperature, and their additive effect for the conditions of Tests 1 and 2 of Table 1. Note that a negative value for surface error implies that the bore diameter is smaller than the nominal diameter.

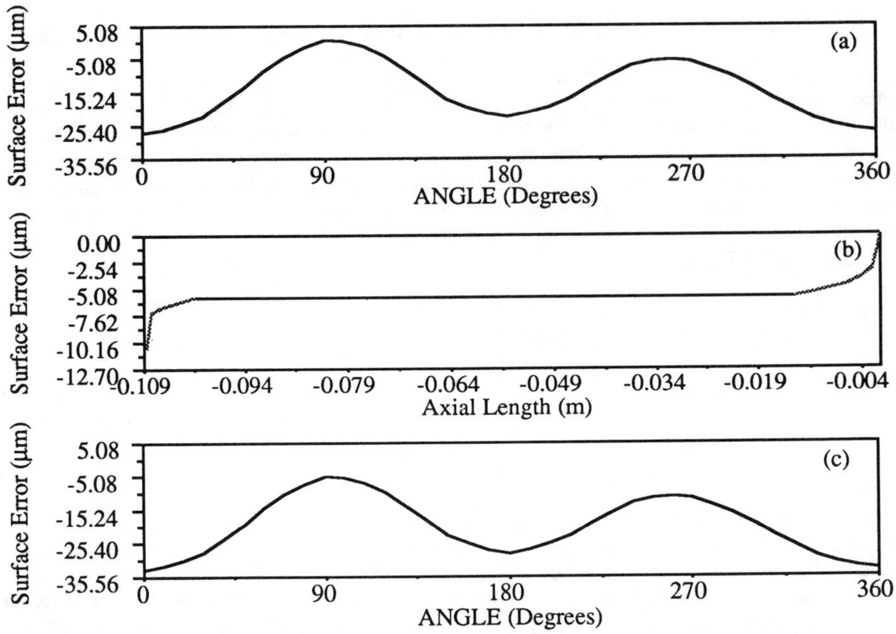

Figure 7 Prediction of Surface Error (Test 1, Table 1): (a) Error at Level 2 due to Cutting Force; (b) Error along the Bore Axis due to Bore Temperature; (c) Additive Effect due to Cutting Force and Bore Temperature at Level 2

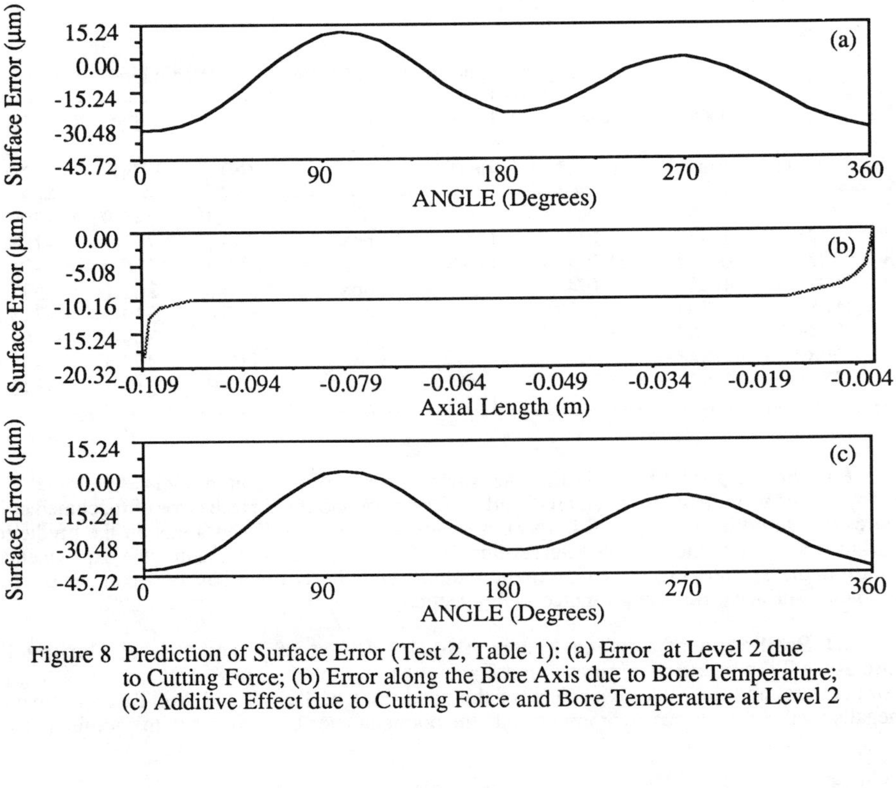

Figure 8 Prediction of Surface Error (Test 2, Table 1): (a) Error at Level 2 due to Cutting Force; (b) Error along the Bore Axis due to Bore Temperature; (c) Additive Effect due to Cutting Force and Bore Temperature at Level 2

For the purpose of comparison three positions along the axis of the bore are identified as Level 1, Level 2, and Level 3. These positions are at 31.7 mm, 61.7 mm, and 91.7 mm from the top deck face of the bore, respectively. In Figures 7 and 8, the figures identified by (a) show the surface error due to cutting force around the bore at Level 2 (computed using Eq.9). As can be seen, the double-sinusiodal shape of the surface error profiles is due to the nature of the bore design, owing to the presence of the water jackets and the Siamese nature of the bore configurations of the engine (see Figure 6), and is reflective of the influence coefficient data shown in Figure 5. It should be pointed out that the actual shape and magnitude of surface error depends not only on the influence coefficient data, but also on the magnitude of the cutting forces (tangential, radial, and axial to the bore). Furthermore, since the design of the engine block is not symmetrical about the x- and y-axes, the "peaks" and "valleys" of the predicted profiles are not of equal magnitude.

In Figures 7 and 8, the figures identified by (b) show the variation of surface error along the axis of the bore due to its thermal expansion (computed from Eq.17). Owing to the azimuthal symmetry assumed for the bore geometry in the model for temperature prediction, the surface error due to thermal expansion is independent of the angular location of the point of surface generation. Furthermore, since the temperature rise at the point of surface generation remains approximately the same over the length of the bore, the surface error due to thermal expansion remains virtually unchanged over most of the length of the bore. However, at the start of cut and at the end of cut, the bore temperature is lower and higher, respectively, than that at the mid-section of the bore, and consequently, the corresponding surface error profile shows the indicated trends. It must be noted that while the the shape of the surface error profile due to thermal expansion is as shown in the figures, the actual magnitude of the error at the two ends of the bore may be different from that shown. This is due to the assumption made in the last section whereby the analysis for thermal expansion was performed along the cross-section to the bore at the point of surface generation. This assumption would not be strictly valid at the two ends of the bore.

The figures identified by (c) in Figures 7 and 8 show the combined effect of surface error due to cutting force and bore temperature at Level 2. The effect of bore thermal expansion is to cause the average diameter of the bore at any axial position to be smaller than the nominal diameter. Therefore, the combined error is obtained by adding together the error caused by cutting force and the error due to thermal expansion at Level 2. Thus, the combined surface error profile has the same shape as that due to cutting force, but its average value is shifted down by the magnitude of surface error due to thermal expansion (-5.842 μm and 10.16 μm in Figures 7 and 8, respectively).

4.2 Experimental Test Facility. A schematic of the layout of the machine tool used for the cylinder boring experiments is shown in Figure 9. The machine tool system consists of a spindle driven by a spindle motor which is controlled by an Indramat spindle motor controller. The spindle and the spindle motor are mounted on a slide whose linear motion provides the feed during the boring operation. The spindle is belt driven by a Siemens-Allis induction motor with an adjustable frequency drive which allows speeds up to 3,600 rpm. The cutter was designed to accept tool cartridges. Tool geometries such as radial and axial rake angles and lead angle could be changed by replacing a tool cartridge. In the tests conducted during the course of this research, several different cartridges were used. These allowed for lead angles of 5 and 15 degrees, and positive and negative rake angles. The inserts used in the tests were indexable, uncoated, and made of either carbide or silicon nitride. The nose radii used in these tests were 2/64 inch or 3/64 inch. The workpiece is mounted on an angle fixture which is mounted on another slide aligned at right angles to the spindle slide. The motion of the workpiece slide is controlled through an Allen-Bradley micro controller.

Cutting tests were performed for test conditions as listed in Table 1 on a single cylinder bore cut out from a cast four-cylinder engine block.The cutter used in these experiments has a diameter of 3.57" and tests were run at a spindle speed of 3000 rpm.

4.3 The Incometer System. The system used for measuring surface error is an Incometer system, which is designed for accurate measurement of the inner contour (distortion) of cylinder bores in engine blocks. The Incometer system consists of a measuring probe which is connected to an electronic control unit and is controlled through a desk-top computer. A reproduction of a photograph of the Incometer set-up showing the measuring probe mounted inside a bore and the computer system is shown in Figure 10.

The measurement of bore distortion is accomplished by mounting the Incometer probe inside the bore to be measured. A cross-sectional view of the Incometer probe mounted inside a cylinder bore is shown in Figure 11. Adjustable clamping jaws are provided on the probe for centering and locking the probe into the cylinder bore. The measuring head of the probe with its pick-up touching the inner surface of the cylinder can be rotated and moved axially on the shaft, thus facilitating the

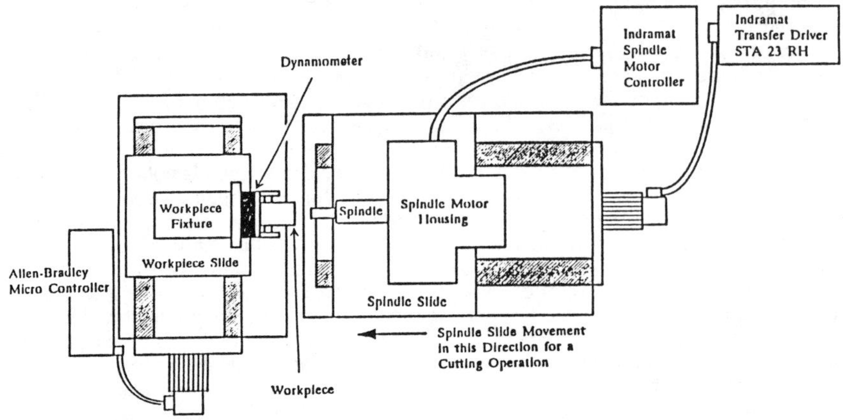

Figure 9 Schematic of Cylinder Boring Process Machine Tool Layout

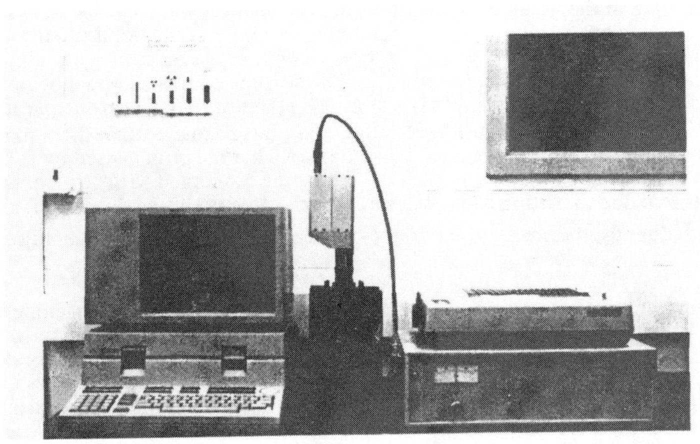

Figure 10 Incometer System for Measurement Bore Cylindricity

measurement of the bore contour. As the pick-up moves over the bore surface, it is displaced radially according to the contour of the bore. The pick-up consists of a leaf spring with strain gages glued on. A pin with a ruby tip which touches the inner surface of the cylinder deflects the leaf spring and the strain gage according to the distortion encountered during scanning the cylinder. The electrical signal generated by the strain gage is transmitted to an electronic control unit and on to the computer as measured data.

The Incometer system is provided with a software package for correction and analysis of the measured data. When making surface error measurements using an Incometer, the axes of the Incometer and the cylinder bore rarely coincide. Consequently, the "raw" data as measured by the Incometer is of little value and has to be processed to compensate for the misalignment error between the axes. This compensation is done by fitting a cylinder to the measured data using the least squared criterion and then referencing the surface error to this fitted cylinder. However, this procedure compensates not only for the misalignment error between the axes, but also tends to eliminate the effect of non-symmetry in the bore design on the magnitude and shape of the surface error. The effect of error compensation is schematically shown in Figure 12.

In order to compare the predicted surface error with the corrected-measured data given by the Incometer, a least squares criterion was employed with the predicted data to determine the location of the center and the radius of a best fit circle (at a given axial location along the bore axis). The predicted surface error values were then recomputed by referencing the surface error to this circle.

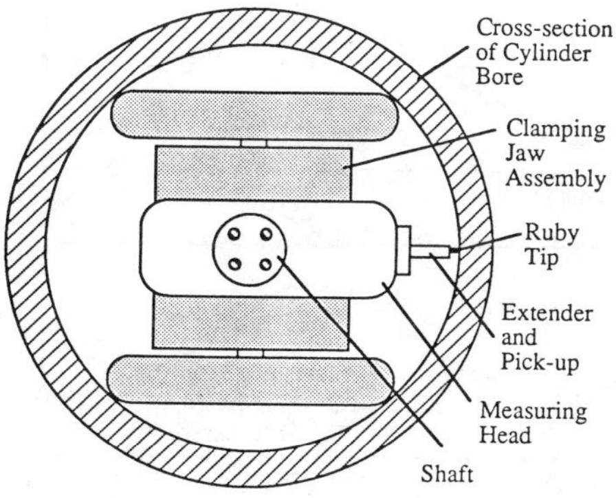

Figure 11 View of an Incometer System Mounted Inside a Cylinder Bore

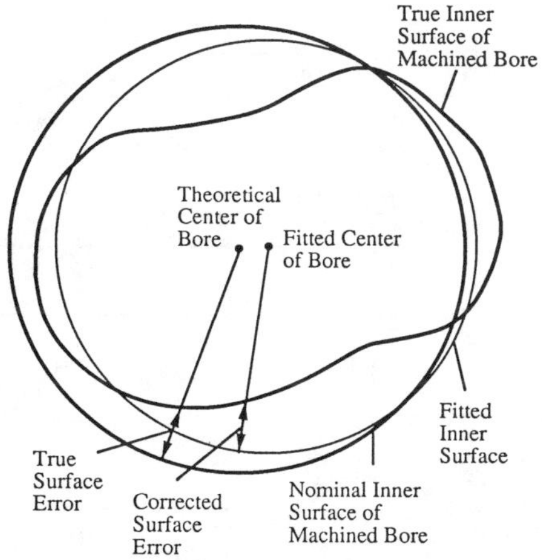

Figure 12 Schematic Representation of True Surface Error and Corrected Surface Error for a Machined Bore

4.4 Comparison Between Measured and Predicted Surface Error. Following the procedure outlined above, the corrected-predicted surface error was computed and compared with the corrected-measured surface error given by the Incometer for a few experiments, and the results are shown in Figures 13-16. In each of these figures, the comparison is shown at the three levels mentioned earlier. From an examination of these figures the following conclusions can be drawn:

(i) The shape of the measured surface error profile is well predicted by the model. As mentioned earlier, the particular shape of the surface error profile is due to the Siamese nature of the bore configurations and the presence of water jackets on the sides of the bores. It must be noted that there exists a phase difference between the measured and predicted error profiles, particularly at the upper levels of the bore. A possible explanation for this is that when the Incometer was mounted in the cylinder bore, the zero degree position of the measurement did not coincide exactly with that of the prediction. Another possible explanation could relate the phase shift to the characteristics of the Incometer response. Different time lags associated with individual frequence compoenets could cause both the phase shift and distortion.

(ii) The magnitude of the corrected-predicted surface error profile is in excellent comparison with the corrected-measured error profile. Since the cylinder bore is fixtured during the boring operation such that top deck face is farthermost away from the fixture, the magnitude of surface error decreases gradually along the axis of the bore from the top deck face due to the decreased flexibility of the bore wall closer to the fixture. This trend is shown both by the predicted and the measured surface error profiles. Furthermore, due to the correction made to the measured and predicted data, the "peaks" and "valleys" of the error profile are of nearly equal magnitude at any given level for a given test condition.

(iii) The effect of increasing the depth of cut is to increase the magnitude of surface error (as can be seen by comparing Figures 13 and 14, and Figures 15 and Figure 16). This is primarily due to the increase in the magnitude of cutting force and temperature rise at larger depths of cut. Similar conclusions can be drawn for an increase in feed also. However, it must be pointed out that if the feed used in an experiment is larger than about 0.254 mm/r, then the measured surface error profile shows a confounding between the surface error and surface waviness components of the machined surface, and hence a comparison between the measured and predicted profiles is not meaningful.

5. SUMMARY AND CONCLUSIONS

In this paper, a model for prediction of surface error arising from elastic deflection of the bore wall due to cutting forces and the thermal expansion of the bore due to the bore temperature field during machining has been developed. Surface error due to elastic deflections is determined using the predicted cutting forces and an influence coefficient of the bore structure, while the error due to

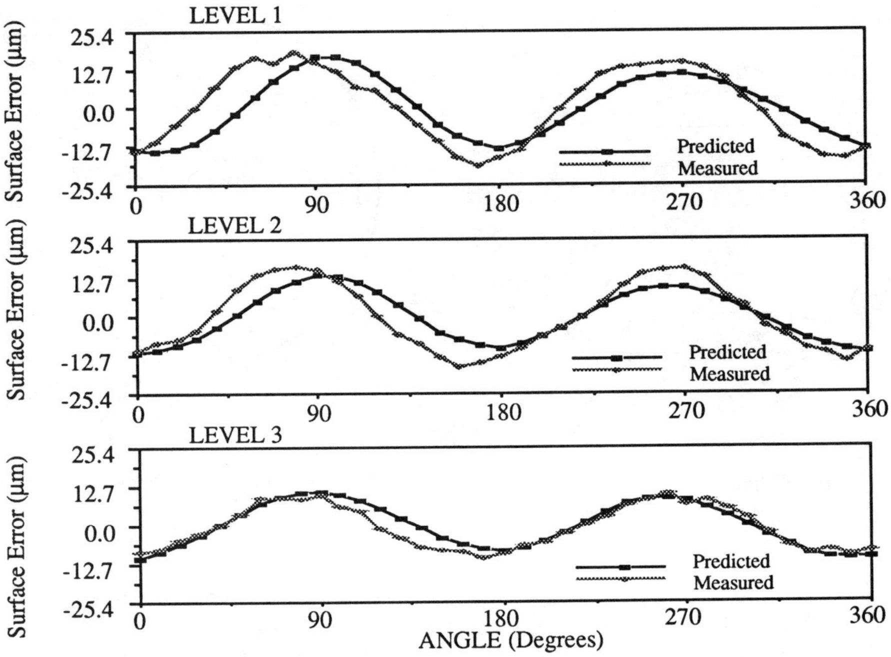

Figure 13 Comparison Between Measured and Predicted Surface Error (Test 1, Table 1)

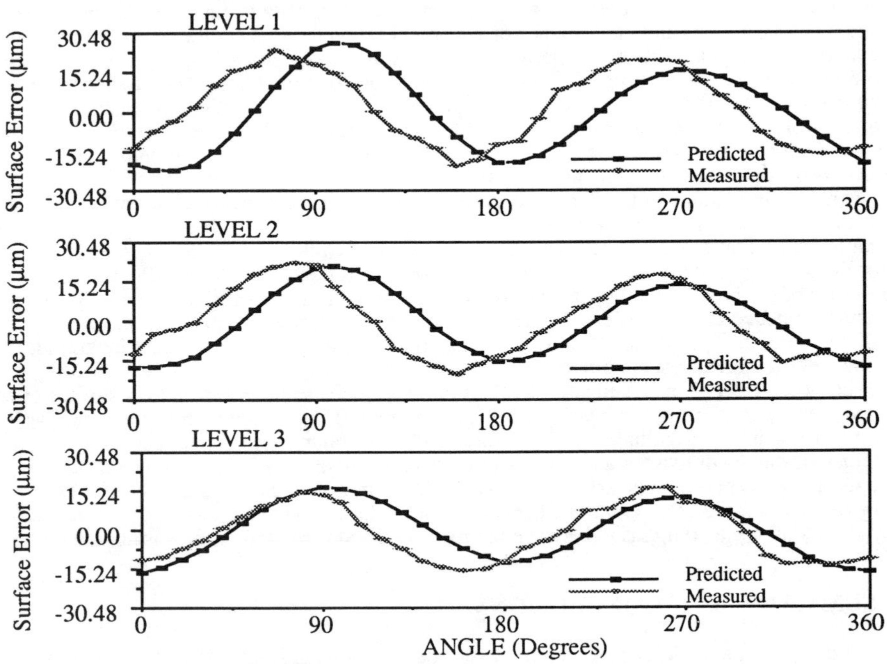

Figure 14 Comparison Between Measured and Predicted Surface Error (Test 2, Table 1)

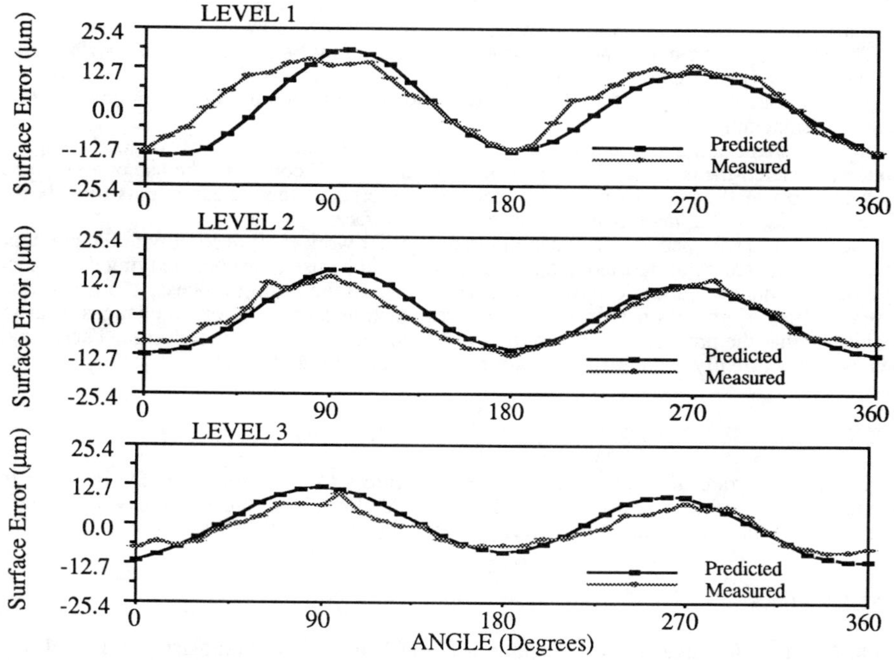

Figure 15 Comparison Between Measured and Predicted Surface Error (Test 6, Table 1)

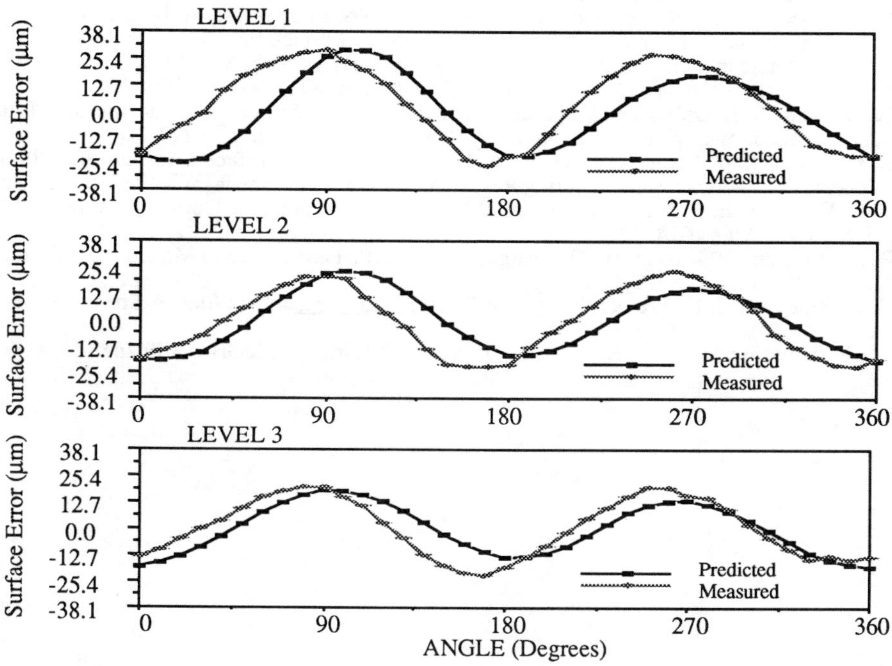

Figure 16 Comparison Between Measured and Predicted Surface Error (Test 8, Table 1)

thermal expansion is found from solving for thermal stresses in the bore due to the heat generated during machining. Specifically, the following conclusions may be drawn:

(i) The shape and magnitude of surface error are strongly affected by the boring process variables and the bore structure design.

(ii) The magnitude of surface error is proportionally affected by the magnitude of cuttiong forces and temperature.

(iii) The shape and magnitude of the correlated-predicted surface error is in good comparison with the correlated-measured surface error obtained using an Incometer. The use of an Incometer may lead to misleading results since the Incometer references the measurements to a "best-fit" cylindrical surface rather than to the "theoretical" bore surface.

The models developed in this paper and in our earlier work [1,2] dealing with the predictions of cutting forces, temperature field and surface error could be used for both studying the impact of a proposed product (engine) design on the performance of the boring process, and for obtaining the sensitivities of process performance measures, such as surface error, to changes in process variables so that the process may be optimally chosen for a given product design. These models when used concurrently could lead to improved quality of cylinder bores.

ACKNOWLEDGEMENTS

The authors gratefully acknowledge General Motors Corporation-BOC Division for their financial support of this research. The Ingersoll Milling Machine Company provided the experimental facility. Computer funds were provided, in part, by IBM Corporation.

REFERENCES

1. G. Subramani, R. Suvada, S. G. Kapoor, R. E. DeVor, and W. Meingast, " A Model for the Prediction of Force System for Cylinder Boring Process," Proc. xv North American Manufacturing Research Conference, Lehigh University, PA, pp 439-446.

2. G. Subramani, M. C. Whitmore, S. G. Kapoor, and R. E. DeVor, " Temperature Distribution in a Hollow Cylindrical Workpiece During Machining: Theoretical Model and Experimental Results, " PED-Vol. 43, Proc. of the Symposium on 'Fundamental Issues in Machining", WAM 90, Dallas, Tx, pp 245-260. (also to appear in the Journal of Engineering for Industry).

3. F. Nassirpour and S. M. Wu, " Statistical Evaluation of Surface Finish and Its Relationship to cutting Parameters in Turning," Int'l. J. Mach. Tool Design Research, Vol. 17, 1977, pp 197-204.

4. T. Sata, " Surface Finish in Metal Cutting," CIRP Annals, XII, 1963, pp 190-197.

5. W. A. Kline, R. E. DeVor and I. Shareef, " The Prediction of Surface Accuracy in End Milling, "Trans. ASME, J. of Engineering for Industry, Vol. 104, No. 3, 1982, pp 272-278.

6. R. E. DeVor, J. W. Sutherland, and W. A. Kline, " Control of Surface Error in End Milling, " Proc. XI North American Manufacturing Research Conference, Madison, WI, 1983, pp 356-362.

7. R. G. Watts and E. R. McClure, " Thermal Expansion of Workpiece During Turning, " ASME Paper No. 68-WA/Prod-24, 1968.

8. R. Ichimiya and H. Kawahara, " Investigation of Thermal Expansion in Machining Operations, " Bull. JSME, Vol. 14, No. 78, 1971, pp 1363-1371.

9. B. A. Boley and J. H. Weiner, Theory of Thermal Stresses, John Wiley & Sons, New York, 1982.

10. S. P. Timoshenko and J. N. Goodier, Theory of Elasticity, McGraw Hill Inc., New York, 1982.

Sensors, Controls, and Quality Issues
in Manufacturing
ASME 1991

THE INFLUENCE OF TECHNOLOGICAL PARAMETERS ON CYLINDRICITY DEVIATION OF DEEP DRAWN CUPS

Lýdia Sobotová
Faculty of Mechanical Engineering
Technical University of Košice
Košice, Czechoslovakia

The papers deals with the influence of technological parameters on accuracy of drawing cups, by means of factor experiment.(factors: lubrication, drawing speed, draw punch radius).

The advantage of factor experiments lies in the possibilities of evaluating numerous objectiv information on influence the character of individual factors and their interactions on the quantities observed.

Factor experiment consists of fewer factors, each of them being at severals levels.

The measured values form the statistical complex which may by arranged in a special combination table. The deviations of measured values are due to the effects of the measured factors X, Y, Z and the uncontrollable and negligible factors (casual influence).

By evaluating the combination tables, it is possible to find the degree of importance of individual factors.

Solving the problems of production quality control is an important area in the process of product manufacturing.

With regard to the fact that with forming we get lower order accuracy than with the cutting machines, the problem of improving accuracy in forming seems to be very important in the present stage of development.

In connection with modern methods of production there has been a requirement to develop and solve the diagnostical methods for production inaccuracies in metal forming, with the purpose to increase the accuracy of deep drawing by means of feedback control. A number of technological parameters may influence the accuracy of pressed pieces in flat forming.

If we want to increase the accuracy of pressed pieces, it can be best achieved by changing the technological factors whose influence is the most remarkable. However, the problem lies in finding the order of importance of technological factors.

This article shows the process of determining the importance of individual technological parameters on the cylindricity deviation of drawn cups.

Problem solution

The process of determining the influence of technological parameters can be divided into two main groups:

- the processes suitable for determining the deviations which can be utilized for feedback purposes by the operator or the computer controlled systems which have been developed recently; the deviations are determined on each drawn cup in the course of in-operation control,
- the processes suitable for determining the influence of individual parameters on the geometrical deviations of drawn cups; these indications are determined a single way and they can serve as the information for the selection of the most suitable technological parameters form the point of view of production and accuracy of drawn cups. They can be utilized in the decision making of monitoring systems (1) or in expert systems (2) as the databank information.

The solutions introduced in this paper present the second group of problems. A method of determining the influence of individual parameters by systematic experiment bassed on factor variance analysis has been developed.

Technological parameters

Two kinds of frequently utilized deep drawn sheet materials have been chosen for the experiment:

- KOHAL 20 E, ČSN 41 1135.21, nominal thickness 0,8 mm; tarnished surface,
- KOHAL 24, nominal thickness 0,8 mm; tarnished surface.

For the experiment flat bottom cylindrical cups have been used with the internal diameter Ø 60 mm. The cups were drawn on a double acting mechanical crank press LTN 25, using press tool with the blank holder for one drawing operation. Two drawing punches have been used:

- with the punch curvature radius $r_{p1} = 2,5$ mm,
- with the punch curvature radius $r_{p2} = 5$ mm

and a drawing die of the radius 6 mm. The drawing tools were developed at the Department of Metallurgy and Technology, Faculty of Mechanical Engineering, Technical Universisty, Košice. The drawing speeds were changed in dependence on the revolutions of the press motor and the gear of press ram (3). The following three drawing speeds were used:

- speed of tool ram $v_1 = 1.94.10^{-1}$ m.s^{-1},
- speed of tool ram $v_2 = 2.49.10^{-1}$ m.s^{-1},
- speed of tool ram $v_3 = 3.15.10^{-1}$ m.s^{-1}.

In the experiment we used three types of lubrication:

- bearing oil B2, viscosity 16 - 22 mm^{2}s^{-1} at 50 ^oC,
- mineral oil J3, viscosity 35 - 42 mm^{2}s^{-1} at 50 ^oC,
- VP lubrication, viscosity 48 mm^{2}s^{-1} at 50 ^oC.

The cups were drawn form the blank of various diameters ranging from Ø 123 mm to Ø 129 mm, graded in 1 mm.

Realization of the experiment

In the systematic experiment the cups were drawn in various combinations of technological parameters. From Tab.1 and Tab.2 it follows that there are various combinations the cups are drawn without any problems, without wrinkles or scratches (crossing fields - a). In some cases drawing has failed, some of the cups were torn and in some other combinations of technological parameters no cups were drawn (empty fields - c).

In each combination of technological parameters six cups were drawn.

Theoretical aspects

The decision in favour of the way the experiment was carried out was based on the demand to obtain as much information as possible from a minimum number of measurements. This can be achieved by proper planning of the experiment (5), (6). We performed the five-factor variance analysis (7). The indicidual influences on technological parameters were analyzed in the cylindricity deviation in various conditions (levels), Tab.3.

The purpose of the experiment was to divide the total dispersion variance of cylindridity deviation into partial dispersion variances caused by individual factors (4), (8).

Tab.3 Identification of technological parameters - factors

Examined parameter	Identification of factor	Number of levels
Lubrication	A	a = 2
Drawing speed	B	b = 2
Blank	C	c = 2
Material	D	d = 2
Radius of drawing punch	E	e = 2

The observation y_{ijklmz} of cylindricity deviation has the form of (7) :

$$y_{ijklmz} = \mu + \alpha_i + \beta_j + \gamma_k + \delta_l + \varepsilon_m + (\alpha\beta)_{ij} + (\alpha\gamma)_{ik} + \dots + (\delta\varepsilon)_{lm} + (\alpha\beta\gamma)_{ijk} + \dots + (\gamma\delta\varepsilon)_{klm} + (\alpha\beta\gamma\delta)_{ijkl} + \dots + (\beta\gamma\delta\varepsilon)_{jklm} + (\alpha\beta\gamma\delta\varepsilon)_{ijklm} + \xi_{ijklmz}, \qquad (1)$$

where $i = 1, .. a$, $j = 1, ..b$, $k = 1,..c$, $l = 1,..d$, $m = 1,..e$, $z = 1, .. r$,

μ - general average value of the deviation of cyl.

$\alpha_i, \beta_j, \gamma_k, \delta_e, \varepsilon_m$ - the effect of the level of factors A, B, C, D, E on the evarage value of cylindricity deviation,

$(\alpha\beta)_{ij}$ - combined effect (interaction) of level combination (A_iB_j) of factors A and B,

$(\alpha\beta\gamma)_{ijk}$ - combined effect (interaction) of level combination $(A_iB_jC_k)$ of factors A, B and C,

$(\alpha\beta\gamma\delta)_{ijkl}$ - combined effect (interaction) of level combination $(A_iB_jC_kD_l)$ of factors A, B, C and D,

$(\alpha\beta\gamma\delta\varepsilon)_{ijklm}$ - combined effect (interaction) of level combination $(A_iB_jC_kD_lE_m)$ of factors A, B, C, D and E,

ξ_{ijklmz} - random observation error.

The evaluation of results was based on the dispersion variance analysis of the observed influences with the residual dispersion variance by means of F tests, for example:

$$F = \frac{S_A}{S_R} \tag{2}$$

where S_A - the dispersion variance due to the influence of A,
S_R - the dispersion variance caused by random influence.

The results of the evaluated experiments are shown in Tab.4. From this table it follows that the influence of statistically significant factors can be determined if $F_{\alpha} > F(\vartheta_A, \vartheta_R, \alpha)$.

The results and discussion

For the evaluation of five factor experiment we used the results of measuring two cups from each group of technological parameters. The cup measurements were carried out on a 3D measurement unit, Ferranti - Merlin 750, available measurement accuracy $\pm (4 + 0.5\ L/100)\ \mu m$.

On each cup 300 points were measured (50 points in 6 cross sections) from which the deviation values were determined by the evaluation program. This article has dealt with the influence of technological parameters on cup cylindricity and diameter deviations. From the results of our experiments it is possible to arrive at the following conclusions:

- the most significant influence on cylindricity deviation is exercised by material properties. More accurate cups can be obtained with the use of KOHAL 24 material Fig.2,
- significant influence on the total dispersion variance; the cylindricity deviations showed interactions DE, ABC, ACE, CDE, ABCE a high value of the testing criterion F was also obtained for the interaction AC.

For more accurate distinction of the influence of individual factors we performed a three factor analysis from the interactions, neglecting the influence of material, based on factors ABC and ACE. It follows from the results that the punch radius is of significant influence, the measured deviations being lower at the smaller punch radii r_{pl} = 2.5 mm, as shown in Fig.1 and Fig.2.

- Significantly better results in both materials are obtained with the use of VP lubrication at the same punch radius r_{pl} = 2.5 mm, Fig.1,
- better results with the blank Ø 123 mm and lubrication VP were achieved with the punch radius r_{pl} = 2.5 mm,

- cylindricity and diameter deviations are related in that the smaller cylindricity deviations also show smaller dispersion variance of diameter deviations, from which it follows that anisotropy is lower and the dimensions are nearer to the theoretical values of diameters.

Conclusion

In connection with the restructuring process in our national economy there has been stress on penetrating into the world market. To achieve this aim it is necessary to produce high accuracy products. The above article has introduces the method of verifying the influence of technological parameters on dimension accuracy and cylindricity deviation of cylindrical cups. The results obtained show the advantages of KOHAL 24 materials from the point of view of accuracy. From some other influences, both analyzed materials KOHAL 24 show the best results at smaller punch radii $r_{pl} = 2.5$ mm, at the drawing speed $v_3 = 3.15.10^{-1}$ m.s^{-1} and with the use of VP lubrication.

The introduced method can be realized in a single-purpose way. Recently there has also been development in the area of metal forming towards expert systems where the results of the introduced analysis can be used for various materials in databanks and serve as comparative values for the application of technological parameters in the pruduction of concrete drawn cups.

References

(1) SCHMID,J.: Sensoren zur automatischen Qualitäts-kontrole bei der Blechumformung, Blech Rohre Profile, 1990, 9, p. 547-551

(2) HELD,H.J.-AMS,L.-FRITZ,J.: SPES - Stufenpressen-Expertensystem, VDI-Z 132, 1990, 2, p. 34-36, 39-40

(3) Blaščík,F. et al.: Výskum parametrov tvarovateľnosti plechov. Výskumná správa III-1-5/09, SjF VŠT Košice, 1985

(4) STAREK,L.-MEČÁRIK,K.: Basics of Engineering Experiments. SjF SVŠT Bratislava, 1982

(5) SOBOTOVÁ,L.: The Influence of Technological Parameters on Accuraci of Drawn cup. SjF VŠT Košice, 1986

(6) SOBOTOVÁ,L.-KAHANCOVÁ,E.: The Influence of Technol. Param. on Accur. of Dr. Cups, IV. ICMF"88 conference Gyor, 1988,p.335-340

(7) SOBOTOVÁ,L.: Vplyv technologických parametrov na presnosť výliskov. (Disertation work), KSMaT, SjF VŠT Košice, 1991,print

(8) ČSN 01 0230 - Applied statistics. Analysis of variance.

Tab. 1 Combinations of technological parameters of drawn cups, material Kohal 20 E, r_{p1} = 2.5 mm

		123	124	125	126	127	128	129
B2	v_1	a	a	a	b	c	c	c
	v_2	a	a	a	b	c	c	c
	v_3	a	a	a	b	c	c	c
J3	v_1	a	a	a	b	c	c	c
	v_2	a	a	a	b	c	c	c
	v_3	a	a	b	c	c	c	c
VP	v_1	a	a	a	a	a	c	c
	v_2	a	a	a	a	a	b	b
	v_3	a	a	a	a	a	b	b

a b c

Tab. 2 Combinations of technological parameters of drawn cups, material Kohal 20 E, r_{p2} = 5 mm

		123	124	125	126	127	128	129
B2	v_1	a	a	a	b	b	b	c
	v_2	a	a	a	a	b	b	c
	v_3	a	a	a	a	b	b	b
J3	v_1	a	a	a	a	b	b	b
	v_2	a	a	a	a	b	b	b
	v_3	a	a	a	a	b	b	b
VP	v_1	a	a	a	a	a	a	c
	v_2	a	a	a	a	a	a	a
	v_3	a	a	a	a	a	a	a

a b c

Tab. 4 Analysis of dispersion variance

$F_{(1,32,0.05)}$, $F_{\alpha} = 4.154$

source of variability	sum of squares	number of degree of freedom	F
A	6,2500	a-1 = 1	0,0068
B	462,2500	b-1 = 1	0,5011
C	Ø	c-1 = 1	Ø
D	100,0000	d-1 = 1	0,1084
E	73305,6875	e-1 = 1	79,47224
AB	115,5625	(a-1)(b-1)	0,1253
AC	3751,5625	(d-1)(c-1)	4,0671
AD	60,0625	(a-1)(d-1)	0,0651
AE	2303,8750	(a-1)(e-1)	2,4977
BC	232,5625	(b-1)(c-1)	0,2522
BD	742,5625	(b-1)(d-1)	0,8050
BE	1260,1250	(b-1)(e-1)	1,3661
CD	68,0625	(c-1)(d-1)	0,0738
CE	1848,8750	(c-1)(e-1)	2,0044
DE	210,1250	(d-1)(e-1)	0,2278
ABC	3844,0000	(a-1)(b-1)(c-1)	4,1674
ABD	30,2500	(a-1)(b-1)(c-1)	0,0328
ABE	1870,6875	(a-1)(b-1)(e-1)	2,0281
ACD	380,2500	(a-1)(c-1)(d-1)	0,4122
ACE	7612,6875	(a-1)(c-1)(e-1)	8,2531
ADE	1958,1875	(a-1)(d-1)(e-1)	2,1229
BCD	2116,0000	(b-1)(c-1)(d-1)	2,2940
BCE	189,1875	(b-1)(c-1)(e-1)	0,2051
BDE	33,1875	(b-1)(d-1)(e-1)	0,0360
CDE	7612,6875	(c-1)(d-1)(e-1)	8,2531
ABCD	3,0625	(a-1)(b-1)(c-1)(d-1)	0,0033
ABCE	6320,1250	(a-1)(b-1)(c-1)(e-1)	6,8518
ABDE	1122,1250	.	1,2165
ACDE	650,1250	.	0,7048
BCDE	6,1250	.	0,0066
ABCDE	2730,1875	(a-1)(b-1)(c-1)(d-1)..	2,9599
R	29517	a.b.c.d.e (r-1) = 32	
sum	150463,4375	a.b.c.d.e (r-1) = 63	

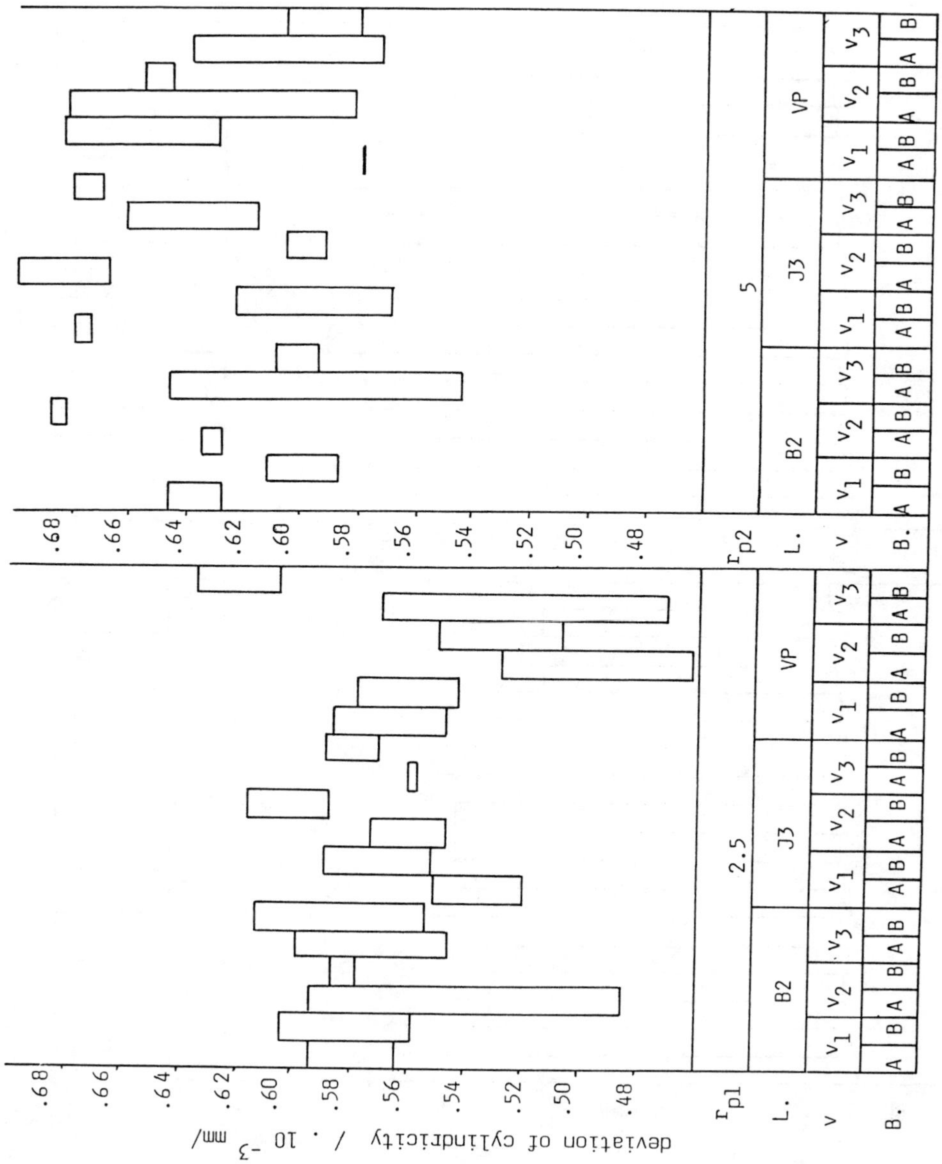

Fig. 1 Deviation of cylindricity

r_p - punch radius, L. - lubrication, v - drawn punch, B. - blank (A = 123 mm, B = 125 mm) , material Kohal 24

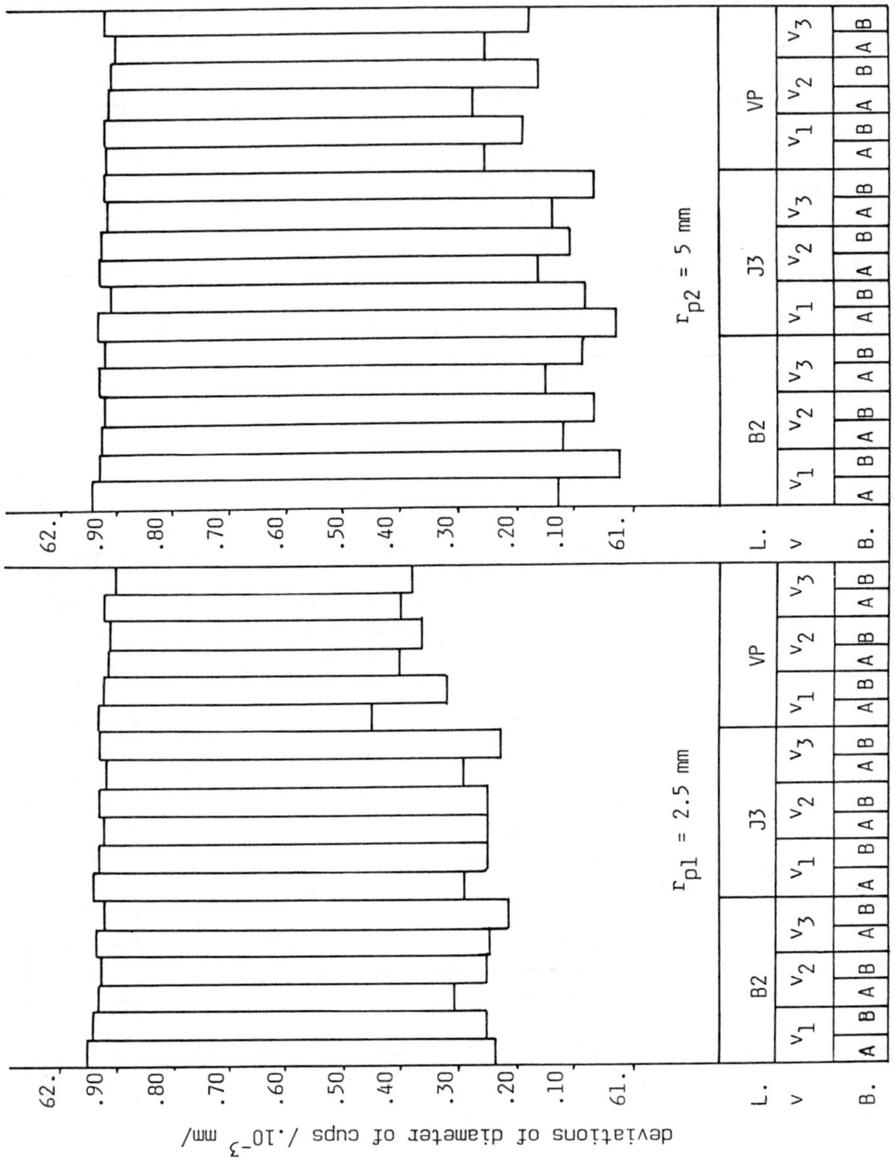

Fig. 2 Deviations of diameter of cups

r_p- punch radius, L.- lubrication, v - drawn speed, B. - blank (A = 123 mm, B = 125 mm), material Kohal 20 E

Sensors, Controls, and Quality Issues
in Manufacturing
ASME 1991

GEOMETRICAL DEVIATIONS OF DEEP-DRAWN CUPS

Ervína Kahancová and Lýdia Sobotová
Faculty of Mechanical Engineering
Technical University of Košice
Košice, Czechoslovakia

The papers deals with the influence of technological parameters (material, diameter of blanks, drawing speed,) on form deviation of deep - drawing cups (circularity deviation, cylindricity deviation).

There will determined the regression dependences.

In the production of drawn cups an unsuitable combination of technological parameters can cause that the required drawn cup is not possible to be drawn. Therefore, and due to relatively higher values of the form and position deviations in pressing than in cutting operations, a more detailed analysis of influences of technological parameters on the accuracy of drawn cups was not the centre of atention of researchers.

With an increase of automation in pressing, for example multiple drawing in production lines, and with growing requirements for accuracy of drawn cups from the point of view of the customers, it has become necessery to find new ways of raisisng accuracy of drawn cups (1).

This article analyzes the influences of technological parameters on geometrical deviations of cylindrical drawn cups made of KOHAL 20 E, (material properties: 0,07 % C, 0,41 % Mn, 0,01 % P, 0,007 % S, 0,067 % Al, Re = 190 - 200 MPa, Rm = 330 - 350 MPa, Re/Rm = 0,582 - 0,594, A_{80} = 37 - 42 %, thickness = 0,8 mm) by Czechoslovak standard ČSN 41 1145.21.

Problem Formulation

The geometrical accuracy of drawn cups can be judged on the basis of kinds of deviations. These deviations can be devided into two groups:

a) the deviations, which are exactly defined by the standard, using specified methods of meassuring and regularly occurring in each technological object (3), (4), (5), whereby in the frame work of one object there are inherent systematically repeated manifestations of their existence or they show specific trend.

In this group it is possible to name the following types of geometrical deviations:

- deviations of dimension,
- deviations of form,
- deviations of position,
- deviations of roughness of surface,

b) the deviations, which are not normalized in the machine industry, they are published in (6), (7), or they are defined in metallurgical standard materials (8). However, these inaccuracies can be changed in the operations of pressing. They occur accidentaly, i.e. they may not occur at all or they may occur only somewhere on the surface of drawn cups. These include:

- the deformations which arise in manipulation and transport, (damaged corners, edges, cavities),
- hair cracks, especially in the zone of strong deformation,
- stampings, from the materials remaining on the press tool,
- the rests of the material on the drawn cup itself,
- scratches.

In this article we want to point out the form and position deviations occurrence given in the point a) in the relation to various technological parameters.

The Results of the Experiment

Implementation of the Experiment and the Results

The conditions of the experiment were fully described in the article Sobotová,L.: The Influence of Technological Parameters on Accuracy of Deep Drawn Cups.

The wide possibilities of measurement evaluations on a 3D measuring unit make it possible to analyze more types of deviations at one time.

Fig.1 shows the record of evaluations of the measurement of the cylindrical drawn cup, drawn from the material KOHAL 20 E, ČSN 41 - 1135.21, with the lubrication - mineral oil J3, from the drawing blank of Ø 125 mm diameter, with the drawing speed $v_1 = 1{,}94.10^{-1}$ $m.s^{-1}$ and with the drawing punch radius $r_{p2} = 5$ mm.

The sum of the results from the whole experiment is shown for individual types of form and position deviations as follows:

- the deviations of cylindricity (Fig.2) of the drawn cup,
- the deviations of roundness of the drawn cup (Fig.3), the highest value of the deviation of roundness from six cutting operations on one drawn cup,
- the deviations of the perpendicularity of surface of the drawn cup (Fig.4),
- the deviation of the flatness of the bottom of the drawn cup (Fig.5).

Evaluation and Discussion

- From the point of view of accuracy of cylindrical drawn cups it is very important to follow up the deviations of cylindricity, where major diferences occur in the values of the deviations of cylindricity (Fig.2) at various technological parameters. The differences are not very profound at the maximum of the deviations roundness (Fig.3). The differences occur only with the lubrication VP and at the drawing speed $v_3 = 3{,}15.10^{-1}$ $m.s^{-1}$, where higher than average deviations of roundness may be found.
- The deviations of perpendicularity of surface (Fig.4) are changed mostly in the dependence on the punch radius: the lower values are at the punch radius $r_{p1} = 2.5$ mm and within this group the lowest values of the deviations of perpendicularity are achieved with the lubrication VP.
- The deviations of flatness (Fig.5) of the bottom drawn cups have not shown the differences in various materials or they have not changed markedly in the dependence on some technological parameters, which is logical and expectable. The bottom of the drawn cup was not so much dependent on punch radius as the cylindrical covering. Fig.5 confirms the correctness of measuring,it means that the differences between the measured deviations in some other kinds of deviations are not the expression of inaccuracy of measuring, but of the real difference between the observed deviation at various combinations of technological parameters.

When comparing the deviations of perpendicularity (Fig.4) and of flatness (Fig.5), at the same technological parameters we can see, that the extreme deviations of perpendicularity arise at higher deviations of flatness. It is due to the fact that at higher deviations of flatness the bottom does not unambiguously sit on the support which causes the axis of drawn cup and therefore the cylindricity surface to turn to side, which may be seen in higher deviation of perpendicularity.

When determining technological parameters in the production of drawn cups, an error - free realisation of the drawn cup is mainly stressed. However, for securing the right function of cylindrical drawn cups, it is often required to meet the demands of geometrical accuracy of the drawn cup. The dimension accuracy of the drawn cup is determined basically by the pressing tool.

The form and the position accuracy is determined by all technological parameters.

The above paper has shown the influence of the material, lubrication, drawing speed, the size of the drawing blank, and the punch radius on the form and position deviations of the cylindrical drawing cup.

From the analysis it follows that in addition to the material, the punch radius exercises great influence in that smaller punch radii produce higher accuracy regardless the material used. With the same pressing tool, higher form and position accuracy was reached with more viscosive lubrication materials and higher drawing speed.

The concrete results may serve as the deciding criterion in optimization of the range of technological parameters and may be used in databanks for further utilization in securing the quality of the drawn cups.

References

(1) SCHMID,J.: Sensoren zur automatischen Qualitätskontrolle bei der Blechformung, Blech Rohre Profile 37 (1990) 9, p. 547-551

(2) SOBOTOVÁ,L.: Vplyv technologických parametrov na presnosť výliskov. (Disertation work), Fak.of.Mech.Eng., TU Košice in print 1991

(3) ČSN 01 4401 - Tolerances of form and position. General.

(4) ČSN 01 4420 - Form and position deviations. Measuring.General.

(5) ČSN 01 4406 - Geometrical tolerances not indicated on drawings.

(6) ALEXY,M. a kol.: Chyby povrchu na oceľových plechoch a pásoch pri spracovaní v Studenej valcovni VSŽ. Oceľové plechy, 1984

(7) MAZUR,E.L. et al.: Pričiny defektov pri štampovke avtomobiľ. detalej, Kuznečno-štampovočnoe proizvodstvo, No. 8, 1978

(8) ČSN 42 0015 - Errors of formed steel products.

PROTOKOL O MERANI
pre:VST-Kosice

VYSKUMNY A VYVOJOVY USTAV ELEKTRICKYCH STROJOV TOCIVYCH - BRNO,
Kombinatny vyskumny ustav - zavod Kosice

MERACIE ZARIADENIE:
SMS fy FERRANTI
Typ:MERLIN 750
VYHODNOCOVACIA JEDNOTKA:
Pocitac fy HEWLETT PACKARD
Typ:9816 S

Bacikova c.5
040 43 KOSICE

ODDELENIE METROLOGIE
Rastislavova c.100
c.t.: 535 11-535 14 kl.202,265

TYP VYROBKU :Plech
NAZOV DIELCA :Vytazok
CISLO VYKRESU :III-1-5/03-2/1
CISLO DIELCA :2 A1 II 4
TEPLOTA DIELCA :20

CISLO PROTOKOLU:790110/09
MERAL :Ing.Peter HULO
NAZOV PROGRAMU :E0027A
DATUM/CAS :18 Jun 1990/21:36:36
TEPLOTA OKOLIA :20

[mm]	Measuring	Drawing	Deviations	
Skupina: 1				Group 1
Odch.valcovitosti :	.661			deviations of cylindricity
Odch.kolmosti :	.365			deviations of perpendicularity
Odch.rovinnosti :	.030			deviations of flatness
Skupina: 2				Group 2
Priemer-vyska 7 :	61.253	61.600	-.347	diameter high 7
Odch.kruhovitosti :	.084			deviations of roundness
Excentricita :	.001			excentricity
Orientacia :	267.113			orientation
Skupina: 3				
Priemer-vyska 13 :	61.710	61.600	.110	
Odch.kruhovitosti :	.215			
Excentricita :	.128			
Orientacia :	241.472			
Skupina: 4				
Priemer-vyska 19 :	61.906	61.600	.306	
Odch.kruhovitosti :	.225			
Excentricita :	.189			
Orientacia :	244.182			
Skupina: 5				
Priemer-vyska 25 :	61.911	61.600	.311	
Odch.kruhovitosti :	.160			
Excentricita :	.186			
Orientacia :	248.952			
Skupina: 6				
Priemer-vyska 31 :	61.884	61.600	.284	
Odch.kruhovitostí :	.243			
Excentricita :	.116			
Orientacia :	255.193			
Skupina: 7				
Priemer-vyska 37 :	61.845	61.600	.245	
Odch.kruhovitosti :	.310			
Excentricita :	.001			
Orientacia :	80.492			

Fig. 1 The record of evaluations of the measurement

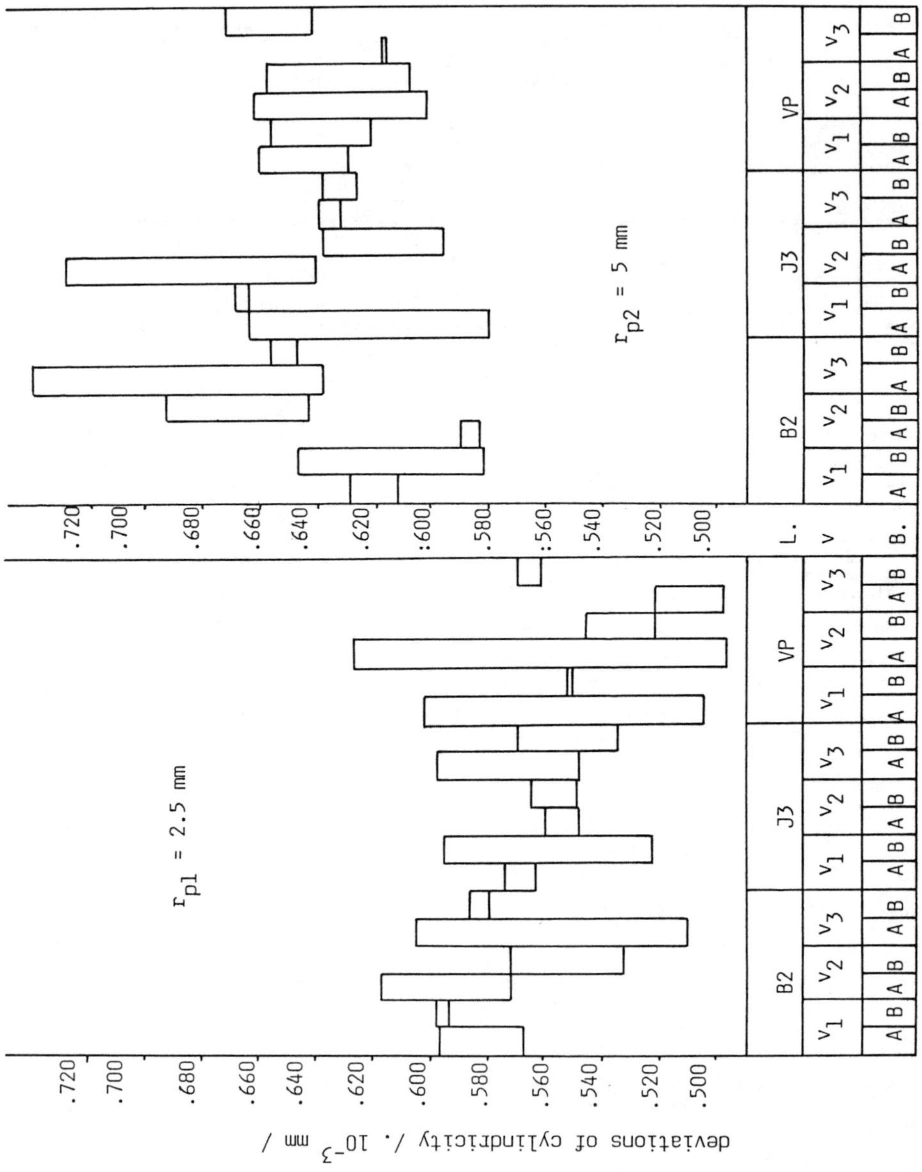

Fig. 2 Deviations of cylindricity
r_p- punch radius, L.- lubrication, v - drawn speed, B.-blank
(A = 123 mm, B = 125 mm), material Kohal 20 E

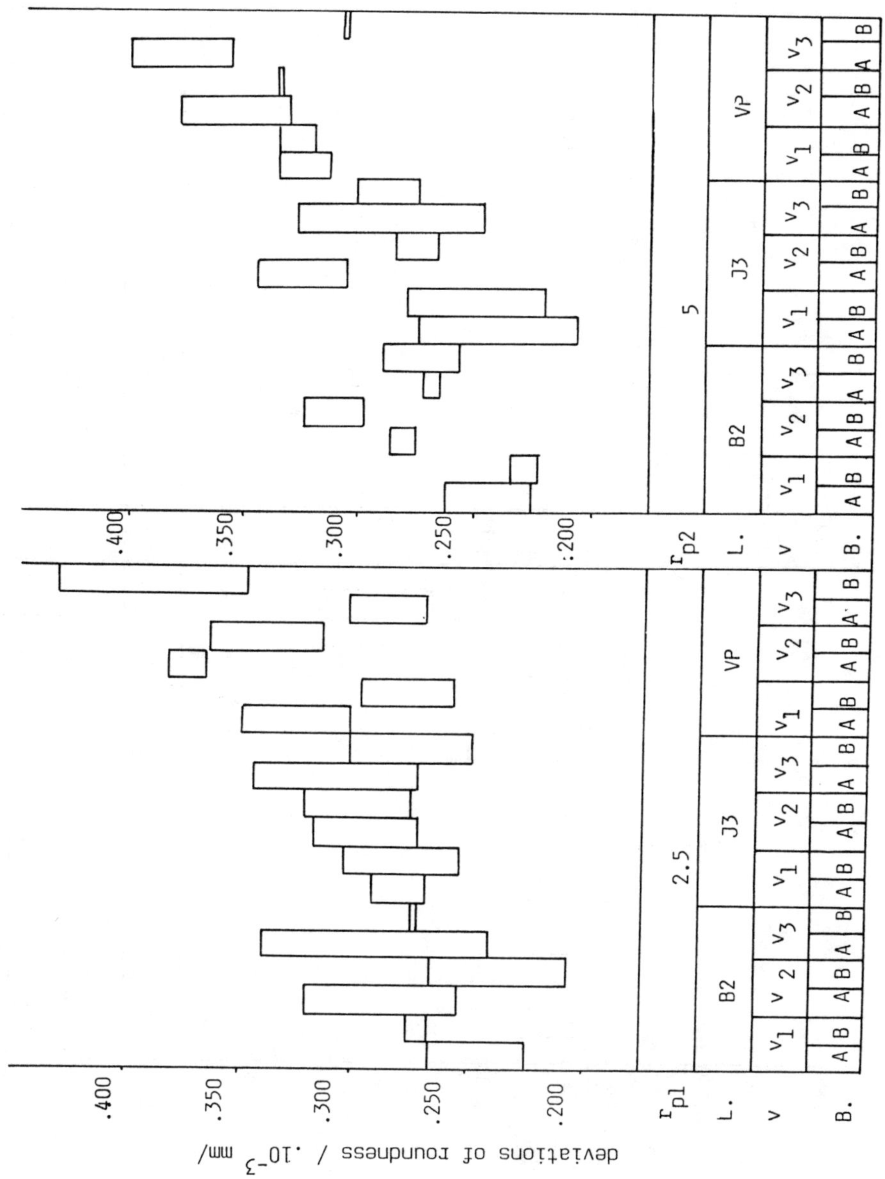

Fig. 3 Deviations of roundness

r_p- punch radius, L.-lubrication, v - drawn speed, B.- blank (A = 123 mm, B = 125 mm), material Kohal 20 E

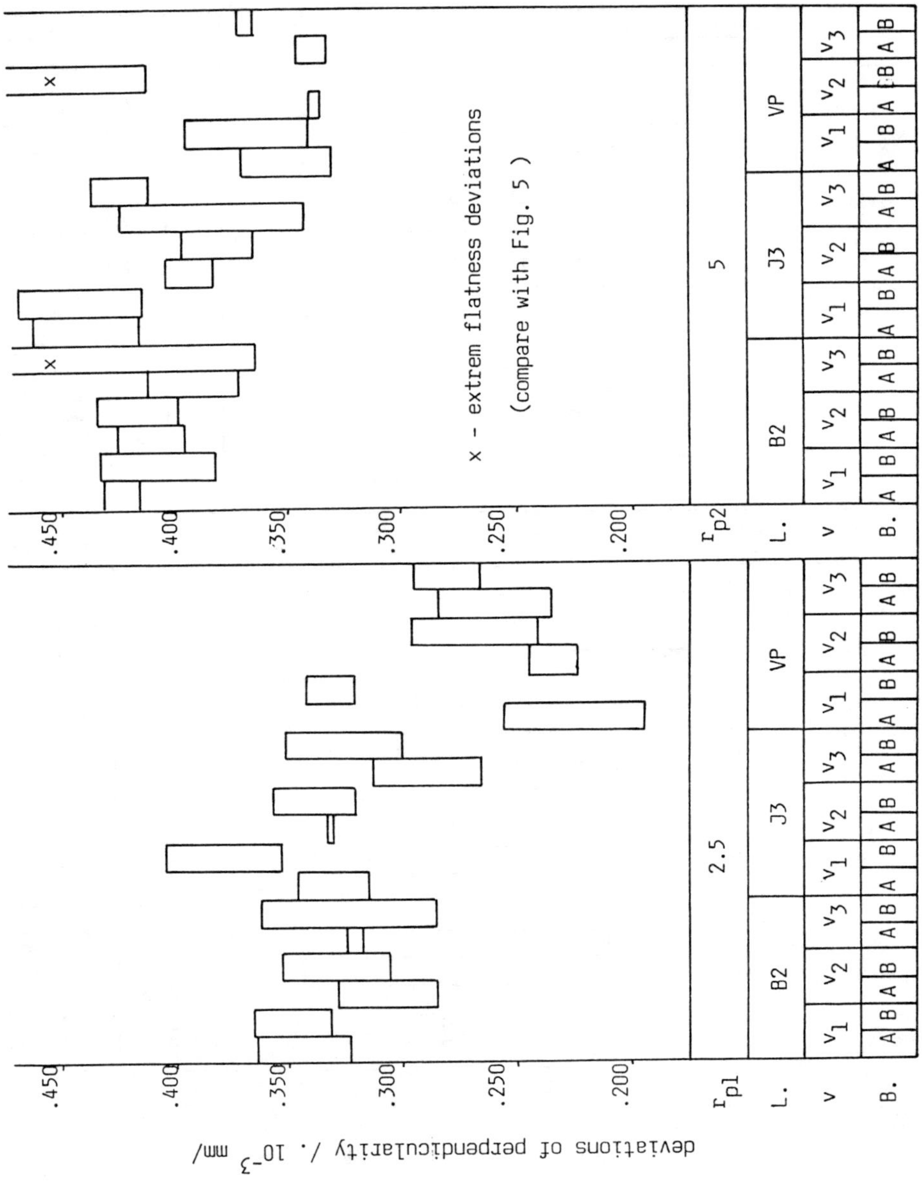

Fig. 4 Deviations of perpendicularity
r_p- punch radius, L.- lubrication, v - drawn speed, B.- blank (A = 123 mm, B = 125 mm), material Kohal 20 E

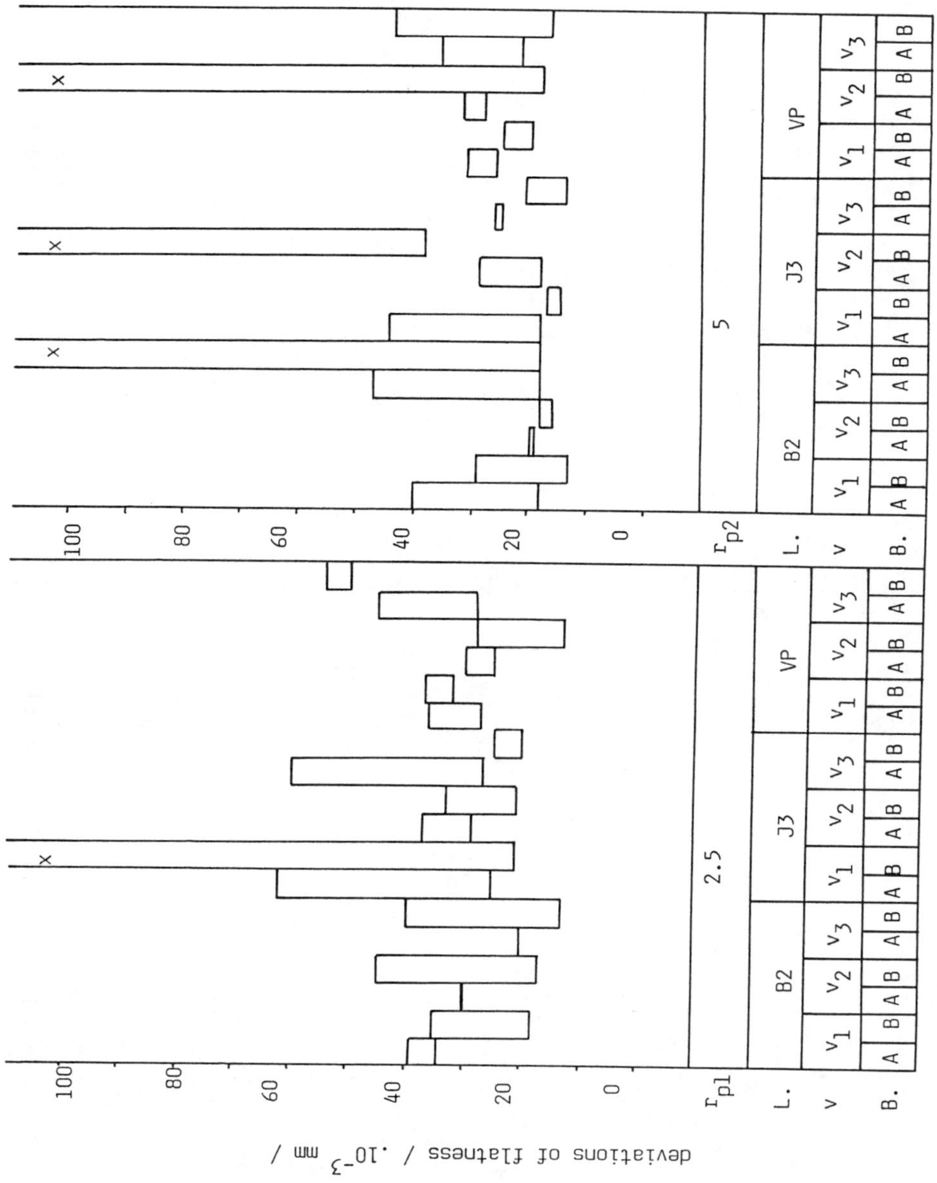

Fig. 5 Deviations of flatness

r_p- punch radius, L.-lubrication, v - drawn speed, B.- blank (A = 123 mm, B = 125 mm), material Kohal 20 E

Sensors, Controls, and Quality Issues in Manufacturing ASME 1991

AN INVESTIGATION OF SURFACE ROUGHNESS CHARACTERIZATION USING AN ART2 NEURAL NETWORK

N. Srinivasa and M. Jouaneh
Department of Mechanical Engineering
University of Rhode Island
Kingston, Rhode Island

Abstract

Surface roughness is an important quality of surfaces. A shortcoming of currently used techniques for characterizing surface roughness is the absence of a single descriptive number that can be used to differentiate the roughness of different surfaces. This paper discusses the development of an Adaptive Resonance Theory (ART2) neural network-based approach for surface roughness analysis. The objective is to develop a unique, more descriptive parameter of surface roughness. ART2 neural networks offer an attractive approach to classify sensory data such as surface roughness profiles, since they have the capability of classifying the data in a self-scaling, unsupervised fashion, and of deciphering the data's invariant properties. In this paper, an ART2 network structure is enhanced to enable the network to address the need of attributing a single numerical value to each classified category. The above methodology was tested on six simulated surfaces and the results obtained showed that the ART2 network is capable of classifying each surface into a distinct category and assigning a roughness index to each category that is more descriptive than the commonly used CLA parameter. However, further work is still needed to obtain the ultimate single number index.

1. Introduction

Surface roughness is an important quality of surfaces. It not only has an important effect on the wear, fatigue, reflection and adhesive properties of many materials, but also affect their appearance and feel. Surface roughness is defined as the macro and micro geometrical properties of a surface. Since it is more difficult to characterize a three dimensional quantity, it has been traditionally characterized by the properties of a two dimensional profile taken from a representative part of the surface. The geometrical properties of this profile are called the surface roughness characteristics.

Recent advances in sensor technology have made available a variety of techniques to measure surface roughness other than the commonly used mechanical stylus [1]. These range from a variety of optical techniques [2] that yield a direct profile of the surface such as in profilometry and interferometry, to techniques that yield a parameter which is related in complex fashion to surface topography as in light scattering [2], speckle pattern analysis [2], ultrasonic [3] and capacitance [4] methods. While providing different choices in complexity, measurement range, and cost for measuring surface roughness, it would be more cost effective to use simple sensing techniques, but increase the wealth of information gained through incorporating advanced computing methods in analyzing sensor data. Artificial neural networks, a dense interconnection of simple computing elements, would provide such a computing approach.

Due to the importance of measuring surface roughness a variety of approaches have been developed to analyze surfaces. These techniques range from purely quantitative ones (basic statistical parameters [5], fractal approach [6], and time-series analysis [7]) that give one or more numbers as a measure of the geometrical properties of the surface to those that yield continuous functions. The latter include time [8,9] and frequency domain techniques [10,11] which describe the properties of the surface as a function of the height variable, the shift distance along the profile, or the frequency content of the profile.

However, none of the techniques yield a single number that is representative of the surface and can be used to compare different surfaces. Rather, each of these techniques measures one attribute of the continuous time profile. The height basic statistical parameters (CLA and RMS), and on which national standards for measuring surface roughness are still based [12], do not uniquely characterize surface roughness. The CLA is defined as the average of the absolute value of profile heights from a center line, while the RMS is defined as the root mean square value of profile heights from a center line. Basic statistical parameters also do not yield any information about the periodic components of surface roughness. On the other hand, time-domain techniques (amplitude density curves, auto-correlation) can only distinguish between a completely random profile and a deterministic one. Similarly, frequency domain techniques can only reveal directly the frequency components of the signal.

A unique surface roughness number that is more descriptive of the surface profile can be used to relate the surface characteristics to the functional usage of the surface. For example, a parameter that can be used to relate the geometrical properties of the surface to the volume of adhesive that is needed to cover the surface of a particle-board to is more appropriate than the CLA or the frequency content of the profile. Moreover, a unique more descriptive parameter of surface roughness offers a rich feedback signal that can be used to control the manufacturing process used to generate the surface.

Furthermore, in many sensory data processing applications, there is a requirement to classify sensory data into different classes based on their invariant properties. This is needed in on-line production environments to automate the classification and inspection of a variety of products such as tiles, carpets, and coated abrasive. Obtaining a unique numerical parameter of each classified entity, which consists of a continuous analog profile, addresses this need also.

In this paper an artificial neural network based approach is developed for surface roughness analysis. The remainder of this paper is organized as follows. Section 2 discusses the principles of unsupervised neural networks; Section 3 discusses the architecture of ART2 network, a special class of neural networks; Section 4 describes the enhancements made to the ART architecture to incorporate the feature of characterization of input data by a single parameter; In section 5, details of computer simulations are presented and discussed; Concluding remarks are given in section 6.

2. Artificial Neural Networks

Artificial neural networks, a dense interconnection of simple computing elements, offer a new way to process and interpret data obtained from sensors by their ability to decipher the data's invariant properties. Two classes of neural networks are available based on their ability to learn with or without supervision. The Back-propagation algorithm [13], the Functional link net [14], Boltzmann machine [15] and Multilayered Perceptrons [27] are examples of neural nets that require supervision from an external agent. On the other hand, the Adaptive Resonance Theory (ART) network [16,17], Kohonen's network [18], Anderson's network [19], and the Bidirectional Associative Memory (BAM) network [20] are examples of neural nets that do not require supervision for training.

Unsupervised learning networks are attractive since they do not need to be trained by exemplars before coding can proceed. The Kohonen and Anderson networks work well only with binary/orthogonal input vectors, and can't learn new associations once they are trained. The BAM network, while conceptually simpler than ART network, is subject to pattern orthogonality constraints and pattern storage problems. On the other hand, the ART network is based on a biologically plausible models and does not have the above limits on its performance. Furthermore, processing of inputs in real-time is possible because learning of a novel pattern can be achieved instantaneously using "fast learning conditions" [17], and because every time a familiar pattern is presented, it directly activates the appropriate stored pattern without any learning. In addition, ART2 networks possess many additional features that would be desirable for the surface roughness analysis problem. These features include:

a) Self Scaling Ability:

This refers to the ability of the network to treat mismatches in input patterns with few features as essential while suppressing the same mismatches as noise in input patterns with many features [21]. For surface roughness analysis, the self scaling ability implies that depending on the scale of the data presented, the network will be selective to certain features of the profile. When surface profiles are compared over the same range of data measurement, the network differentiates the profiles based purely on their surface characteristics alone and not on the relative amplitudes of the individual profiles.

b) Self Stabilizing Ability:

This refers to the ability of the network when presented by an input pattern that is highly irrelevant to its established codes, to slip past all its established codes and codes this input into a new category. This implies that the network is not restricted to certain types of profiles.

c) Self Organizing ability:

This refers to the ability of the network to perform its coding process without an external teacher. Thus the network is not instructed how to classify surface roughness data, and would be able to code complicated input data from non-profile type sensors.

Currently ART2 networks have been used successfully to classify analog input profiles in engineering applications [22-25] such as speech recognition, radar detection and invariant pattern recognition. However, no work has been done to attribute a numerical parameter to each classified category. Thus, there is need to analyze and enhance the network architecture to obtain a numerical measure of each classified input. This is necessary in order to use an ART2 network to provide a more descriptive measure of surface roughness.

3. Structure of an ART2 Neural Network

3.1 Working Principle

An ART network is defined as a neural network that can classify an arbitrary sequence of input patterns into distinct categories in real-time without a teacher and without destroying the previously coded categories that have been stored in its memory [16]. Two types of ART networks called ART1 and ART2 have been developed to classify input patterns into stable recognition codes. The ART1 architecture classifies binary input patterns, while the ART2 architecture can classify both binary and analog input patterns.

The architecture of the ART2 neural network is shown in Figure 1. The network has two processing layers F_1 and F_2. Input from the environment in the form of an analog input pattern is received by the F_1 layer. The F_1 layer performs the function of normalizing the input to achieve high sensitivity and of filtering it from very small magnitude signals. This is accomplished by creating an activation pattern in the nodes of F_1. The activation pattern describes the amount of activity or excitation at the nodes of each activity center due to the input presented to it. This activation pattern is stored temporarily in the memory of the F_1 layer called the short term memory (STM) until the input pattern has been coded by the network. The activity of the nodes in the F_1 layer are then transmitted to the nodes in the next processing layer F_2. During this transmission or bottom-up activation (BUA), the signals from F_1 are multiplied or gated by certain weight factors or long term memory traces (LTM).

The BUA process activates the nodes in the F_2 layer. The F_2 layer performs the function of selecting an exemplar from the activation pattern of its nodes and of storing the LTM traces of the coded input pattern. The nodes of the F_2 layer are competitive in nature in that the node with the maximum activity suppresses activities from all other nodes in F_2. The maximally active node stores its activity in the STM of F_2 and sends back a gated signal or top-down expectation (TDE) to the F_1 layer through a different set of LTM traces connecting the nodes of F_2 and F_1.

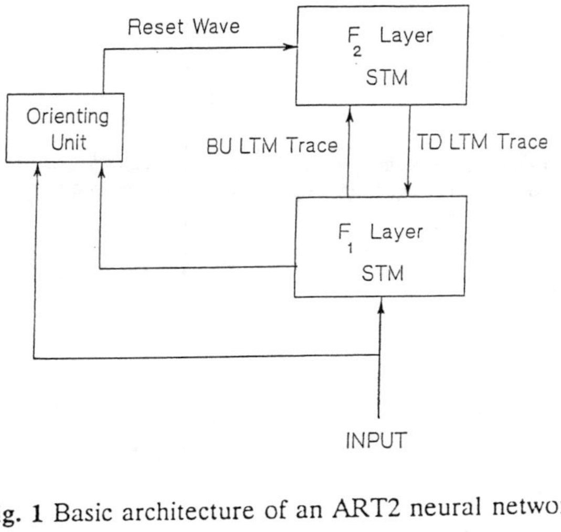

Fig. 1 Basic architecture of an ART2 neural network

After preliminary processing at F_2, the matching process begins. The process compares the TDE with the input activation pattern. The degree of matching can be tuned according to the user's requirements by means of a parameter called the vigilance parameter. High value of the vigilance parameter imply that the network is vigilant to very small mismatches between the TDE and the input pattern and vice versa. If there is a high amount of mismatch, the activity level of the F_1 nodes is attenuated. The attenuation excites another unit of the ART architecture called the orienting unit. This unit sends an arousal signal to F_2 so as to inhibit the activity of its maximally active node (reset) and this inhibition process prolongs until the current input pattern is finally coded. The inhibition process is followed by the selection of the next maximally active node at F_2. The TDE process is again initiated and the process continues until a previously stored pattern is matched with the input pattern (occurrence of resonance) or a new node at F_2 is activated. The resonance of the network results in the LTM traces being updated (learning) according to the choice made by the F_2 layer. The above steps constitute a complete cycle in the process of coding of an input pattern by the ART neural network.

3.2 STM and LTM Equations of an ART2 Neural Network

The detailed processing in an ART2 network can be illustrated with the aid of Figure 2. In the figure, the filled circles represent activity centers. Each activity center represent M nodes that process the inputs from all the M elements of the input pattern in parallel. In each sub-figure solid lines represent signal paths that are active at that processing stage while the dashed lines represent inactive signal paths.

The processing can be divided into four stages: Bottom-up activation, top-down expectation, matching, and vigilance test and reset. Each of the processing stages is detailed below.

I Bottom-Up Activation:

In this stage, an analog input pattern I_i (i=1,...,M) is received by activity center 1 of the preprocessor (labelled loop 1) of the ART2 network as shown in Figure 2(a). After processing in center 2, it is normalized and filtered in activity centers 3 & 4. The filtering is controlled by the threshold parameter θ set by the user. The equations in this processing stage are:

For (i = 1,....,M)

a) At Activity Center 1:

$$e_i = I_i$$

b) At Activity Center 2:

$$f_i = \frac{e_i}{\|e\|} \quad where \; \|e\| = \sqrt{e_1^2 + \ldots\ldots\ldots + e_M^2}$$

c) At Activity Center 3:

$$g_i = 0 \quad if \quad 0 \le f_i \le \theta$$

$$= f_i \quad if \quad f_i > \theta$$

where θ is the threshold or cut-off for the filter.

d) At Activity Center 4:

$$h_i = \frac{g_i}{\|g\|}$$

The normalized and filtered input stream h_i is then fed to activity center 5. The processing of the signals in loop2 is exactly similar to that in loop1. The equations in loop2 can be summarized as:

a) At Activity Center 5:

$$w_i = h_i$$

b) At Activity Center 6:

$$x_i = \frac{w_i}{\|w\|}$$

c) At Activity Center 7:

$$v_i = x_i \quad if \; x_i > \theta$$

$$= 0 \quad if \; 0 \le x_i \le \theta$$

where θ is the threshold parameter as defined before.

d) At Activity Center 8:

$$u_i = \frac{v_i}{\|v\|}$$

After computing u_i, the signals are fed to activity center 9.

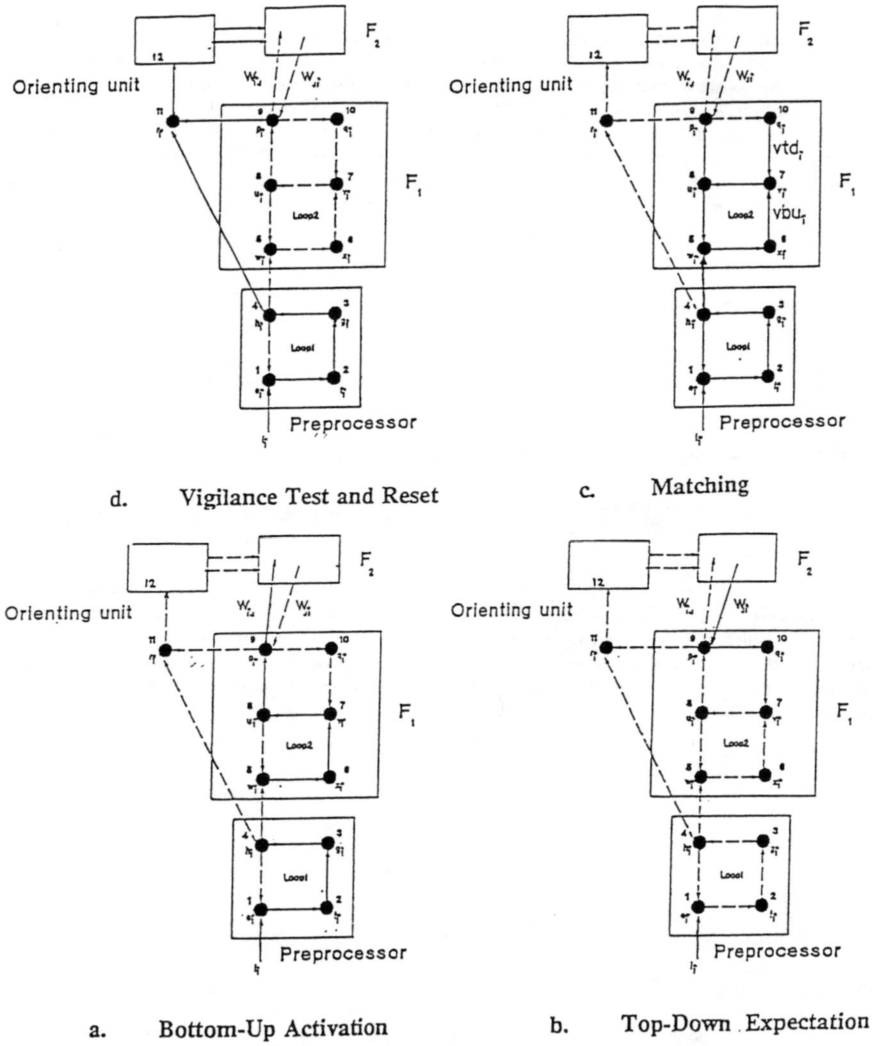

Fig. 2 Processing stages of an ART2 network. All solid lines indicate active signal paths while all dashed lines indicate inactive signal paths during each processing stage.

e) At Activity Center 9:

$$p_i = u_i$$

The p_i signals are then multiplied by the weight vector w_{ij} and are transmitted to the nodes in the F_2 layer as T_j where

$$T_j = \sum_{i=1}^{M} p_i w_{ij} \qquad j = M+1, \ldots, N$$

The w_{ij}'s are initialized to the value $(1-d)\sqrt{M}$ in order to ensure that the ART2 network stabilizes [17]. The set $\{e_i, f_i, g_i, h_i, w_i, x_i, v_i, u_i, p_i\}$ constitute the STM activations during the BUA process.

II Top-Down Expectation

This process is shown in Figure 2(b). The nodes in the F_2 layer are now activated due to T_j signals. The F_2 layer selects the maximally active node J on the basis of the magnitudes of T_j's. By this selection, the F_2 layer essentially makes a hypothesis that the maximally active node is probably the best exemplar for the input pattern. The node J transmits its activity to the F_1 layer through a gating of its signals by the LTM traces w_{Ji} connecting the Jth node in the F_2 layer to all the nodes of activity center 9 in the F_1 layer. The w_{ji} values are all initialized to zero to ensure that no reset occurs when an uncommitted F_2 first becomes active [17]. The gated signal is normalized and then multiplied by a factor 'b' at activity center 7 and is fed to the middle layer of F_1. This marks the end of the top-down expectation process. The equations of this process are as follows:

a) At Activity Center 9:

$$p_i = u_i + d w_{Ji}$$

b) At Activity Center 10:

$$q_i = \frac{p_i}{\|p\|}$$

c) At Activity Center 7:

$$vtd_i = b q_i \quad \text{if } q_i > \theta$$

$$= 0 \quad \text{if } 0 \le q_i \le \theta$$

III Matching

This process is shown in Figure 2(c) and occurs in the F_1 processor. The signal u_i from activity center 8 is amplified by a factor 'a' and is added to the input pattern h_i at activity center 5. The resulting signal w_i is then fed to activity center 6 which generates the vbu_i signals or the bottom-up processing signals. The signals vtd_i due to TDE are matched with the vbu_i signals at activity center 7. The resultant signals are then fed to activity center 9 via activity center 8. This marks the end of matching process. The processing equations in this stage are:

a) At Activity Center 5:

$$w_i = h_i + a u_i$$

b) At Activity Center 6:

$$x_i = \frac{w_i}{\|w\|}$$

c) At Activity Center 7:

$$vbu_i = x_i \quad \text{if } x_i > \theta$$

$$= 0 \quad \text{if } o \le x_i \le \theta$$

and

$$v_i = vbu_i + vtd_i$$

d) At Activity Center 8:

$$u_i = \frac{v_i}{\|v\|}$$

e) At Activity Center 9:

$$p_i = u_i + d w_{Ji}$$

IV Vigilance Test and Reset

After the matching process is complete, the F_1 layer sends its signals to an orienting unit to check for the degree of match between the input pattern and the TDE. The degree of match is decided by the user through the vigilance parameter. If the vigilance is high, the network is highly sensitive to very small differences between the input pattern and the top-down expectation and vice versa. When the mismatch is greater than the limit set by the vigilance parameter, the F_2 layer is reset wherein the current maximally active node is inhibited. The reset also initiates a parallel search in F_2 to find the next most highly active node and create a new TDE. If the degree of match is within the limit of the vigilance parameter, the current maximally active node in F_2 is considered to have coded the input pattern. The process of vigilance testing and reset is shown in Figure 2(d). The equations involving this stage are summarized below.

a) At Activity Center 11:

$$r_i = \frac{h_i + cp_i}{\|h\| + \|cp\|}$$

b) At Activity Center 12:

$$If\ \frac{\rho}{\|r\|} > 1\ then\ reset\ process\ begins$$

$$If\ \frac{\rho}{\|r\|} < 1\ the\ current\ maximally\ active\ node\ J\ codes\ the\ input.$$

and the LTM traces are updated using fast learning conditions[17] by the equations

$$w_{iJ} = \frac{u_i}{(1.0 - d)}$$

and

$$w_{Ji} = \frac{u_i}{(1.0 - d)}$$

The STM and LTM equations in this section made use of four parameters a, b, θ and ρ. The 'a' and 'b' parameters are called the F_1 feedback parameters. They are critical in helping the learned TDE from being recorded in an unstable manner by the bottom-up inputs [21]. The threshold parameter θ is critical in maintaining the non-linearity in the F_1 layer. If θ is set to zero, the F_1 layer is made completely linear and hence loses its property of contrast enhancement and noise suppression [21]. The importance of the vigilance parameter ρ has already been discussed before. The entire operation of the ART2 network is summarized in the flow chart shown in Figure 3.

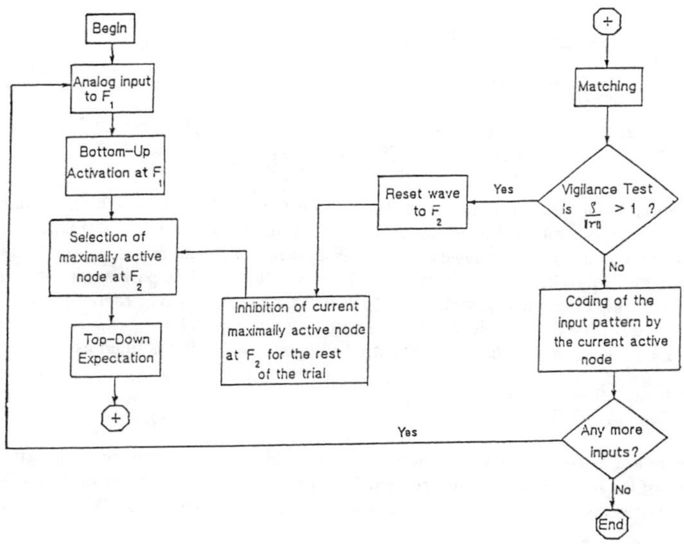

Fig. 3 Flow chart showing the sequence of operations in an ART2 neural Network

A common objective of surface roughness characterization methods is to assign a numerical value for each surface profile to allow comparison between different surfaces. Hence in the next section, we will show how to obtain a numerical value for each classified category of an ART2 network.

4. Identification of a Parameter for Quantification of Surface Profiles

A key to the use of an ART2 network to quantify sensory data is the identification of a parameter in the network that has this ability. This parameter should be based on a comparison between the input pattern presented and the coded categories in the F_2 layer. A possible candidate for this purpose is the $\|r\|$ parameter, discussed in the previous section, since it gives a measure of the degree of match between the input pattern and the TDE of the categories already stored in the F_2 layer. Each component of the r vector is given by

$$r_i = \frac{h_i + cp_i}{\|h\| + \|cp\|} = \frac{h_i + cp_i}{1 + \|cp\|} \qquad (\|h\| = 1)$$

Since

$$\|r\| = \sqrt{r_1^2 + r_2^2 + \ldots\ldots + r_M^2}$$

this implies that

$$\Rightarrow \|r\| = \frac{\sqrt{\|h\|^2 + \|cp\|^2 + 2c(h_1p_1 + \ldots\ldots + h_Mp_M)}}{1 + \|cp\|}$$

or

$$\Rightarrow \|r\| = \frac{\sqrt{1 + \|cp\|^2 + 2\|cp\|\cos(h,p)}}{1 + \|cp\|} \qquad (1)$$

Equation (1) can be expanded to relate the input pattern vector(h_1,...,h_M) to the LTM traces (w_{J1},....,w_{JM}) which store the features of the coded exemplars. Writing $\|p\|$ as:

$$\|p\| = \sqrt{1 + \|dw_J\|^2 + 2\|dw_J\|\cos(u,w_J)} \qquad (\|u\| = 1) \qquad (2)$$

The dot product between h and p can be written as

$$\|h\|\|p\|\cos(h,p) = \|p\|\cos(h,p) = h_1p_1 + \ldots\ldots + h_Mp_M$$

This implies

$$\|p\|\cos(h,\ p) = \cos(h,\ u) + d\|w_J\|\cos(h,\ w_J) \qquad (3)$$

Substituting (2) and (3) into (1) we obtain:

$$\|r\| = \frac{\sqrt{1+c^2+c\|cdw_J\|^2+2c\|cdw_J\|\cos(u,w_J)+2c\cos(h,u)+2\|cdw_J\|\cos(h,w_J)}}{1+\sqrt{c^2+c\|cdw_J\|^2+2c\|cdw_J\|\cos(u,w_J)}} \qquad (4)$$

From equation(4) we can see that $\|r\|$ is dependent on the three angles cos(h,u), cos(u,w_J) and cos(h,w_J) in addition to the parameters c and d. For given values of c and d, the vectors u and h that appear in cos(h,u) are perfectly parallel when coding occurs in a new node. Also, during direct access of a previously stored code, the vectors u and h are almost parallel and cos(h,u) $> \rho$ where ρ is the vigilance parameter. So in general, for given values of c and d and a high vigilance parameter ρ ($\rho \geq 0.95$), cos(h,u) $\approx$ 1 and therefore cos(h,w_J) $\approx$ cos(u,w_J). Thus, the only term in equation (4) that clearly changes with different inputs h is cos(h,w_J). This term represents the angle between the input vector h and the LTM or stored categories w_J and is defined as:

$$\cos(h,w_J) = \frac{h \cdot w_J}{\|h\|\|w_J\|}$$

Since the two vectors are of constant length, cos(h,w_J) gives a measure of "match" or "similarity" between the two vectors by definition of the inner product. Since cos(h,w_J) parameter was found to show more spread when compared with the $\|r\|$ parameter, it was decided to use cos(h,w_J) as an index for measuring surface roughness.

To obtain a numerical value for each input that is classified by the ART2 network, it is necessary to have a reference pattern to which other patterns can be compared. In the case of characterizing surface roughness profiles, this reference profile would be a perfectly flat surface. A perfectly flat surface is an ideal smooth surface and has no roughness.

To implement the quantification task in the ART2 network, the network architecture is enhanced by adding three units as shown in Figure 4. These units are:
i) An auxiliary unit
ii) A reference layer
iii) A comparator

The auxiliary unit is exactly identical to the F_1 layer of the ART2 network and the equations governing this unit are similar to those discussed in the section on bottom-up activation. The auxiliary unit activates the reference layer which contain the learned code of the perfectly smooth surface as defined above. The TDE of the reference layer is matched with the BUA of the input. The result of this matching is the activation of the comparator. The comparator evaluates the unique parameter $\cos(h,w_1)$ which compares the input pattern h with the perfectly smooth surface coded in the LTM trace vector w_1 and gives a measure of the degree of match between the two.

5. Computer Simulation and Results

To illustrate the ideas presented in this paper, six surface profiles resembling those obtained from machined surfaces [26] were used as input patterns. The profiles are shown in Figures 5(a)-5(f). Each profile, represented by 1000 data points, is constructed over a 25mm length. The profile heights range over the interval of 0 to 1. As a reference profile, a perfectly flat surface with height equal to 0.5 is chosen. The network parameters used in this simulation are:

i) $a = b = 10$
ii) $c = 0.1$; $d = 0.9$
iii) $\rho = 0.95$
iv) $\theta = 0.98/\sqrt{M}$
These parameters were chosen to enable the network to code the surfaces with all its special properties as described earlier.

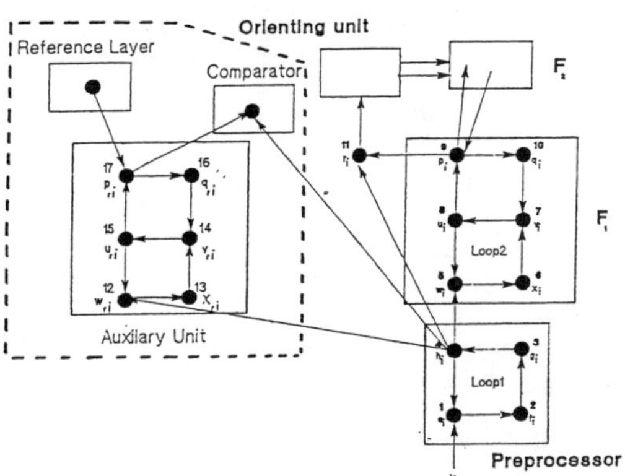

Fig. 4 An enhanced ART2 network showing three additional units incorporated to assign a numerical value to each classified category

In the simulation the modified ART2 network is first presented with the reference profile. The ART2 network codes the reference profile into the first node of F_2 and the LTM traces are updated to store this recognition code. Concurrently, the LTM traces connecting the reference layer node to activity center 17 of the auxiliary unit are initialized with the stored LTM values of the reference profile to be used later for quantifying other surfaces. For each of the six surface profiles, the modified ART2 network codes each profile into a new node or category in F_2. Each surface is also processed by the auxiliary unit to perform a comparison between the reference profile and the current input pattern. The matching process results in quantifying the current input pattern with respect to the reference pattern in terms of the parameter $\cos(h,w_1)$ as described earlier. For illustration, we define the roughness index (RI) of a surface profile as,

$$RI = 1.0 - \cos(h, w_1)$$

The RI is dimensionless and has a value between 0 and 1. For a perfectly smooth surface, the RI is 0. Thus, the closer the value of the RI to 1.0, the rougher the surface is supposed to be. The RI parameter values for the six surfaces along with the CLA and RMS values are given in Table 1. The data shows that the RI seems to be proportional to the extent of height and spatial differences between major crests in the profile. Figure 6 displays the relationship between the RI and CLA for the six surfaces. The figure shows that the relationship is non-linear. This implies that the RI parameter is not merely a function of the height features of the profile and hence is more sensitive than the CLA values. Also, the RI is displayed in Table 2 for the two surfaces that have the same CLA (a periodic and a random profile), and the data shows that the RI is unique for profiles that have the same CLA values. However, these results are not conclusive in our effort to obtain a unique roughness index. Further investigation is in progress to study this parameter and to relate it to other measures of the profile such as the mean crest excursion [26].

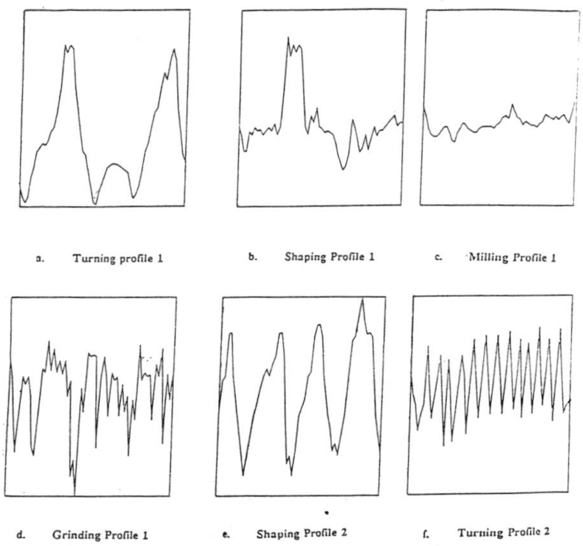

Fig. 5 Six simulated test profiles. All profiles are within the range 0 to 1.

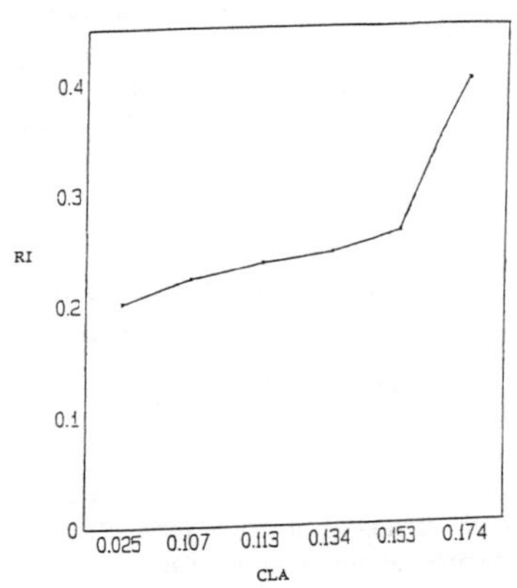

Fig. 6 A plot of RI vs CLA for the six profiles shown in Figure 5.

Profile Number	CLA	RMS	RI
Turning profile 1	0.1737	0.2542	0.4003
Shaping profile 1	0.1527	0.2235	0.2641
Milling profile 1	0.1344	0.1967	0.2452
Grinding profile 1	0.1125	0.1646	0.2347
Shaping profile 2	0.1071	0.1567	0.2231
Turning profile 2	0.0251	0.0367	0.2016

Table 1: Comparison between the CLA, RMS and RI values for six different simulated surface profiles taken over a 25mm length with a 1000 data points.

Profile Number	CLA	RI
1	0.4013	0.6789
2	0.4013	0.5231

Table 2: Comparison of CLA and RI values for two profiles with same CLA.

6. Conclusions

Adaptive Resonance Theory Neural Networks (ART2) have the capability of self organizing continuous input patterns into stable categories in real-time. In this paper an ART2 neural network scheme is studied and applied to the problem of surface roughness characterization. In order to attribute a numerical value to the classified surface profiles, the network is modified by adding three additional units: a reference layer, an auxiliary unit and a comparator. Furthermore, $\cos(h,w_1)$ was identified as a possible parameter that could characterize surface roughness provided the network has a high vigilance parameter. The quantification technique is applicable to any characteristic of an input pattern provided that a pattern with the desired characteristic can be used as a reference. The results obtained show that the network not only has the capability of classifying surfaces into different categories but also attributing a number to each category that is more descriptive than the CLA measure. However, further work is needed to understand the properties of this number. Current work is focussed on relating the obtained RI number to other measures of the profiles. Attention is also focused on addressing the problem of incorporating intelligence into the network so as to be invariant to the direction of traverse on a surface during data acquisition.

7. References

1. Jouaneh, M., "Review of Surface Roughness Measurement and Characterization," Report No. 36.01.134, Forest Products Laboratory, University of California, Berkeley, September 1989.
2. Sherrington, I., and Smith, E., "Modern Measurements Techniques in Surface Metrology: Part II; Optical Measurements," Wear, vol. 125, no. 3, pp. 289-308, August 1988.
3. Blessing, G., and Eitzen, D., "Ultrasonic Sensor for Measuring Surface Roughness," Surface Measurement and Characterization, Jean M. Bennet, Editor, Proc. SPIE 1009, pp. 281-289, 1989.
4. Brecker, J., "A Capacitance-Based Surface Texture Measuring System," Annals of the CIRP, vol. 25, no. 1, pp. 375-377, 1977.
5. King, T., and Spedding, T., "On the Relationships between Surface Profile Height Parameters," Wear, vol. 83, pp. 91-108, 1982.

6. Gagnepain, J., and Roques-Carmes, C., "Fractal Approach to Two-Dimensional and Three-Dimensional Surface Roughness," Wear, vol. 109, pp. 119-126, 1986.
7. DeVries, W., "Autoregressive Time Series Models for Surface Profile Characterization," Annals of the CIRP, vol. 28, pp. 437-440, 1979.
8. Peklenik, J., "Investigation of Surface Topology," Annals of the CIRP, vol. 15, no. 4, pp. 381-385, 1967.
9. Pesante, M., "Determination of Surface Topology by Means of Amplitude Density Curves," Annals of the CIRP, vol. 12, no. 2, pp. 61-68, 1964.
10. Ber, A., and Braun, S., "Spectral Analysis of Surface Finish," Annals of the CIRP, vol. 16, pp. 53-59, 1968.
11. Church, E., and Berry, H., "Spectral Analysis of the Finish of Polished Optical Surfaces," Wear, vol 83, pp. 189-201, 1982.
12. American Standard, ASA B46.1-1962, Surface Texture, ASME, NY, 1962.
13. Rumelhart, D. E., Hinton, G. E., and Williams, R. J., "Learning Internal Representations by Error Propagation," Parallel Distributed Processing, vol. 1, Chap. 8, MIT Press, Cambridge, MA., 1986.
14. Pao, Y. H., Adaptive Pattern Recognition and Neural Networks, Addison-Wesley Publishers, Reading, MA., 1989.
15. Sobajic, D. J., Artificial Neural Nets for Transient Stability Assessment of Electric Power Systems, Ph.D. Thesis, Systems Engineering Department, Case Western Reserve University, Cleveland, Ohio, 1987.
16. Carpenter, G. A., and Grossberg, S., "The ART of Adaptive Pattern Recognition by a Self-Organizing Neural Network," Computer, vol. 21, pp. 77-88, 1988.
17. Carpenter, G. A., and Grossberg, S., "ART2: Self-Organization of Stable Category Recognition Codes for Analog Input Patterns," Applied Optics, vol. 26, pp. 4919-4930, 1987.
18. Kohonen, T., "Self-Organized Formation of Topologically Correct Maps," Biological Cybernetics, vol. 43, pp. 59-69, 1982.
19. Anderson, J., "A simple Neural Network Generating an Interactive Memory," Mathematical Sciences, vol. 14, pp. 197-220, 1972.
20. Kosko, B., "Adaptive Bidirectional Associative Memories," Applied Optics, vol. 26, pp. 4947-4960, 1987.
21. Carpenter, G. A, Grossberg, S., " A Massively Parallel Architecture For a Self-Organizing Neural Pattern Recognition Machine," Computer Vision, Graphics, and Image Processing, vol. 37, pp. 54-115, 1987.
22. Carpenter, G. A., and Grossberg, S., "ART2: Stable self-organization of Pattern Recognition Codes for Analog Input Patterns," Proceedings, First International Conference on Neural Networks, San Diego, 1987.
23. Carpenter, G. A., and Grossberg, S., "Invariant Pattern Recognition and Recall by an Attentive Self-Organizing ART Architecture in a Nonstationary World," Proceedings, First International Conference on Neural Networks, San Diego, 1987.
24. Hartley, K., "Seeing the Need for ART," Science News, 14, pp. 132, 1987.
25. Kolodzy, P., "Multidimensional Machine Vision Using Neural Networks," Proceedings, First International Conference on Neural Networks, San Diego, 1987.
26. Osman, M. O. M., Sankar, T. S., "Profile Characteristics of Manufactured Surfaces Using Random Function Excursion Technique Part 2: Application," Journal of Engineering for Industry, Trans. ASME, Series D, 1974.
27. Lippmann, R. P., "An Introduction to Computing with Neural Nets," IEEE ASSP Magazine, pp. 4-22, 1987.

**Sensors, Controls, and Quality Issues
in Manufacturing
ASME 1991**

THE CHAOTIC CHARACTERISTICS OF THE THREE DIMENSIONAL TURNING

Ibrahim Nur Tansel
Department of Mechanical Engineering
Florida International University
Miami, Florida

Cahit Erkal
Department of Physics
Central Michigan University
Mount Pleasant, Michigan

Fillios Karamidas
Department of Mechanical Engineering
Tufts University
Medford, Massachusetts

ABSTRACT

The present machine tools are designed very conservatively to have linear characteristics and to work in a stable region. During finite amplitude instability, metal cutting operations have nonlinear characteristics and vibration stays within certain limits. In this paper, the chaotic nature of finite amplitude instability is studied on simulated and experimental turning signals. Lyapunov exponents and Poincare maps of the signals were calculated to investigate the chaotic nature of the systems. The results of this study agreed with all the previous cutting simulation and experimental studies. In the future, the chaotic nature of metal cutting can be used effectively to design new machine tools to obtain very large metal removal rates.

I. INTRODUCTION

Machine tool designers have been designing machine tools to assure the stability of cutting operations. Machine tool structures are traditionally designed bigger and stronger to increase metal removal rates without creating chatter. On the other hand, nature uses nonlinear systems that demonstrate chaotic characteristics. Chaotic characteristics allow these systems to avoid catastrophic unstable responses (severe weather conditions calm down) and perform their tasks better (chaotic heartbeats avoid clotting of blood). In metal cutting, we know that under unstable cutting conditions the amplitude of vibrations increases until the tool starts to leave the workpiece and the amplitude of the vibrations reaches a limit. Subsequently, the system will be in finite amplitude unstable cutting conditions. Previously, an analytical model of orthogonal cutting was studied and it was found to be chaotic. In this paper, finite amplitude instability of 3-D cutting is studied on semi-experimental and experimental data.

During the cylindrical turning of long slender bars, radial (thrust direction) cutting force and relative tool-workpiece vibrations are the most critical [1,2,3,4]. These operations can be simulated by considering only the thrust direction vibrations [4] and cutting forces for the simulations. The transfer functions of cutting and structural dynamics have to be determined experimentally (preferably) or analytically. The cutting [5,6,7] and structural [1,2,3,4,8,9] dynamics can be represented with a linear transfer function in oblique and orthogonal cutting.

The uncut chip area variation can be represented with a linear gain during orthogonal cutting under stable cutting conditions. However, it becomes nonlinear [10,11,12,13] when the cutting tool starts to leave the workpiece during severe chatter. In oblique cutting, the uncut chip area variation and tool vibration have a nonlinear relationship [14,15]. When the tool starts to leave the workpiece, the nonlinearity increases [4]. In this study, simulation of cylindrical turning was prepared by using experimentally identified cutting and structural dynamics, and a non-linear uncut chip area model.

Most of the previous studies on turning operations have focused on orthogonal cutting. Estimation of the threshold of instability [3,8,16,17], the nonlinear nature of the uncut chip area when the tool leaves the workpiece [10, 13], and the amplitude of finite amplitude instability [10,13] were studied in these research investigations. Grabec [18] has demonstrated the chaotic dynamics of the displacement signals of orthogonal cutting. In that study, analytical models were derived to represent cutting and structural dynamics. The chaotic nature of the displacements were indicated by using a Poincaré return map.

To develop chatter detection [19,20] or active suppression algorithms [21-22], the characteristics of the cutting force and displacement signals must be studied. During the growth and finite amplitude instability stages of chatter, the nonlinear model of the system is very complex and cutting force fluctuations are very high. It is very difficult to study such a complex system analytically. Similarly, under the influence of such large forces, experimental data collection is dangerous and may damage the instrumentation.

The purpose of this paper is to study the characteristics of the simulated cutting force signals at severe chatter conditions. The simulation data will be generated by using experimentally identified cutting and structural dynamics to obtain more realistic results. Also, the most commonly used oblique cutting operations were considered instead of the orthogonal cutting operations in Ref. [18]. The predictability (chaotic nature) of the signals will be investigated by using the Lyapunov spectrum. The results of the study will be useful in the development of chatter detection and suppression systems.

The theoretical background of the study (the block diagram of the turning operation and calculation of the Lyapunov exponents) will be discussed in the next section. Simulation of the turning operation and analysis of the signals will be presented in the following sections.

II. THEORETICAL BACKGROUND

In this section, the block diagram of the cylindrical turning operation and the Lyapunov exponents will be very briefly outlined. For more information on the block diagram of turning, readers should see Refs. [3] and [4]. For identification of the chaotic behavior of experimental data, the Lyapunov exponents will be used in this paper. Section II.2 will introduce the Lyapunov exponents.

II.1. Block Diagram of Cylindrical Turning:

The block diagram of the simplified cylindrical turning operation is presented in Fig. 1 [4]. Through cutting dynamics (G_{ix}, G_{iy}, G_{iz}, G_{ox}, G_{oy}, G_{oz}), the uncut chip area variations ($dA(x_i,y_i)$ and $dA(x_o,y_o)$) are converted into the F_x, F_y, F_z cutting force fluctuations. Direct (G_{xx}, G_{yy}) and cross (G_{yx}, G_{zx}, G_{xy}, G_{zy}) receptances of the tool and workpiece structure generate relative vibrations (x_i, x_o) in the X (feed) and (y_i,y_o) in the Y (thrust) directions under the influence of cutting force fluctuations. The present (x_i) and previous (x_o) vibrations generate the $dA(x_i,y_i)$, $dA(x_o,y_o)$ uncut chip area variations. In this study only the thrust direction vibrations G_{iy}, G_{oy} and G_{yy} transfer functions will be considered.

II.2. Lyapunov Exponents from Time Series:

A reliable procedure for recognition of the chaotic behavior of experimental data has been developed using Lyapunov exponents [26]. The Lyapunov exponents can be calculated for nonlinear systems from the differential equations that govern the time evolution of the path [27]. Lyapunov exponents are, on the average, the rate of convergence or divergence of the nearby orbits in the phase space (x,x) where x and x are velocity and displacement, respectively. Calculation of Lyapunov exponents using the phase space, requires a knowledge of the equations of motions [27, 28]; however, in the last decade, a number of algorithms have been developed to analyze the chaotic characteristics of a system using experimental data [27,28,29,30,31,32,33]. These algorithms utilize calculation of the Lyapunov exponents spectrum. For a given time series, $\{x(t_k), x(t_k+\tau), \ldots x(t_k+(m-1)\tau)\}$, where $t_k = kT$, $k=1, \ldots, \infty$, an m dimensional phase portrait can be reconstructed with a delay coordinate, τ, provided almost arbitrarily [34,35,36]. The distance, $L(t_1)$, between the initial point, $\{x(t_1), x(t_1+\tau),$

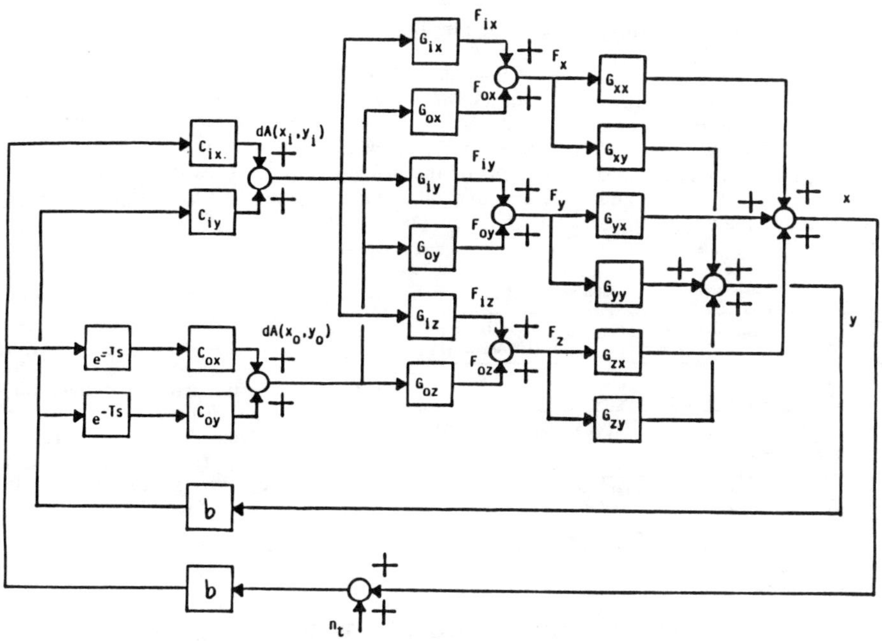

Fig. 1. Block diagram of the turning operation.

....$x(t_1+(m-1)\tau)\}$, and the nearest neighbor, evolves in time. Time evolution of $L(t_1)$ is monitored through the data (see Ref. [24] for details), and the largest Lyapunov exponents are calculated according to the following measure:

$$\Gamma = \frac{1}{t_m - t_o} \sum_{k=1}^{M} \log_2 \frac{L^1(t_k)}{L(t_{k-1})} \tag{1}$$

where t_m is the replacement time, M is the number of replacement steps, and $L^1(t_k)$ is the evolved length [26]. In the equation, $T = t_{k+1} - t_k$ is used for fixed time intervals. The results of Eq. (1) are given in bits of information per iteration (bits/iteration). This is the conventional unit adapted from "information theory" and commonly used in various analyses.

The sign and the magnitude of the largest Lyapunov exponent, Γ, determine the type and the dimension of the attractor of the dynamical system represented by the time series, x(t). A negative exponent implies a periodic cycle (periodic attractor); a positive exponent, on the other hand, signals the existence of a chaotic attractor. It has been shown that the method outlined has correctly identified the simulated data of various chaotic systems [27,28]. The identification of an attractor and its structure is very important, since the predictability of further steps of the system is strongly dependent on the type of attractor the system settles down to. From an engineering point of view, a prediction of the onset of unpredictability (chaos) plays an important role in design and performance of monitoring and control systems.

III. SIMULATION OF TURNING OPERATIONS

To simulate turning operations, the transfer functions of cutting and structural dynamics (Fig. 1) must be identified. However, identification of all the transfer functions is extremely time consuming, expensive and sometimes unnecessary. For example, cylindrical turning of long slender bars is one of the critical operations in which chatter is encountered and can be simulated with the simplified version of Fig. 1. In the following sections, the simplification of the block diagram in Fig. 1 for this special case, identification of the transfer functions of this simplified system, calculation of uncut chip area variation, and preparation of the simulation will be very briefly outlined. For more information on the preparation of the simulation, readers should see Ref. [4].

III.1. Cylindrical Turning of Long Slender Bars:

During the cylindrical turning of long slender bars (approximately L/D > 12), chatter is mainly created by the thrust direction vibrations of the workpiece. The feed direction relative tool–workpiece vibrations are minimum. In such cases, the critical vibrations are in the thrust direction (Y) of the workpiece. For this case, the general block diagram in Fig. 1 can be modified by eliminating the feed (X) direction vibrations of the machining process, as shown in Fig. 2. The transfer functions of cutting and structural dynamics (Fig. 2) have to be identified with separate experiments and the uncut chip area variation must be calculated.

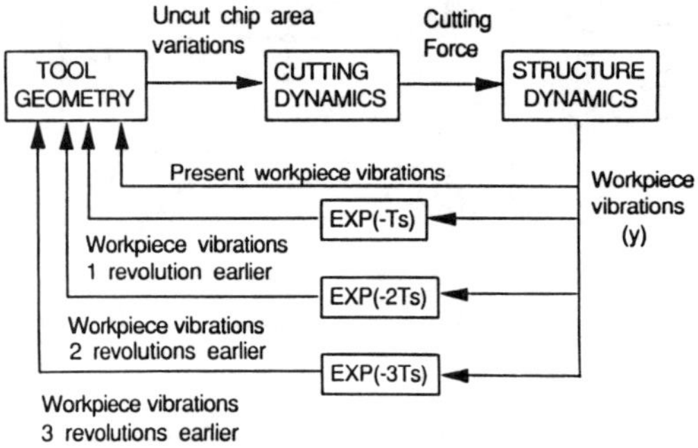

Fig. 2. Simplified block diagram of the turning operation. The block diagram in Fig. 1 is simplified by considering the cylindrical turning of long slender bars.

III.2. Identification of Cutting and Structural Dynamics:

The discrete transfer functions of the cutting process dynamics were identified using a special test rig with one degree of freedom, Fig. 3 [37]. A tool was attached to the test rig and vibrated in the x direction (parallel to the axis of the workpiece) by using an electro–hydraulic exciter. The cutting forces and tool displacements were measured with a dynamometer and proximity–type transducers, respectively. The contribution of the inertia forces to the cutting force measurements was compensated by subtracting properly amplified acceleration signals. During the experiments, the y direction tool displacements were negligible in comparison to the x direction displacements. The G_{ix} and G_{ox} transfer functions (in Fig. 1) of the cutting process were identified using the Recursive Multichannel Maximum Entropy Method (RMMEM) [38] after the present and one revolution earlier tool positions were separated.

The cutting dynamics experimental data were collected while a 1045 steel bar was being cut at 105 m/min cutting speed with a 0.13 mm/rev feed. The magnitude and phase of the characteristics of the G_{ix} and G_{ox} transfer functions are shown in Fig. 4.

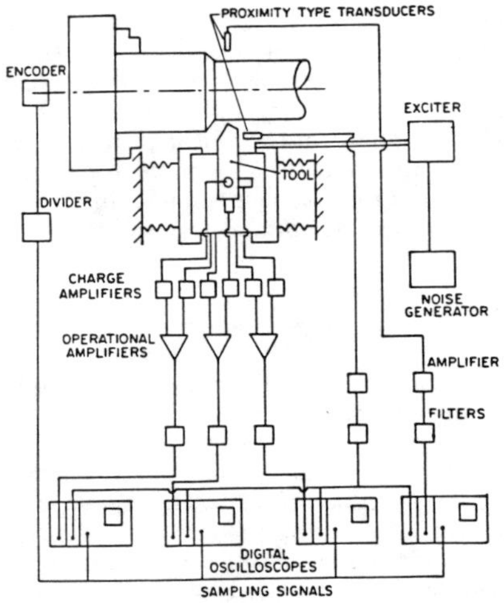

Fig. 3. Experimental setup to identify the cutting dynamics.

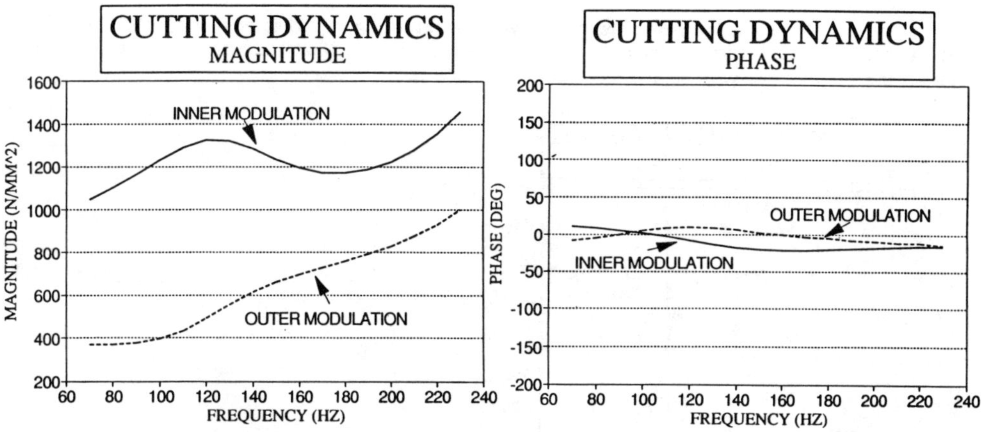

Fig. 4. Characteristics of the cutting dynamics.

To identify the structural dynamics of a long slender bar, a 1045 steel workpiece with 760 mm length and 65 mm diameter was attached to a lathe between the spindle and tailstock. The workpiece was excited from the middle, perpendicular to its axis, by using an electro–hydraulic exciter. The excitation force and workpiece displacements were measured by using a load cell and velocity pick–up (the absolute displacements were obtained by integrating the velocity signals by an analog circuit). A diagram of the test rig is presented in Fig. 5.

After collecting the displacement data from structure identification tests and filtering with a band–pass filter with cut–off frequencies at 100–200 Hz, ARV models were estimated with the RMMEM to obtain the discrete transfer function of the structural dynamics. The frequency response characteristics of the workpiece are presented in Fig. 6.

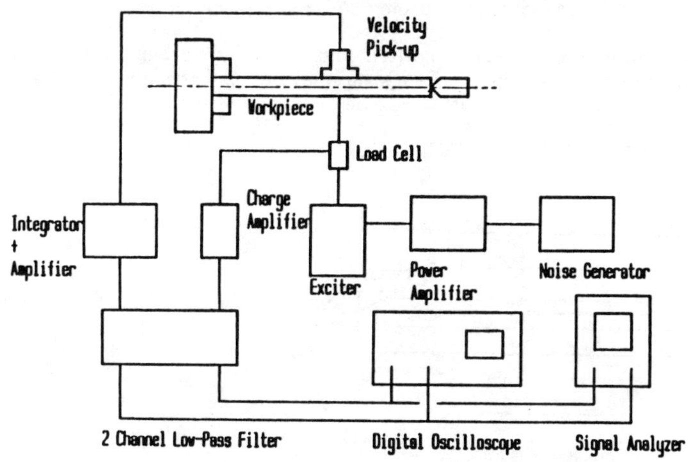

Fig. 5. Experimental setup to identify the structural dynamics.

III.3. Simulation of Cylindrical Turning Operations:

The cylindrical turning of long slender bars was simulated according to the block diagram in Fig. 2. Experimentally identified cutting (G_{iy} ($G_{oy} \approx G_{iy}$)) and structural (G_{yy}) dynamics transfer functions were used. The uncut chip area was calculated using a simulation subroutine that calculates the coordinates of the cutting tool by considering the side and end cutting edge angles and nose radius of the tool. The uncut chip area was calculated by considering the tool position in four consecutive rotations. Detailed information on the uncut chip area calculations is given in Ref. [4].

The simulations in this paper represent cylindrical turning of a 1045 steel bar, at 105 m/min cutting speed with 0.13 mm/rev feed. The workpiece has a length of 760 mm and a diameter of 65 mm. In the simulations, various depths of cuts were selected.

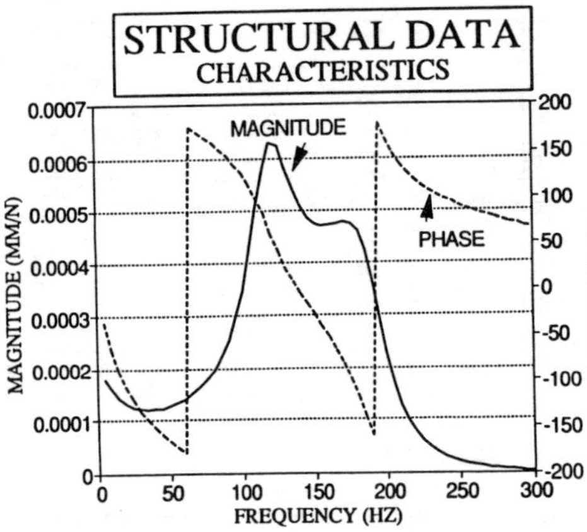

Fig. 6. Characteristics of the structural dynamics.

IV. EXPERIMENTAL DATA COLLECTION FOR FINITE AMPLITUDE INSTABILITY

The machine tool vibrations were monitored by measuring the acceleration signals at the tailstock of the Clausing (Colchester 15) lathe. The length of the workpiece was 600 mm and its diameter was 70 mm. A piezoelectric accelerometer (B&K accelerometer type #4333) was attached to the tailstock of the lathe on a magnetic base in the horizontal plane perpendicular to the axis of the workpiece. The signals of the accelerometer were converted to voltage variation with a charge amplifier (B&K #2635) and digitized by using an IBM PS2/50 personal computer with a Megabyte DAS-16G analog to digital (A/D) converter after the signal was conditioned by an electronic band-pass filter with the cutoff frequencies at 50 and 500 Hz.

The acceleration signals were digitized and stored at various cutting conditions. A representative example of experimental turning data at severe chatter conditions (collected at 470 rpm cutting speed, 0.119 in/sec feed rate with a carbide insert tool of radius 0.031") is presented in Fig. 7.

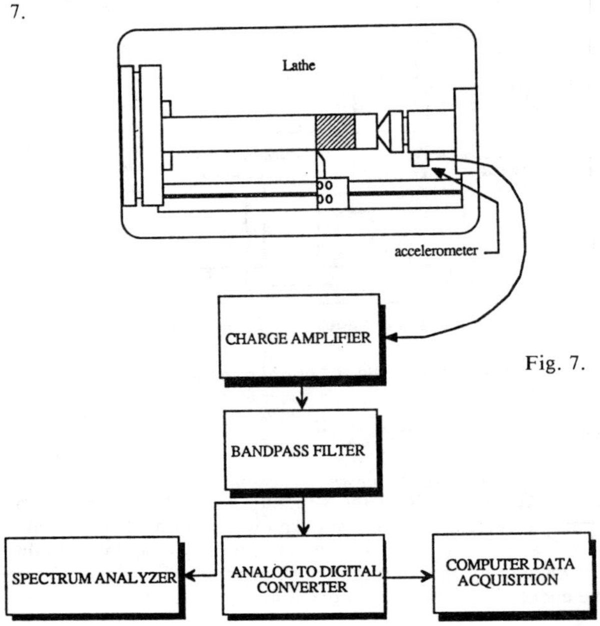

Fig. 7. Experimental setup to digitize acceleration signals during chaotic cutting.

V. RESULTS AND DISCUSSION

The cylindrical turning of the long slender bars was simulated and studied at many different depths of cuts. For each case, the cutting force and displacement data were created until the system reached finite amplitude instability and finished its transition. The first nine seconds of chatter were simulated and the characteristics of the signal were studied after the first four seconds when the signal settles. Under stable cutting conditions, the cutting force and thrust direction workpiece vibration signals quickly diminished, Fig. 8.

Under unstable cutting conditions, the cutting force variations and workpiece vibrations increased until they reached a certain amplitude. Later, the frequency of the signal almost stayed constant; however, the amplitude continuously varied (Fig. 8). The thrust direction workpiece displacements of all the simulated turning signals had a dominant frequency (Fig. 10).

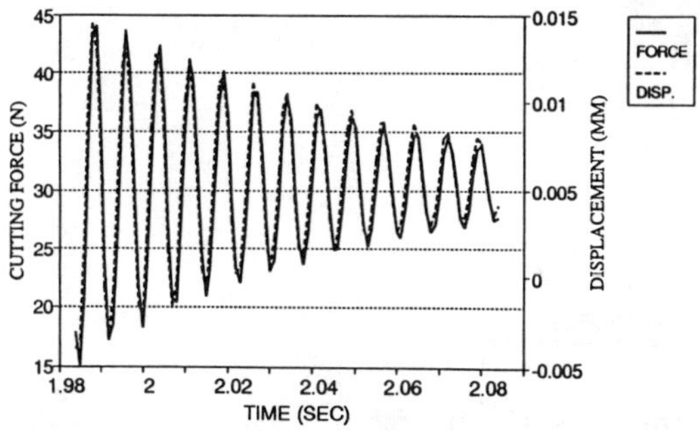

Fig. 8. Simulated cutting force and thrust direction workpiece vibration signals. Depth of cut is 0.5 mm, stable cutting.

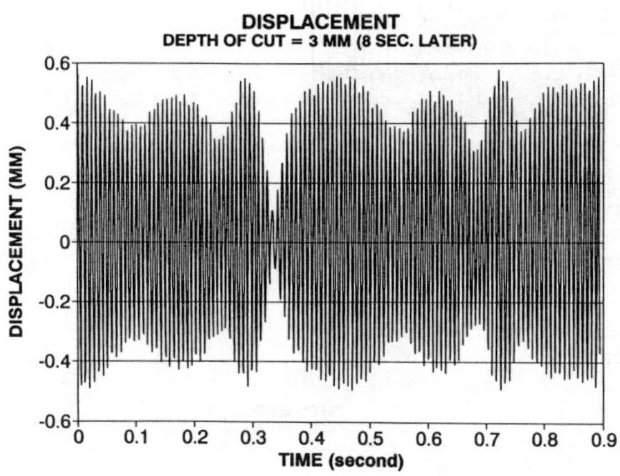

Fig. 9. Thrust direction workpiece vibrations.

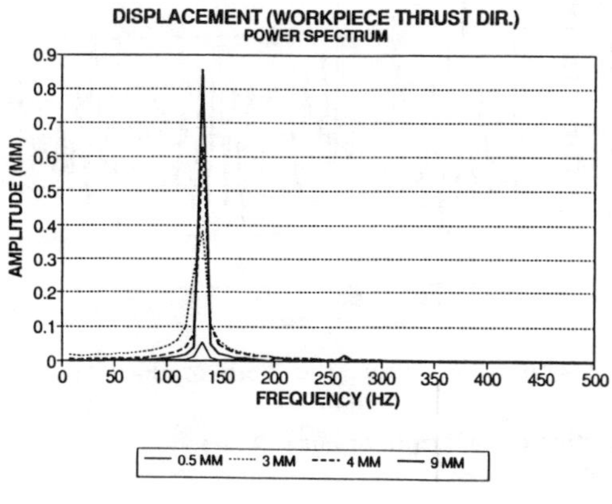

Fig. 10. The power spectrums of workpiece direction displacement data.

The variation of the thrust direction cutting forces was simulated with the 3 mm and 9 mm depths of cut and presented in Figs. 11 and 12, respectively. The continuous change of the characteristics of the signal can be seen in Fig.11. The characteristics of the machining operation have been outlined in Section III.3. When Figs. 12.b, and 12.c are inspected one by one, a constant pattern can be considered; however, there is no continuity between the patterns of Figs. 12.b and 12.c.

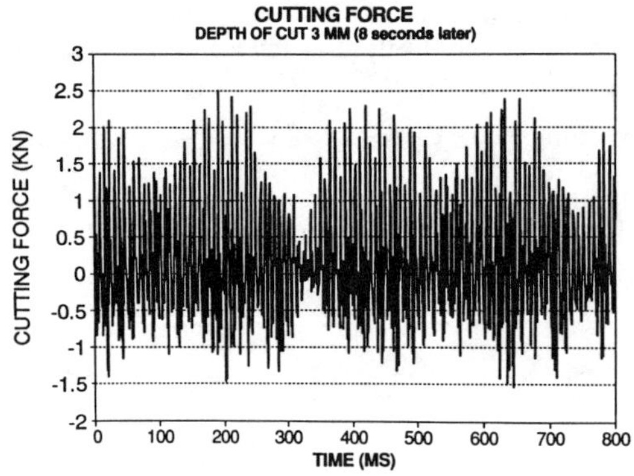

Fig. 11. The thrust direction cutting force. Depth of cut is 3 mm.

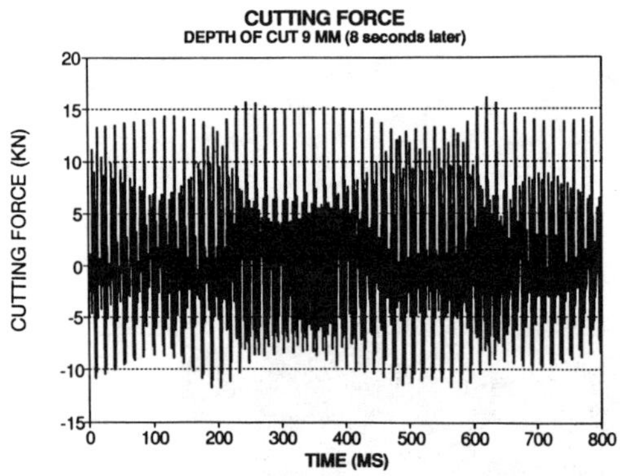

Fig. 12.a. From the 8th to the 9th second.

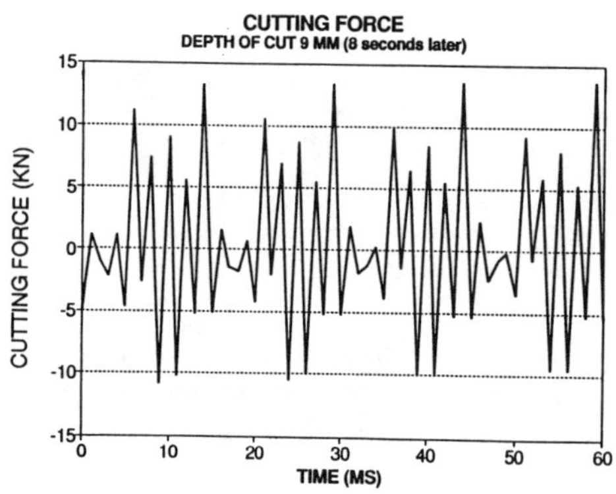

Fig. 12.b. From the 8th to 8.1st second.

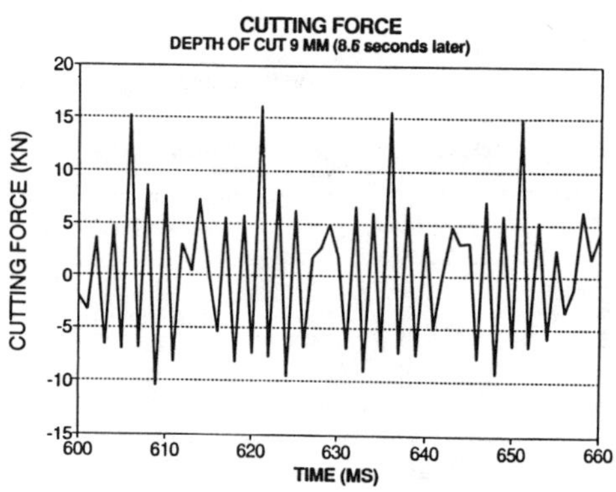

Fig. 12.c. From the 8.6th to the 8.7th second.

Fig. 12. The thrust direction cutting force. Depth of cut is 9 mm.

To investigate the chaotic nature of the cutting force signals, Lyapunov exponents were calculated [27] from the simulated time series data with fixed evolution time. A chaotic system can be recognized from the Lyapunov exponents, since it would have at least one positive Lyapunov exponent. Fig. 13 shows the Lyapunov exponents calculated for two particular sets of data versus replacement steps. Lyapunov exponents stabilized as the algorithm approached the end of the data and reached 4.7 and 12.7 for 3 mm and 9 mm depths of cuts, respectively. The Lyapunov exponents of the system always stayed positive and indicated the cutting force signals could be chaotic.

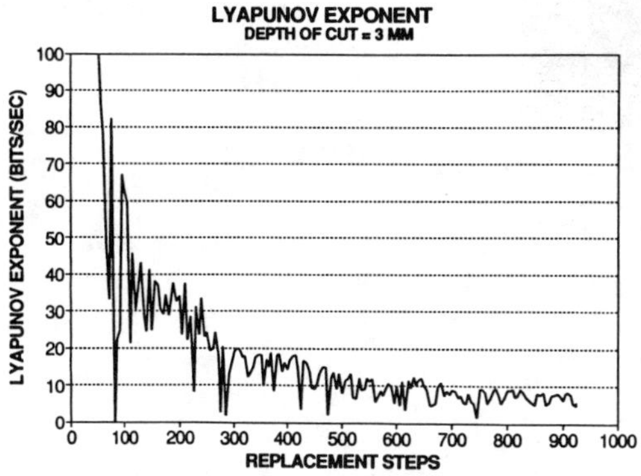

Fig. 13.a. Depth of cut is 3 mm.

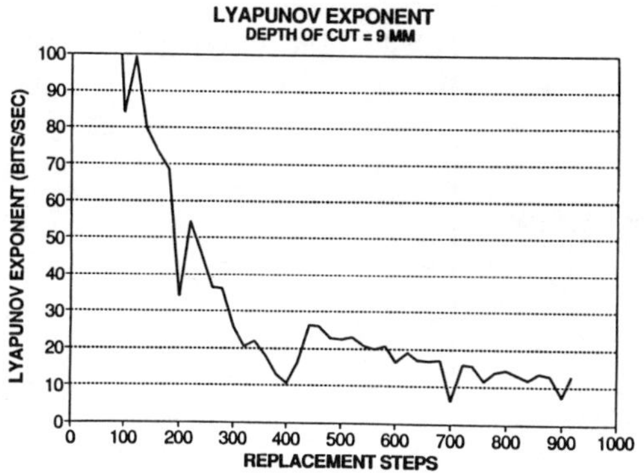

Fig. 13.b. Depth of cut is 9 mm.

Fig. 13. The Lyapunov exponents of the cutting force signal.

The chaotic nature of the experimental cutting signals were also investigated to support our results on the semi-experimental data. The Lyapunov exponents of the experimental acceleration signals (Fig. 14.a) of finite amplitude unstable cutting were calculated and presented in Fig. 14.b. The Lyapunov exponents of experimental acceleration also confirmed the chaotic nature of the turning operations. Another tool for recognition of the chaotic signals is the Poíncaré maps. The periodic signals create a single point on the Poíncaré map while the random signals are indicated with randomly distributed points in a two-dimensional plane. The chaotic signals generate a pattern that indicates a kind of regularity. The Poíncaré map of the acceleration signal displayed a pattern that supported the chaotic nature of the signal(Fig. 14.c).

The results of this study also agree with the previous simulation studies and experimental results. In Parthimos and Ehmann's [39] recent study, the simulated horizontal motion of tool and cutting force signals was very similar to the experimental results of Kondo et al. [40] while chatter grows. The simulated and experimental signals had periodic characteristics until vibrations reached a maximum amplitude. During finite amplitude instability, Kondo et al.'s [40] cutting force signals were demonstrating an unpredictable amplitude variation.

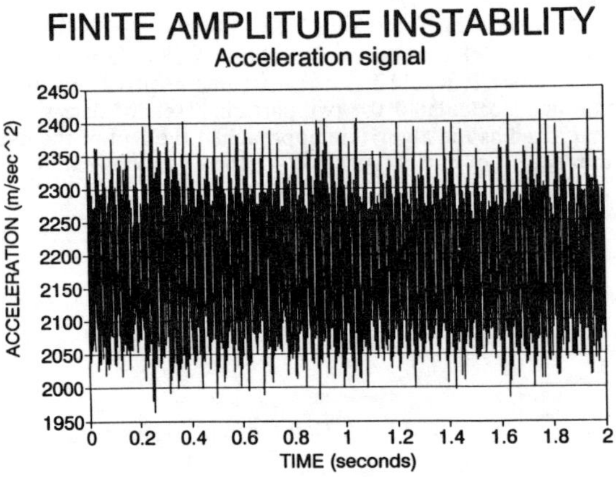

Fig. 14.a. Experimental acceleration signal.

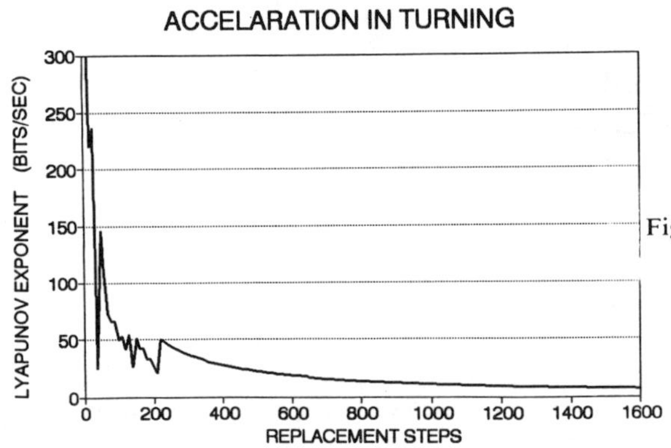

Fig. 14.b. The Lyapunov exponent of Fig. 14.a.

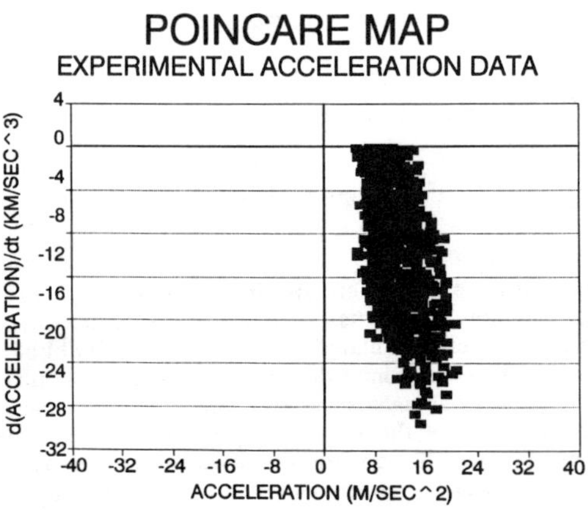

Fig. 14.c. The Poincaré map of Fig. 14.a.

Fig. 14 Investigation of the chaotic nature of experimental signals.

These results indicate that the thrust direction workpiece vibrations are a reliable source to detect chatter or to make a forecasting chatter suppression type controller, even at extremely nonlinear working conditions (at finite amplitude instability). Previously, Eman and Wu [20] and Tsai [6] used acceleration signals to detect chatter. Both approaches are expected to detect chatter without any problem, even at finite amplitude instability.

On the other hand, the thrust direction cutting force was found to be chaotic during finite amplitude instability. This means the cutting force signal will lose its predictability when the tool starts to leave the workpiece. In this study, alternating periodic-chaotic transition sequences (similar to ones reported in Refs. [41,42]) were not encountered during finite amplitude instability. In actual cutting conditions, the marks of the previous pass, slight variation of the cutting parameters, ongoing tool wear, and material characteristics will reduce the predictability of the signals further. Researchers should be very careful when they design chatter detection and suppression systems based on cutting force measurements. The chaotic cutting force signals (during finite amplitude instability) may have several spikes or a high background noise that may deceive a chatter detection system. Similarly, active chatter suppression systems may expect to have larger prediction errors if a forecasting compensatory control (FCC) [43] type algorithm is going to be used to suppress chatter by monitoring the cutting forces. Of course, researchers should also consider the advantages of the cutting force measurement (such as relatively simpler instrumentation [22,23]) based chatter suppression systems.

VI. CONCLUSION

The nonlinear dynamics of the cylindrical turning of long slender bars are studied on semi-experimental and experimental data of 3-D cutting when the system reaches finite amplitude instability. Semi-experimental data were generated using experimentally identified cutting and structural dynamics transfer functions, and an uncut chip area simulator. The thrust direction cutting force and displacement data were created by using the simulation program.

After the chatter reached to finite amplitude instability, the cutting force signal showed a non-periodic characteristics. An inspection of the Lyapunov exponents of the simulated cutting force variations and experimental acceleration data indicated that the signals may be chaotic. Also, the poincare maps supported the findings. In a future study the dimensional aspect of the signals will be discussed to scrutinize the results of this study. The study found the use of the displacement signal advantageous to detect chatter development, or suppression systems, to avoid the problems related to the probably chaotic nature of cutting forces during finite amplitude instability.

The amplitude of the vibrations during finite amplitude instability is limited. This feature can be effectively used to design a new generation machine tools with very high metal removal rates.

REFERENCES

1. T. Moriwaki, K. Iwata, "In-Process Analysis of Machine Tool Structure Dynamics and Prediction of Machining Chatter," *Journal of Engineering for Industry*, Trans. of ASME, Feb. 1976, pp. 301-305.

2. B.H. Lu, Z.H. Lin, X.T. Hwang, C.H. Ku, "On-Line Identification of Dynamic Behavior of Machine Tool Structures During Stable Cutting," *Annals of the CIRP*, Vol. 32/1, 1983, pp.315-318.

3. I.N. Tansel, "Stability Analysis of Long Slender Bars by Using the Unified Transfer Functions," *Symposium on Product and Process Design, Manufacturing International 1988*, (edited by G. Chryssolouris and R. Komanduri), ASME, 1988, pp.345-351.

4. I.N. Tansel, "Simulation of Turning Operations," *Int. J. of Mach.Tools and Manuf., Des., Res. and Applic.*, Vol.30, No.4, 1990, pp.535-547.

5. M.M Nigm, M.M. Sadek and S.A. Tobias, "Determination of Dynamic Cutting Coefficients from Steady State Cutting Data," *Int. J. Mach. Tool Des. Res.*, Vol. 17, 1977, pp. 19-37.

6. J. Tlusty and B.S. Goel,"Measurement of the Dynamic Cutting Force Coefficient," J. Tlusty and B.S. Goel, *Proceedings of the Second NAMR Conference*, 1974, pp. 649-665.

7. I.N. Tansel, K.F. Eman, "Transfer Function of Cutting Dynamics in Three Dimensional Cutting," *Proc. of NAMR Conf.*, May 1985, pp. 476-481.

8. F. Koenigsberger, J. Tlusty, *Machine Tool Structures*, Pergamon Press, Oxford, England, 1970.

9. Y.C.Shin, K.F.Eman, S.M. Wu, "Experimental Complex Modal Analysis of Machine Tool Structures," Trans. of ASME, *Jour. of Eng. for Ind.*, Vol.11, 1989, pp. 116-124.

10. H.M. Shi, S.A. Tobias, "Theory of Finite Amplitude Machine Tool Instability," *Int. J. Mach. Tool Des. Res.*, Vol.24, No.1, 1984, pp. 45-69.

11. J. Tlusty, F. Ismail, "Basic Non-Linearity in Machining Chatter," *Annals of CIRP*, Vol.30, No.1, 1981, pp. 299-304.

12. K. Jemielniak, A. Widota, "Numerical Simulation of Non-Linear Chatter Vibration in Turning," *Int. J. Mach. Tools Manufact.*, Vol.29, No.2, 1989, pp. 239-247.

13. F. Jiulian, Y. Zhejun, Y. Yingxue, "A Unified System Model of Cutting Chatter and Its Transformation Function," Vol.29, No.2, 1989, pp. 239-247.

14. K. Srinivasan, C.L. Nachtigal, "Investigation of the Cutting Process Dynamics in Turning Operations," *Journal of Engineering for Industry*, Trans. of ASME, Vol.100, 1978, pp.323-331.

15. H. Takeyama, "Study on Chatter Vibration of Cutting Tool," *Proc. Int. Conf. on Prod. Eng.*, Tokyo, 1974, pp. 185-190.

16. S.A. Tobias, *Machine Tool Vibration*, Blackie and Son, 1965.

17. M.M. Nigm, "A Method for the Analysis of Machine Tool Chatter," *Int. J. MTDR*, Vol. 21, 1981, pp. 251-261.

18. I. Grabec, "Chaotic Dynamics of the Cutting Process," *Int. J. Mach. Tools Manufact.*, Vol. 28, No. 1, 1988, pp. 19-32.

19. S.Y. Tsai, "On-Line Identification and Control of Machining Chatter in Turning Through Dynamic Data Systems Methodology," Ph.D Thesis, University of Wisconsin-Madison, 1983.

20. K.F. Eman, S.M. Wu, "A Feasibility Study of On-Line Identification of Chatter in Turning Operations," *Journal of Eng. for Ind.*, Trans. of ASME, Vol.102, 1980, pp. 315-321.

21. H.Y. Hwang, J.H. Oh, and K.J. Kim, "Modelling and Adaptive Pole Assignment Control in Turning," *Int. J. Mach. Tools Manufact.*, Vol.29, No.2, 1989, pp. 275-285.

22. K. Srinivasan, C.L. Nachtigal, "Analysis and Design of Machine Tool Chatter Control Systems Using the Regeneration Spectrum," *Jour. of Dynamic Systems, Measurement and Control*, Trans. of ASME, Vol.100, 1978, pp. 191-200.

23. C.L. Nachtigal, N.H. Cook, "Active Control of Machine-Tool Chatter," *Journal of Basic Eng.*, Trans. of ASME, June 1970, pp. 238-244.

24. T.R. Comstock, F.S. Tse, J.R. Lemon, "Application of Controlled Mechanical Impedance for Reducing Machine Tool Vibrations," *Journal of Eng. for Ind.*, Trans. of ASME, November, 1969, pp. 1057-1062.

25. E.E. Mitchell, E. Harrison, "Design of a Hardware Observer for Active Machine Tool Control," *Jour. of Dynamic Systems, Measurement and Control*, Trans. of ASME, Dec. 1977, pp. 227-232.

26. V.V. Nemyfskii and V.V. Stepanov, *Qualitative Theory of Differential Equations*, Edt. Marston Morse and A.W. Tucker, *Princeton Mathematical Series*, 1960, pp. 307-420.

27. A. Wolf, J.B. Swift, H.L. Swinney and J.A. Vastano, "Determining Lyapunov Exponents from a Time Series," *Physica 16D*, 1985, pp. 285-317.

28. I. Shimada and T. Nagashima, "A Numerical Approach to Ergodic Problem of Dissipative Dynamical System," *Progress in Theoretical Physics*, Vol. 61, No. 6, June 1979, pp. 1605-1616.

29. J.P. Gollub, E.J. Romer and J.E. Socolar, "Trajectory Divergence for Coupled Relaxation Oscillators: Measurements and Models," *J. Stat. Phys.*, **23**, 1980, pp. 321-333.

30. J.L. Hudson and J.C. Mankin, "Chaos in the Belousov-Zhabotinskii Reaction," *J. Chem. Phys.* **74**, 1981, pp. 6171-6177.

31. H. Nagashima, "Experiment on Chaotic Response of Forced Belousov-Zhabotinskii Reaction," *J. Phys. Soc. Japan*, **51**, 1982, pp. 21-22.

32. A. Wolf and J. Swift, "Progress in Computing Lyapunov Exponents from Experimental Data" in: *Statistical Physics and Chaos in Fusion Plasmas*, C.W. Horton Jr. and L.E. Reichl, eds. (Wiley, New York, 1984).

33. J. Wright "Method for Calculating a Lyapunov Exponent," *Phys. Rev.*, **A29**, 1984, pp. 2923-2927.

34. J.C. Roux, R.H. Simoyi and H.L. Swinney, "Observation of a Strange Attractor," *Physica*, **8D**, 1983, pp. 257-266.

35. N.H. Packard, J.P. Crutchfield, J.D. Farmer and R.S. Shaw, "Geometry from a Time Series," *Phys. Rev.*, **lett - 43**, 1980, pp. 712-716.

36. F. Takens "Detecting Strange Attractors in Turbulence" in lecture notes in *Mathematics*, Vol. 898, D.A. Rand and L.S. Young, eds. (Springer, Berlin, 1981) pp. 366-381.

37. I.N. Tansel and K. F. Eman, "Transfer Function of Cutting Dynamics in Three Dimensional Cutting," *Proc. of the 13th NAMRC*, 1985, pp. 476-481.

38. M. Morf, A. Vieira, D.T. Lee, T. Kailath, "Recursive Multichannel Maximum Entropy Spectral Estimation," *IEEE Transaction on Geoscience Electronics*, Vol. GE-16, No. 2, 1978, pp. 85-94.

39. D.T. Parthimos and K.F. Ehmann, "The Phase-Shift Between Successive Waves in Dynamic Metal Cutting," *Modeling of Machine Tools*, Edited by P.M. Ferreira, S.G. Kapoor, A.C.-Y. Wang, ASME, 1990, pp. 101-120.

40. Y. Kondo, O. Kawano, H. Sato, "Behavior of Self-Excited Chatter Due to Multiple Regenerative Effect," *Jour. of Eng. for Ind.*, Trans. of ASME, Vol. 103, No. 3, 1981, pp. 324-329.

41. J.C. Roux, J.S. Turner, W.D. McCormick, H.L. Swinney, "Experimental Observations of Complex Dynamics in a Chemical Reaction," *Nonlinear Problems: Present and Future*, edited by A.R. Bishop, D.K. Campbell, B. Nicolaenko, North-Holland, Amsterdam, 1982, p. 409.

42. J.S. Turner, J.C. Roux, W.D. McCormick, H.L. Swinney, "Alternating Periodic and Chaotic Regimes in a Chemical Reaction-Experiment and Theory," *Phys. Lett.*, **85A**, 1981, pp. 9-12.

43. C.W. Park, K.F. Eman, S.M. Wu, "Forecasting Compensatory Control (FCC) of Machining Flatness," *Int. J. Mach. Tools Manufact.*, Vol. 28, No.1, 1988, pp. 59-67.

Sensors, Controls, and Quality Issues
in Manufacturing
ASME 1991

THE DIMENSION OF ATTRACTORS ASSOCIATED WITH METAL CUTTING DYNAMICS

B. S. Berger
Department of Mechanical Engineering
University of Maryland
College Park, Maryland

I. Minis
Department of Mechanical Engineering
University of Maryland
Baltimore, Maryland

M. Rokni
Department of Mechanical Engineering
University of Maryland
College Park, Maryland

ABSTRACT

The dynamics of orthogonal turning are investigated through an analysis of tool-workpiece relative acceleration measurements for three post transient cutting cases corresponding to depths of cut $h = 1.0$ mm, 2.2 mm and 2.3 mm. These cases range from static cutting, $h = 1.0$ mm, to mild chatter, $h = 2.3$ mm. The information dimension, $D(0)$, has been determined for these cases. It was found that the dynamics in case $h = 1.0$ mm are dominated by noise. However, cases $h = 2.2$ mm and 2.3 mm are characterized by attractors of low dimension, $2.2 < D(0) < 2.6$. The non-integer values of $D(0)$ for cases $h = 2.2$ mm and 2.3 mm suggest that the cutting dynamics on the attractor may be chaotic. If the box dimension is approximately equal to $D(0)$, then it follows that embeddings for the attractors of cases 2.2 mm and 2.3 mm are of dimension greater than 4.76 and 5.02, respectively. It can be shown that the post transient tool-workpiece relative motions of these cases are described by not more than 5 and 6 dependent variables, respectively.

1 INTRODUCTION

The dynamics of metal cutting affect significantly both the quality of machined parts and the productivity of machining operations. The relative oscillations between the tool and the workpiece, which are induced by the dynamic cutting force, are detrimental to part dimensional accuracy and surface finish. Furthermore, machine tool chatter, i.e. self-excited oscillations that develop under high values of the depth of cut, limit the productivity of machining, promote tool wear and may lead to tool fracture and damage of the workpiece or the machine tool.

The desire to design efficient machine tools that produce parts of high quality and require minimum human intervention has prompted numerous research efforts towards understanding the dynamics of the machining system, which includes the structural dynamics of the machine tool coupled to the dynamics of the cutting process. Linear mathematical models have traditionally been proposed to describe these major sub-systems. However, there is strong experimental evidence that both the structural and process dynamics contain non-linearities.

Although the structure of the machine tool is a complex distributed parameter system, in metal cutting research it is sufficient to consider the relationship between the cutting force and the relative tool-workpiece displacement. Many investigators have assumed this relationship to be linear and examined the frequency response of the structure-workpiece system [1,2,3]. Other studies have introduced non-linear stiffness terms to the structural models to account for intrinsic non-linearities [4,5]. The dynamics of the cutting process have been described by experimentally obtained linear models that contain time delays [6]. On the other hand, attempts to derive the governing equations of cutting analytically usually result in non-linear cutting equations [7,8,9]. Recently, non-linear cutting equations were combined with linear structural models to show that the coupled machining system may exhibit chaotic oscillations [10,11]. A computational study based on a non-linear model of cutting dynamics was conducted by the authors of the present paper and showed that the information dimension of the machining system assumes non-integer values [17] (see also Appendix A).

The fact that the machining system is both non-linear and infinite dimensional suggests complex dynamical behavior, which may not be adequately described by conventional linear models. Measurements of the tool-workpiece accelerations during orthogonal turning of mild steel workpieces were analyzed. The depth of cut, h, was varied while all other parameters were held constant. The information dimension, $D(0)$, was found for data sets corresponding to h = 1.0 mm, 2.2 mm and 2.3 mm, which are referred to as cases 1.0, 2.2 and 2.3, respectively. Case 1.0 may be described as static cutting, while case 2.3 may be described as mild chatter. All three cutting states studied are post transient. The measurements of case 1.0 were found to be dominated by noise. However, both cases 2.2 and 2.3 are characterized by low dimensional attractors. For these cases $D(0)$ is a non-integer suggesting that the cutting dynamics on the attractor may be chaotic. These results are of both theoretical and practical significance. They reveal attributes of metal cutting dynamics not determined previously and provide an upper bound on the number of dependent variables required to determine the post transient motion. This information is of use in modeling cutting dynamics. In addition they may eventually provide a means for the characterization of the machining state and detection of deleterious cutting conditions.

The paper is structured as follows. Section 2 describes the experimental apparatus and the procedures followed during the cutting experiments. Section 3 outlines the theoretical background upon which the data analysis is based. Section 4 presents the results of the data analysis and discusses their significance. The conclusions of the study are summarized in Section 5.

2 EXPERIMENTAL METHODOLOGY

This investigation of machining dynamics was exclusively based on the analysis of data obtained from orthogonal turning experiments. The single-point turning process was selected as the most fundamental, yet simple, machining operation. The experimental set-up, shown in Fig. 1, is comprised of a CNC lathe, a special force dynamometer and its associated electronics, accelerometers and electronics, and a data acquisition system that provides real time analysis capabilities.

The lathe (Hardinge Superslant AB) is driven by a 10 hp motor and provides maximum travel of 419 mm (16.5 in) and 120 mm (4.75 in) along the Z and X axes, respectively. It is equipped with a custom force dynamometer that measures all three components of the dynamic cutting force. The dynamics of the lathe's structure have been thoroughly studied (see [12]). It has been found that the mode of oscillation which dominates the chatter phenomenon has a natural frequency of approximately 112 Hz.

The three components of the tool-workpiece relative acceleration were measured by two sets of three accelerometers each. The first set was mounted on the cutting tool as closely to the cutting edge as possible (see Fig. 1). The second set was mounted on the cutting center of the lathe, which during cutting was pressed against the workpiece with a predetermined load [4540 N (1000 lb)]. It has been found that this arrangement provides reliable acceleration

Fig. 1 The experimental set-up. (Reprinted from [12]).

data for the stubby workpieces used in this study [12]. The two sets of acceleration measurements were subtracted from each other by an analog circuit to obtain the relative tool-workpiece acceleration. The time wave forms of the appropriate force and/or acceleration signals were recorded by a spectrum analyzer (HP 3562A) operated in the time capture mode.

Right-handed turning of solid cylindrical workpieces was performed using a positive rake 0.5 in. tool shank (Kennametal), which provided $5°$ rake angle and $0°$ lead angle. The tool holder carried Kennametal TPG inserts with nominally zero nose radius and a clearance angle of $6°$. Special care was taken to keep the cutting conditions as uniform as possible throughout the experiments. Consequently, all workpieces were cut from the same bar stock of mild steel AISI 1018. Furthermore, the width of the tool's flank wear zone was kept less than 0.1 mm. The feedrate of the tool and the surface speed of the workpiece were 0.127 mm/rev (0.005 in/rev) and 90 m/min (295.3 ft/min), respectively. Preliminary investigations had shown that the right-handed configuration at this particular surface speed yielded the lowest value of the critical depth of cut for chatter-free turning.

The sole parameter varied during the experimental investigation was the depth of cut h, which greatly affects the dynamics of machining. Cutting experiments were conducted starting from h = 1 mm (0.039 in.), increasing up to $h = 3.5$ mm (0.138 in.) in steps of 0.1 mm (0.0039 in.). The initial and final value of h corresponded to static cutting and mild chatter, respectively. At each value of h the measurements were obtained after the transient had settled, since the present analysis is only concerned with post transient cutting cases. The time wave forms of the acceleration component that is normal to the surface being cut (opposite to the feed direction) as well as the corresponding thrust force component were filtered, sampled and recorded by the analyzer. The sampling frequency was 1536 Hz, which corresponded to a cutoff frequency for the anti-aliasing filters of 600 Hz. For each signal up to 143,360 samples were recorded during a measurement time of up to 93.33 sec. The raw data from the entire set of experiments were transferred from the mass storage of the analyzer to a workstation (DECstation 3100) for analysis. The computation of $D(0)$ for a data set of n =143,360 points with m =3000 and $0 < E \leq 21$ (see Section 3) required 14 hours of DEC3100 CPU time. Both the force and acceleration measurements exhibit the same characteristics and, therefore, only the latter are discussed here.

3 DIMENSION FUNCTION

The qualitative features of the time evolution of many physical systems are identical to those of the solution of a typical differential equation of the form

$$\frac{d\vec{x}(t)}{dt} = \vec{f}(\vec{x}(t), \vec{\mu}) \tag{1}$$

where $\vec{\mu}$ is a vector of experimental parameters. In a phase space, E_n, the n components of $\vec{x}$ are taken as coordinates of points in an n-dimensional Euclidian space. For a dissipative system volume elements of E_n contract under the action of (1) and there exists a set of points, U, of E_n, initial conditions, which asymptotically evolve to a compact set of points, A, called an attractor.

Dimension is a measure of the scaling properties on the attractor as the observational resolution is increased [14,15]. In [14] a dimension function, $D(\gamma)$, is defined as a function of the Renyi dimensions in terms of suitable entropies. For applications $D(\gamma)$ may be expressed as [16],

$$D(\gamma) = -\lim_{n\to\infty} \log_b(n/k) / \left[(1/\gamma)\log_b(1/m) \sum_{j=1}^{m} \delta_j^{\gamma}(k, E, n) \right] \tag{2}$$

where m =number of reference points $\vec{y}_j$, $j = 1, \ldots, m$ on the attractor, E = dimension of the delay-coordinate space, $\gamma = (1-q)D_q$, where D_q = Renyi dimension [14], $\delta_j^{\gamma}(k, E, n)$ = distance between the reference points $\vec{y}_j$ and the k-th nearest neighbors for an E dimensional delay-coordinate space and n =number of generic points of the attractor. $D(0)$ = information dimension [16].

A relationship between $D(0)$, and the Lyapunov exponents has been conjectured in, [15],

$$D(0) = K - \left(\sum_{i=1}^{K} \lambda_i \right) / \lambda_{K+1} \tag{3}$$

where K is the largest value for which $\sum_{i=1}^{K} \lambda_i \geq 0$, [18]. For dissipative systems for which the conjecture is valid it follows that a non-integer value of $D(0)$ implies the existence of positive Lyapunov exponents and thus chaotic dynamics. Of perhaps more practical importance are the dimension function properties for characterizing the structure of experimental time series and acting as a guideline in the construction of mathematical models.

The continuous mass distribution of the tool-workpiece system together with the time delays introduced by regenerative cutting indicate the possibility of a system of infinite dimension. Computational studies based on a theoretical cutting model have indicated that this system has a low dimensional attractor (see Appendix A). For post transient motion the number of dependent variables needed to determine a trajectory on the attractor and thus model the system is equal to the embedding dimension, [19]. It is shown in [20] that an embedding is given by any $E > 2\times$ the box dimension of the attractor. The embedding dimension is defined as the minimum integer value of E satisfying the inequality.

The determination of the box dimension is in most instances computationally impractical requiring very large data sets and excessive CPU time, [21]. Instead, a more accessible dimension, $D(0)$, is studied. For systems of ordinary differential equations $D(0) \cong$ box dimension, [19,20].

4 $D(\gamma)$ FROM CUTTING DATA

$D(0)$ was computed for three sets of measurements of the relative tool-workpiece acceleration, $\ddot{x}(t)$, which is normal to the surface being cut (opposite to the feed direction). These sets were identified by visual and auditory observations of the machine operator as no-chatter, pre-chatter and mild chatter states and corresponded to values of the depth of cut $h = 1.0$ mm (0.039 in.), 2.0 mm (0.079 in.) and 2.3 mm (0.091 in.), respectively. For all cases the determination of $D(0)$ from (2) used reconstruction of the attractor based on delaying $\ddot{x}(t)$ by $i = 1, \ldots, E$ multiples of Δt, $\ddot{x}_i(t) = \ddot{x}(t + (i-1)\Delta t)$. The embedding dimension associated with any of the cases 1.0 through 2.3 is unknown. The evaluation of $D(0)$ from (2) was therefore carried out for a sequence of delay-coordinate spaces, $E = 1, 2, \ldots, 21$ with all other parameters held constant.

In the absence of noise and for sufficiently large values of m and n, $D(0)$ for an existing attractor would be approached by $D(0)$ vs. E, [13,14,15,16]. If high levels of noise are present then $D(0) \cong E$ for a range of small values of E depending on n. $D(0)$ increases with E for larger values of E and the slope of $D(0)$ vs. E is monotonically decreasing, Figs. 2 and 8. The presence of low levels of noise on a low dimensional attractor typically introduces a plateau or reduction in the slope of $D(0)$ vs. E, which is then followed by an interval of increased slope, Figs. 4 and 6. An estimate of $D(0)$ is furnished by the range of values of $D(0)$ associated with the plateau.

For case 1.0, $D(0)$ is shown in Fig. 2 as a function of E with $m = 3000$ and $n = 100{,}000$. The upper, middle and lower curves correspond to $k = 100$, 200 and 300, respectively. For noise $D(0) \cong E$. This relationship is approximately satisfied for $0 < E \leq 4$. Accurate determination of $D(0) > 4$ requires larger data sets, [16]. The slope of $D(0)$ vs. E decreases monotonically with E and no plateau exists. This overall behavior is characteristic of a noise dominated system, [16]. The dominant presence of noise is also evident from the time wave form of $\ddot{x}(t)$ shown in Fig. 3. Case 1.0 is typical of static cutting, in which the amplitude of the tool-workpiece acceleration (0.75 m/sec^2), and, consequently, the amplitude of the corresponding displacement, are low.

$D(0)$ is shown in Fig. 4 for the case 2.2 data set with $m = 3000$ and $n = 100{,}000$. For $E > 8$, $D(0)$ increases with E. This behavior is characteristic of noise. However, for $2 < E < 8$ a plateau emerges for which $D(0) = 2.38 \pm 0.10$. This estimate is the average and mean deviation of $D(0)$ for E = 4, 5, 6 and k = 100, 200, 300. For noise-free data the curves corresponding to k = 100, 200, 300 converge to the same value for large E. The presence of noise prevents the realization of the limit, [16]. Some structure is apparent in the time wave form of $\ddot{x}(t)$ shown in Fig. 5. In this case the amplitude of the tool-workpiece acceleration has increased to approximately 1.6 m/sec^2.

For case 2.3, $D(0)$ is shown in Fig. 6 as a function of E = with $m = 3000$, $n = 100{,}000$ and k = 100, 200 and 300. For $E > 0$ the behavior of $D(0)$ is characteristic of noise. However, for $2 < E < 8$ a plateau exists for which $D(0) = 2.51 \pm 0.08$. This estimate is the average and mean deviation of $D(0)$ for E = 4, 5, 6 and k = 100, 200, 300. Case 2.3 exhibits the typical wave form pattern of mild chatter (see Fig. 7). The maximum amplitude of $\ddot{x}(t)$ is approximately 5 m/sec^2. It is noted that cutting in this case was accompanied by the characteristic sound emitted during machine tool chatter.

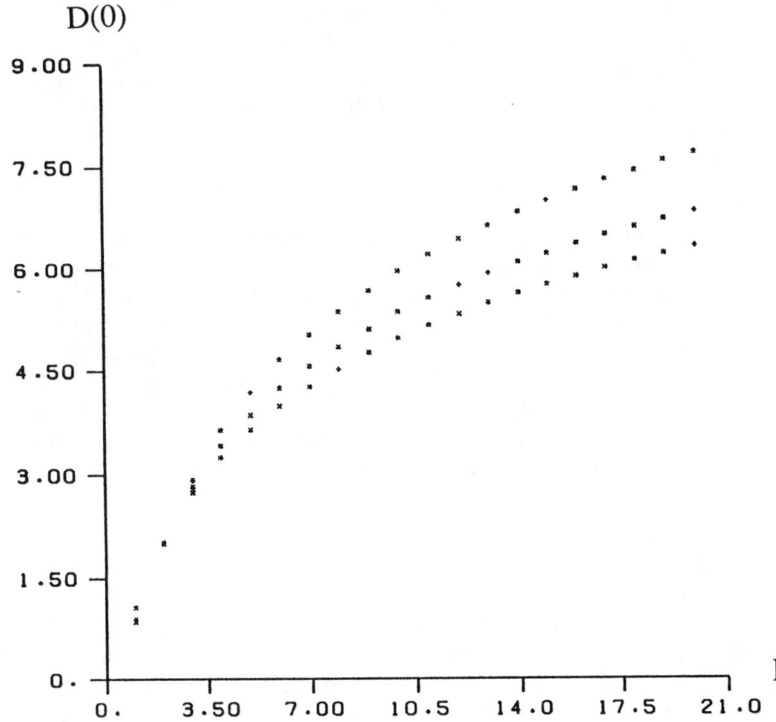

Fig. 2 $D(0)$ vs E for $h = 1.0$ mm. $m = 3000$, $n = 100{,}000$ and $k = 100$ (upper curve), 200 (middle curve) and 300 (lower curve).

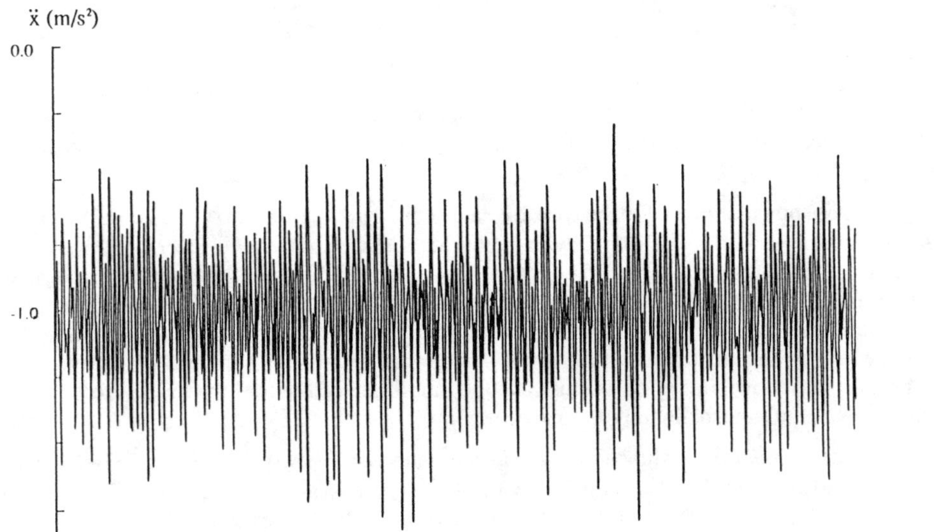

Fig. 3 $\ddot{x}(t)$ for $h = 1.0$ mm.

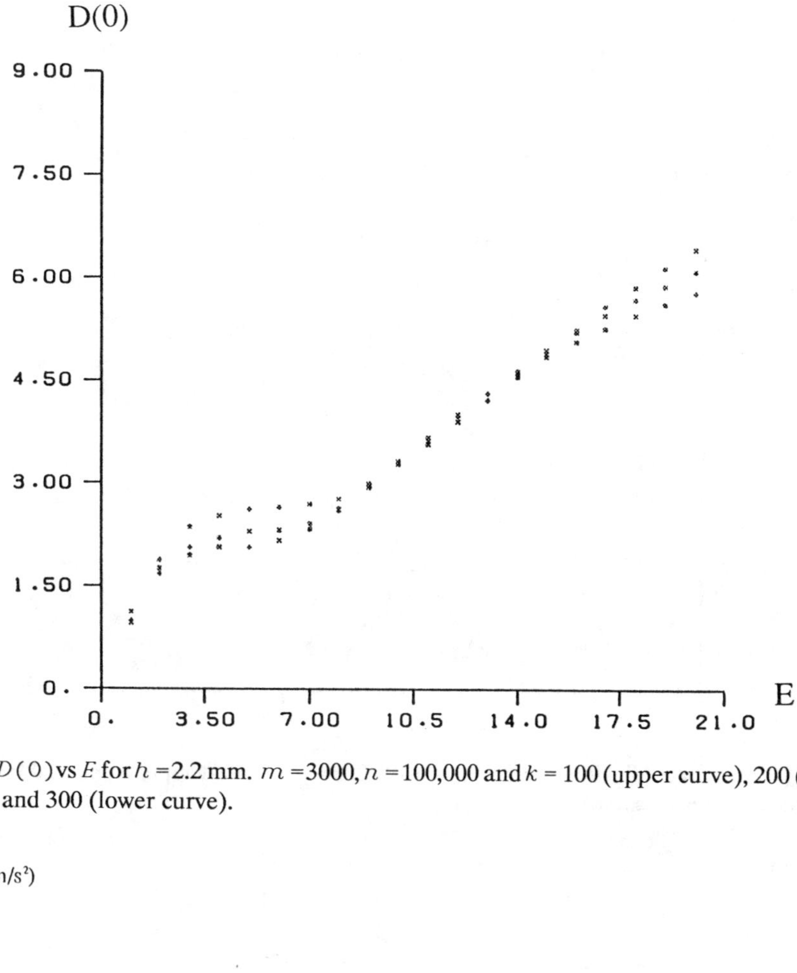

Fig. 4 $D(0)$ vs E for h =2.2 mm. m =3000, n =100,000 and k = 100 (upper curve), 200 (middle curve) and 300 (lower curve).

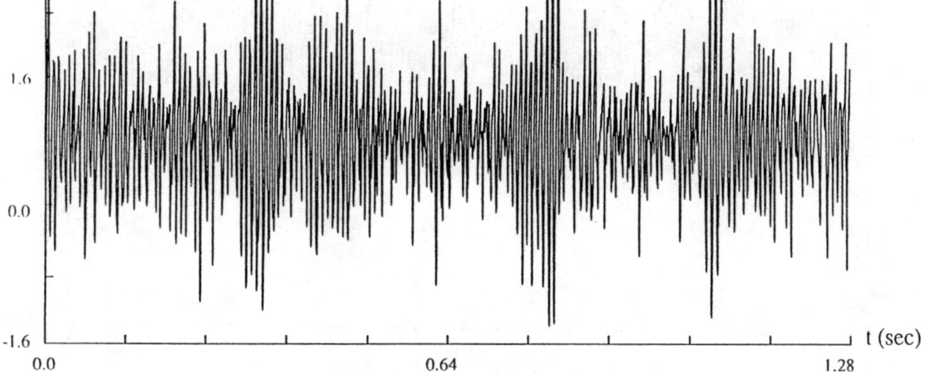

Fig. 5 $\ddot{x}(t)$ for h =2.2 mm.

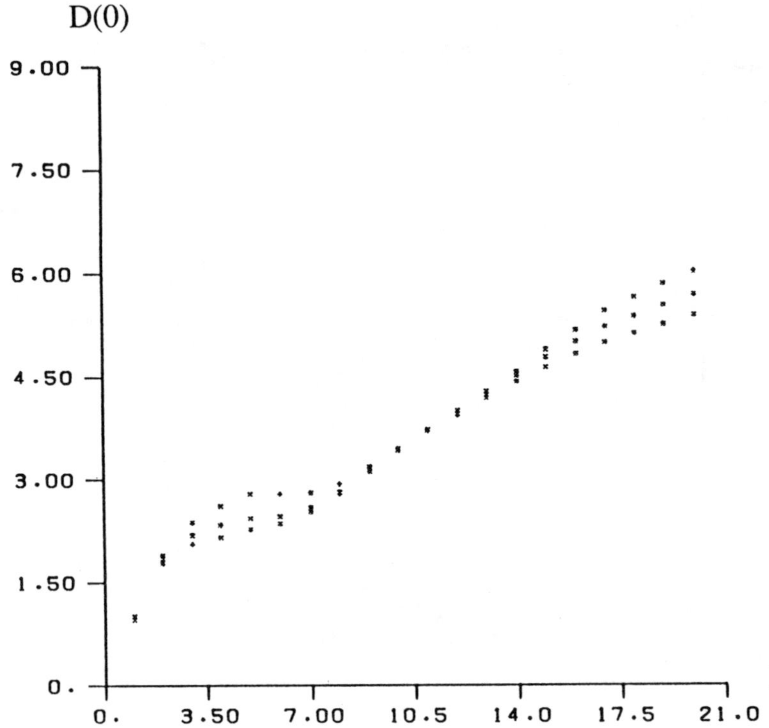

Fig. 6 $D(0)$ vs E for h =2.3 mm. m =3000, n =100,000 and k = 100 (upper curve), 200 (middle curve) and 300 (lower curve).

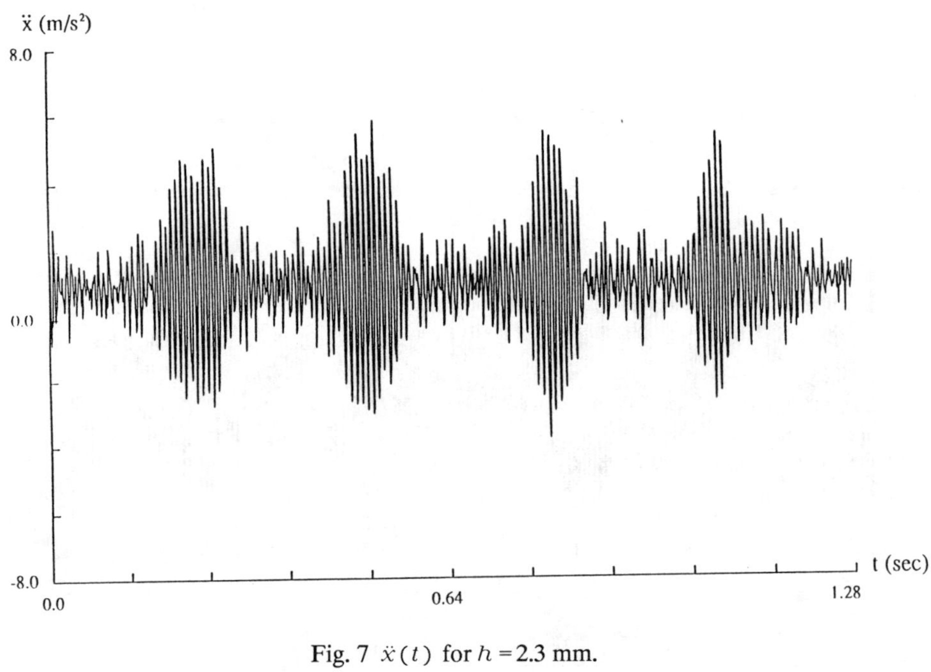

Fig. 7 $\ddot{x}(t)$ for h =2.3 mm.

$D(0)$ is shown in Fig. 8 for case 1.0 with m =3000, n = 143,360 and k = 10, 20, ..., 300. The characteristics of noise dominance are evident over the entire range of E and k. This result is consistent with the determination of $D(0)$ vs. E shown in Fig. 2.

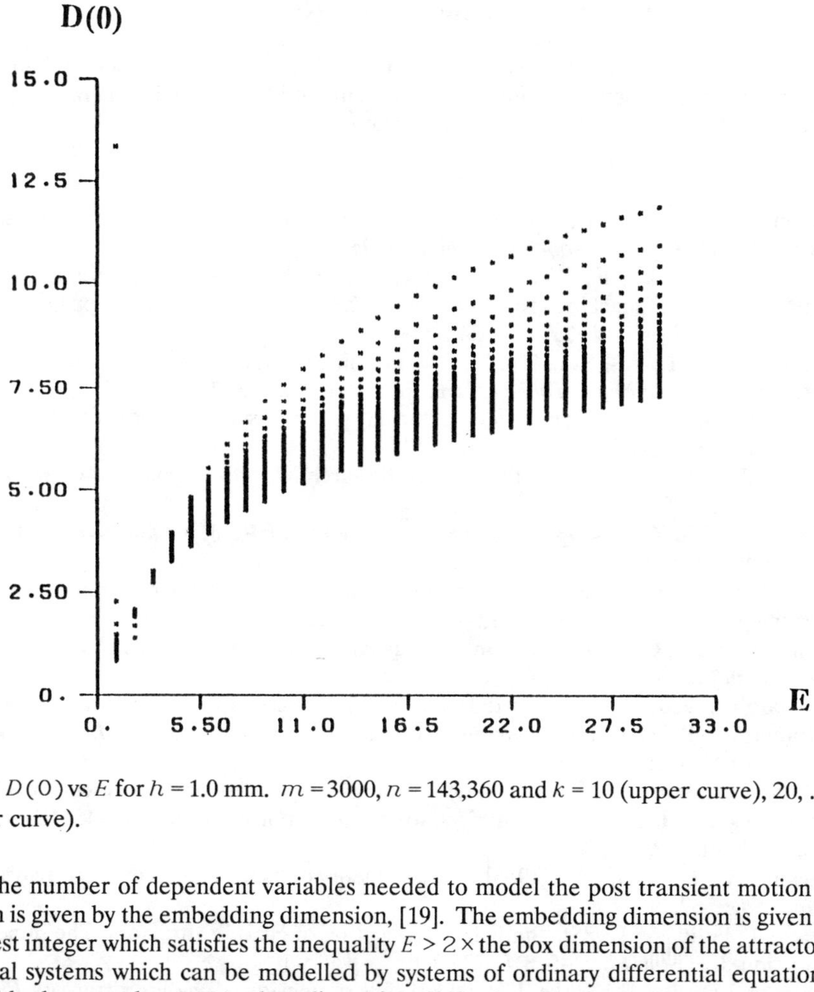

Fig. 8 $D(0)$ vs E for h = 1.0 mm. m =3000, n = 143,360 and k = 10 (upper curve), 20, ... , 300 (lower curve).

The number of dependent variables needed to model the post transient motion of the system is given by the embedding dimension, [19]. The embedding dimension is given by the smallest integer which satisfies the inequality $E > 2 \times$ the box dimension of the attractor. For physical systems which can be modelled by systems of ordinary differential equations it is probably the case that $D(0) \cong$ box dimension, [20]. Apparently no rigorous mathematical proof of this approximation is known. For cases 2.2 and 2.3 the corresponding embedding dimensions are 5 and 6, respectively.

5 CONCLUSIONS

The information dimension, $D(0)$, has been found for three data sets corresponding to depths of cut h = 1.0 mm, 2.2 mm and 2.3 mm. These cases range from static cutting, h = 1.0 mm, to mild chatter, h =2.3 mm. The 1.0 case has been shown to be dominated by noise which completely obscured any possibly existing low dimensional attractor. Attractors for which $D(0) \cong 2.38$ and 2.51 were found to be associated with cases 2.2 and 2.3, respectively. The existence of low dimensional attractors for orthogonal turning has been demonstrated from an analysis of experimental data. The fact that $D(0)$ is fractional suggests that the dynamics on the attractor are chaotic for cases 2.2 and 2.3. Implications for modelling of the metal cutting process have been noted.

ACKNOWLEDGEMENT

The authors wish to acknowledge the support of the Computer Science Center, University of Maryland and the computer facility of the Department of Mechanical Engineering. The assistance of C.J. Berger and Dr. G. Broggi is greatly appreciated.

REFERENCES

1. Koenigsberger, I., Tlusty, J., [1971]. *Structures of Machine Tools,* Pergamon Press.
2. Weck, M., [1985]. *Handbook of Machine Tools, Vol. 4,* Wiley.
3. Kim, K.J., Eman, E.F., Wu, S.M., [1984]. Identification of natural frequencies and damping ratios of machine tool structures by the dynamic data system approach, *Int. J. of Mach. Tool Des. and Res.,* Vol. 24, 161-170.
4. Hanna, N.H., Tobias, S.A., [1969]. The non-linear dynamic behaviour of a machine structure, *Int. J. of Mach. Tool Des. and Res.,* Vol. 9, 293-307.
5. Klamecki, B.E. [1989]. On the effects of turning process asymmetry on process dynamics, *ASME J. of Eng. for Ind.,* Vol. 111, 193-198.
6. Tlusty, J., [1978]. Analysis of the state of research in cutting dynamics, *CIRP Annals,* Vol. 27, 583-589.
7. Wu, D.W., Liu, C.R., [1985]. An analytical model of cutting dynamics. Parts 1 and 2, *ASME J. of Eng. for Ind.,* Vol. 107, 107-118.
8. Wu, D.W., [1989]. A new approach to formulating the transfer function for dynamic cutting processes, *ASME J. of Eng. for Ind.*, Vol. 111, 37-47.
9. Lin, J.S., Weng, C.I. [1990]. A nonlinear dynamic model of cutting. *Int. J. Mach. Tools and Manufact.*, Vol. 30, 53-64.
10. Grabec, I., [1986]. Chaos generated by the cutting process, *Phys. Lett.,* Vol. 117, 384-386.
11. Grabec, I., [1988]. Chaotic dynamics of the cutting process, *Int. J. Mach. Tools and Manufact.*, Vol. 28, 275-280.
12. Minis I., Magrab, E., Pandelidis, I. [1990]. Improved methods for the prediction of chatter in turning. Part I: Determination of the structural response parameters, *ASME J. of Eng. for Ind.,* Vol. 112, 12-20.
13. Eckmann, J.P., Ruelle, D., [1985]. Ergodic theory of chaos and strange attractors, *Rev. Mod. Phys.*, 57, 3.
14. Badii, P, Politi, A., [1985]. Statistical description of chaotic attractors: The dimension function, *J. of Stat. Phys.* 40, 516, 725-750.
15. Farmer, J.D., Ott, E., Yorke, J.A., [1983]. The dimension of chaotic attractors, *Physica,* 7D, 153-180.
16. Broggi, G., [1988]. Evaluation of dimensions and entropies of chaotic systems, *J. Opt. Soc. of America*, B, 5, 5, 1020-1028.
17. Berger, B.S., Rokni, M., Minis, I. [1991]. Complex dynamics of metal cutting, accepted for publication, *Quarterly of Applied Math., American Math. Society*.
18. Lichtenberg, A.G., Liberman, M.A. [1983]. *Regular and Stochastic Motion*, Springer Verlag.
19. Farmer, J.D. [1982]. Chaotic attractors of an infinite-dimensional dynamical system, *Physica* 4D, 366-393.
20. Sauer, T., Yorke, J.A., [1991]. Embedology, *DARPA Report*, Dept. of Math. Sciences, George Mason Univ., Fairfax, VA 22030.
21. Greenside, H.S. [1982]. Impracticality of a box-counting algorithm for calculating the dimensionality of strange attractors, *Physical Review* A, 25, 6, 3453-3456.

APPENDIX A
Computation of $D(0)$ using the Wu, Liu Cutting Model

A non-linear mathematical model of cutting dynamics, proposed by Wu and Liu in [7], was used in reference [17] to investigate the non-linear dynamics of the machining system. Considering orthogonal cutting and a two degree-of-freedom machine tool structure, the system equations of motion given in [12] are:

$$m_x\ddot{x} + b_x\dot{x} + k_x = 2h\tau(x_o - x)\left[(A_x - C_x v_0) + \frac{B_x}{2}(\dot{x} - \dot{x}_o) - \frac{C_x}{2}(\dot{y} - \dot{y}_o)\right] - \frac{kh}{v_0}\dot{x} \quad (A-1)$$

$$m_y\ddot{y} + b_y\dot{y} + k_y = 2h\tau(x_o - x)\left[(A_y - C_y v_0) + \frac{B_y}{2}(\dot{x} - \dot{x}_o) - \frac{C_y}{2}(\dot{y} - \dot{y}_o)\right] \quad (A-2)$$

where x and y represent the displacement components of the tool along the directions X and Y that are opposite to the feed and surface speed, respectively, m_x, b_x, k_x, m_y, b_y and k_y are the modal parameters of the two modes under consideration, h is the depth of cut, τ is the shear strength of the workpiece material, v_0 is the nominal surface speed, $A_x, B_x, C_x, A_y, B_y, C_y$ and k are constants that depend on the nominal cutting conditions. The study in [17] used $x_o(t) = x(t-T)$ and $y_o(t) = y(t-T)$, where T is the period of the workpiece revolution. It is noted that the non-linearity of the cutting force terms in Eqs. (A-1) and (A-2) is due to the dependence of the shear angle of cutting on the instantaneous speed of the tool $(\dot{x}, \dot{y})$ [7].

The values of the modal parameters in the computational study were experimentally determined in [12] for the two dominant modes of the machine tool used in the experiments described in Section 2. The values of the cutting process parameters, however, were taken from reference [7] and correspond to low carbon steel ASE 1020. Equations (A.1) and (A.2) were numerically integrated to yield long trajectories of the tool motion. Phase plane plots and Poincare sections revealed complex dynamics. The information dimension of the attractor was computed to be $D(0) \cong 2.7$. This value is close to those obtained from the experimental cutting data used in the present paper (see Section 4).

Sensors, Controls, and Quality Issues
in Manufacturing
ASME 1991

A HIDDEN MARKOV MODEL APPROACH TO THE STUDY OF RANDOM TOOL MOTION DURING MACHINING

Guanming Zhang and Chien Lin
Department of Mechanical Engineering and Systems Research Center
University of Maryland
College Park, Maryland

ABSTRACT This paper presents a new approach to the study of random tool motion during machining. Theory of the hidden Markov model is applied to formulate a comprehensive random excitation system present during machining. Based on the microstructural analysis, characteristics of the hardness distribution in the material being machined are identified for analyzing the cutting dynamics in microscale. The machining action within one revolution of the workpiece and the relation between the machining actions in consecutive revolutions are interpreted as a double stochastic process. Computer simulation based on the hidden Markov model approach is used to predict values of surface roughness characterization indices under given machining conditions. The predictions are compared with the data obtained from direct measurements, showing good agreements. The developed approach has brought new light on a better understanding of vibration control during machining.

1. INTRODUCTION

Today's technological innovations in manufacturing are driven by demands to maintain a consistent, high level of product quality in the modern manufacturing environment. Quality of manufactured products such as functionality and reliability depends on finish quality of machined surfaces. Manufacturing engineers are now faced with the difficult problem of improving finish quality without compromising productivity. As a result, research on the control of finish quality of machined surfaces has attracted great attention.

It has been well known that the roughness profile of a machined surface contains periodic and nonperiodic components. The geometry of machining action is responsible for the generation of the periodic components. During a turning operation, the spindle speed, machining feed and tool geometry determine a spiral tool path. However, random tool motion during machining aggravates the surface texture formation. Factors, which cause the tool vibration in a random manner, could be built-up edges, tool wear, and nonhomogeneity of basic material properties. Evaluation and control of tool vibration, for both deterministic and stochastic parts, have been the focus for improving finish quality of machined surfaces.

Results from previous research have strongly suggested the use of precise spindle and feed mechanisms for machining error reduction (Blaedel, 1980 and Kohno et al., 1989). Work on machine tool accuracy enhancement has been successfully improving the rigidity of machine tool structures or/and compensating the deterministic part of tool path errors during machining (Donmez, 1986, Tlusty and Ismail, 1983, Tobias, 1961). Study on measurement and assessment of topography of machined surfaces has been extensive (Nayak, 1973 and Zhou et al., 1990). Methods like wavelength decomposition of surface roughness (Pandit and Revach, 1981), time series modeling of machined surfaces (Kapoor and Wu, 1978), and surface models from light scattering measurements (Vorburger and Scace, 1990) were developed for establishing the relationship between surface parameters and pertinent machining parameters. A recent study indicated that the nonhomogeneity of basic material properties, such as hardness, could be one of the major sources to excite random tool motion during machining (Zhang and Kapoor, 1991). Research has been conducted to establish a mapping function between the variability of standard material properties and the formation of surface irregularities.

The work presented in this paper represents a continuation of the research effort for evaluating random tool motion during machining. A statistic method of hidden Markov modeling is applied to describe the random excitation phenomenon. The developed model employs two stochastic processes to formulate the mechanism of random excitation. The first stochastic process characterizes the random tool motion within a single revolution of the workpiece. The second stochastic process quantifies the interrelationship of random tool motion between the consecutive revolutions of the workpiece. Analysis of the cutting dynamics in microscale under the formulated random excitation provides rich information about possible strategies for controlling the quality of machined surfaces.

The paper is organized as follows. Section 2 presents the modeling process of random excitation. The nonhomogeneity of hardness distribution in the workpiece material is assumed to be a major random excitation source. Concepts of the segment excitation and overall excitation are defined to form a basis for the development of a hidden Markov model to describe the random excitation observed during machining. Section 3 presents a technical detail about determination of the states and state transition matrix. Computer implementation is also discussed in this section. A case study is presented in section 4 to demonstrate the procedure to apply the developed hidden Markov model for the study of random tool motion during machining. Discussion of results is given in section 5 where results from the experimental work are compared with the predictions based on the hidden Markov modeling approach. Effects of the state selection and state transition matrix on random tool motion are investigated through a two-level factorial design. Section 5 contains concluding remarks.

2. BASIC METHODOLOGY

Random tool motion, a phenomenon observed in the production line, has been blamed as a major element to deteriorate product quality. Research on a quantitative identification of the random excitation source still remains as a challenge to the manufacturing community. In this research, we assume that the nonhomogeniety of basic material properties, such as material hardness, is of primary importance to cause random tool motion during machining.

2.1 Random Tool Motion

Figure 1 presents an intuitive view to explain the mechanism that a nonhomogeneous distribution of the material being machined could excite random tool motion during machining. The individual blocks in Fig. 1 represent volumes of the material being removed at time intervals Δt. From the machining point of view, the cutting force generated at these time intervals is related to the hardness of the volume material. It is evident that the cutting force varies as the hardness value of an individual block varies. The variation of the cutting force, which is of random nature, introduces random tool vibration during machining. It becomes necessary to characterize the hardness distribution of material for the evaluation of random tool motion.

2.2 Microstructural Analysis

It has been known that the hardness distribution of material in microscale is characterized by the size, shape, and segregation of the microstructures present in material. Figures 2a and 2b are photographs of the microstructures in an AISI 1020 carbon steel bar. Figure. 2a illustrates the microstructural distribution observed in a cross-section perpendicular to the bar axis, and Fig. 2b the microstructual distribution observed in the cross-section along the bar axis. The dark and white parts stand for pearlite and ferrite

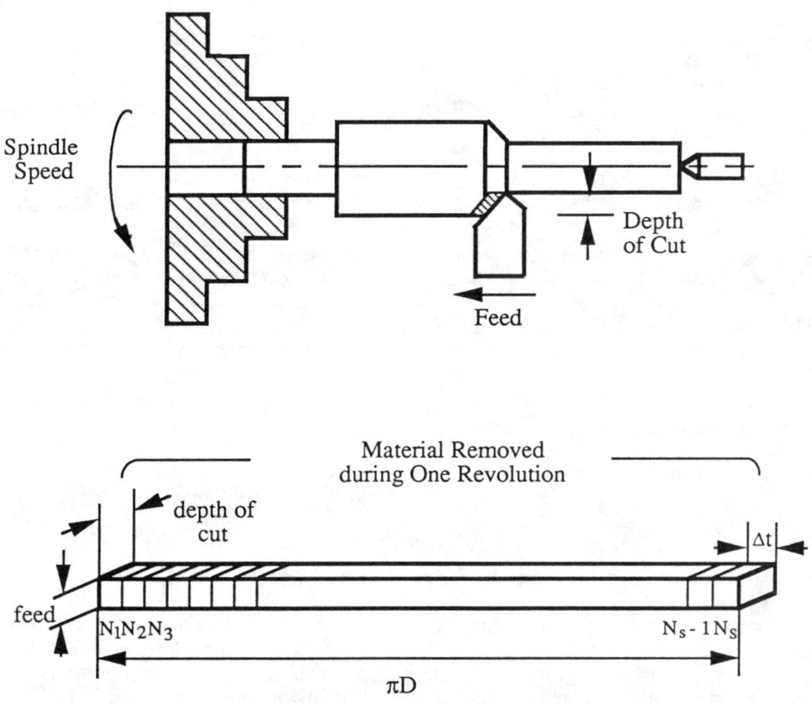

Figure 1 Mechanism of Random Excitation

structures, respectively. Comparing the microstructures displayed in Fig. 2a and Fig. 2b, the sizes of pearlite structure shown in Fig. 2a are relatively small and their shapes look irregular. On the other hand, the pearlite structures in Fig. 2b are shown as narrow strips. These narrow strips are distributed in parallel, depicting a distribution pattern. This narrow and parallel strip pattern results from the rolling process when the AISI 1020 carbon steel was fabricated. Due to the fact that the hardness of pearlite structure is much higher than the hardness of ferrite structure, patterns of a nonhomogeneous distribution of microstructures lead to patterns of a nonhomogeneous distributions of hardness correspondingly. During the machining of carbon steel materials, the presence of a nonhomogeneous distribution of hardness gives rise to variation of the cutting force, which causes the tool to randomly vibrate (Rabiner and Juang, 1986).

2.3 Hidden Markov Modeling

Theoretically speaking, a hidden Markov model is a doubly stochastic process with an underlying stochastic process that is not observable (it is hidden), but can only be observed through another set of stochastic processes that produce the sequence of observed events.

The strategy of modeling random excitation is to view the random excitation during the machining of carbon steel materials as a direct concatenation of "short time" random excitation segments. Each of "short time" random excitation segments represents the random excitation action during one revolution of the workpiece. The statistical model for the segment excitation is subjected to the hardness distribution observed in the cross-section of the material being machined, i.e., the one shown in Fig. 2a. The overall excitation action is a synchronous sequence of these segments. Therefore, it can be described by a hidden Markov model. The statistical model to represent the transition process from one segment to the consecutive segment is subjected to the hardness distribution in the machining feed direction, i.e., the one shown in Fig. 2b. As a result, the random excitation during the entire machining process is considered to be a doubly embedded stochastic process.

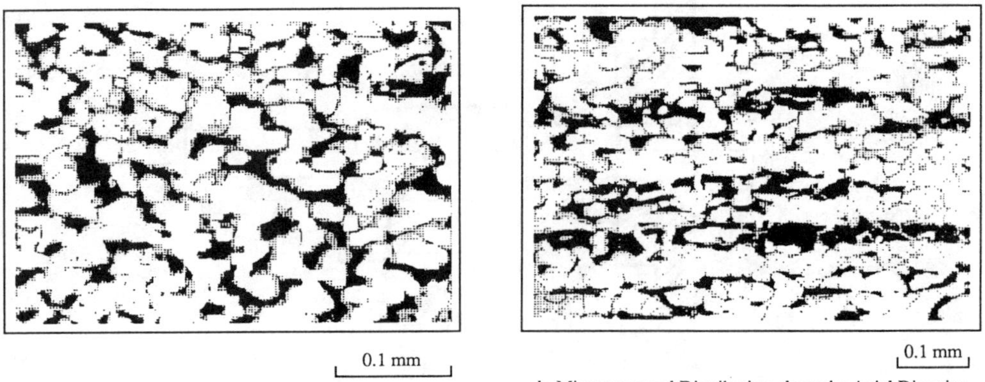

a. Microstructural Distribution in the Cross-Section

b. Microstructural Distribution along the Axial Direction

Figure 2 Microstructural Analysis

2.3.1 Modeling of Segment Excitation

Imagining that the machining process is to take individual blocks away from the workpiece material, the ratio of pearlite structure to ferrite structure within the block being removed varies from one block to another. Meanwhile, the cutting force follows the ratio variation to vary, forming a microscale random excitation.

A statistical model to describe this excitation action was developed and reported in (Zhang and Kapoor, 1991). Development of the reported statistical model was based on an assumption, i.e., the excitation action at a specific machining instant was determined by the mean hardness value of the block being removed. Due to the nonhomogeneous hardness distribution, the parameter of block mean hardness was treated as a random variable, which was assumed to obey a normal distribution because of the central limit theorem in statistics. The mean and variance of the normal distribution for modeling the variation of mean hardness values were calculated through microstructural analysis, as briefly discussed in section 2.2.

In this research, this reported statistical model is used to characterize the segment excitation action, or the excitation action within one revolution of workpiece. For given mean and variance values of the statistical model and the number of blocks divided along the workpiece circumference, one can construct a data set, using a random number generator on a computer, to represent the block mean hardness variation along the workpiece circumference, or within one revolution of the workpiece.

2.3.2 Modeling of Overall Excitation

As discussed above, the segment excitation model characterizes the effect of the nonhomogeneous distribution of hardness on the variation of the cutting force during one revolution of the workpiece. If examining the distribution pattern shown in Fig. 2b, which is viewed along the machining feed direction, the concurrence of different microstructures appearing in the neighboring areas, or in the consecutive revolutions of the workpiece, can be observed. This concurrence is directly related to the narrow strips of pearlite and characterized by

1. the presence of a hard (or soft) spot on a specific circumferential location is correlated with the presence of a hard (or soft) spot on its consecutive, or neighboring locations, and
2. in a general case, the changing process, which is from a hard (or soft) spot to the soft (or the hard) spot is gradually proceeding, although the abrupt jumps between an extreme hard and an extreme soft spots, or vice versa, occasionally happen.

To further clarify the two observations, Fig. 3 illustrates a state-transition representation. The entire area of the microstructural photograph shown in Fig. 2b was divided into 12 equal gridblocks. Each gridblock represented an area of 0.14 x 0.14 mm^2. Using electronic scanning, the numbers of pearlite and ferrite within each of the 12 gridblocks were identified and listed in the corresponding gridblock. The mean hardness values of the 12 gridblocks were calculated where the hardness values of pearlite and ferrite structures were assumed to be 150 BHN and 38 BHN, respectively. It is evident that the mean hardness value varies from one gridblock to another. If we assume that gridblock A-1 (μ_{A-1} = 122 BHN) represents a soft spot location and gridblock B-1 (μ_{B-1} = 132 BHN) a hard spot location, an observation can be made that the variation of the mean hardness values on row A ranges from 122 BHN to 121 BHN and the variation of the mean hardness values on row B ranges from 132 BHN to 121 BHN. This observation again confirms that along the machining feed direction, the appearance of a hard (or soft) spot location on the circumferential direction is constrained by its neighboring hardness status. In order to evaluate the overall excitation action during machining, there is a pressing need to formulate a statistical method to describe the distribution pattern of the mean hardness values in the consecutive locations along the machining feed direction. In this research, theory of hidden Markov modeling is used to extend the segment excitation model into a comprehensive random excitation model.

3. DETERMINATION OF STATES AND STATE TRANSITION MATRIX

In the theory of hidden Markov modeling, the possible outcomes of a stochastic process are classified as events. The basic principles of hidden Markov modeling are using an observable stochastic process to toss the outcome (one of the classified events) and using an underlying stochastic process to add a specific constraint on the likelihood of the concurrence of different states. This constraint is called "state transition matrix".

3.1 Determination of States of Segment Excitation

In this research, the segment excitation model is considered to be the observable stochastic process. If the number of blocks on the circumferential direction is assumed to be N_s, the N_s mean hardness values, i.e., $\mu_1, \mu_2, \mu_3, \ldots, \mu_{Ns-1}, \mu_{Ns}$, obey a normal distribution shown in Fig. 4. Now we classify the N_s mean hardness values into *n* states and each of the *n* states covers a certain range of the mean hardness values. As indicated in Fig. 4, the probability associated with each of the *n* states is given by ratio of the number of those mean hardness

	1	2	3	4
A	No. of Pearlite =88093 No. of Ferrite = 28347 Mean Hard-ness Value = 122.7	No. of Pearlite =87533 No. of Ferrite = 28907 Mean Hard-ness Value = 122.2	No. of Pearlite =86824 No. of Ferrite = 29616 Mean Hard-ness Value = 121.5	No. of Pearlite =86907 No. of Ferrite = 29533 Mean Hard-ness Value = 121.6
B	No. of Pearlite =98549 No. of Ferrite = 17891 Mean Hard-ness Value = 132.8	No. of Pearlite =93815 No. of Ferrite = 22625 Mean Hard-ness Value = 128.3	No. of Pearlite =93382 No. of Ferrite = 23058 Mean Hard-ness Value = 127.8	No. of Pearlite =86685 No. of Ferrite = 29755 Mean Hard-ness Value = 121.4
C	No. of Pearlite =97866 No. of Ferrite = 18574 Mean Hard-ness Value = 132.1	No. of Pearlite =91869 No. of Ferrite = 24571 Mean Hard-ness Value = 126.4	No. of Pearlite =95741 No. of Ferrite = 20699 Mean Hard-ness Value = 130.1	No. of Pearlite =85026 No. of Ferrite = 31414 Mean Hard-ness Value = 119.8

Figure 3 The Block Mean Hardness Values Calculated from Photograph Fig. 2b

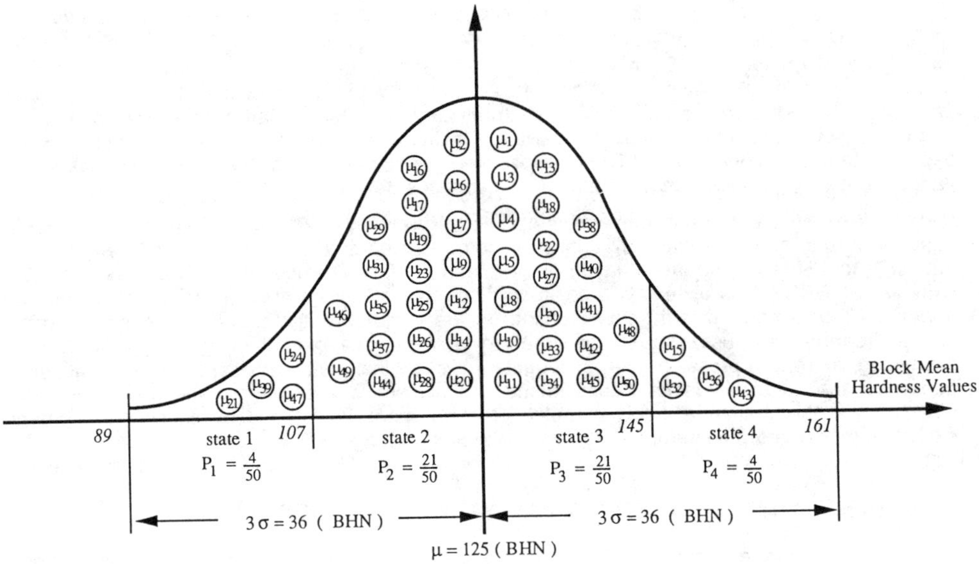

Figure 4 Normal Distribution N(125, 12^2) for Segment Excitation (Block Number = 50)

values falling into its corresponding interval to the block number N_s, and the probabilities associated with the *n* states for the N_s blocks, or within one revolution of the workpiece, are governed by

$$\sum_{i=1}^{n} P_i = 1$$

As an example to demonstrate the formation of a four (4) state segment excitation , we assume that the material being removed from the workpiece circumference during one revolution is divided into 50 blocks. The mean (μ) and standard deviation (s) of the normal distribution associated with the 50 block mean hardness values are 125 BHN and 12 BHN, respectively. Figure 4 actually depicts this normal distribution. Within the range [μ - 3s, μ + 3s] or [89, 161], four cells with an equal span $\frac{3s}{2}$ are formed. As illustrated, state 1 covers the mean hardness values from 89 BHN to 107 BHN. There are 4 mean hardness values falling into this cell. The probability associated with state 1, P_1, is 0.08. In a similar manner, we determine P_2 = 0.42, P_3 = 0.42, and P_4 = 0.08.

3.2 Determination of State Transition Matrix

Serving as the second, or the hidden, stochastic process, the state transition matrix describes both the neighboring and global state transitions. In formulating this matrix, the following two assumptions are made in this research.

1. The next state entered in the hidden stochastic process depends only on the current state, not on the previous states.

2. The transition that can occur is within two adjacent states, implying that the transition is bi-directional. However, the two states on the extremely right and left boundaries are exceptional. Their transitions are unidirectional.

Figure 5a is the state diagram illustrating the model for the neighboring state transition where state 1 represents the state on the left boundary and state 2 is a bi-directional transition state. As illustrated in Fig. 5a, parameter a_1 denotes the transition probability from state 1 to state 1 (itself), parameter a_2 denotes the transition probability from state 2 to state 2 (itself), parameter $(1-a_1)$ denotes the transition probability from state 1 to state 2, and parameter b_1 denotes the transition probability from state 2 to state 1. As a demonstration example, let us assume that $a_1 = 0.7$, $(1 - a_1) = 0.3$, $a_2 = 0.4$, $b_1 = 0.3$, and $(1 - a_2 - b_1) = 0.3$. Physical interpretations of these given transition probabilities during the transition process are as follows.

1. There are only two possibilities for state 1 to transit, either to remain its original state, or to become state 2. The likelihood for state 1 to stay in its original state is 0.7 and the likelihood for state 1 to transit to state 2 is 0.3.

2. There are three possibilities for state 2 to transit, either to remain its original state, or to become state 1, or to become state 3. The likelihood for state 2 to stay in its original state is 0.4, the likelihood for state 2 to transit to state 1 is 0.3, and the likelihood for state 2 to transit to state 3 is 0.3.

After having developed the model for the neighboring state transition, an *n*-state model for the global state transition can be established. Figure 5b is the state diagram illustrating the global *n*-state transition. In Fig. 5b, parameter a_i denotes the transition probability from state i to state i (itself), parameter b_i denotes the transition probability from state i to state (i-1), and parameter $(1 - a_i - b_{i-1})$ denotes the transition probability from state i to state (i+1). The neighboring and global state transition diagrams shown in Fig. 5a and 5b can be explicitly represented in a matrix form. Such a matrix is called state transition matrix. Its basic structure is provided below.

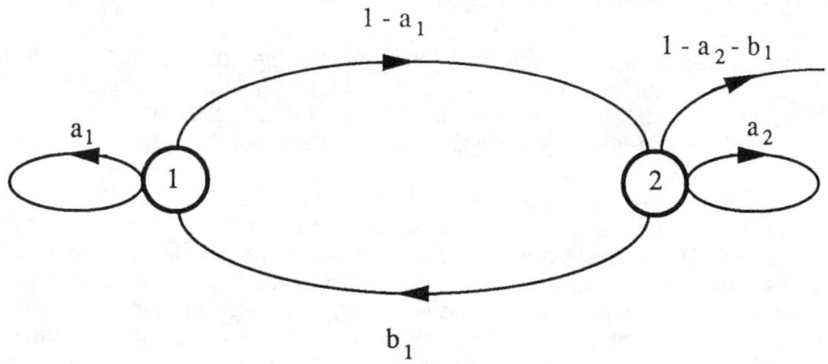

(a) Neighboring state transition diagram

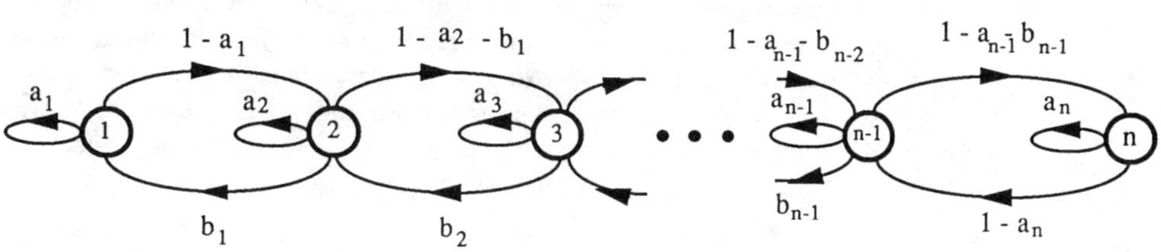

(b) Global n-state transition diagram

Figure 5 State Transition Diagrams

$$
\begin{array}{c|ccccccc|}
 & 1 & 2 & \cdots & & n\text{-}1 & n \\
\hline
1 & a_1 & 1-a_1 & \cdots & & 0 & 0 \\
2 & b_1 & a_2 & 1-a_2-b_1 & \cdots & & 0 \\
\cdot & 0 & b_2 & a_3 & \cdots & & \\
\cdot & 0 & & & & & \\
\cdot & & & & & & 0 \\
n\text{-}1 & 0 & & \cdots & b_{n-2} & a_{n-1} & 1-a_{n-1}-b_{n-2} \\
n & 0 & 0 & \cdots & 0 & 1-a_n & a_n
\end{array}
$$

In the state transition matrix, the elements on the diagonal represent the transition probabilities of remaining their original states, and the off-diagonal elements represent the probabilities of transiting from one state to another. The bandwidth of the transition matrix, which is equal to three (3) in the above matrix, reflects the assumption made for the neighboring state transition. Therefore, the transition matrix characterizes the unobservable, or the hidden, stochastic process.

3.3 Determination of Transition Probabilities

As discussed in section 2.4.1, to determine P_i for i = 1, 2, ..., n, we may group the mean hardness values of $\mu_1, \mu_2, \mu_3, \ldots, \mu_{Ns-1}, \mu_{Ns}$ into their corresponding state cells. An experimental procedure is developed to determine the state transition probabilities, i.e., a_i, b_i, and $(1 - a_i - b_{i-1})$ for i = 1, 2, ..., n. The procedure consists of three steps.

Step 1. Select a representative sample along the machining feed direction from the material to be machined. Use a computer-based scanning and imaging process to identify the type of microstructure at each location in the representative sample. Measure the microhardness value for each of the identified microstructures.

Step 2. Construct the gridblocks. The gridblock dimension should be determined by the number of blocks assumed on the workpiece circumference and the machining feed to be used during machining. Count the numbers of pearlite and ferrite structures within each of the formed gridblocks. Calculate the mean hardness value for each gridblock. Examine the variation among the calculated mean hardness values to confirm the number of states which was selected in the modeling of segment excitation. Otherwise, a proper adjustment for the number of states should be taken. At this stage, a state transition description, which is similar to the one shown in Fig. 3, is completed.

Step 3. Determine the transition probabilities, i.e.,a_i, b_i, and $(1 - a_i - b_{i-1})$ for i = 1, 2, ..., n.. In this initial research, we assume that transition probabilities, a_i, b_i, and $(1 - a_i - b_{i-1})$ for i = 2,3, ..., n-1, are equal to each other. The transition probabilities associated with the two extreme boundary states are defined as $a_1 = (a_2 + b_1)$ and $a_n = (a_{n-1} + b_{n-1})$ because these two states are unidiretional states. Referring to the state transition representation shown in Fig. 3, we may select a_i = 0.4 for i = 2,3,...,n-2,n-1, b_i = 0.3 for i = 1,2,...,n-1, a_1 = 0.7, and a_n = 0.7.

3.4 Implementation on Computer

The basic methodology for modeling a comprehensive random excitation is based on the hidden Markov modeling theory. A computer program was developed to implement this methodology on a computer. The block diagram shown in Fig. 6 outlines the simulation strategy, which is built upon four modules, i.e., the user input module, the segment excitation module, the state transition module, and the output module. Note that the data from the output module represents the random excitation action, which will be used as an input function to the program called "evaluation of random tool motion", which was developed in [14]. The descriptions of these four modules are given below.

1. The user input module. To initiate the computer simulation, the following parameter values should be specified through the designed user interface.

   (1). The mean and standard deviation of the segment excitation model, i.e., μ and s.
   (2). The block number selected for the segment excitation model, N_s, and the

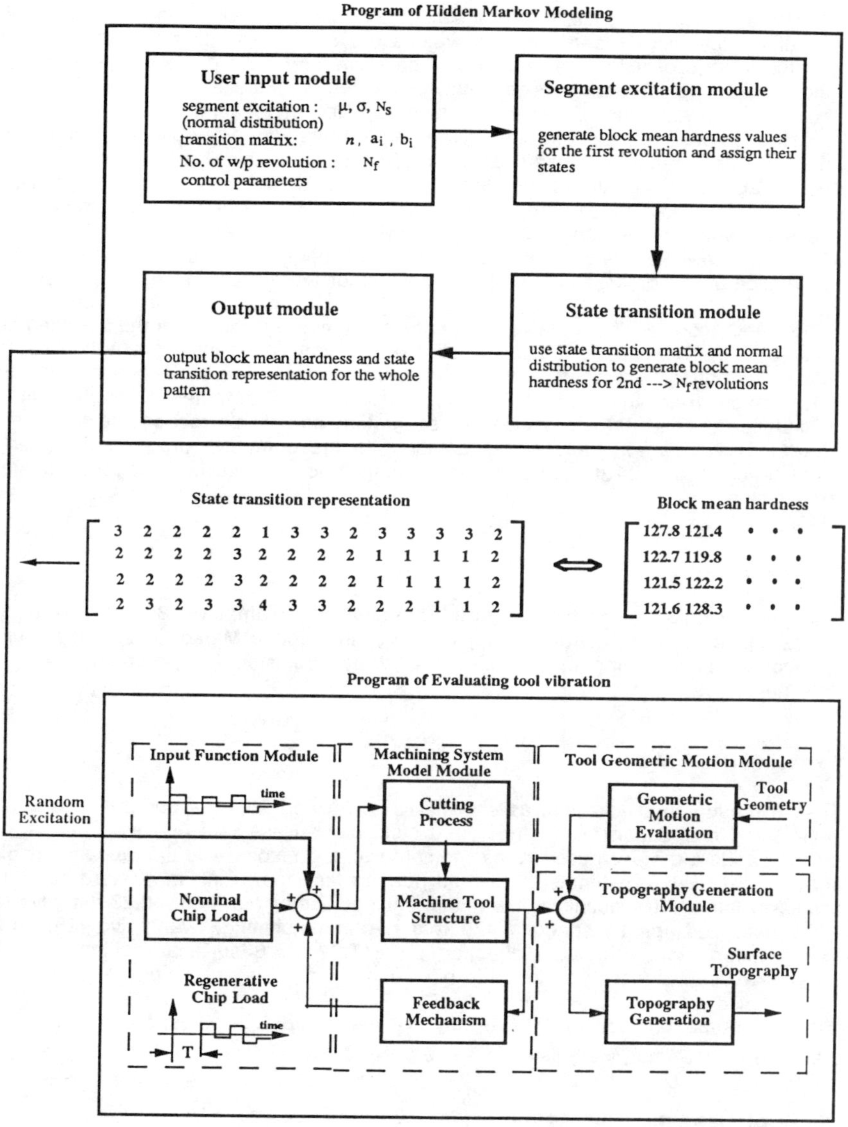

Figure 6 Methodology to Simulate the Topography of a Machined Surface

number of states selected for the state transition matrix, *n*.

(3). The transformation probabilities, i.e., a_i for i = 1,2,3,...,n-1,n, and b_i for i = 1,2,...,n-1.

(4). The number of workpiece revolutions to be simulated.

(5). The control parameters such as the variation levels of μ and s for each revolution of the workpiece.

2. The segment excitation module. A random number generator is built to construct the normal distribution, $N(\mu,s^2)$, which is associated with the block mean hardness values. These generated values are then distributed to individual blocks resembling the workpiece material being removed at specific instants. A built-in divider forms the *n* state groups and assigns the state number to each block. This completes the random excitation simulation for the first workpiece revolution.

3. The state transition module. For the second workpiece revolution and those revolutions to follow, the block mean hardness values are generated using the state transition matrix, instead of using the random number generator built in the segment excitation module. Using the state assignments for the first workpiece revolution as the initial condition, an algorithm is developed to assign the state number to the neighboring location in accordance with the given state transition probabilities. In order to maintain the mean and standard deviation of the assigned block mean hardness values, which also obey a normal distribution, a control scheme is used to regulate the selection of a block mean hardness value from its group. When the assignment for the second revolution is completed, these assigned values will serve as the initial condition to assign new block mean hardness values for the third workpiece revolution. The reiteration of this process completes the second, or the hidden, stochastic process.

4. The output module. It calculates the mean and standard deviation of the assigned block mean hardness values for each revolution of the workpiece, check the normality of the data distribution, and prints out the simulation results. The table attached to Fig. 6 was taken from the state transition representation obtained through computer simulation. This state transition representation and its corresponding block mean hardness values will be used as an input to the computer program designed for simulating tool vibration during machining and the surface topography after machining.

4. A CASE STUDY

A turning operation to machine an AISI 1020 steel bar (diameter = 80 mm and length = 300 mm) was studied to demonstrate an application of the hidden Markov model approach for controlling tool vibration during machining. The study consists of experimental work and computer simulation.

4.1 Experimental Work

First, microstructural analysis, as shown in Fig. 2a and 2b, was performed to quantify the microstructural distribution in the AISI 1020 steel bas. Based on the experimental data, the mean and standard deviation used in the segment excitation model and the probabilities of the state transition matrix were estimated. Second, turning test shown in Fig. 1 were carried out on a Placemaker lathe. The machining parameter settings used were feed = 0.23 mm/rev, depth of cut = 0.5 mm, and spindle speed = 470 rpm. After machining, twenty-five (25) surface profiles were taken along the axial direction on a Talysurf 6 surface profilometer. The measured surface profiles were positioned in parallel to form a surface topography. Figure 7a presents the experimental results. The twenty-five measured R_a values, together with $\overline{R_a}$ = 4.22 μm and s_{Ra} = 0.32 μm, were listed in Table 1.

4.2 Study of Random Tool Motion

Following the experimental work, the developed computer program was used to study

Table 1 Ra values of the Measured and Simulated Surface Topographies

	Measured	Segment Excitation	Hidden Markov Model (0.3,0.4,0.3)		Hidden Markov Model (0.2,0.6,0.2)	
	(μm)	(μm)	state=4 (μm)	state=10	state=4 (μm)	state=10
	4.54	4.19	4.17	3.69	3.75	3.65
	4.38	3.73	4.42	3.72	3.89	3.66
	4.60	3.72	4.22	3.67	3.92	3.66
	4.54	4.00	4.43	3.75	3.72	3.65
	4.21	3.78	4.19	3.66	3.76	3.70
	4.03	3.97	4.14	3.72	3.86	3.68
	4.10	3.73	4.62	3.67	3.84	3.70
	3.65	3.79	3.97	3.71	3.86	3.67
	4.25	4.07	3.95	3.69	3.70	3.66
	4.40	4.03	4.09	3.73	4.00	3.66
	3.80	3.78	4.49	3.68	3.82	3.69
	3.74	3.92	4.19	3.69	3.73	3.67
	4.28	3.80	4.05	3.69	3.70	3.66
	4.51	4.04	4.34	3.71	3.89	3.69
	4.73	3.88	4.10	3.70	3.76	3.68
	4.64	3.82	4.47	3.69	3.85	3.69
	4.08	3.89	4.09	3.70	3.87	3.65
	4.10	3.98	4.15	3.65	3.77	3.69
	3.91	4.00	4.15	3.73	4.48	3.66
	4.54	3.81	4.08	3.71	4.37	3.66
	4.52	3.93	4.50	3.71	3.87	3.69
	3.64	4.11	4.10	3.70	3.81	3.64
	4.01	4.11	4.41	3.70	3.84	3.69
	4.06	3.84	4.42	3.69	3.86	3.70
	4.18	3.95	4.31	3.68	3.89	3.70
Mean	4.22	3.92	4.24	3.70	3.87	3.67
Std.	0.32	0.13	0.17	0.02	0.18	0.02

random tool motion and its effect on the surface texture formation during machining. The block diagram shown in Fig. 6 illustrates how the basic methodology discussed in section 2 is implemented for the evaluation of random tool motion. As illustrated in Fig. 6, the program named as "evaluation of random tool motion" takes the nominal chip load, the residual chip load due to overlapping machining, and the random excitation as the system inputs to evaluate the tool vibratory response. It combines the evaluated tool motion with the spiral trajectory tool path to generate the data base for constructing the topography of a machined surface (Zhang and Hwang, 1990, Zhang and Kapoor, 1991).

In the computer simulation , the machining parameters used were set identical to those used in the experimental work. In the first simulation run, the input of random excitation only takes the segment excitation into consideration. This means that the random number generator in the segment excitation module was used to generate the block mean hardness values for the first workpiece revolutions, which obeyed the specified normal distribution. Also, the segment excitation module was used to generate the block mean hardness values for the consecutive workpiece revolutions, and the state transition module was skipped during the first simulation run. As a result, there were no constraints which were imposed between the neighboring block mean hardness values along the machining feed direction. In fact, the first simulation run duplicated the research work reported in (Zhang and Kapoor, 1991) where random excitation was considered, but only accounted for the nonhomogeneity of the hardness distribution in the cross-section perpendicular to the bar axis. Effect of the nonhomogeneity of the hardness distribution along the machining feed direction was neglected in the previous work (Zhang and Kapoor, 1991). Figure 7b presents the simulation results, showing the surface topography through computer visualization. The corresponding R_a values calculated from the simulated surface profiles, together with $\overline{R_a}$ = 3.92 μm and s_{Ra} = 0.13 μm, were listed in Table 1.

In the second simulation run, the input of random excitation takes both the segment excitation and overall excitation into consideration. During the simulation process, the segment

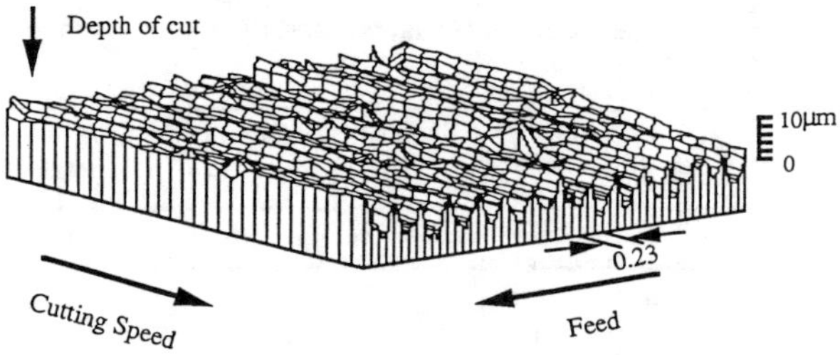

(a) Measured

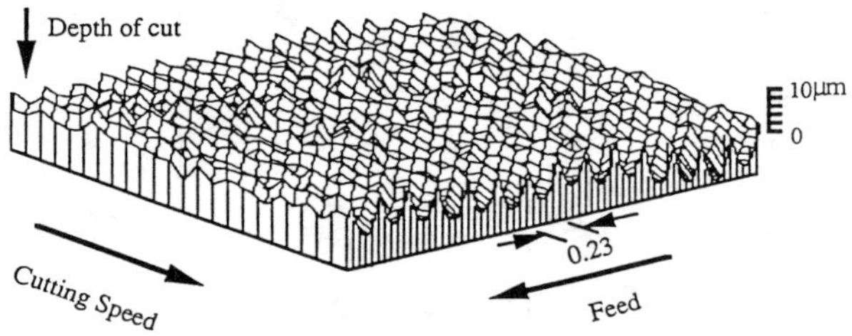

(b) Simulated (segment exitation only)

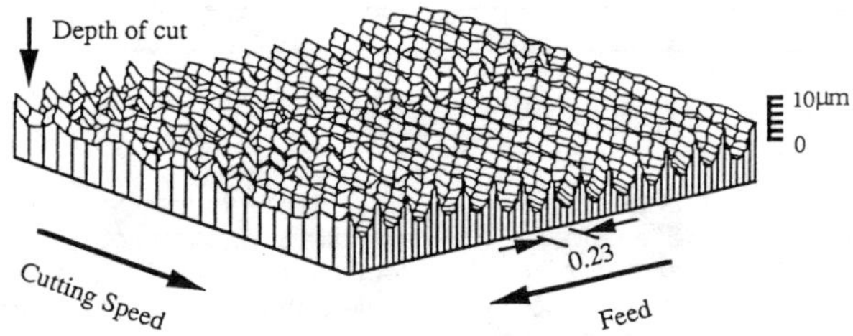

(c) Simulated (hidden Markov modeling.)

Figure 7 Surface Topography Comparison

excitation module was used to generate the block mean hardness values only for the first revolution. Afterwards, the state transition matrix was used to generate the block mean hardness values for the rest of workpiece revolutions. The defined transition probabilities, serving as an imposed constraint, supervise the assignments of the block mean hardness values. The assignment process simulates a direct concatenation of the local segment excitation actions, resembling a hidden Markov chain action. This entire simulation process represents a comprehensive study of the effect of the nonhomogeneity of the hardness distribution of the material being machined on the surface texture formation during machining. Figure 7c presents the simulation results. The corresponding R_a values calculated from the simulated surface profiles, together with $\overline{R_a}$ = 4.24 μm and s_{Ra} = 0.17 μm, were listed in Table 1.

5. DISCUSSION OF RESULTS

5.1 Comparison between the Measured and Simulated Topographies.

The two surface topographies shown in Figs. 7b and 7c provide a qualitative visualization of the effect of random tool motion on the surface irregularity generation. In this research, nonhomogeneous distributions of the basic material properties, such as hardness, were assumed to be the major excitation source. A main difference between the two surface topographies can be observed if examining them carefully. The difference is that the surface topography shown in Fig. 7c displays a gentle variation of the profile heights. Such a gentle trend comes from the cohesive relationship of random excitation actions between the consecutive revolutions during machining. This cohesive relation indicates that the likelihood for the occurrence of strong (or a weak) excitation actions at neighboring locations is higher than the likelihood of the mixed occurrence of strong and weak excitation actions at neighboring locations. Therefore, the simulated Markov chain characterizes the effect of the narrow and parallel pearlite strip pattern of the microstructural distribution on random tool motion during machining. In fact, the gentleness of profile height variation can also be seen from the measured surface topography shown in Fig. 7a. The two surface profiles shown in Fig. 8a and Fig. 8c, which were taken from the two corresponding surface topographies, clearly depict a slow variation pattern of the profile heights. The similarity between the measured topography (Fig. 7a) and the simulated surface topography (Fig. 7c) confirms the validity of proposing the hidden Markov modeling approach for the study of random tool motion during machining.

On the other hand, a high level of variation among the profile heights shown in Fig. 7b can be observed. This is due to the fact that random excitation action was evaluated on a single revolution basis. The random excitation actions between two consecutive workpiece revolutions are considered to be independent of each other. Consequently, the occurrence of a strong (or weak) excitation action at a given location does not impose a condition that a rather strong (or weak) excitation action will occur in its neighboring locations. In fact, the likelihood of having a strong, or a weak, excitation action, in its neighboring location, is solely governed by the segment excitation model which is characterized by a normal distribution. The surface profile shown in Fig. 8b, which was taken from its surface topography, depicts a rather significant variation of the profile heights if compared with those shown in Figs. 8a and 8c. Although more discrepancies can be observed if comparing with the measured surface topography shown in Fig. 8a, the surface topography shown in Fig. 8b would still be a good representation for machining certain carbon steel materials having relatively small and spherical pearlite structures as opposed to the narrow strips of pearlite in the present case study.

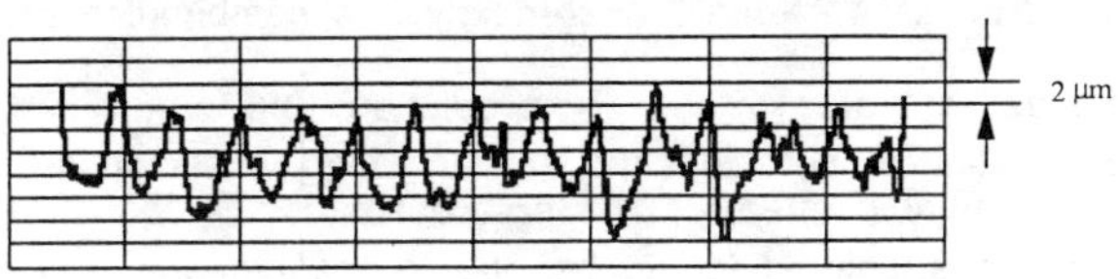

(a) Measured (Ra = 4.11 µm)

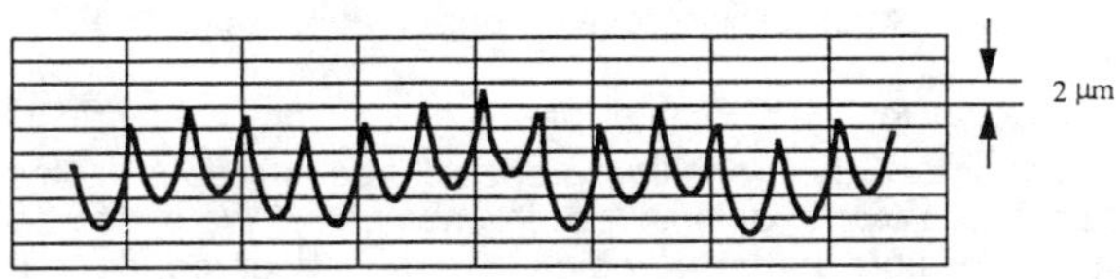

(b) Segment Excitation Model (Ra = 4.07 µm)

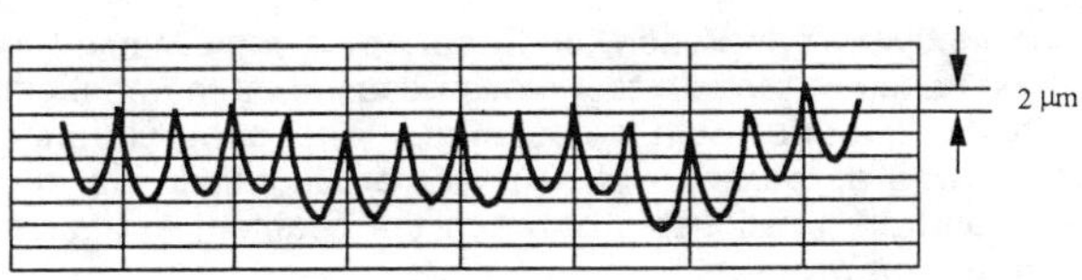

(c) Hidden Markov Model (Ra = 4.09µm)

Figure 8 Surface Profile Comparison

5.2 Effects of State Selection and Transition Probabilities

As discussed above, the block mean hardness values within one workpiece revolution are grouped into the state cells for implementation of the state transition matrix. The number of states and the associated transition probabilities were determined mainly based on the results from microstructural analysis. In order to extend our understanding of using the hidden Markov modeling method to formulate the random excitation observed during machining, a two-level factorial design was utilized to investigate the effects of the state selection and transition probability setting on the evaluation of random tool motion during machining. The LOW and HIGH levels of the state selection were set at 4 and 10, respectively. The low and high levels of the transition probability setting used were (a_i = 0.4 for i = 2,3,...,n-2,n-1, b_i = 0.3 for i = 1,2,...,n-1, a_1 = 0.7, and a_n = 0.7), and (a_i = 0.6 for i = 2,3,...,n-2,n-1, b_i = 0.2 for i = 1,2,...,n-1, a_1 = 0.8, and a_n = 0.8) where the number of the select states is equal to either 4 or 10. Four computer simulation runs were performed using the two-level factorial design. The simulation results in terms of the predicted $\overline{R_a}$ and s_{Ra} values were given below.

Predictions μm	Run 1 (LOW, low)	Run 2 (LOW, high)	Run 3 (HIGH, low)	Run 4 (HIGH, high)
$\overline{R}_a$	4.24	3.87	3.70	3.67
σ_{Ra}	0.17	0.18	0.02	0.02

Examining the simulation results, the combination of a LOW level of the state selection and a low level of the transition probability setting resulted in the highest value of the $\overline{R_a}$ prediction, which is listed as 4.24 μm. Accordingly, the combination of the HIGH and high levels yielded the smallest value of the $\overline{R_a}$ prediction (3.67 μm). Regarding the s_{Ra} estimations, it is interesting to note that the low and high levels of the transition probability settings have an unobservable effect on the s_{Ra} estimation. From the viewpoint of hidden Markov modeling, this phenomenon could be due to the fact that the transition probability setting is related to the hidden stochastic process, which generates patterns along the machining feed directions. At a given state selection, the random hardness values generated by the two transition probability settings maintain the R_a variation at a similar level. This phenomenon deserves a further study. Another interesting observation is that the change of the state selection from a LOW level to a HIGH level has a more significant effect on the $\overline{R_a}$ and s_{Ra} predictions than the change of the transition probability setting does. This indicates that great attention should be paid to a proper selection of the number of the states because the number of states determines the possible patterns which can be observed during the state transition process.

6. CONCLUSIONS

1. As a new and innovative approach, the theory of hidden Markov modeling is applied to study the phenomenon of random tool motion caused by the nonhomogeneity of hardness distribution in the material being machined. The general random excitation during machining is viewed as a series of segment excitation during individual revolutions of the workpiece, and the relationship between the segment excitation actions is governed by the state transition process.

2. The segment excitation and the state transition process are modeled as two stochastic processes. The stochastic process to describe the segment excitation, serving as the observable stochastic process, is characterized by a normal distribution with its mean standing for the material hardness in general and its standard derivation representing nonhomogeneity of the hardness distribution. The distribution pattern of

microstructures is used to establish the state transition matrix, which characterizes the cohesive relationship between random excitation actions in the consecutive workpiece revolutions. The state transition matrix maneuvers the likelihood of concurrence of different states of the random excitation action, resembling a so-called hidden stochastic process.

3. A prototype system to implement the hidden Markov model on a computer is developed for the study of random tool motion and its effect on the surface texture formation during machining. To confirm the validity of the proposed approach, predictions through computer simulation are compared with the results from the experimental work. A general match between them has been observed.

ACKNOWLEDGEMENTS

The authors acknowledge the support of the Systems Research Center under Engineering Research Centers Program: NSFD CDF 8803012, and the support of Minta Martin funding from the College of Engineering. They wish to thank Professor W. L. Fourney, Chairman of the Mechanical Engineering Department, and Professor D. K. Anand, Director of the Advanced Design and Manufacturing Laboratory for their valuable support in conducting this work. The technical assistance from the Precision Engineering Division of the National Institute of Standards and Technology and Mr. T. W. Hwang at the University of Maryland is appreciated.

REFERENCES

Blaedel, K.L., "Error Reduction," in Technology of Machine Tools: Machine Tool Accuracy, Vol. 5, Lawrence Livermore National Laboratory, Livermore, CA, 1980.

Donmez, M.A., Bolmquist, D.S., Hocken, R.J., Liu, C.R., and Barash, M.M., "A General Methodology for Machine Tool Accuracy Enhancement by Error Compensation", Precision Engineering, No. 4, 1986.

Kapoor, S.G., and Wu, S.M., "A Stochastic Approach to Paper Surface Characterization and Printability Criteria," Journal of Physics, Vol. 11, 1978, pp. 83-96.

Kohno, T., Okazaki, Y., Pzawa, N., Mitui, K., and Omoda, M., 1989, "In-Process Measurement and a Workpiece-Referred Form Accuracy Control System (WORFAC): Concept of the Method and Preliminary Experiment," Precision Engineering, Volume 11, Number 1, January 1989, pp. 9-14.

Nayak, P.R., "Some Aspects of Surface Roughness Measurement," Wear, Vol. 26, 1973, pp. 165-174.

Pandit, S.M., and Revach, S., "A Data Dependent System Approach to Dynamics of Surface Generation in Turning," Journal of Engineering for Industry, ASME Trans., Vol. 103b, 1981, pp. 437-445.

Rabiner, L. R. and Juang, B. H.,"An Introduction to Hidden Markov Models," IEEE ASSP Magazine, January 1986, Vol. 3, pp 4-16.

Tlusty, J. and Ismail, F., "Special Aspects of Chatter in Milling," Journal of Vibration, Acoustics, Stress, and Reliability in Design, ASME Trans., Vol. 105, 1983, pp. 24-30.

Tobias, S., "Machine Tool Vibration Research," International Journal of Machine Tool Design and Research, Vol. 1, 1961, pp. 1-14.

Vorburger, T.V., and Scace, B., "Progress Report of the Quality in Automation Project for FY89," NISTIR 4322, National Institute of Standards and Technology, May 1990.

Zhang, G. M. and Hwang, T. W., "Analysis of the Cutting Dynamics in Microscale," Symposium on Fundamental Issues in Machining, 1990 ASME Winter Annual Meeting, PED-Vol. 43, pp. 25-37.

Zhang, G. M. and Kapoor, S. G., "Dynamic Generation of the Machined Surface, Part I: Mathematical Description of the Random Excitation System," Journal of Engineering for Industry, Transactions of ASME, May 1991, pp. 137-144.

Zhang, G. M., and Kapoor, S. G., "Dynamic Generation of the Machined Surface, Part II: Mathematical Description of the Tool Vibratory Motion and Construction of Surface Topography," Journal of Engineering for Industry, Transactions of the ASME, May 1991, pp. 145-153.

Zhou, G.Y., Leu, M.C., and Dong, S.X., "Measurement and Assessment of Topography of Machined Surfaces," ASME PED.-Vol. 46, pp. 89-100, 1990

Sensors, Controls, and Quality Issues
in Manufacturing
ASME 1991

STUDY ON THE DRILL SKIDDING MOTION IN ULTRASONIC VIBRATION MICRODRILLING

De-Yuan Zhang, Xiu-Juan Feng, Wei Zhang, and Ding-Chang Cheng
Department of Manufacturing Engineering
Beijing University of Aeronautics and Astronautics
Beijing, People's Republic of China

ABSTRACT:

In this paper, a new axial ultrasonic vibration microdrilling technology has been developed to restrain the skidding motion of miniature drill at the beginning of penetration. The causes and mechanisms of the initial skidding motion of drill in conventional drilling have been analyzed. The mechanisms of intermittent cutting and initial penetration of the chisel edge, and the dynamic stability of drilling machine have been investigated analytically. An experimental study on the influence of ultrasonic vibration amplitude on the radial deflection of drill point at commencement of drilling and hole accuracy and drill life has been conducted. Finally a new variable–amplitude vibration drilling method has been presented.

NOMENCLATURE:

A	vibration amplitude
C_d	equivalent dumping coefficient of the drill
C_s	equivalent dumping coefficient of the spindle
f	feed
F	vibration frequency
K_d	equivalent bending rigidity of the drill
K_s	equivalent axial rigidity of the spindle
L	drill life
M_d	equivalent mass of the drill
M_s	equivalent mass of the spindle
n	rotation speed of the drill
P	unbalanced radial force
V_n	rotation speed of the spindle
W_F	vibration times per revolution
X_v	displacement of radial vibration of the drill
Z_v	displacement of axial vibration of the spindle
α	inclined angle of the workpiece surface
α_0	measured relief angle of the major cutting edge
α_{ev}	working relief angle of the major cutting edge
γ_ψ	measured rake angle of the chisel edge
$\gamma_{\psi v}$	working rake angle of the chisel edge

P_a	thrust force of the drill	θ_c	angle of rotation in each cutting time
P_r	radial pulse cutting force	θ_i	angle of rotation in each idle time
r	radius on a certain point of the cutting edge	ξ_d	damping ratio of the drill
		ξ_s	damping ratio of the spindle
r_c	half of chisel edge length	ω	radian frequency of ultrasonic
T	vibration cycle	ω_d	bending natural frequency of the drill
t_c	cutting time of each cycle		
t_i	idle time of each cycle	ω_s	axial natural frequency of the spindle
V_{ec}	cutting speed of the chisel edge in conventional drilling	φ	vibration phase at the cutting beginning
V_{ev}	cutting speed of the chisel edge in vibration drilling	σ	mean square deviation of diameter of holes
V_f	feed speed		
V_F	vibration speed of ultrasonic		

1 INTRODUCTION

Because miniature drills are of poor rigidity, if unbalanced radial forces have occurred, the chisel edge will skid easily on the surface of the workpiece when penetration has begun (Yasuhko,1978), causing the hole positional error and size error. Although some research papers address initial penetration mechanism, most were limited to study drill whirling vibration and the distorted polygon shape at entrance (Ema et al., 1988, Fuji et al., 1986, Lee et al., 1987), few investigated the drill skidding motion in microdrilling. Galloway(1957) investigated hole roundness and alignment by measuring the radial deflection of a drill point at commencement of drilling. It has been found that hole accuracy is markedly dependent on the relative lip height of the edges. Lee (1987) reported that the chisel edge skidding motion is produced because of factors such as chisel edge eccentricity and a lack of drill and spindle rigidity. He also observed that because the multifacet drills have a considerably smaller chisel edge, a quicker stabilization of the skidding motion is caused.

This paper deals with not only how the drill skid in conventional drilling, but also how the drill skidding is reduced effectively by axial ultrasonic vibration. The five characteristics of vibration drilling were proposed (Wang, 1990), namely, impacting, separating, changing speed, reciprocating pressing and changing angle. The authors also present a new precise penetration characteristic.

2 INITIAL PENETRATION IN CONVENTIONAL DRILLING

The chisel edge continuously cuts metal along the track $\overline{C_0C_1}$ with a large negative rake angle $-\gamma_\psi$ to form a chip $\widehat{C_1FG}$ as shown on a cylindrical profile of r – radius in Fig.1. The chisel edge cutting speed on the point C_1 is given by:

$$V_{ec} = \sqrt{V_f^2 + V_n^2} = \sqrt{\left(\frac{nf}{60}\right)^2 + \left(\frac{2\pi rn}{60}\right)^2} \tag{1}$$

The cutting ability of the chisel edge becomes very poor due to a large negative rake angle and a low cutting speed.

It is difficult that the tangent plane of workpiece surface at the entry point is perpendicular to the spindle rotation axis in microdrilling because of a variety of errors, so that the chisel edge will easily skid on the surface of the workpiece when penetration

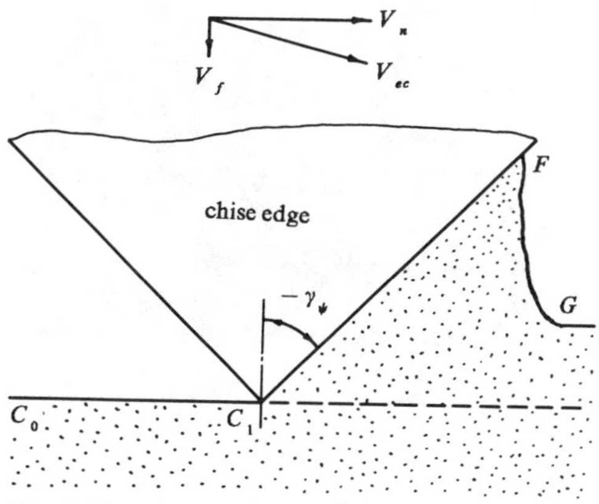

Fig.1 Continuously cutting process of the chisel edge

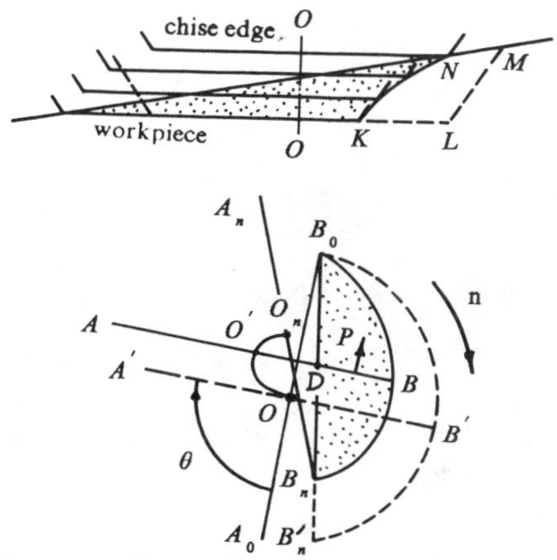

Fig.2 Skidding of the chisel edge in conventional drilling

has begun, leading to hole positional error. Let's take a typical case – drill an inclined plane to analyze the skidding of the chisel edge in conventional drilling. As shown in Fig.2 , when penetration has begun, only part chisel edge cuts metal, and this part is gradually lengthened in pace with the increment of cutting deep. The drill skidding mostly occurred before the chisel edge completely penetrates into workpiece. Therefore we focus attention on how the skidding motion develops in this process. Fig.2 is the skidding motion of conventional drilling. The chisel dege center O had coincided with the spindle rotation axis before the chisel edge cut metal from A_0B_0. If the drill were a rigid body, the chisel edge would be at A'B' after revolving the angle θ. However, because only the $\overline{DB}$ part of the chisel edge cuts metal, a radial unbalanced force P which makes the drill bend and produces radial deflection $\overline{OO'}$ is generated. Meanwhile the metal in the part BB_0B' can not be removed. Conventional drilling is a successive cutting process, so that the radial deflection of chisel edge center should be increased continuously. Finally, when the chisel edge leaves the metal, the part $B_0BB_nDB_0$ is removed, but the part $B_0B'B'_nB_nBB_0$ is left. The drill deflection of per revolution should be summed. When the full chisel edge has penetrated into the metal, the total deflection is e, and the metal in the part KLMN is left. Therefore the skidding motion of the chisel edge on the inclined plane is unavoidable in conventional drilling.

3 INITAL PENETRATION IN ULTRASONIC VIBRATION DRILLING

3.1 Cutting Mechanism of Chisel Edge

In axial ultrasonic vibration drilling, the motion locus of each point of the chisel edge is a sine wave, but it is similar to a sawtooth wave when the rotation rate of the spindle is not too high and the vibration amplitude is relatively large. The function of impact extrusion of rake face causes an intermittent cutting process when the vibration amplitude is relatively large. The intermittent cutting process of the chisel edge is shown on a cylindrical profile of r – radius in Fig.3. The chisel edge moves along the locus $M_0C_0M_1C_1M_2C_2$. When the chisel edge cuts toward point C_0, the metal which is extruded by the rake face forms an extruded face $\overline{C_0F_0}$ and a swollen face $\widehat{F_0G_0}$ due to plastic deformation. The chisel edge is idle along the locus $\overline{C_0M_1N_1}$, and then begins the next cutting process with a relatively large positive working rake angle from the

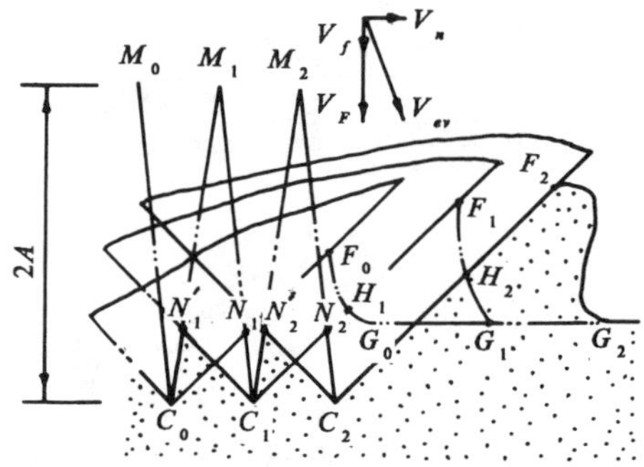

Fig.3 Intermittent cutting process of the chisel edge

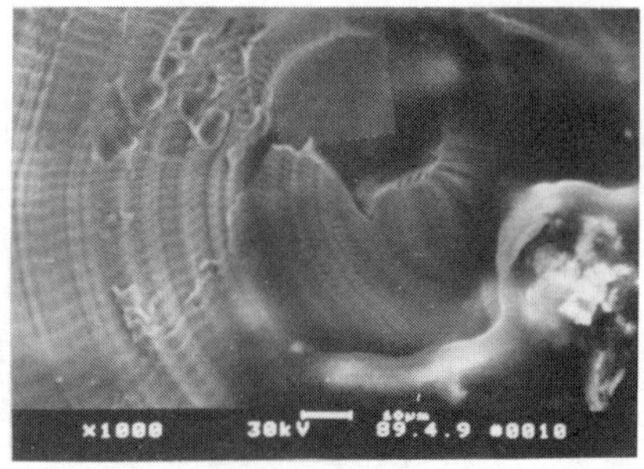

Fig.4 Undular cutting surface and chip

point N_1. While the chisel edge cuts along the locus $\overline{N_1C_1}$, the metal in the part $N_1C_1H_1F_0$ which is extruded by the rake face forms a new extruded face $\overline{C_1F_1}$ and a new swollen face $\widehat{F_1G_1}$ due to plastic deformation, at the same time, the metal in the part $N_1C_0C_1$ which is extruded by the flank face becomes a remained metal $N'C_0C_1$.The rest may be deduced by analogy. In pace with continuation of the cutting process, the swollen part becomes so large that a chip is formed, and the remained metal forms an undular cutting surface. In Fig.2, on the left is the undular cutting surface, on the right is the chip.

As shown in Fig.3, in the intermittent cutting process, the working rake angle $\gamma_{\psi v}$ and the cutting speed V_{ev} at the points from where every cutting begins (such as points N_1, N_2) are given by:

$$\begin{cases} \gamma_{\psi v} = -\gamma_{\psi} - tan^{-1}\dfrac{AW_F cos\varphi}{r} \\ V_{ev} = \sqrt{(V_F + V_f)^2 + V_n^2} = \sqrt{\left(2\pi AFcos\varphi + \dfrac{nf}{60}\right)^2 + \left(\dfrac{2\pi rn}{60}\right)^2} \\ \varphi = 2\pi\left(\dfrac{3}{4} - \dfrac{1}{1 + \dfrac{2A}{\pi r}W_F tan\gamma_{\psi}}\right), \; W_F = \dfrac{60F}{n} \end{cases} \tag{2}$$

where φ is the phase of the points from where every cutting begins. From formular (1) and (2), it can be seen that: $\gamma_{\psi v} > \gamma_{\psi}$ and $V_{ev} > V_{ec}$ because of relatively large vibration frequence. Therfore the ability of the chisel edge in vibration drilling is stronger than in conventional drilling.

Fig.5 shows the motion locus ($\overline{M_0C_0M_1C_1}$) of the chisel edge on a cylindrical profile of r – radius. If the line $\overline{C_0D}$ on the rake face intersects the line $\overline{M_1C_1}$ when the chisel edge moves to the lowest point C_0 (the intersection point is N_1), i.e., the slope of $\overline{C_0D}$ ($1/tan\gamma_{\psi}$) is smaller than that of $\overline{C_0M_1}$ ($2AF/\pi rn$), the intermittent cutting will occur. Also when the cutting parameters meet the following formular, the cutting of the chisel edge will become intermittent cutting:

$$A > \frac{\pi r_c}{2W_F tan\gamma_{\psi}} \tag{3}$$

Within the one vibration cycle ($\overline{C_0M_1C_1}$), the chisel edge is idle along the section

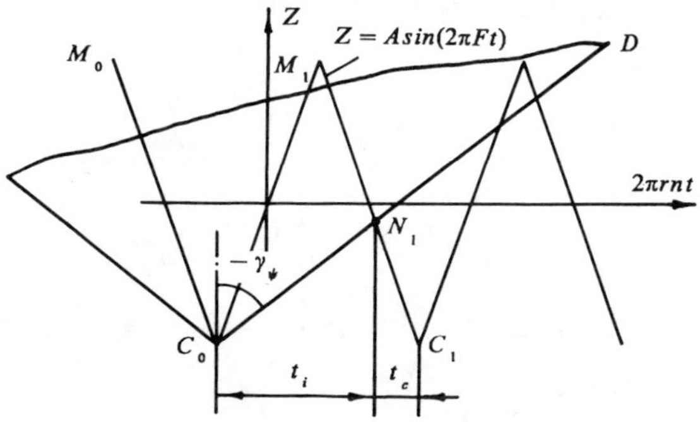

Fig.5 Cutting time and idle time

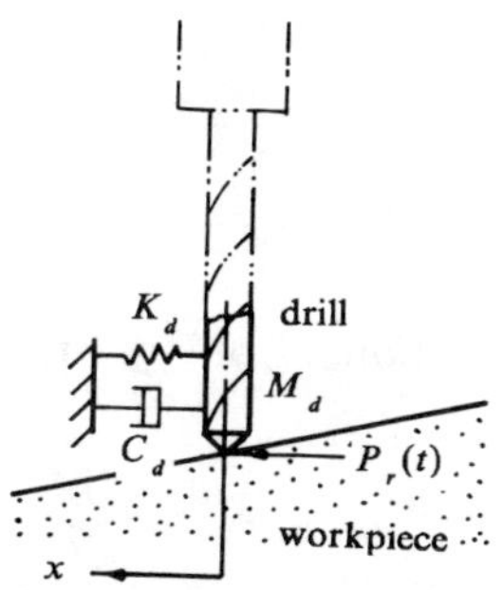

Fig.6 Radial vibration of drill

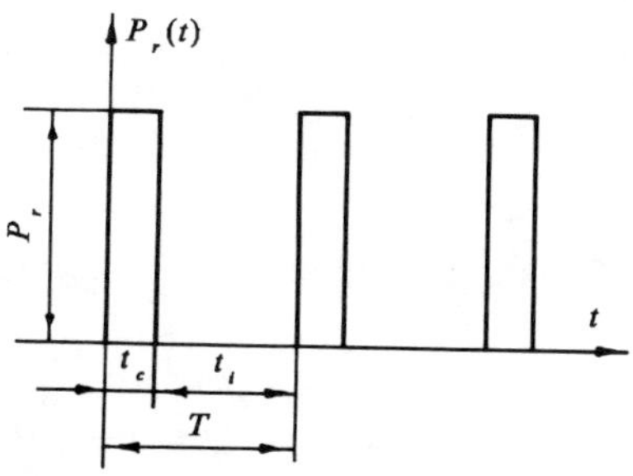

Fig.7 Radial pulse cutting force

$\overline{C_0M_1N_1}$ in which motion time is t_i, and then it cuts metal along the section $\overline{N_1C_1}$ in which the motion time is t_c. t_c and t_i are given by the following equations:

$$t_c = T\left(\frac{1}{1+\frac{2A}{\pi r}W_F tan\gamma_\psi}\right),\quad t_i = T - t_c,\quad T = 1/F \tag{4}$$

3.2 Radial Vibration of Drill

The radial vibration of a drill point at commencement of drilling directly affects the hole accuracy. Here gives a study on the influence of radial cutting force P_r on the dynamic displacement X_v. When penetration has begun, the radial cutting force P_r is pulse force when the vibration amplitude A is relatively large as shown in Fig.6. A vibration system is composed of a drill, a workpiece and a clamping head as shown in Fig.7, The motion equation of the drill is given by:

$$M_d\frac{d^2x}{dt^2} + C_d\frac{dx}{dt} + K_d x = P_r(t) \tag{5}$$

The pulse cutting force $P_r(t)$ is expanded to a Fourier series as the following:

$$P_r(t) = \frac{t_c}{T}P_r + \frac{2P_r}{\pi}\sum_{n=1}^{\infty}\frac{1}{n}sin\left(\frac{t_c n}{T}\pi\right)cos(n\omega t) \tag{6}$$

By solving the equation (5), the vibration displacement of the drill is given by:

$$X_v = \frac{t_c}{T} \cdot \frac{P_r}{K_d} + \sum_{n=1}^{\infty} \frac{P_r}{K_d} \cdot \frac{\left(\frac{2}{n\pi}\right) sin\left(\frac{t_c \pi}{T}\right)}{\sqrt{\frac{4\xi_d^2 n^2 \omega^2}{\omega_d^2} + \left(1 - \frac{n^2 \omega^2}{\omega_d^2}\right)^2}} \cdot sin\left(n\omega t + tan^{-1} \frac{1 - \frac{n^2\omega^2}{\omega_d^2}}{\frac{2n\xi\omega}{\omega_d}} \right) \tag{7}$$

where $\omega \gg \omega_d$, so the formular (7) is approximate to:

$$X_v \approx \frac{P_r}{K_d} \cdot \frac{t_c}{T} \tag{8}$$

here P_r / K_d is the static radial displacement of conventional drilling. From (4), it can be seen that:

$$\frac{t_c}{T} = \frac{1}{1 + \frac{2A}{\pi r_c} W_F tan\gamma_\psi} \tag{9}$$

Normally the value of t_c / T is in the range of $1/50 \sim 1/20$, as an example, when $A = 15\mu m$, $r_c = 0.05mm$, $F = 16000HZ$, $n = 7100r/min$, $\gamma_\psi = -45°$, the value of t_c / T is equated to $1/27$. In other words the vibration displacement of the drill in ultrasonic vibration drilling is approximate to one twentieth of that in conventional drilling. So that the vibration displacement tend to zero when the vibration amplitude A is relatively large.

3.3 Initial Skidding Motion of the Drill

As stated above, the intermittent cutting process is caused when the vibration amplitude is relatively large in ultrasonic vibration drilling. In order to study the initial penetration mechanism of the chisel edge, the process of skidding motion of the chisel edge is analyzed in one vibration cycle such as the cycle which is from AB to A″B″ as shown in Fig.8. The chisel edge begins to cut metal from AB, and then rotates an angle θ_c in the time t_c. The largest possible deflection ($\overline{OO'}$) of the chisel edge is caused by the drill rotating an angle θ_c around the cutting part of the chisel edge because of the radial unbalanced force P. The $\overline{OO'}$ is given by:

$$\overline{OO'} < \theta_c \cdot r_c = \frac{2\pi r_c}{\left(1 + \frac{2A}{\pi r_c} W_F tan\gamma_\psi\right) W_F} \tag{10}$$

The chisel edge skids $\overline{OO'}$ toward A′B′ in the time t_c, next, idles from A′B′ to A″B″ in the time t_i and then begins the next cutting process from A″B″. The large W_F of

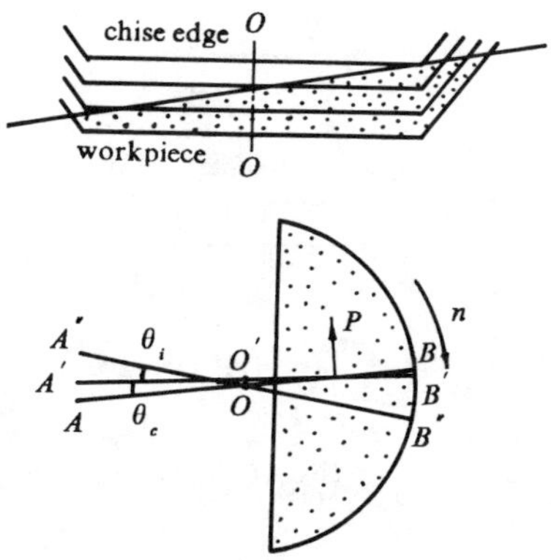

Fig.8 Micro–skidding of the chisel edge in ultrasonic vibration drilling

ultrasonic vibration drilling makes that each deflection $\overline{OO'}$ is very small. The ratio of the idle time t_i to the cutting time t_c is given by:

$$\frac{t_i}{t_c} = \frac{2A}{\pi r_c} W_F \cdot tan\gamma_\psi \qquad (11)$$

The value of the parameter W_F is very large, so that t_i is much larger than t_c for a large vibration amplitude A. During a long time t_i, the deflection $\overline{OO'}$ can be reduced to zero, and the center of the chisel edge can be reset to the center of rotation (the point O) from point O′. In brief, the drilling process in each vibration cycle is: chisel edge cutting and a simultaneous micro–skidding—chisel edge idling and a simultaneous center resetting. The repetitional process ensures that each micro–skidding is not accumulated , and the metal is removed completely in each rotation when the inclined angle of workpiece surface is relatively small. The ability of micro–skidding and center resetting of the chisel edge indicates that the ultrasonic vibration drilling technology has precise penetration characteristic.

4 FULL PENETRATION IN ULTRASONIC VIBRATION DRILLING

When the major cutting edge starts to cut matal, i.e., after the full chisel edge of the drill has penetrated into the workpiece,the smallest working relief angle of the major cutting edge is given by:

$$\alpha_{ev} = \alpha_0 - tan^{-1}\frac{AW_F}{r} \qquad (12)$$

If the vibration amplitude A is relatively large, the working relief angle of the major cutting edge will be a negative value, this causes that the flank wear is aggravated. If $A > f/2$, the intermittent cutting process of the major cutting edge will occur, this causes that the breakage of the major cutting edge is aggravated. So that after full penetration, the major cutting edge must be made to cut continuously and have not overly large negative working relief angle (see Fig.9). The thrust drilling force is equal to $P_a + p_a \sin\omega t$ in the continuous cutting as shown in Fig.10. Now let's analyze the influ-

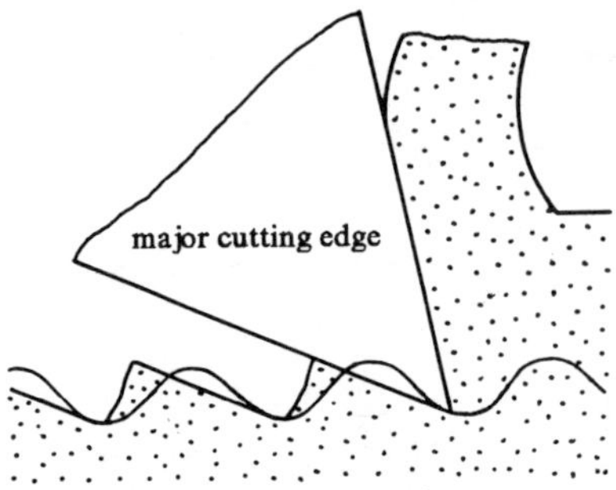

Fig.9 Continuously cutting process of the major cutting edge

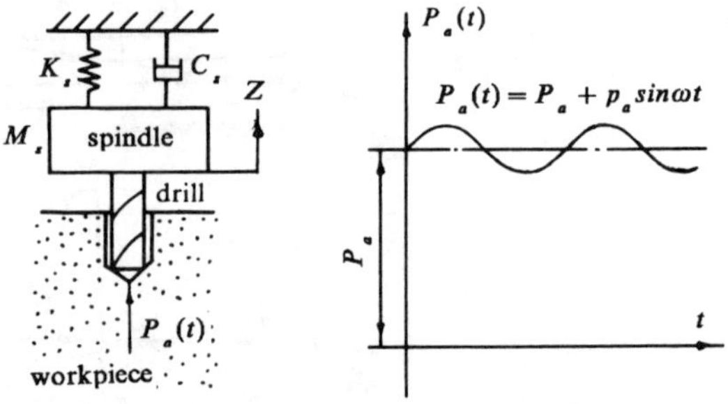

Fig.10 Vibration system and thrust force

ence of the thrust force on the stability of the cutting system. The vibration system is compose of a workpiece, a spindle and a machine tool. The motion equation of the spindle can be expressed as following:

$$M_s \frac{d^2 z}{dt^2} + C_s \frac{dz}{dt} + K_s z = P_a + p_a sin\omega t \tag{13}$$

The displacement of the spindle along z direction is given by:

$$Z_v = \frac{P_a}{K_s} + \frac{p_a}{K_s} \bullet \frac{1}{\sqrt{\left(1 - \frac{\omega^2}{\omega_s^2}\right)^2 + \frac{4\xi_s^2 \omega^2}{\omega_s^2}}} sin(\omega t + \varphi_0) \tag{14}$$

where: $\omega \gg \omega_s$, So that the displacement Z_v is approximate to:

$$Z_v \approx \frac{P_a}{K_s} \tag{15}$$

So in the ultrasonic vibration drilling, the vibration displacement of the spindle almost does not change with time, i.e., it is approximate to static displacement. Therefore the ultrasonic vibration drilling can maintain the dynamic stability of spindle.

5. EXPERIMENTAL ANALYSIS

A common bench drilling machine was used in the experiments. A 16000HZ ultrasonic vibration acted on the workpiece. A tool microscope was used to measure the drill point skidding and the diameter of drilled holes. The workpiece material was low–carbon alloy (18Cr2Ni4WA). The drills were Φ0.32mm HSS twist drills. The depth of drilled holes was 2mm.

5.1 Radial Deflection of Drill

In order to analyze the ability of restrainning radial skidding in the ultrasonic vibration drilling, the workpiece surface is inclined at an angle α . Radial deflection of

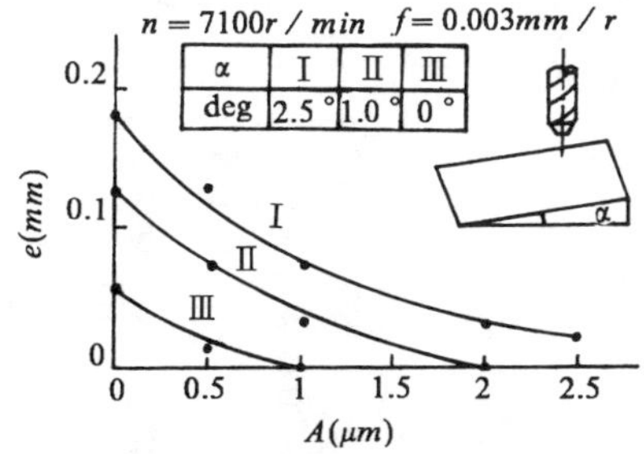

Fig.11 Variation of radial deflection with amplitude and inclined angle

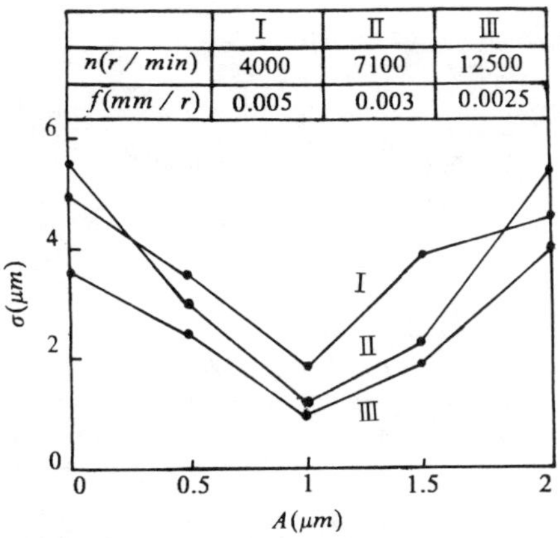

Fig.12 Variation of hole diameter error with amplitude

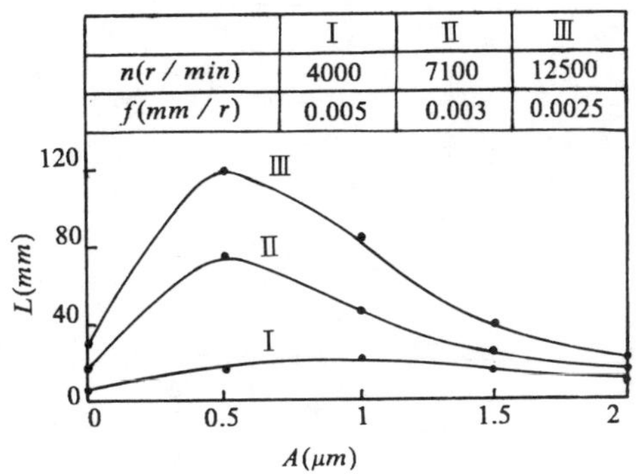

Fig.13 Variation of drill life with amplitude

drill point is measured when the full chisel edge has penetrated into the metal. Fig.11 shows the variation of the radial deflection e with amplitude A and inclined angle α. From the curve, it can be found that, under the different inclined angles of workpiece surface, the radial deflection e in the vibration drilling gradually decreases with the amplitude increasing, and it is smaller than the radial deflection in the conventional drilling ($A = 0$). The reason for this is that with increasing A, t_c will be decreased and t_i / t_c will be increased, moreover the ability of micro–skidding and the center resetting of the chisel edge will be increased, and the radial vibration displacement will be decreased. Also it can be found that the larger the inclined angle is, the larger the vibration amplitude is required to reduce the skidding of the chisel edge. In brief, the larger the amplitude is, the better the hole positional accuracy is.

5.2 Hole Accuracy and Drill Life

In order to analyze the cutting ability after full chisel edge has penetrated into the metal, and to decrease the influence of radial skidding on cutting process after the full penetration, the workpiece surface is set level. Fig.12 and Fig.13 respectively show the variation of hole diameter error and drill life with the vibration amplitude A. The hole diameter error is expressed by the mean square deviation σ of hole diameter. The drill life is expressed by the total length L of drilling before the drill is snapped. From the curves, it can be seen that hole diameter accuracy and the drill life do not always increase with A. The optimum amplitude of hole accuracy curves is $1\mu m$, but that of drill life curves is $0.5\mu m$.

Table 1 The smallest working relief angle of the major cutting edge α_{ev} and the ratio 2A / f

α_{ev} , 2A / f \ n, f, A	r / min	4000		7100		12500	
	mm / r	0.005		0.003		0.0025	
0.5	μm	−17 ° ,	0.2	−3 ° ,	0.3	7 ° ,	0.4
1	μm	−36 ° ,	0.4	−20 ° ,	0.7	−6 ° ,	0.8
1.5	μm	−46 ° ,	0.6	−32 ° ,	1	−16 ° ,	1.2
2	μm	−52 ° ,	0.8	−39 ° ,	1.2	−24 ° ,	1.6

Table 2 The largest working rake angle of the chisel edge $\gamma_{\psi v}$ and the ratio t_i / t_c

$\gamma_{\psi v}$, t_i / t_c \ n, f, A	r / min	4000		7100		12500	
	mm / r	0.005		0.003		0.0025	
0.5	μm	22 ° ,	1.5	8 ° ,	0	−7 ° ,	0
1	μm	33 ° ,	3.1	25 ° ,	1.7	12 ° ,	0
1.5	μm	36 ° ,	4.6	31 ° ,	2.6	22 ° ,	1.5
2	μm	37 ° ,	6.2	34 ° ,	3.4	27 ° ,	2.0

In order to analyze the regularity of these curves under various cutting condition, the smallest working relief angle of the outer corner of the major cutting edge (set $\alpha_0 = 20°$, r=0.16mm, in (12)) α_{ev} and the ratio 2A / f are computed, the results are shown in table 1., and the largest working rake angle of the corner of the chisel edge (set $r = r_c = 0.005$mm, $\gamma_\psi = -45°$, if $\varphi < 180°$ or (3) is not true, set $\varphi = 180°$ in (2)) $\gamma_{\psi v}$ and the ratio t_i / t_c are also computed, the result are shown in table 2. Table 1 and tableb 2 are compared with Fig.12 and Fig.13, it can be seen that: ① when a=0.5 μm, t_i / t_c is reletively small, the ability of restrainning the skidding of the chisel edge is not strong,this cause that the hole diameter error is not obviously smaller in vibration drilling than in conventional drilling (A=0). On the other hand, the cutting speed (it was not computed in the table) and the working rake angle of the chisel edge are obviously large, at the same time, overly large negative relief angle of the chisel edge does not occur, these cause that the drill life is obviously longer in vibration drilling than in conventional drilling, the drill life can be increased about four times at most. ② when A=1μm, the drill skidding is removed due to large t_i / t_c (see Fig.11), and because the working rake angle of the chisel edge is obviously increased, the cutting ability of the chisel edge is improved, at the same time, the inflence of the negative working relief angle of the major cutting edge on hole accuracy is still relitively small, so that the hole accuracy is the highest in this case. But such a large negative working relief angle of the major cutting edge increases the flank wear and causes drill life is lower than the optimum value. ③ when A> =1.5μm, although the initial skidding is removed due to relitively large t_i / t_c, very large negative relief angle of the major cutting edge occurs, and the intermittent cutting process of the major cutting edge occurs (because 2A / f>1), these cause that the bending deformation of the drill is increased, the chip is fragmental and the wear and breakage of the major cutting edge are increased. Therefore hole accuracy and drill life are decreased. To sum up, the new technology of axial ultrasonic vibration microdrilling mainly improves the cutting ability of the chisel edge and restrains the skidding of the chisel edge, under not overly large negative working relief of the major cutting edge improves the hole accuracy and prolongs the drill life.

6 DISCUSSION AND CONCLUSION

The above theoretical analysis and experimental results indicate that a large amplitude ultrasonic vibration is beneficial to improve the cutting ability of chisel edge and restrain the skidding of chisel edge. When the curvature of workpiece surface is large, it is difficult that the tangent plane of workpiece surface at entry point is perpendicular to the spindle rotation axis, and radial disturbing forces are easily produced. So that a large vibration amplitude is required to make the initial skidding of a drill tend to zero. But an overly large amplitude will worsen the cutting ability of major cutting edge, increase hole size errors and shorten drill life. In order to solve these problems, a variable–amplitude vibration drilling method should be used, that is: before full chisel edge completely penetrate into workpiece, the skidding of chisel edge is restrainned with large vibration amplitude to keep high positional accuracy of holes, then vibration amplitude is reduced to the optimum value to keep longer drill life. This method can be further bringed into play the effect of axial ultrasonic vibration cutting.

References

Ema,S., Fujii,H., and Marui,E., 1988, "Whirling Vibration in Drilling. part 3, "ASME Journal of Engineering for Industry, Vol.110, No.11, pp.315–321.

Fujii,H., Marui,E., and Ema,S., 1986, "Whirling Vibration in Drilling. Part 1–2," ASME Journal of Engineering for Industry, Vol.108, No.3, pp.157–168.

Galloway,D.F., 1957, "Some Experiments on the Influence of Various Factors on Drill Performance," Transactions of ASME, Vol.79, pp.191–231.

Kanaya,Y., and Arai,K., 1987, "Small Hole Drilling: The Japanese way," Circuits Manufacturing, pp.33–35.

Lee,S.J. Eman, K.F. and Wu,S.M., 1987, "An Anlysis of the Drill Wandering Motion," ASME Journal of Engineering for Industry, Vol.109, No.11, pp.297–305.

Wang,L.J., Zhang,D.Y., and Qiu,S.B., 1990, "Investigation on Zero–Phase–Difference Vibration Chip Breaking, "Chinese Journal of Mechanical Engineering, Vol.3, No.1, pp.50–57.

Sensors, Controls, and Quality Issues in Manufacturing
ASME 1991

INTEGRATED SIMULTANEOUS ENGINEERING: AN EXPERT SYSTEM

D. S. Ermer and S. Araj
Department of Mechanical Engineering
University of Wisconsin, Madison
Madison, Wisconsin

ABSTRACT

An approach called Integrated Simultaneous Engineering (ISE) is presented which makes use of an export system to combine the application on-line and off-line quality control for continuous quality and productivity improvements. It is a methodology for establishing product robustness and manufacturability at the very early stages of product design and process development. The methodology employs engineering statistical tools to implement the approach, and stresses the use of Statistical Experimental Design in all phases of product/process design and development.

INTRODUCTION

The growing need for continuous product and process design improvement, and the emerging role of off-line-quality control is the major thrust behind the development of a expert system to assist design and manufacturing engineers in meeting these goals. This expert system possesses a knowledge domain in both on-line and off-line quality control and provides effective information and direction for solving product or process quality and design related problems. The method of integrating both on-line and off-line quality control is called Integrated Simultaneous Engineering [1]. This ISE method provides complete communications between all components of the design and manufacturing stages to contribute to a never ending cycle of improvement.

INTEGRATED SIMULTANEOUS ENGINEERING (ISE)

Integrated Simultaneous Engineering (ISE) is a methodology to build robustness and manufacturability at the initial stage of product design and process development. The idea is to also maintain the product/process at its target value - both in terms of the nominal value and also with minimum variation, through the use of statistical process control (SPC) methods. Although this is a two part process, the ISE approach stresses the successful joining together of the design and manufacturing functions to produce a "stalwart" design. Hence, the ISE methodology is a way of connecting off-line quality control methods with on-line quality control methods [2]. The Statistical Experimental Design (SED) method is used to identify product parameters or process control parameter settings that reduce the product or process sensitivity to use or manufacturing variation and therefore should be controlled. In order to achieve the optimal functional characteristics, a series of factorial experiments are conducted on the proposed product design. Whereas the role of SPC methods is to feed back information about the product and process to the design and manufacturing departments so as to determine how the product or process can be revised or improved when unexpected variations are observed. Therefore, SPC is the quality control technique needed to reduce the variability within the production phase. Hence, the production process is diagnosed at appropriate times, and if any production process factor is statistically found to have deviated from its nominal (functional characteristics) a corrective action is taken immediately. This action could be a simple adjustment, or sometimes it may be required to conduct a SED to be able to learn now to adjust the process to remove the cause of the increased variability.

THE ISE EXPERT SYSTEM

Currently, off-line and on-line quality control techniques are mostly being used separately, but they should complement each other in the field of design and manufacturing. Therefore, these techniques need to be integrated by ISE for effective implementation. The ISE approach can only be implemented when both engineering and statistical knowledge and know-how exist and is combined for robust design. Therefore, the ISE approach needs engineers with an extensive background in engineering statistical methods -- this includes SPC and SED - in order to successfully integrate product and process design. Since many engineers do not have an extensive background in engineering statistics, an expert system called Integrated Simultaneous Engineering Expert System was developed which has the expert knowledge and know-how in the field of engineering statistics. That is, the role of the expert system for ISE is to efficiently furnish the expertise needed by design/process engineers to implement the strategy of ISE. In addition, the expert system will help engineers avoid using some of the misleading ideas which are being reported in the literature on the use of statistical experimental designs in industry. Thus, the expert system will provide the knowledge and know-how necessary at the early stages of design in order to achieve the desired product robustness. At the same time, since it is not possible to provide a completely perfect design, the expert system will have the ability to implement on-line control.

Off-Line Consultation

This consultation session is conducted to solve design and manufacturing problems at the early design stage. Figure 1 gives a graphical representation of this session where it is shown that the consultation starts after the system design phase, where the engineering concepts of the product/process design are developed. Quality Function Deployment (QFD) principles - where the voice of customer is taken into consideration - are used, and after system design the consultation session has two steps:

Parameter Design-Step 1. The parameter design consultation session starts by extracting information from the user (designer or manufacturing engineer) about the product/process to be developed. The needed information includes design and manufacturing requirements, material requirements and working conditions. In general, the expert system diagnoses the given information and provides a SED recommendation. At this stage the consultation session is to be interrupted so that the user performs the recommended experiment.

Next, the off-line quality control process is continued by the user performing the recommended experiment. The analysis of the experimental results are analyzed by an easy to use computer program which is developed as part of the expert-system external programs. Finally, the parameter-design-consultation stage is resumed by feeding back the results of the experimental analysis. If the resulting product/process design is not robust, a sequential process of SED's may be implemented to reach a robust design. However, if the design is found to be robust then two cases may arise:

•*Case One:* User is satisfied by the design and hence the consultation is terminated.

•*Case Two*: User needs to refine the tolerances around the designed parameters and hence the consultation continues.

Tolerance Design - Step 2. The tolerance design consultation session is based on a newly developed approach called Integrated Simultaneous Engineering Tolerancing (ISET) [1]. The ISET concept integrates the activities of both design and manufacturing engineering in a cycle of continuous improvement. Furthermore, Taguchi's loss function is an important tool in the ISET process, and is used to relate customer expectation with design and manufacturing requirements. The main goal of tolerance design is to set tolerances around the process/product parameters so that the final specifications of the design are met. This design stage is also a sequential process and should be repeated to achieve the targeted specifications.

The first part of tolerance design is to estimate the specifications of a certain product/process, and this is done by a market study and engineering considerations using QFD. Then Taguchi's loss function principle is used to connect both the market study and engineering considerations to find the required specification. If the required specification is $\pm\Delta$, then for a capable process the standard deviation of the process should be less than or equal to $\Delta/3$. At this point, engineering statistics are used to estimate the standard deviation of a certain process/product. If this standard deviation is greater than $\Delta/3$ then the process is incapable and the tolerances of the parameters that contribute to the variance of the functional characteristics should be reduced. The reduction of the parameter tolerances is a function of the degree of contribution to the total variance and the degree of control implied by the tolerances of each specified parameter.

On Line Consultation

On-line consultation is implemented during manufacturing. This stage starts by using statistical control charts. If quality problems are detected, a problem diagnosis feature in the expert system can be actuated to help determine the cause of the problem. The next

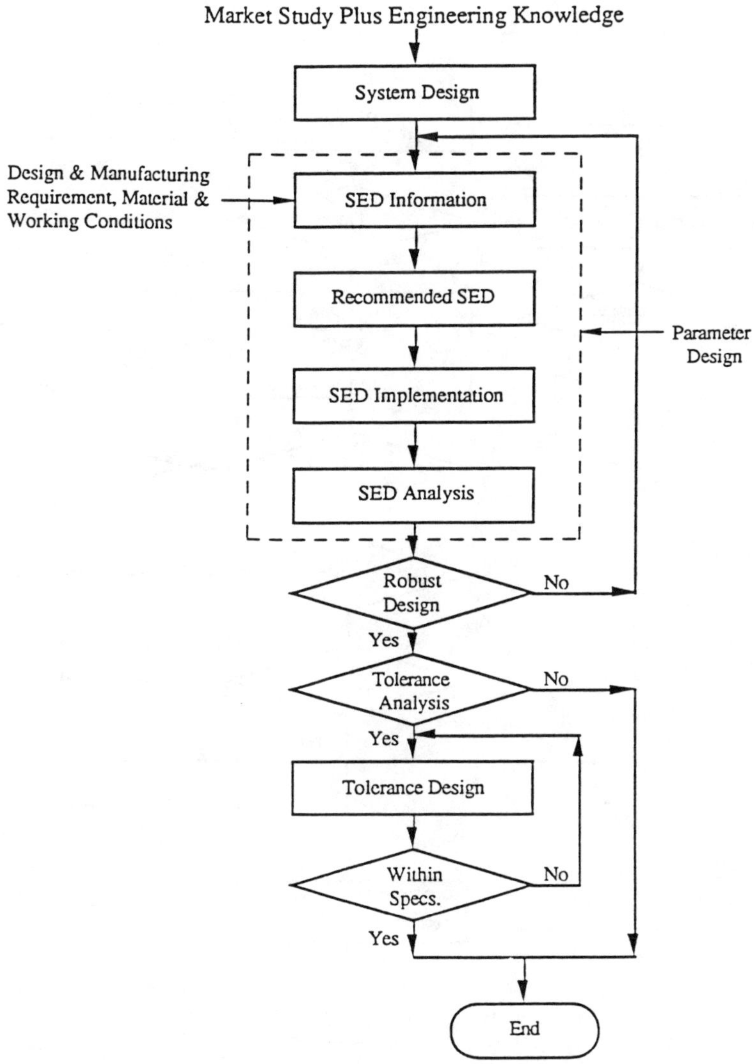

Figure 1 Off-Line Quality Control Consultation.

step is to correct the problem and to feed back the correct information to the expert system. Figure 2 shows a more detailed representation of the on-line quality control consultation loop. This figure shows that the expert system can handle the two situations that may arise during any production operation.

• *Case One:* In this cased rational sampling procedures are being followed to establish SPC charts. These samples are used in the development of Shewhart control charts or Cumulative Sum Control (CUSUM) Charts to "diagnose" and monitor the production process for stability. The expert system analyzes these data by plotting the required control charts. Then it calculates the variability of the data and checks for trends, seasonality, or out-of-control points. At this stage one of the following situations will be displayed to the user by the expert system:

- Process is in control and capable
- Process is in control but not capable
- Process is out of statistical control

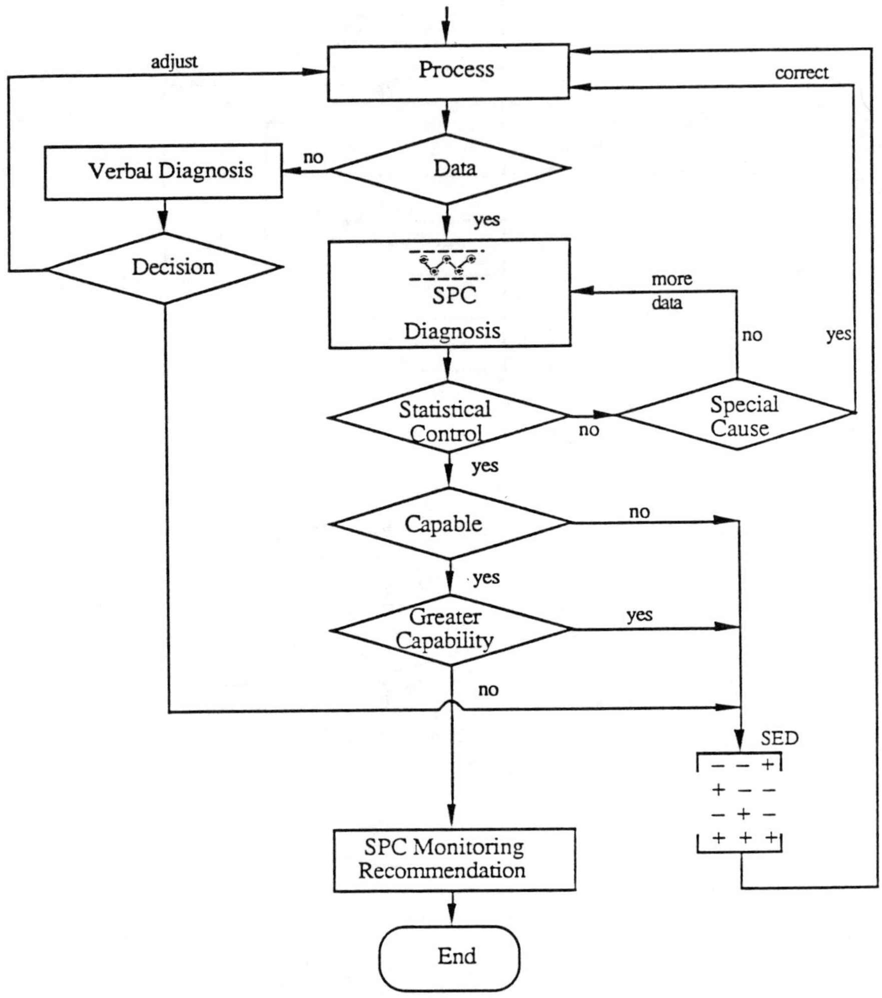

Figure 2 On-line Quality Control Consultation.

If the process is in control and capable, then no further actions are needed but recommendation will be made to monitor the process with reduced sampling. If the process is in control but not capable, then the expert system will extract information from the user and recommend a specific action to be taken such as executing a statistically designed experiment. The purpose of the proposed experiment will be to help the user understand the process and reduce its variability. However, if the process is out of control this means that the process might also be incapable. At this stage the expert system will also extract information from the user in order to diagnose the problem. After diagnosing that a special cause is present, the problem can be solved locally by adjusting or removing the process fault or system problem. However, if no special cause is found, a statistical designed experiment will again be provided. This experiment will help the user uncover the cause of the fault or the cause of excessive variation. It should be noted that histograms and normal probability plots are also considered to ensure that the process is normally distributed if an analytical capability study is to be carried out.

• *Case Two*: The expert system may be used in the case where no reliable or scientific data is available. That is, for some manufacturing process it is sometimes possible to decide whether the process is doing well or not, without any data. In such situations, and when a special cause is detected, it may be possible for an experienced person to adjust the process to eliminate the error. However, if there is no experienced person or if the experienced person could not determine the cause of the error, then the expert system will be used. In such a case the expert system will extract information from the user about the type of problem and manufacturing situation, and after diagnosing the problem a recommendation will be issued. The recommendation could be a direct adjustment to the process in the case of a special cause of variation or to recommend a statistically designed experiment in the case of large-common cause variability.

A SUMMARY OF THE MAIN FUNCTIONS OF THE EXPERT SYSTEM

1. Collect information about the product or process in order to:
    a. Work towards ISE
    b. Optimize a new product or process
    c. Solve an existing manufacturing/design problem
    d. Improve an existing product or process
    e. Investigate the robust design of a new product or process
2. The information will be collected in one or more of the following forms:
    a. By asking multiple choice questions
    b. By asking direct questions like "how much"
    c. As a set of data when it is available (dependent or independent)
3. Analyze the given information and recommend solution(s). This analysis includes:
    a. Analysis of verbal information
    b. Analysis of data (when available) by using formulas
    c. Analysis of data (when available) by plotting graphs
4. After analysis, one of the following situations may be recommended:
    a. Special cause is found: provide long term solution to the problem
    b. No special cause is found: provide statistical designed experiment to find a long-term solution
    c. Process in control but need to reduce variability: Provide a statistically designed experiment to help determine how to reduce the variation
    d. Design a new product or manufacturing process: provide a statistically designed experiment to get a robust product or process
    e. For designing a system: provide a SED to be used in a simulation analysis
5. Provide recommendations for preparing, executing and collecting the data of the designed experiment. Also provide recommended steps to follow when analyzing the data or proposing a simulation.
6. Complete analysis of the experimental data. This program includes:
    a. Lambda Plot: A tool used to help choose the correct scale for data transformation.
    b. Effect estimation and representation in graphs such as two factor interaction plots and cube plots.
    c. Normal probability and dot plots for residuals.
    d. Model fitting analyses.
    e. Model checking tools, which includes:
        i. Normal Probability and histogram of residuals
        ii. Plot of residuals versus run order
        iii. Plot of residuals versus predicted values
7. Summary of results

EXPERT SYSTEM KNOWLEDGE BASE

An expert system building tool called "VP-Expert" [3] was used for developing the ISE expert system. This means that the VP-Expert program (shell) provides the inference engine, the user interface and everything needed to create a working knowledge for the ISE expert system. More importantly, the VP-Expert is a rule-based tool with forward and backward chaining, and it provides the developer with options of entering a set of cases or examples that the system converts into rules. The key thing that a knowledge engineer has to do is to encode knowledge of the problem into the knowledge base. In addition, this VP-Expert shell has the ability to execute external programs that are needed for an integrated system. Some of the external programs provided are:

1. Program for designing a statistically designed experiment
2. Program for analyzing a two level factorial experiment.
3. Program for analyzing a three level factorial experiment.
4. Program for regression analysis.
5. Programs for generating control charts such as:
   - $\bar{X}$-R Charts
   - X-Moving Range Charts
   - P-Charts
   - C-Charts
   - U-Charts
   - CuSum Control Charts

Figure 3, which follows, summarizes the structure of the knowledge base for the ISE Expert System.

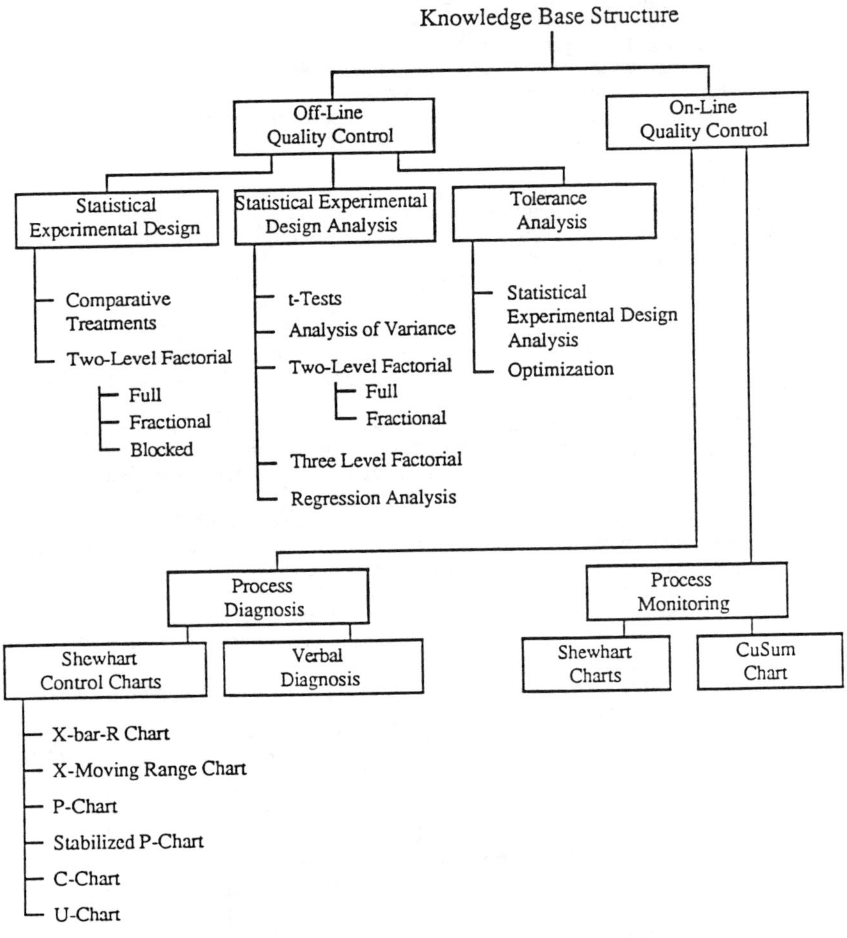

Figure 3 The Knowledge Base Structure Contained in the ISE Expert System.

UPDATING AND EXPANDING THE ISE KNOWLEDGE-BASE

The on-going quality revolution is strengthened by the concept of continuous and never ending quality improvement. The ISE methodology is based on this concept, and therefore it is receptive to on-going system improvement. Accordingly, the knowledge base of the ISE Expert System is designed for dynamic changes, i.e., continuous updating and expansion of the knowledge base is the key to the Expert System usefulness in the long run.

The ISE Expert System knowledge-base is a collection of two types of knowledge-base structure, i.e., rule-base structure and external computer programs. In the former additional 'If-Then-Rules' can be added to the rule-base structure for the purpose of enhancing, updating and expanding this structure. However, the expansion of the rule-base structure is restricted to the available computer memory. On the other hand, this restriction can be overcome by dividing the knowledge base into smaller knowledge-bases and connecting these knowledge bases through a chaining command. Similarly, additional external computer programs can be added to enhance and improve the capability of the Expert System. These computer programs can be written in any of the high level computer languages and then integrated with the ISE Expert System knowledge-base through a set of 'If-Then Rules'.

HOW TO USE THE ISE EXPERT SYSTEM FUNCTION

Implementation of the ISE methodology by the expert system is based on the knowledge-base and the domain-knowledge which is contained in a rule-based computer program and in other computer programs. These programs are used in the ISE Expert System to perform five different functions.

1. Problem Diagnosis
2. Process Monitoring
3. Statistical Experimental Design
4. Statistical Experimental Data Analysis
5. Tolerance Design Consultation

A consultation session with the ISE Expert System will trigger one or more of the above functions which are described as follows:.

1. Problem Diagnosis Function

The diagnostic function of the ISE Expert System is utilized when quality characteristics of a certain product/process needs investigation. The diagnostic procedures could be implemented in two ways, verbal with no data diagnoses or analytical with data diagnoses:

a. Verbal (No Data) Diagnosis. If this situation is triggered because the user has no data, then the ISE Expert System diagnoses the problem through a question and answer session with the user. Questions about the operators and their job skills, working conditions, supervisor strengths, raw material used, etc. are asked. By the interaction of the user with the questions asked by the ISE Expert System the problem is traced to its roots. However, if the problem cannot be diagnosed the ISE Expert System issues a recommendation for a statistically designed experiment. If this recommendation is accepted, then the diagnostic session is integrated with the third function (SED) of the ISE Expert System and the consultation session is continued.

b. Analytical (Data) Diagnosis. Process diagnosis is achieved by data analysis through the use of statistical control charts. The major source of information needed for process analysis is the data, and hence a data reliability investigation is carried out first by the ISE Expert System. To narrow down the investigation, information about the type of data is given to the ISE Expert System. The user needs to be able to differentiate between the two major data types (i.e., variable measurements or attribute information), and give this information to the ISE Expert System when requested.

For variable-measurement data the ISE Expert System needs to know whether the data was collected in subgroups or not. If the data is collected in subgroups an investigation of the validity of these subgroups is carried out. Rational sampling procedures are checked to make sure that special causes of variation are not mixed-up within common causes of variation. Data and/or rational subgroup verification will lead to one of the following three cases:

i. Data is not reliable: Recommendations for proper rational sampling are displayed to the user.

ii. Data mixture within the subgroups: For this case a separation of the subgroup data is recommended if possible. For example, if the subgroups contains data from different operators, it might be possible to separate the data in the subgroups with respect to each of the operators. In such case SPC charts for individual measurements (X-Moving range charts) are recommended. If not, the subgroups are declared unreliable and a recommendation for collecting rational subgroups is issued.

iii. Data is reliable: For this case the $\bar{X}$-R charts are actuated for data analysis. Similarly, if the collected data are individual measurements, rational sampling procedures are investigated. If the data is found to be reliable, then X-Moving range charges are actuated for data analysis. In addition, rational sampling procedures are recommended for the case of incorrect data. On the other hand, if the collected data are attribute measurements, the ISE Expert System asks questions to verify whether the data is a fractional defective type or a nonconforming (defect) type of attribute data. Accordingly, the Expert System recommends a P-Chart or C-Chart analyses if rational sampling procedures are not violated. If the data is rejected then rational sampling procedures are recommended.

The next step in process diagnosis is to use the SPC tools to get the needed information about the process. Shewhart Control Charts are used for this purpose, while the CuSum control chart is not used for diagnoses, since it is used for monitoring of the process average.

The ISE Expert System uses the rules and theories for out-of-control and pattern checking:

1. Process problem rules:
   - Point outside of control limits
   - Ad hoc zone or sequencing rules
   - Trend and cycle rules
2. Sampling problem rules:
   - For mixtures
   - For stratification
   - Webb products

It should be noted here that not all of these rules are applicable to all the control charts available; hence the ISE Expert System is programmed to use the applicable rules for the specified control chart.

For the $\bar{X}$-R charts analyses the ISE Expert System adopts the strategy of not attempting to diagnose the $\bar{X}$ Chart unless the R-Chart is put in statistical control first. Hence, starting with the R-Chart diagnoses, any point that is found to violate any of the out-of-control rules is circled and a general list of possible special causes for this violation is displayed. If the special cause for this violation is identified and removed, then the circled point is removed from the chart. A recalculation of the control limits will then take place. This process is repeated until the process is put in a state of statistical control. However, if the point is circled due to a mixture or stratification, the data diagnostic procedure is terminated, since this is a sampling problem.

If the R-chart is put in statistical control, the standard deviation of the data is calculated from the average range ($\bar{R}$), is denoted by sigma and is equal to $\bar{R}/d_2$. The control limits on the $\bar{X}$ chart are analyzed through the use of the sequencing rules, and any point found to violate these rules is circled and a list of possible special causes for the violation is displayed. If a special cause is identified and removed, then the point or points in question are removed from both the $\bar{X}$ chart and the R-chart. Next, the analysis procedure starts again, and the value of sigma is recalculated if necessary. Also the new control limits are recalculated and the checking procedure repeated.

After completing the statistical charting diagnosis, the ISE Expert System enters the capability analysis phase. No capability analysis is conducted if the corresponding control charts are not in statistical control. Moreover, if the control charts are put into statistical control after taking out points found to have special causes of variation, and the remaining points (samples) are less than twenty five, then no attempt is made to perform the capability analysis.

If the process is in statistical control and 25 or more subgroups are plotted on the chart, then the two types of capability analysis, i.e., traditional and Taguchi methods are conducted. Accordingly, one of the following situations might describe the analyzed data and its respective process:

1. Process can't be diagnosed because of sampling problem.
2. Not enough data to establish proper control limits and therefore can't draw any useful conclusions
3. Process is out of control.
   - Special cause is found and needs to be removed
   - Special cause is not found
4. Process is in statistical control:
   - Process is capable
   - Process is not capable and variation needs to be decreased.
   - Specification limits calculated by the loss function do not contradict with the traditional manufacturing specification and hence they could be used for future analyses.
   - Specification limits calculated by the loss function do not match the traditional manufacturing specifications and hence the specifications needs modification.

After issuing one of the above diagnostic situations, the ISE Expert System asks the user whether to:

1. Stop consultation
2. Proceed to adjust process mean
3. Proceed to reduce process variance
4. Proceed to adjust process mean and reduce variance

If item one is chosen then the consultation session is terminated. However, if any other item is chosen then the consultation session continues, which means that the third function of the ISE Expert System, namely, Statistical Experimental Design (SED), is triggered. The SED function is employed to further study the sources of variation and their magnitudes.

2. Process Monitoring Function

Process monitoring is one function of the ISE Expert System which requires a stable and capable process, since only then can the on-line monitoring function be used efficiently for the process control. Once again, rational sampling procedures are needed for data gathering. The next step in process monitoring is to use Statistical Process Control to get the needed information out of the data. The SPC monitoring tools available in the ISE expert system are the same as those used for process diagnosis plus the CuSum chart. The rules used to check for out-of-control process violations or sampling violations on the Shewhart Control Charts are also used for process monitoring, whereas the V-mask is used to monitor the process average on the CuSum charts. There are two strategies:

a. Shewhart Control Charts Monitoring Strategy. The on-line monitoring starts by a process declaration, i.e., the ISE Expert System needs to know the process situation:

1. If the process has been diagnosed by the ISE Expert System and found to be stable and capable, then the user should provide the file name where the process past history is stored. Accordingly, the process control limits stored in the file for this particular process, are used for process monitoring.
2. If the process has been diagnosed by other programs and found to be stable and capable, then the user is asked to input the control limits that are going to be used for process monitoring.
3. If this monitoring consultation is a continuation of a previous process monitoring consultation, then the ISE Expert System asks for the name of the file used for this process.

For example, if $\bar{X}$-R charts are used the general strategy for using Shewhart Control Charts and the expert system is as follows. The current reading is plotted on the R-chart, and if a point beyond the control limits or a run of eight successive points are above or below the average is detected then the last point in question is circled; and the ISE Expert System issues a warning and a list of possible special causes for the violation. The point(s) in question is/are removed from the chart if a special cause(s) is identified. If no special cause is identified the monitoring procedure will continue without taking out the circled points; for keeping these points on the chart will not affect the monitoring procedure of the $\bar{X}$-R charts since these control limits are calculated from historical data. Moreover, the circled points will serve as a reminder for future modification of the control limits if these out-of-control violations continue to persist. Trends and cycles are checked for, but no stratification or mixture checking will take place since rational sampling procedures are established at the diagnostic step.

b. CuSum Charts Monitoring Strategy. When CuSum charts are invoked the R-Chart monitoring procedure described above is also used first. After monitoring the R-chart, the V-mask of the CuSum Chart is used to detect small changes in the process average. The mask angle is analogous to the control limits of the $\bar{X}$-R Chart, and the monitoring procedure starts by putting the mask on the second point of the data and checking for any significant shift in the process average. If the first point is located outside the V-mask a shift in the process average is considered significant; and if a special cause is found and corrected then the monitoring procedure is continued. If no shift is detected, than the V-mask is put at the third point and the two proceeding point are checked, and accordingly the mask continues moving along for monitoring purposes. Finally, if a shift in the process average is detected, but a special cause is not found, then the user is encouraged to study the process and decide whether the shift could be controlled/eliminated or a modification of the process average is needed.

In summary, the monitoring function of the ISE Expert system is applied to detect any significant shifts or process changes that might occur in both the process mean and/or variability. It also, issues a warning if such a shift is detected. The type of SPC control chart used for monitoring depends on the type of data that represents the quality characteristics of the product under study, and must be carefully defined.

3. Statistical Experimental Design (SED) Function

The SED function or consultation could be started in two situations. The first situation is after data analysis by SPC is terminated and the SED function is employed to further investigate the improvement of the process/product. The second situation is for off-line quality control consultation. The purpose in this latter case is either to study and learn about improving an existing product or process or to help design a new process or product. The ISE Expert System requires information about the general type of experiment to be run and the general form of the experiment is determined accordingly. The following two subsections describe the two forms of experiments that could be designed and recommended by the ISE Expert System.

a. Experiments for Comparison of Treatments. These types of experiments are employed when two or more treatments need to be compared and the main interest of the investigation is to look for a significant difference between the treatments. A general inquiry about the different treatments under consideration is requested by the ISE Expert System. Such information includes:

- Number of treatments.
- Variables that must be kept constant
- Variables that must be blocked due to their nuisance nature

According to the information supplied, the ISE Expert System suggests a suitable independent or paired t-test, or an analysis-of-variance test. The experiment design contains the random order of experimental runs and also includes the blocking arrangement, when blocking elements are specified.

b. Factorial Experiments. This type of experiment is conducted to study and investigate a process/product in order to learn how to improve quality characteristics. In this regard the ISE Expert System issues inquires about:

- Number and names of independent variables
- Possible levels of independent variables
- Variable nature (i.e., can they be adjusted and controlled precisely?)

If a factor can be independently adjusted, then this variable is called a controllable or independent variable; on the other hand if an independent variable is difficult or expensive to control or adjust, then it is called a noise (independent) variable. Considering the above information the expert system suggests a preliminary design for the experiment. The design could be either a:

- Full factorial design with replication and no outer array design if no noise variables are present.
- Full factorial design with replication, and an outer array design for the noise variables.

To confirm the suggested design the ISE Expert System needs information about the cost and time needed to run the experiment. The user is then informed about the type of design to be used and the time and cost needed for the experiment is also given. Now, a decision is to be made by the user on whether to:

1. Accept the preliminary design.
2. Block the design because of time or nuisance variable limitations.
3. Reduce the number of replications.
4. Reduce the number of experimental runs by performing a fractional factorial experiment.
5. Consider situations 3 and 4 together.

If situation 1 is chosen then the design of the preliminary experiment is confirmed and stored in a file so that the user can call it when running the experiment. However, if any of the other four situations are executed, a new design is determined, and then a new cost and time estimate is carried out. The design resolution is also recalculated if situations 4 or 5 are actuated. Again, the user is asked whether to accept the new design or suggest any modification. This procedure is repeated until the user confirms the design, and then this latest design is stored in a file for later experimental use. Finally, a recommendation on the experiment operation is given to the user in order to facilitate a successful execution of the experiment. The SED execution recommendation includes:

- Team planning for carrying out the experiment is recommended, and should include the operators that are going to run the experiment.
- Provide the operators and supervisors with detail instructions about how the experiment will be conducted
- Recommendation on when to start data collection, i.e., after steady state conditions are reached if there is a transient period.
- Note-taking recommendation includes;
  - Time of the experiment.
  - Which variables are held constant.
  - Record any change or unusual observation that occurs during execution.
- Go back to the ISE Expert System for data analysis.

4. Analysis of Statistical Experimental Design Function

The Statistical Experimental Design Analysis Function available in the ISE Expert System is capable of analyzing two forms of experimental design

1. Experiments for comparative treatments.
2. Factorial type experiments.

After the ISE Expert System asks the user about the general form of the experiment under consideration, the user is then given access to the analysis procedure. The following subsections describe the general forms of the consultation procedures available for the analysis of the experiments.

a. Analysis of Comparative Treatments. Under this heading three types of analyses could be performed by the ISE Expert System. The analysis type depends on the number of treatments and on the presence of blocking variables. After obtaining the appropriate information, one of the following methods is chosen for the analysis of the experimental results:

1. Independent t-test. If two independent treatments are to be compared, then the hypotheses testing procedure is employed. The user is asked to choose the confidence level or alpha level for the analysis. Accordingly, the result of the comparison analysis is displayed.

2. Paired t-test. This test is performed when the effects of uncontrollable factors in the experiment need to be blocked in order to minimize inflation of the standard error. Once more, hypotheses testing procedures are recommended and performed and the user is informed about the result of the analysis for a given level of confidence α.

3. Analysis of Variance. This method is used when three or more treatments are to be tested. The sums of squares of each treatment are calculated and an F-test is conducted for a user specified confidence level. Any significant deference between the treatments is then identified.

4. Analysis of Factorial Experiments. The ISE Expert System has the ability to analyze 2^k and 3^k factorial type experiments. The analysis procedures for these two major types of factorial experiments are not similar due to the complexity of analyzing three-level factorial experiments. The procedural analysis for these two types as adopted by the ISE Expert System is described below:

b. Two Level Analysis. The forms of the two level factorial are limited to those designed by the SED Function of the Expert System. These forms are:

1. Full factorial design.
2. Fractional factorial design.
3. Inner-outer array design.
4. Fractional inner-outer array design.

The above designs could include replication and/or blocking. The two types of data analyses included in the ISE Expert System are:

1. Analysis of the mean (location effect analysis)
2. Analysis of the standard deviation (dispersion effect analysis).

The procedures for the two types of analyses are quite similar, with minor modifications in data treatment for the dispersion effect analysis. The experimental data are entered for the analysis in the standard order of the design. Also, the design matrix with facts about the number of fractions, blocks, replications and the design of the outer array as determined in the SED Function are utilized. The following describes the analysis procedures adopted in the ISE Expert System.

Analysis of Location Response

The step by step analysis of the location effects is as follows:

Lambda Plot [3]

This plot is used as a tool to help the user choose the proper scale for data transformation. The vertical axes for this plot is the t-ratio of the effects and is a function of the estimate standard error of the effects. The ISE Expert System adopts the following strategy in estimating the standard error of the effects.

- For a SED with replication, the standard error of the effects is calculated from the variability between true replications.
- For a SED without replication:
  - Inner-outer array design: The outer array design is regarded as a replication and hence, the standard error is calculated from the variability between these replications. However, it should be pointed out in this case that the estimated value of the experimental error will be inflated due to the variations in the experimental setting of the assumed replications. Therefore, the user of the ISE Expert System should be aware of this inflation of the standard error when determining the significant effects.
  - Full factorial design: The user is asked to choose higher order interactions that might be considered as noise, which will be used for the estimation of the standard error.
  - Fractional factorial designs: The following procedure is used for the estimation of standard error:
    i. Effects are calculated for the original data.
    ii. A Normal plot of these effects is displayed to the user.

iii. A list of the effects names, their alias structure and calculated values are displayed; and the user is asked to determine which of these effects are to be regarded as insignificant (noise).

iv. The standard error of the effects is estimated from the noise effects chosen in step 'iii' above.

After establishing a method for estimating the standard error of effects the Lambda plot is displayed to the user with the following statement on how to choose the best scale for data transformation:

> "The best 'Lambda' value for transformation is the value at which the greatest separation and simplification of the t-ratios of the effects is located. Separation means that wide and high t-ratios and their corresponding effects are distinguishable from noise, whereas simplification means the use of a minimum number of significant effects and interactions are used in establishing the process model."

The user is asked to choose a suitable value from the displayed lambda values for data transformation according to her/his best interpretation.

Main and Interactions Effect Calculation

The main and interaction effects are calculated using the correct scale for data transformation and displayed to the user. Also, the normal plot of these effects and their interactions are shown. The standard error of the effects is calculated according to the strategy explained in the previous section.

Mathematical Model Building

A mathematical model that represents the data is estimated at this stage. One of two strategies is used for model building:

1. Significant effects are determined by user input. If this strategy is adopted a list of the effect names, their aliases and estimated values are displayed to the user; and the user is asked to choose the significant effects. Hence, the model is constructed from the data mean and the chosen effects.
2. Significant effects are determined according to a confidence interval ($1-\alpha$) chosen by the user. In this case the user is asked to choose a confidence interval ($1-\alpha$), to be used for significant effect testing. Next, an effect is considered significant if its t-ratio is equal to or greater than that value obtained from a t-distribution for a confidence interval alpha (α) and degrees of freedom equal to the number of error terms used in the estimation of the standard error.

Interactions representation and Model Testing

Significant interaction effects are shown in two forms:

1. Two-way plots for two-factor interactions.
2. Cube plots for three-factor interactions;

And the following tools are used to test the fitted model:

1. Normal plot of residuals
2. Plot of residuals vs run order
3. Plot of residuals vs predicted values

After testing the fitted model the user is asked whether the model is confirmed or a different one should be fitted. Model fitting and testing procedures are repeated upon user request.

Analysis of Standard Deviation (Dispersion Response)

This analysis is conducted when the user desires to study the variability of the response represented by the experimental data. This analysis could be conducted if the experimental data comes from any of the following designs:

1. Any Replicated SED.
2. Any Inner-Outer Array Design.

The standard deviations for the different experimental settings are calculated from replication or the outer-array design. These standard deviations are Chi-square distributed, but the logarithms of these standard deviations can be approximated by a normal distribution. Hence, the first step in the analysis of standard deviations is to log-transform these data in order to facilitate the analysis. The procedure for analyzing this data follows the same pattern as that used for the analysis of the means. However, since the standard-deviation responses are not replicated, the estimation of the standard error of the effects is carried out by using the same method used for the location effects for the case of a fractional factorial without replication, i.e., by a normal probability plot.

c. Analysis of Three Level SED. The analysis of three-level factorial experiments are limited to full factorial designs. Two types of analysis can be performed on the experimental data: i. Analysis of the mean, and ii. Analysis of the standard deviation. Similar procedures are used for both types of data analysis, which include:

1. *Data Transformation*

The Lambda plot cannot be used for three level factorial experiments because of the fact that the t-ratios used in this plot can only be obtained from a comparison of two levels. Thus, choosing the correct scale for data transformation at the initial stage of data analysis is not easy. In this regard, the ISE Expert System asks the user to input a suggested value of lambda to be used for data transformation. It should be pointed out that the suggested value can be revised after model fitting and checking. In other words, patterns that might show up on the plots of residuals versus predicted values are an indication of a wrong scale for data transformation and therefore another value for lambda should be tried.

2. *Significant effect determination and model building and testing*

The variances of the three-level factors are used to provide inferences about the means of the factors, and the analysis of variance method is used to calculate these variances. The significant factors and interactions are determined by comparing the F-ratios of these factors and interactions with that of the critical $F_{v1,v2;\alpha}$, where v_2 is the degree of freedom of the effect and $1-\alpha$ is the significant confidence level. The ISE Expert System also allows the user to choose the significant effects without regard to a significant confidence level.

A mathematical model is then fitted from the determined significant factors and interactions by using regression analysis. The fitted model can be tested by:

1. Normal plot of residuals
2. Plot of residuals vs run order
3. Plot of residuals vs predicted values

Finally, the user is asked to confirm the fitted model or to go back and fit another model or choose another scale for data transformation. This process continues until the user confirms the fitted models by residual analysis.

5. Tolerance Design Consultation.

This function of the ISE Expert System can be employed in two ways:

1. After experimental data analysis: In this case the Tolerance Design Consultation Function is invoked to find optimal settings for the process variables and to set the specification limits on the quality characteristics in question.
2. General case of tolerance analysis: In this case, if a mathematical model for the process is not available, then the SED function of the Expert System is consulted first, followed by the SED analysis procedures, and then the tolerance analysis consultation. Otherwise, tolerance analysis consultation procedures are utilized.

The theory of tolerance design is given in Reference [1], where a new methodology for tolerance design analysis was developed. This new methodology is called Integrated Simultaneous Engineering Tolerancing (ISET) and implementation of this methodology within the ISE Expert Systems is also possible. Following is an illustration on how ISET is implemented.

This example is depicted from an experiment reported by Pignatiello and Remberg [4]. In this experiment the heat treatment of leaf springs for trucks was studied and the free length of the spring with a target value of eight inches was investigated. As reported by Pignatiello and Remberg, this experiment was carefully designed and analyzed. However, for the purpose of illustrating the new concept of ISE, it is assumed that the investigation is to be conducted by engineers with limited background in statistics but using the Expert System. A brief presentation of a consultation session is described below, where "Q" denotes the question requested by the program, "A" represents the user's response to the inquiry, and "R" represents an additional remark with regard to "Q" and "A".

Q: Choose the type of consultation you want to perform:
   1. Process monitoring.
   2. Process diagnosis
   3. Statistical experimental design
   4. Statistical experimental design analysis
   5. Tolerance design consultation

A: No. 2

R: The engineers wanted to diagnose their process and hence, they chose item 2.

Q: Do you have data?

A: Yes

Q: What is the data type?
   1. Variable measurement data
   2. Attribute measurement data
   3. Individual measurement data

A: No. 1

Q: For variable measurement data the following charts are available
   1. $\bar{X}$-R charts
   2. $\bar{X}$-Moving Range Charts

   Please choose one of the above charts.

A: No. 1

R: According to the original example, variable measurement data in the proper form for charting was recorded and hence, the user chose item 1. At this point SPC control charts for $\bar{X}$-R are invoked and the user is asked to input his/her data. For the present data the $\bar{X}$-R charts show no out of control points, which means that the process is in statistical control. The process mean and standard deviation were also calculated, and are equal to 7.9074 " and 0.1625" respectively.

Q: What is the Upper (and Lower) specification limits on the product/process?

A: 8 ± 0.2

Q: What is the tolerance value of the product/process above which the product/process stops functioning properly (i.e., customer tolerance Δ_0)?

A: $\pm$ 0.3

Q: What is the loss (Δ_0) associated with this customer tolerance?

A: $15.00

Q: What is the loss to the manufacturer (A_1) if the product is scrapped or disabled at the processing stage?

A: $4.00

R: The input to the above four inquiries are assumed values, and are needed to determine whether the manufacturing tolerance is sufficient or not. Now, a capability study is performed and the value of C_p is found to be equal to 0.41 and $C_{pk} = 0.357$, i.e., process not capable and not centered. Also, by using the relation $\Delta = \Delta_0 \sqrt{A_1/A_0}$, the required specification is equal to ± 0.154. The user will be notified that his/her process is not capable and that the specification limits need modification. The following question will be asked:

Q: According to data analysis choose one of the following:

a. The process is acceptable. End consultation.
b. The process mean needs to be adjusted.
c. The process variance needs to be reduced.
d. Adjust process mean and reduce process variance

A: d

R: At this point the engineers realized that they need to adjust the process mean and reduce the process variance, and hence they chose item 'd' to continue. Thus, by choosing item d the SED function of the ISE Expert Systems is actuated, and the consultation proceeds.

Q: Input the total number of factors that have an effect on the free length of the springs?

A: 5

Q: Do you think that some of these factors are difficult or expensive to control?

A: Yes

Q: Input the number of these factors.

A: 1

R: As reported in the original experiment five factors are known to have an effect on the product quality characteristics and are listed in Table 1. However the engineers believe that the oil quench temperature is difficult to control and therefore, they classified this variable as a noise factor. Due to the existence of such a noise factor an Inner-Outer array design was recommended for this consultation. Next, the expert program prompted the user to input the name of these factors and the low and high levels to be used for an experiment. Table 1 lists the input to this inquiry where the first four variables constitute the inner-array and the fifth factor constitutes the outer-array of the design.

Table 1 Factors Involved in the Leaf Spring Experiment

	Factors	Symbol	Low Level	High Level
A	High heat temperature (F°)	TEMP	1840	1880
B	Heating time (Sec)	HEAT	25	23
C	Transfer time (Sec)	TRANS	12	10
D	Hold down time (Sec)	HOLD	2	3
E	Quench oil temperature (F°)	QUENCH	130-150	150-170

Q: How much time does it take on the average to change the setting of these factors from one level to another and start running a stable process?

A: 1/2 hr

Q: How much time does it take to run an average trial?

A: 1/4 hr

Q: Give a cost estimate of the amount needed to change to another set of trial condition
A: $5.00
Q: Give a cost estimate to operate the process at steady state and take duplicate samples
A: $0.5
R: The above values are needed to estimate the time and money needed for the experiment. The above inputs to the inquiries are assumed values. Now, the user will be informed of the approximate total time and money needed to perform the experiment. These values are calculated from the above information and assuming a full factorial design for the inner array with three replication, for this example. The next inquiry is:
Q: Do you think that you need to run the experiment in blocks due to any external conditions that might change during the experiment.
A: No
R: To remind the user to use a blocked design if needed.
Q: According to the above estimation and your knowledge of the process, choose one of the following:
a. Accept the initial recommended setup
b. Reduce trials by having less replication.
c. Reduce trials by using a fractional factorial design
A: c
Q: Choose the number of fractions: 0 or 1
A: 1
R: According to the original experiment the engineers believe that some of the two factor interactions might not have a significant effect on the process and hence they did not need to run a full factorial design. For the inner design a choice of fraction of 1 means a half factorial design, i.e., only 8 trials while 0 means a full factorial design. Now the resolution of the inner design will be calculated and displayed to the user.
Q: The resolution of this inner design is 4.
R: In a resolution 4 design, two factor interactions are confounded with other two factor interactions.
Q: Do you want to get a higher resolution design?
A: No
R: As noted before the engineers believe that they know the type of interaction factors and hence they accepted the resolution of the design. Now, the new estimate of time and money will be displayed and the following question is raised.
Q: According to the new estimates of time and money choose one of the following:
a. Accept the proposed new design.
b. Go back to the original menu
A: a
R: At this stage the design of the experiment will be sent to a file which shows the sixteen unique trial combinations, the random order of the actual runs and the standard order of the design. The user will be prompted to use this file which is shown in Figure 4 in order to perform his experiment. Also he/she will be advised to come back to the Expert System program for data analysis.
Assuming that the experiment is performed and that the results of this experiment are the same as those reported by Pignatiello and Ramberg, and also that the random order shown in Figure 4 represents the order of runs for the experiment, then the last column of this figure represents the responses of this assumed run order.
Now, the user goes back to the ISE Expert System and starts from the beginning. In this case the user will choose item four of the first question, which is the SED analysis function of the Expert System. The user is prompted to input the results.
Q: What type of analysis do you need to perform?
1. Analysis of location effects
2. Analysis of location effects
3. End analysis procedures
A: No. 1
R: This type of analysis is conducted to show the significant factors that have an effect on the process mean.
R: The following analysis procedure is conducted:
1. Lambda Plot
A Lambda plot is displayed to the user, where the main purpose of this plot is to check whether the data needs transformation. Figure 5 shows this plot with a brief help statement on how to use this plot. The best scale for a transformation

STANDARD ORDER DESIGN SETUP
WITH 3 REPLICATIONS

Run #	Temp	Heat	Trans.	Hold	Quench	Response
1	1840	25	12	2	140	7.78
3					160	7.50
13	1880	25	12	3	140	8.15
31					160	7.88
36	1840	23	12	3	140	7.50
14					160	7.50
28	1880	23	12	2	140	7.59
27					160	7.63
8	1840	25	10	3	140	7.94
37					160	7.32
43	1880	25	10	2	140	7.19
26					160	7.56
47	1840	23	10	2	140	7.56
32					160	7.18
46	1880	23	0	3	140	7.56
2					160	7.81
35	1840	25	12	2	140	7.78
22					160	7.25
39	1880	25	12	3	140	8.18
41					160	7.88
4	1840	23	12	3	140	7.56
16					160	7.56
17	1880	23	12	2	140	7.56
19					160	7.25
5	1840	25	10	3	140	8.00
38					160	7.44
29	1880	25	10	2	140	8.09
23					160	7.69
44	1840	23	10	2	140	7.62
7					160	7.18
18	1880	23	10	3	140	7.81
45					160	7.50
40	1840	25	12	2	140	7.78
33					160	7.12
30	1880	25	12	3	140	7.88
12					160	7.44
21	1840	23	12	3	140	7.50
24					160	7.50
9	1880	23	12	2	140	7.75
25					160	7.56
20	1840	25	10	3	140	7.88
34					160	7.44
42	1880	25	10	2	140	8.06
15					160	7.62
48	1840	23	10	2	140	7.44
10					160	7.25
11	1880	23	10	3	140	7.69
6					160	7.59

RANDOM ORDER
DESIGN SETUP

Run #	Temp	Heat	Trans	Hold	Quench	Response
1	1840	25	12	2	140	7.78
2	1880	23	10	3	160	7.81
3	1840	25	12	2	160	7.50
4	1840	23	12	3	140	7.56
5	1840	25	10	3	140	8.00
6	1880	23	10	3	160	7.59
7	1840	23	10	2	160	7.18
8	1840	25	10	3	140	7.94
9	1880	23	12	2	140	7.75
10	1840	23	10	2	160	7.25
11	1880	23	10	3	140	7.69
12	1880	25	12	3	160	7.44
13	1880	25	12	3	140	8.15
14	1840	23	12	3	160	7.50
15	1880	25	10	2	160	7.62
16	1840	23	12	3	160	7.56
17	1880	23	12	2	140	7.56
18	1880	23	10	3	140	7.81
19	1880	23	12	2	160	7.75
20	1840	25	10	3	140	7.88
21	1840	23	12	3	140	7.50
22	1840	25	12	2	160	7.25
23	1880	25	10	2	160	7.69
24	1840	23	12	3	160	7.50
25	1880	23	12	2	160	7.56
26	1880	25	10	2	160	7.56
27	1880	23	12	2	160	7.63
28	1880	23	12	2	140	7.59
29	1880	25	10	2	140	8.09
30	1880	25	12	3	140	7.88
31	1880	25	12	3	160	7.88
32	1840	23	10	2	160	7.18
33	1840	25	12	2	160	7.12
34	1840	25	10	3	160	7.44
35	1840	25	12	2	140	7.78
36	1840	23	12	3	140	7.50
37	1840	25	10	3	160	7.32
38	1840	25	10	3	160	7.44
39	1880	25	12	3	140	8.18
40	1840	25	12	2	140	7.78
41	1880	25	12	3	160	7.88
42	1880	25	10	2	140	8.06
43	1880	25	10	2	140	7.69
44	1840	23	10	2	140	7.62
45	1880	23	10	3	160	7.50
46	1880	23	10	3	140	7.56
47	1840	23	10	2	140	7.56
48	1840	23	10	2	140	7.44

Figure 4 Inner-Outer Array Design. (Response is Free Height)

should be chosen at the point where the greatest separation and simplification of the t-ratio occur, i.e., the point at which minimum significant effects and interactions occur. It is clear from this figure that no transformation is needed for the data since the Lambda plot does not show a clear preference for any lambda value, i.e., a transformation is not necessary.

2. Effects analysis:
   The names of the effects for the inner $2^{(4-1)}$ SED and their alias structure and estimated values are displayed to the user as shown in Table 2. Next, the normal plot of these effects is displayed (see Fig. 6) where it is clearly seen that effects A, B, and D can be regarded as significant.

Table 2 Estimated Effects and Their Alias Structure

Alias structure	Estimated effect
A'BCD	0.221
B'ACD	-0.176
C'ABD	-0.029
D'ABC	0.104
AB'CD	-0.017
AC'BD	-0.020
AD'BC	-0.035

Key: A - Temp. B - Heat C - Trans. D - Hold

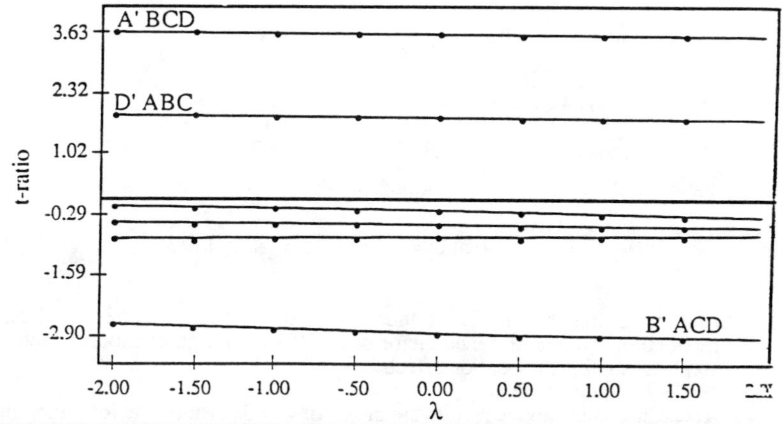

The best 'lambda' value for tranformation is the value at which the greatest separation and simplification of the t-ratios of the effects is located. <u>Separation</u> means that wide and high t-[illegible] and their corresponding effects are distinguishable from noise, whereas <u>Simplification</u> means [illegible] of a minimum number of significant effects and interactions are used to establish the process [illegible].

Figure 5 Lambda Plot

3. Significant effects analytically determined.
The user is asked to choose the significant level for the analysis from a set of different α values. For this example $\alpha = 0.1$ is chosen. Now, the t-ratios of the effects will be calculated and displayed as shown in Table 3 and the user is prompted to choose the effects that she/he considers significant. The user can also let the program decide the significant effects based on the level of significance chosen at the start of this step.

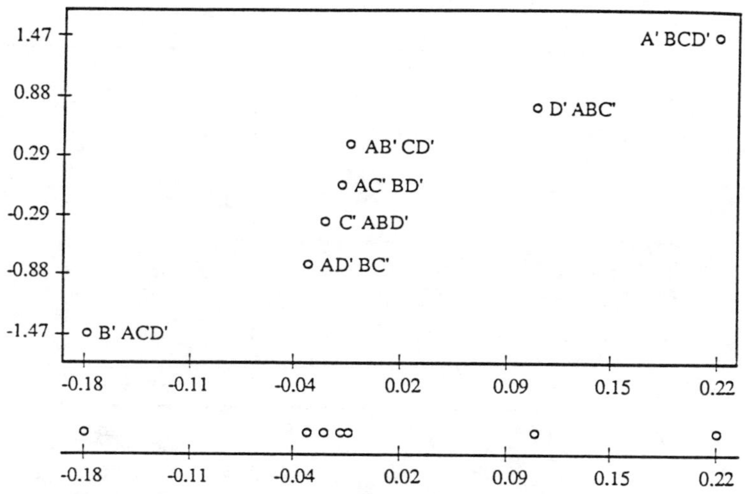

Figure 6 Normal Plot of Effects.

Table 3 The Estimated Effects and Their t-Ratios

Alias structure	Effect	Estimated t-value
Average	7.636	
A'BCD	0.221	3.57
B'ACD	-0.176	-2.84
C'ABD	-0.029	-0.46
D'ABC	0.104	1.67
AB'CD	-0.017	-0.28
AC'BD	-0.020	-0.32
AD'BC	-0.035	0.57

Note: The t-ratio at $\alpha = 0.1$ and DF = 40 is 1.3

Assuming that the choice of the significant effects is to be based on a true replication and a confidence factor of $\alpha = 0{,}1$ the program chooses factors A, B, and D as significant location effects.

4. Model building and testing:
According to the above results the program will determine the following model to represent the process:

$$\hat{Y} = 7.636 + 0.1105*(\text{TEMP}) - 0.088*(\text{HEAT}) + 0.052*(\text{HOLD}) \quad (1)$$

5. Residuals Calculated for Checking Model Assumptions:
Now, the estimated values at the eight distinct combinations of process setting will be calculated and are called the predicted values. The differences between the predicted values and the observed experimental values are also calculated and are called the residuals. The normal plot of these residuals is shown in Figure 7 which shows that these residuals comes from an approximate normal distribution and hence suggest that the model is adequate. Figure 8 shows the plot of the residuals vs the run order and is shown here for the sake of presenting a complete picture about the analysis program. However, this figure does not give the true plot since the original run order is not known. However, in an actual experimental analysis, this plot should not show any clear relationship between the residuals and the run order if it is to be used as n indication of the model's acceptability. Figure 9 shows the plot of residuals versus the predicted values where no pattern can be identified, which also supports the hypothesis of the adequacy for the proposed model. Finally the user is asked if whether she/he wants to repeat the model fitting procedure or not. For the present case the fitted model appears to be very adequate and there is no need to repeat this procedure.

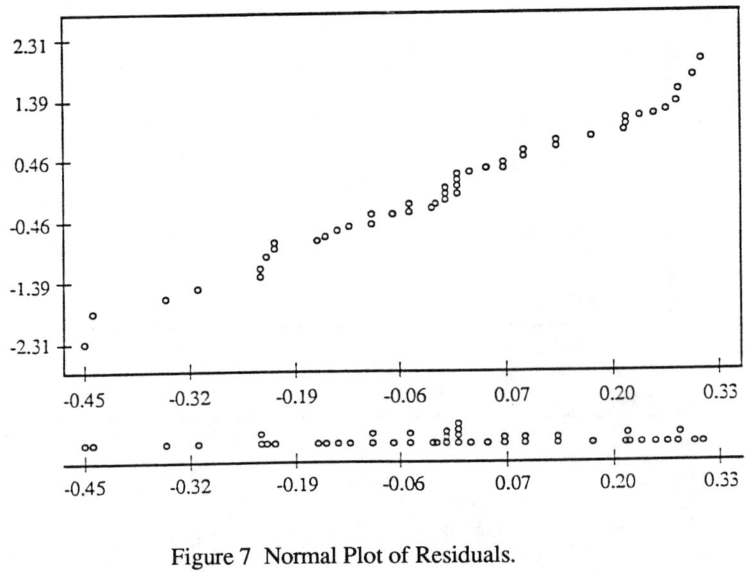

Figure 7 Normal Plot of Residuals.

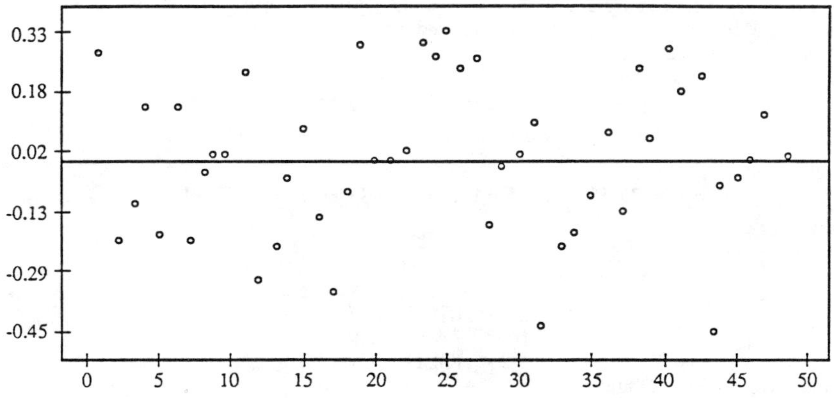

Figure 8 Plot of Residuals vs Predicted Values.

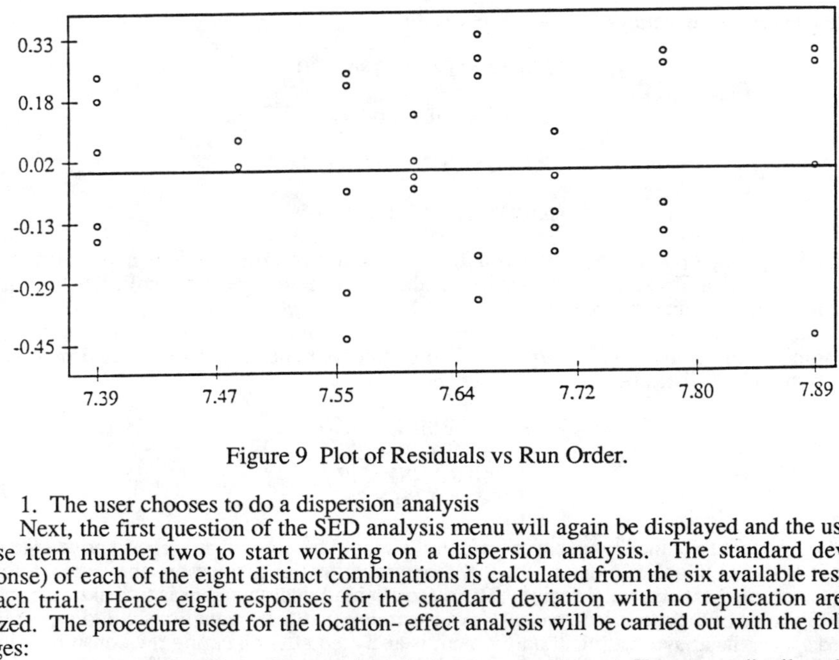

Figure 9 Plot of Residuals vs Run Order.

1. The user chooses to do a dispersion analysis

Next, the first question of the SED analysis menu will again be displayed and the user will choose item number two to start working on a dispersion analysis. The standard deviation (response) of each of the eight distinct combinations is calculated from the six available responses for each trial. Hence eight responses for the standard deviation with no replication are to be analyzed. The procedure used for the location- effect analysis will be carried out with the following changes:

a. Data handling: The standard deviation responses are Chi-square distributed which means they are not easy to analyze. However, the logarithms of these values can be approximated by a normal distribution The ISE Expert System adopts this theory for the dispersion analysis.
b. Standard Error estimation: This value is estimated from effects considered to be not significant.

The fitted model for the standard deviation is

$$\hat{\sigma} = \left(10^{(-0.49-0.135 \cdot \text{Heat})} \right)^{\left(\frac{-1}{2.5}\right)0.5} \qquad (2)$$

2. Determining the optimum settings

The next step of the data analysis is to set the variables at their optimal values. To set up for this situation the user is asked to write a production cost relationship. The assumed production cost relationship is represented by Eq. (7) below. The optimal setting of the variables are to be determined using the following optimization procedure:

Minimize the total cost

$$\left(\frac{A_o}{\Delta_o^2}\right) \bullet (S^2 + (\hat{y} - T)^2) + C \qquad (3)$$

subject to

$$8 = \hat{y} = 7.636 + .1105 \bullet \text{Temp} - .088 \bullet \text{Heat} + 0.052 \bullet \text{Hold} \qquad (4)$$

$$D > 3*S \qquad (5)$$

$$1820 < \text{TEMP1} < 1880 \qquad (6a)$$
$$23 < \text{HEAT1} < 25 \qquad (6b)$$
$$10 < \text{TRANS1} < 12 \qquad (6c)$$
$$2 < \text{HOLD1} < 3 \qquad (6d)$$

Where S is given by Eq. (2) and

$$C = 1 + 0.00025*\text{TEMP1} - .02*(4*\text{HEAT1} + 5*\text{TRANS1} + 1.5*\text{HOLD1}) \qquad (7)$$

and

$$\Delta = \Delta_o \sqrt{\frac{A_1}{A_o}} \qquad (8)$$

and where the coding equations for the variables are:

$$\text{TEMP} = (\text{TEMP1} - 1860)/20 \qquad (9a)$$

$$\text{HEAT} = (\text{HEAT1} - 24)/1 \qquad (9b)$$

$$\text{TRANS} = (\text{TRANS1} - 11)/1 \qquad (9c)$$

$$\text{HOLD} = (\text{HOLD1} - 2.5)/0.5 \qquad (9d)$$

In addition $A_1 = C$, $A_0 = 15$, $A_0 = 0.3$ and $T = 8$. The above problem does not converge when using a computer optimization routine, i.e., it does not yield a feasible solution because the constraints in Eqs. (4) and (5) are not met.

Thus, the user will be prompted to help solve this problem. The user is asked to try to loosen some of the constraints of Eq. (6). The engineers believe that they could modify their specification limits as listed below:

$$1820 < \text{TEMP1} < 1890 \qquad (10a)$$
$$22 < \text{HEAT1} < 27 \qquad (10b)$$
$$10 < \text{TRANS1} < 12 \qquad (10c)$$
$$2 < \text{HOLD1} < 3 \qquad (10d)$$

where the first two tolerance ranges were widen. It should be noted that a good knowledge of the process is essential in order to determine the above modifications, since the above constraints go beyond the original experimental region. The reason behind this statement is to make sure that Eq. (1) and (2) hold for the modified space. Now, assuming that these equations could represent the modified space, the above optimization problem is analyzed after changing the constraints to Eqs. (10). A feasible solution is then found, and the following values are obtained from this solution:

1. High heat temperature (TEMP1) = 1887.5 F°
2. Heating time (HEAT1) = 27 Sec
3. Transfer time (Trans1) = 10 Sec
4. Hold down time (HOLD1) = 2 Sec

From Eq. (2) the standard deviation is 0.04468, and the process mean is centered at eight inches.

After setting the process at the above optimum values, the process could be operated, sequential samples obtained, and the process monitored by statistical control charts for confirmation. This step could be achieved by implementing the SPC monitoring option of the Expert System. It is important to note that if the optimization problem does not achieve a feasible solution, the Tolerance analysis procedures should be followed as outlined by Araj and Ermer [1].

SUMMARY

The implementation of the emerging role of two complementary concepts, off-line quality control and on-line quality control in the fields of product/process design and manufacturing is the major thrust behind the expert system. Off-line quality control is a methodology for designing quality and robustness into the product/process at early design stages, through Taguchi's cycle of system design, parameter design and tolerance design. At the same time, on-line quality control is a methodology for monitoring the manufacturing processes in real time so that any process variation can be detected and corrected. A new methodology that integrates both of these concepts in a cycle of never ending improvement for producing robustness has been developed and is called Integrated Simultaneous Engineering.

The developed ISE methodology employs engineering statistical tools to implement the approach, and stresses the use of Statistical Experimental Design at the early stages of product/process design and development. Statistical Experimental Design is used to study the independent process/product variables (parameters) during the parameter design stage so that the process/product can be put on target with minimum variation. Statistical experimental results combined with Taguchi's Loss Function Principle and non-linear optimization techniques are used to put the product/process on target with not only minimum variation, but with minimum cost while meeting customer requirements. At the same time, since manufacturing and environmental variations cannot be removed completely, statistical process control tools are used during the manufacturing stage so that any process variation can be detected and removed if a special cause is found, or reduced in the case of common cause variation.

A new methodology for tolerance analysis called ISE tolerancing for tolerance design and allocation is also included in the expert system. The main goal of this methodology is to allocate the individual tolerances of product/process variables so that the final manufacturing tolerance can meet the customer requirements. This method is a more refined stage of experimental design and optimization techniques, where Taguchi's inner-outer array experiments are used to study the effect of slight variations of the variables nominal values on the total tolerance of the product/process. This approach is combined with non-linear optimization methods so that optimal tolerances of the process and product variables can be allocated. This tolerance design methodology was illustrated by a processing example, and it can be used for any general application in product or assembly design and process development.

The main advantages that can be gained from implementing the ISE methodology are:

1. Effectively and efficiently combines off-line and on-line quality control.
2. Helps integrate the design and manufacturing functions.
3. Combines statistical control charts and statistical experimental design for continuous improvement of product design and quality.
4. Integrates Taguchi's Loss Function with statistical experimental design and non-linear optimization to produce robustness and manufacturability at low cost, while meeting customer requirements and satisfaction.
5. Provides an easy to implement method for tolerance design and allocation that can be used for quality problems as product design, assembly and processing.

An expert system that contains the knowledge base of the full scale ISE methodology is called the ISE Expert System, and can provide five consultation functions: 1. Process diagnosis, 2. Process monitoring, 3. Planning Statistical Experimental Designs 4. Analysis of Statistical Experimental Design and 5. Implementation of Tolerance Analysis. These five functions can be used separately or they can be integrated to implement the full scale ISE methodology. The main advantages that can be achieved from using the Expert System are:

1. Promotes the implementation of Integrated Simultaneous Engineering.
2. Provides an easy to use, but effective tool, for engineers with minimum knowledge and experience in applying statistics.
3. Integrates new and old ideas of statistical experimental design, e.g., handles both dispersion and location analyses.
4. Prevents engineers from using misleading methods currently reported in engineering and quality publications on the application of statistical methodologies.
5. Disseminates rare and costly expertise for more effective and efficient use of human experts, which will save resources by preparing and conducting proper experiments.

It should also be pointed out that the ISE Expert System can be improved and enhanced by continuous upgrading and expansion of its knowledge-base. This can be achieved by improving the present structure of the knowledge base and by adding new rules or codes to this knowledge-base structure.

Thus the ISE Expert System developed covers the full methodology of ISE. The five integrated functions of this Expert System are designed to deal with the broad issues of product/process design and process diagnosis/monitoring, and a major advantage of this Expert generality. On the other hand, this generality has its disadvantage in regards to one of the Expert System functions, namely, the process diagnosis function. This function is composed of two parts, the verbal and the statistical process control charts diagnoses. The diagnostic knowledge base of these two parts deals with the general problems that might affect the manufacturing processes. Hence if a special processing problem is detected it cannot be traced to its roots but rather to its general source. Therefore, a major enhancement to the system can be achieved by tailoring the knowledge base of this diagnostic function to fit a certain manufacturing process, as for example, a plastic injection modelling process. A greater advantage can be gained by expanding this knowledge base to contain much wider applications in industry, such as polymer processing in general.

Overall the existing ISE Expert System can handle many types of quality decisions, and has the capability to be expanded as experience with its use is gained. Thus the cycle of never ending design and quality improvement can go on by applying the Integrated Simultaneous Engineering Methodology developed.

REFERENCES

1. Araj, S. and Ermer, D.S., "Integrated Simultaneous Engineering Tolerancing," PED-Vol 42, p. 97, Winter Annual Meeting, December 1989.
2. Taguchi, G., *Introduction to Quality Engineering*, Asian Productivity Organization, Tokyo, 1986.
3. Box, G. and C.A. Fung, "Some Consideration in Estimating Data Transformation," MRC Technical/Summary Report #2609, Mathematics Research Center, University of Wisconsin-Madison, 1983.
4. Pignatiello, Jr., J.J. and Rember, J.S., "Discussion on Taguchi Methods," Journal of Quality Technology, Vol. 17, No. 4, pp. 198-206, October 1985.
5. Dessouky, M.I., Kapoor, S.G. and DeVor, R.E., "A Methodology for Integrated Quality Systems," ASME Journal of Engineering for Industry, Vol. 109, pp. 241-247, 1987.
6. DeVor, R.E., "The Role of Parameter Design in Simultaneous Engineering of Products and Processes," Quality: Design, Planning and Control, PED-Vol. 27, pp. 131-135, New York, ASME WAM 1987.
7. Ermer, D.S., "Proper Bases for Establishing SPC Charts," ASME WAM, Dec. 1987, PED-Vol. 27, p. 55.
8. Cox, N.D., "Tolerance Analysis by Computer," Journal of Quality Technology, Vol. 11, No. 2, p. 80-87, April 1979.
9. Cox, N.D., "How to Perform Statistical Tolerance Analysis" ASQC Quality Congress Transactions, pp. 223-226, Minneapolis, 1987.
10. Liao, Y.S., Wu, S.M. and Ermer, D.S., "A Time Series Approach to Quality Assurance," Inspection and Quality Control in Manufacturing Systems, ASME, PED, Vol. 5, pp. 25-32, November, 1982.
11. Notohardjono, B.D. and Ermer, D.S., "Time Series Control Charts for Correlated and Contaminated Data," ASME Journal of Engineering for Industry, Vol. 108, pp. 219-226, August 1986.
12. Phadke, Madhav S., *Quality Engineering Using Robust Design,* Prentice Hall, Englewood Cliffs, N.J., 1989.

**Sensors, Controls, and Quality Issues
in Manufacturing
ASME 1991**

A STATISTICAL PROCEDURE FOR DEFECT CONTROL IN HIGH QUALITY MANUFACTURING

T. N. Goh
Department of Industrial and Systems Engineering
National University of Singapore
Singapore

Abstract

Advanced technologies today are such that it is possible to keep the occurrence of defects in manufactured products at very low levels. The use of the conventional c-chart for statistical control of defects in high quality products would encounter serious practical difficulties because the small number of defects would render invalid the assumptions behind the construction of the chart. Based on reasoning with fundamental probability distributions, this paper offers a reliable solution that is particularly suited to on-line inspection and testing operations such as those found in an automated manufacturing environment.

INTRODUCTION

The statistical approach to quality control has long been known to production engineers. The application of statistical techniques aims at alerting the user any variations in the properties of a manufactured product that are considered originating from causes other than random fluctuations inherent in the parameters of the physical process generating the product: see, for example, Duncan (1974) or Grant and Leavenworth (1988) for an account of the theoretical foundation and implementation of such techniques.

With the advent of automation and advanced manufacturing technologies, it is now possible to reduce the disturbances of extraneous parameters in manufacturing processes to an extent that instances of non-conformities to product specifications are literally very few and far between. This has given rise to a situation where some well-established techniques in statistical quality control, such as lot acceptance sampling plans and process control charts, are becoming difficult to apply since defects in products are non-existent most of the time. This paper addresses the difficulties faced by the users of one common type of process control charts, namely the *c*-chart, for the statistical control of the number of defects found in a continuous stream of manufactured goods. The problem is solved by the application of fundamental probability considerations which replace the statistical approximations implicitly used in the construction of such charts by production or quality control engineers.

CONVENTIONAL *c*-CHART

The *c*-chart is applicable in cases where there is a continuous output of identical product units that can be sampled for inspection. Product defects defined by various functional or cosmetic descriptions are detected and counted from on-line samples. A sample may consist of one or more discrete units of production, for example six gearboxes, or it may comprise a given amount of the product, say 10 m of a chain or 100 m^2 of a rolled metal sheet: whatever the definition, the sample size remains constant from sample to sample.

A production process is considered to be statistically *under control* as long as the number of defects c per sample fluctuates in a random manner. If c is found to be excessive, then the process is deemed *out of control* and must be halted for necessary rectifications. A c-chart is constructed first by inspection of a number, say k, of preliminary samples collected from a steady-state process. With each sample containing c_i defects, where $i = 1,2,...,k$, the average number of defects per sample is given by

$$\bar{c} = \Sigma c_i / k \tag{1}$$

Since the occurrence of defects is random in nature, c can be characterized by the Poisson distribution. This distribution has a mean and a variance that share a value that can be readily estimated by $\bar{c}$. The c-chart can then be set up with a central line (CL), a lower control limit (LCL), and an upper control limit (UCL) defined by

$$CL = \bar{c} \tag{2}$$

$$LCL = \bar{c} - 3(\bar{c})^{1/2} \tag{3}$$

$$UCL = \bar{c} + 3(\bar{c})^{1/2} \tag{4}$$

respectively. Any preliminary sample which yields a value of c_i that does not satisfy

$$LCL < c_i < UCL \tag{5}$$

is investigated and removed, and calculations with Eqs (1) to (4) repeated so that eventually all c_i values fall between the current LCL and UCL, thereby establishing the final values of CL, LCL and UCL for the chart. Any new value of c_i, $i > k$, from subsequent sampling must fall between the control limits, as illustrated in Fig. 1; otherwise the process is considered out of statistical control.

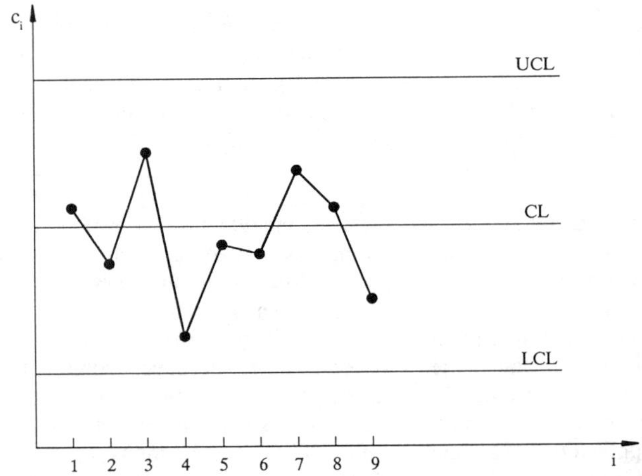

Fig. 1 Typical *c*-chart for control of defects in a product

The operation of the c-chart is based on the assumption that values of c_i, having a Poisson distribution, can be approximately described by a normal distribution with both mean and standard deviation estimated by $\bar{c}$. As indicated by Eqs (3) and (4), the *LCL* and *UCL* are each three standard deviations away from the mean; when the process is under control and the normal approximation of the Poisson distribution is justified, the probability that any single c_i value falling either below the *LCL* or exceeding the *UCL* is only 0.00135. Since this is a very small probability, whenever a c_i value does fall beyond the *UCL*, it is highly likely that the process is out of statistical control, *i.e.* the mean number of defects per unit is now significantly more than $\bar{c}$, and an investigation of the underlying causes is called for. In the event that the mean has actually remained unchanged, then a value of c_i above the *UCL* is a false alarm, known as *Type I error* in statistical inference: its probability of 0.00135 is deemed tolerable in practice.

PROBLEMS AT LOW DEFECT LEVELS

Although effective most of the time, the c-chart becomes inappropriate when the number of defects per sample is too small, as then the normal approximation to the Poisson is no longer valid. To illustrate this situation, consider a process that has improved to such an extent that there is now only an average of 0.04 defects per sample. According to Eqs (2) to (4), $CL = 0.04$, $LCL = -0.56$, $UCL = 0.64$ respectively. Since a negative value has no physical meaning, the *LCL* is often set to 0. This leads to a chart illustrated in Fig. 2.

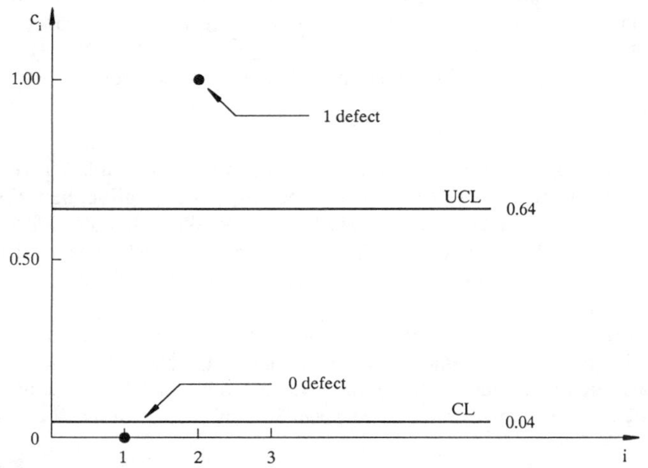

Fig. 2 Example of c-chart when a very low number of defects are encountered

From several points of view, the use of c-chart for process control at such a low value of $\bar{c}$ is hardly valid. First, the control limits are not symmetrical about the central line, which suggests that the idea of three standard deviations from either side of the mean as the limits of natural variation is no longer applicable. Secondly, at low levels of defects, the c-chart is sluggish in its response to process shifts. Suppose the average number of defects per unit has increased to 0.10, for which the *UCL* as given by Eq (4) is 1.05. Since the probability that any individual sample will carry more than one defect is only 0.005, no c_i value will exceed the *UCL* for a long time: according to the theory of geometric distribution, it will take an average of 1/0.005 or 200 samples for the first of such samples to appear. Thus any subsequent action to investigate the deteriorated process would not be timely enough, as a situation classified as *Type II error* in statistical decision making has long taken place.

A more serious drawback of the c-chart for a low $\bar{c}$, such 0.04, is one that would be immediately apparent to the chart user. It can be seen from Fig. 2 that under such circumstances, the only way the process can be deemed under control is to keep generating samples containing no defects at all. One defect found in a sample will yield a c_i of 1.0, which exceeds the *UCL* and causes the process to be considered out of control. This is tantamount to an absolutely "zero defect" requirement which, with the process stabilized at 0.04 for $\bar{c}$, not only cannot be expected to be met in practice, but also runs counter to the very concept of statistical process control, namely elimination of systematic shifts but tolerance of random fluctuations in product quality. This situation surfaces whenever

$$\bar{c} + 3(\bar{c})^{1/2} < 1, \tag{6}$$

or $\bar{c}$ less than 0.092. Thus it is seen that for a low $\bar{c}$, the c-chart will lose its intended purpose since it tends to give rise to false alarms while the process is still under control, yet may not be capable of giving timely warnings when the process is truly out of control.

DEVELOPMENT OF ALTERNATIVE DECISION RULES

The difficulties described in the last section arise primarily from the fact that the state of control of a process is determined on the basis of defects found in single samples. An alternative set of decision rules will now be developed based on the Poisson distribution for the number of defects in single samples as well as the binomial distribution for the occurrence of defects in successive samples. The decision rules are devised such that the risk of unjustifiably stopping a process is kept at about the same level as that associated with a usual c-chart, namely 0.00135. This represents an unavoidable risk when decisions are based on imperfect information, *i.e.* results from inspection of samples.

The probability that d defects are found in a single sample is given by

$$P(d) = e^{-\mu}\mu^{d}/d! \tag{7}$$

where μ, the mean number of defects per sample, is estimated by $\bar{c}$. Table 1 gives the $P(d)$ values for μ up to 0.15; for this range of μ, $P(d)$ for d exceeding 1 is negligible, being considerably less than 0.01. Thus the detection of a defect in a sample, which makes a sample *defective*, is a binomial event, and the behavior of a number of consecutive samples can be analyzed by the binomial distribution with a view to detecting systematic shifts in process behavior.

The general framework for the control of a low defect process is based on the frequency of occurrences of defective samples, rather than the number of defects in individual samples. Let n be the number of consecutive samples inspected, and r be the number of defective samples found therein. The occurrence of a given pattern, defined by a specific combination of n and c, is governed by the probability

$$P(n, r) = {}^{n}C_{r}\,[P(1)]^{r}\,[1 - P(1)]^{r} \tag{8}$$

A series of easily recognizable patterns of occurrence of defective samples can be drawn up, such as "two consecutive defective samples", "two defective samples out of three consecutive samples", etc. When the process is still under control, the probability of occurrence of each pattern can be set at 0.00135, the same probability that a value of c_i will exceed the *UCL* of a c-chart. Thus with

$$P(n, r) = \alpha \tag{9}$$

where α is allowed to vary within 0.00135 ± 0.00020 to accommodate the constraints posed by integer values of n and r, the value of $P(1)$ can be solved for each pair of (n, r) combination. Table 2 summarizes the patterns and the associated probability values.

Table 1. Probability of occurrence $P(d)$ of d defects in a single sample

μ (estimated) by $\bar{c}$	d	
	0	1
0.01	0.990	0.010
0.02	0.980	0.020
0.03	0.970	0.029
0.04	0.961	0.038
0.05	0.951	0.048
0.06	0.942	0.056
0.07	0.932	0.065
0.08	0.923	0.074
0.09	0.914	0.082
0.10	0.905	0.090
0.11	0.896	0.098
0.12	0.887	0.106
0.13	0.878	0.114
0.14	0.869	0.122
0.15	0.861	0.129

Table 2. Probability of a given pattern of occurrence of defective samples

Pattern of occurrence	Binomial probability $P(1)$
1 out of 2	0.001
2 out of 5	0.012
2 out of 4	0.015
2 out of 3	0.021
2 consecutive	0.037
3 out of 5	0.053
3 out of 4	0.072
3 consecutive	0.110

Use can now be made of Tables 1 and 2 for the formulation of the appropriate process control procedure. For example, when $\bar{c} = 0.04$, Table 1 gives $P(1) = 0.038$. The closest $P(1)$ value in Table 2 is 0.037, which corresponds to a pattern of two consecutive defective samples. This means that one defective sample does not constitute sufficient evidence of an out-of-control situation, but if two defective samples occur successively, the process should be considered out of control. In the use of the two tables, if no close match can be found, then the computed $P(1)$ from a given $\bar{c}$ can be rounded down to the nearest lower available value in Table 2 for a more sensitive test pattern. For example, with $\bar{c} = 0.08$, Table 1 gives $P(1) = 0.074$; since the closest lower value in Table 2 is 0.072, the process can be monitored for 3 defective samples out of 4 inspected consecutively.

PRACTICAL GUIDES

Tables 1 and 2 are necessary for a full understanding of the reasoning behind the alternative process control principle. For the convenience of practitioners, ranges of values of $\bar{c}$ can be calculated and directly defined for the various patterns of sample behavior, as shown in Table 3. In actual applications, the value of $\bar{c}$ is first obtained from a steady-state process; process control can be effected by the conventional *c*-chart for values of $\bar{c}$ that exceed 1.5. The "pattern recognition" approach may be instituted for lower values of $\bar{c}$, and becomes a requirement when $\bar{c}$ falls below 1.0.

Table 3. Direct reference table for process control

$\bar{c}$	Pattern
0.001 to 0.012	1 out of 2
0.012 to 0.016	2 out of 5
0.016 to 0.022	2 out of 4
0.022 to 0.039	2 out of 3
0.039 to 0.056	2 consecutive
0.056 to 0.078	3 out of 5
0.078 to 0.125	3 out of 4
0.125 to 0.15	3 consecutive

With a given value of $\bar{c}$, an appropriate pattern is first obtained directly from Table 3, and a record of occurrence of defects maintained. The process being monitored is out of control whenever two or more defects are found in a single sample, but if only one defect is found, the process is not necessarily out of control; it is considered so only when, together with subsequent samples, the specified pattern is generated. For example, if *X* represents a sample carrying a defect and *O* represents one that is defect free, then after a long series of *O*'s, the generation of the pattern *X*-*O*-*O*-*X* is acceptable, *i.e.* the process considered under control, for $\bar{c} = 0.03$ (for which the out-of-control pattern is "2 out of 3") but not when $\bar{c} = 0.02$ (for which the pattern is "2 out of 4"). Thus whenever a new sample is inspected, the result is interpreted together with those of the immediate past few samples for pattern detection. The overall decision making procedure for the monitoring and control of defects in a product is summarized in Fig. 3.

CONCLUSIONS

The procedure described in this paper is useful for the monitoring and control of high quality production processes that give rise to only a very small number of defects in the product. It has the following advantages:

1. Simplicity in application: no process-specific calculations are needed, and the decision rules, based on patterns of defect occurrences, can be clearly laid out in a single reference table.

2. Compatibility with automation: the decision rules are particularly suited to automated sensing, testing, and record keeping as no intervention with human judgment is required.

3. Rigor in theoretical foundation: the decision rules are based on exact probability calculations rather than approximations of probability distributions as is the case with the conventional *c*-chart.

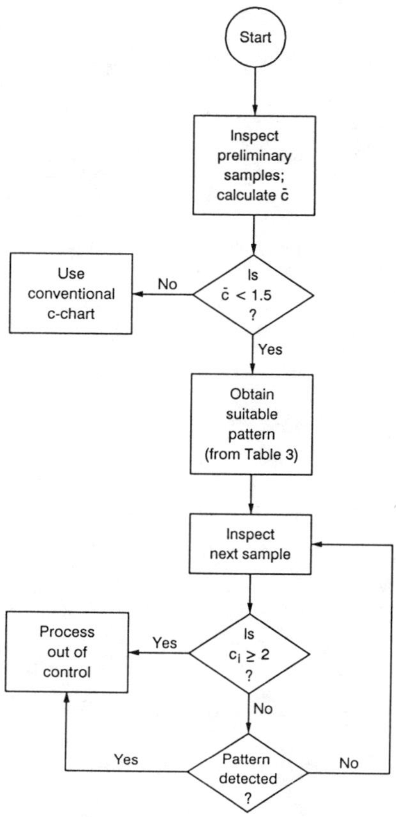

Fig. 3 Decision making in statistical monitoring and control of defects

4. Robustness in statistical inference: the procedure permits the occasional exhibition of random quality variations even in a very low defect situation, thus minimizing Type I error or false alarms; yet it is sensitive to actual shifts in process performance thereby reducing Type II error in process control.

It is clear that with the increasing sophistication in manufacturing and production systems, the above advantages are realized even more readily with on-line information processing and real-time systems of performance tracking and decision making. The ease of interpretation and presentation of product quality data would in fact make the procedure extremely attractive to industry personnel, ranging from management to shop-floor operators.

References

[1] Duncan, A.J., 1974, *Quality Control and Industrial Statistics*, 4th edition, Irwin.

[2] Grant, E.L., and Leavenworth, R.S., 1988, *Statistical Quality Control*, 6th edition, McGraw-Hill.

Sensors, Controls, and Quality Issues in Manufacturing
ASME 1991

DEVELOPMENT OF OPERATIONAL CHARACTERISTICS AND DIAGNOSTIC PROCEDURES FOR $\overline{X}$ CONTROL CHARTS

Gregory E. Rahn, S. G. Kapoor, and R. E. DeVor
Department of Mechanical and Industrial Engineering
University of Illinois, at Urbana-Champaign
Urbana, Illinois

ABSTRACT

Although control charts have had a tremendous impact on quality improvement, their inability to precisely measure performance has hindered their full potential of functional performance. Measures of performance are presented for traditional "three-sigma" charts and for $\overline{X}$ charts which incorporate four rules. "Three-sigma" charts are evaluated for power at a single subgroup, as well as for cumulative power over m subgroups. Power of $\overline{X}$ charts that employ four rules is formulated on the single-subgroup level using a rule dependency approach and is compared to the operational characteristics of "three-sigma" charts. Applications of these models to simulated data demonstrate their accuracy of power prediction. In addition, a diagnostic methodology is described which utilizes the derived performance measures to predict the mean and variance of a shifted distribution.

INTRODUCTION

Since their inception in 1931, Shewhart control charts have been a driving force towards an enhanced competitive position through quality improvement. The primary function of Shewhart control charts is to model the variability pattern of the process in order to characterize the level of common cause variability and to signal the presence of abnormal behavior due to special cause variability, thereby providing a basis for corrective action and subsequent process improvement. When monitoring off-line, control charts serve two purposes: (a) to establish initial process control by eliminating special causes of variability (b) to provide continuous process improvement through the removal of common causes of variability. When monitoring on-line, Shewhart charts serve as a tool at the machine in the decision-making process of whether to adjust the process or simply leave it alone (DeVor, 1991). Whether monitoring on-line or off-line, control charts provide a feedback system necessary for their successful application to a variety of processes. Although control charts clearly have withstood the test of time, some concerns regarding their performance have arisen as the proliferation of charting has evolved over the past ten years.

Shewhart used "three-sigma" limits on the $\bar{X}$ chart primarily to minimize false indications of lack of control (Shewhart, 1931) and (Wheeler, 1983). When all of the statistical assumptions associated with $\bar{X}$ charts are satisfied, the utilization of control charts will exhibit precise rates of false indications of lack of control (α) and false indications of control (β). The probability of a false alarm using Shewhart charts has been precisely determined to be .0026998 ≈ .0027. One criticism of using "three-sigma" limits is that their extreme confidence ($1-\alpha=.9973$) has resulted in moderate insensitivity to small shifts in the process. Consequently, three additional rules have been implemented over the years on the $\bar{X}$ chart to increase sensitivity toward various shifts in process location. The resulting rules are:

(1) a single subgroup falling beyond $\pm 3\sigma$
(2) two out of three consecutive subgroups falling beyond $\pm 2\sigma$
(3) four out of five consecutive subgroups falling beyond $\pm 1\sigma$
(4) eight consecutive subgroups falling strictly above or below the midpoint (DeVor, 1991).

All four of these rules are commonly accepted in practice, yet the probability of a false alarm (α) and the probability of a false indication of control under a given shift (β) have not been quantified, i.e. the behavior of $\bar{X}$ charts when all four rules are implemented is virtually unknown. α errors may lead to needless downtime, maintenance, and possibly equipment replacement while β errors tend to result in lost opportunities to improve the underlying process. Because of the severe consequences when observing false indications (α and β), the rate of their occurrence should be known precisely. The primary purpose of this paper is to derive exact α and β errors for control charts that employ all four rules.

This paper also attempts to expand the role of control charting procedures beyond the mere detection of abnormal behavior to include the diagnosis of faults entering the process. Exact probability determination is essential to obtain a model of normal behavior in terms of α and abnormal behavior in terms of β. However, when compared to α, the role of β in the implementation of control charts has been insignificant. For example, β, the probability of a false indication of control, has been used to compare different charting procedures or possibly to adjust control limits for specific applications, but it has had no bearing on real-time control chart monitoring. A methodology will be introduced which utilizes the characteristic behavior of derived β errors to estimate the mean and variance of the shifted distribution given observed performance.

The remainder of this paper will be devoted to the development of performance measures for $\bar{X}$ charts that incorporate four rules and the introduction of diagnostic procedures founded on these performance measures. Specifically, this investigation will: (a) state the definitive assumptions of $\bar{X}$ charts as well as specific assumptions required of this methodology (b) develop performance measures for "three-sigma" charts as well as for charts which use four rules (c) implement a diagnostic procedure based on these derived performance measures to predict the true shifted distribution in terms of its mean and variance. An application example will be discussed to demonstrate the utilization of this diagnostic procedure.

MATHEMATICAL ASSUMPTIONS FOR SPC CHARTING

Ultimately, the effectiveness of control charts stems directly from their ability to predict behavior under certain process conditions; therefore, performance measures must precisely be determined for $\bar{X}$ charts. In order to obtain precise probability estimates, several assumptions have been made regarding (1) distribution of sample means (2) charting procedures (3) manifestation of shifts.

Distribution of Sample Means

Exact probabilities can only be determined using **known distributions.** Specifically, $\bar{X}$ charts adhere to a standard normal distribution which assumes that (a) samples are identically and independently distributed (b) sample means are independently normally distributed (c) σ^2 is known and (d) μ is known. The standard normal transformation for sample means is $Z = \sqrt{n}\,(\bar{X}_i-\mu_0)/\,\sigma_0$. Upon examination of the deviates $[\bar{X}_i-\mu_0]$:

$$
\begin{aligned}
E[\bar{X}_i-\mu_0] &= \mu_i-\mu_0 \\
VAR[\bar{X}_i-\mu_0] &= VAR[\bar{X}_i] + VAR[\mu_0] - 2*COV[\bar{X}_i,\mu_0] \\
&= VAR[\bar{X}_i] \\
&= \sigma^2_0/n \qquad \text{since } \mu_0 \text{ is a constant.}
\end{aligned}
$$

Charts which use σ^2_0/n as the $VAR[\bar{X}-\mu_0]$ intuitively assume that μ_0 (the centerline of the $\bar{X}$ chart) is a constant rather than a random variable. Since the distribution of $\bar{X}_i$ is normal, the distribution of $[\bar{X}_i-\mu_0]$ is also normal.

Charting Procedures

All derived probabilities will be formulated under the policy of "control with respect to a given standard" which means that the estimates of μ and σ^2 (μ_0,σ^2_0) are not functions of the monitored data but are constants obtained from previously monitored data (King, 1954),(Olds, 1961),(Scheffe', 1959). The assumed mean, μ_0, which typically is the average of the sample means obtained from a previously monitored data set, is a realized constant with respect to the current data set: therefore, the deviations $[\bar{X}_i-\mu_0]$ are independent for all subgroups. This may seem trivial, yet monitoring processes that utilize $\bar{\bar{X}}$ as the centerline will **always** exhibit a reduced rejection rate due to the dependency of deviations $[\bar{X}_i-\bar{\bar{X}}]$.

Manifestation of Shifts

The behavior of $\bar{X}$ charts with respect to various shifts in process mean and variance was the primary focus of this investigation. In agreement with previous studies, this investigation observed the behavior of power under the condition that a shift occurred prior to actual monitoring of the process, such that all of the subgroups collected follow the same "shifted" distribution (King, 1952_a),(Olds, 1961),(Scheffe', 1959),(Wheeler, 1983). As a result, the null and alternative hypotheses for the $\bar{X}$ chart are: H_0: $\mu_i=\mu_0$, $\sigma^2_i=\sigma^2_0$ $\forall$ i and H_A: $\mu_i=\mu_A$, $\sigma^2_i=\sigma^2_A$ $\forall$ i (i=1,• •,m), respectively. This investigation did not incorporate shifts during the monitoring process primarily because the subsequent operational characteristics would have to be evaluated with respect to numerous shifts of various duration, and timing, as well as shift type and magnitude. However, the derivation of OC with all four rules can easily be modified to account for multiple shifts(distributions) throughout the monitoring process.

DEVELOPMENT OF PERFORMANCE MEASURES

Although control charts have had a tremendous impact on quality improvement, their inability to precisely measure performance has hindered their full potential of functional performance. Performance measures must be developed in order to determine the ability of control charts to signal the presence of abnormal behavior. Performance has typically been evaluated by using its Operational Characteristics (OC) curve, which yields the probability of rejection {or "acceptance"} of a single subgroup. Thus, a performance estimate can be obtained at each subgroup i. The probability of rejection at subgroup i will be referred to as OC on the **single-subgroup level**. Two conditions arise where the probability of rejection is of prime importance: (1) when the null hypothesis is true (2) when the null hypothesis is false. The null hypothesis for $\bar{X}$ charts on the single-subgroup level is $H_0: \mu_i=\mu_0$, $\sigma^2_i=\sigma^2_0$, where μ_0 and σ^2_0 are implied to be known constants. When the null hypothesis is true, i.e. the process is stable, a low probability of rejection(α) is ideal. However, when the null hypothesis is false, i.e. a shift of a given magnitude does occur, a high probability of rejection (power) is ideal, where power is equivalent to $(1-\beta)$. The OC curve on the single-subgroup level for the "three-sigma" chart has been thoroughly covered for shifts in the mean (King, 1954),(Olds, 1961),(Scheffe',1959).

When control charts are used to monitor m subgroups over a period of time, performance measures become cumulative in nature. In effect, one measure of performance can be used to characterize the entire m subgroups. For example, if one or more subgroups are rejected, the process does not constitute a constant system of common causes. The performance criterion on the **multiple-subgroup level** thus becomes the probability of one or more violations in m subgroups. Although the OC curve on the multiple-subgroup level appears to be a significant measure of control chart performance, it has received very little attention. When examining cumulative performance, the null hypothesis for $\bar{X}$ charts is $H_0: \mu_i=\mu_0$, $\sigma^2_i=\sigma^2_0 \ \forall\, i$ $(i=1,\bullet\bullet,m)$, where m is the number of subgroups monitored. Just as in the single-subgroup case, when the null hypothesis is true, a low probability of rejection is ideal, and when the null hypothesis is false, a high probability of one or more violations is desirable. These probabilities of rejection, on the single-subgroup level and on the multiple-subgroup level, must be determined to evaluate control chart performance.

Established performance estimates include: (a) the probability of rejection of "three-sigma" charts on the single-subgroup level for shifts in mean (King, 1954),(Olds, 1961) and for shifts in mean and variance (Kanazuka, 1986),(Scheffe', 1969) (b) the probability of rejection of "three-sigma" charts on the multiple-subgroup level for shifts in mean (Wheeler, 1983). This paper will specifically develop (**a**) the probability of rejection of "three-sigma" charts on the multiple-subgroup level for shifts in both mean and variance (**b**) the probability of rejection of the additional rules (2,3,4) and of (any rule) for $\bar{X}$ charts using four rules on the single-subgroup level.

$\bar{X}$ Chart (Rule 1): Single-Subgroup Level

The probability of detecting a rule 1 violation at subgroup i has been well documented and is given by the following formula:

$$P_{\bar{x}(1)_i} = \Pr[\bar{X}_i \geq \mu_0 + 3\sqrt{\sigma^2_0/n}\,] + \Pr[\bar{X}_i \leq \mu_0 - 3\sqrt{\sigma^2_0/n}\,] \qquad (1)$$

where μ_0 and σ^2_0 are the mean and variance under the null hypothesis. Note that this is a two-tailed test in that the null hypothesis can be rejected if the sample mean falls above the upper limit or below the lower limit. The subscript (1) of $P_{\bar{x}(1)_i}$ indicates that only rule (1) was implemented. The probability of rejection of rule 1 at subgroup i equals α under the null hypothesis and $(1-\beta)$ under the alternative hypothesis. Typically, the power of $\bar{X}$ charts has been portrayed strictly with respect to changes in process mean. However, Scheffe'(1959) and Kanazuka(1986) did acknowledge the significance of shifts in mean and variance on the power of "three-sigma" charts on the single-subgroup level. Since the $\bar{X}$ chart is a function of both $\bar{X}$ and S^2, shifts in process variance may in fact be manifested as a violation on the $\bar{X}$ chart. Therefore, the OC curve of the $\bar{X}$ chart should be viewed with respect to shifts in both mean and variance. When the process shifts, the data is assumed to come from a $N(\mu_A, \sigma^2_A)$ population. Upon conversion to a standard normal distribution where

$Z=(\bar{X}-\mu_A)/\sqrt{\sigma^2_A/n}$ (Kanazuka, 1986), equation (1) can be expressed as:

$$P_{\bar{x}(1)_i} = \Pr[Z \geq (\mu_0-\mu_A)/\sqrt{\sigma^2_A/n} + 3\sqrt{\sigma^2_0/\sigma^2_A}\,] + \Pr[Z \leq (\mu_0-\mu_A)/\sqrt{\sigma^2_A/n} - 3\sqrt{\sigma^2_0/\sigma^2_A}\,] \quad (2)$$

The probability of rejection can now be calculated for any shift combination of mean and variance. In order to efficiently portray these combinations, standardization measures must be developed because an infinite number of mean/variance combinations exist with respect to the alternative distribution $\{N(\mu_A,\sigma^2_A)\}$. These standardization measures are X_i and V_i where:

$$X_i = (\mu_A-\mu_0)/\sqrt{\sigma^2_0/n} \quad (3)$$

$$V_i = \sigma_A/\sigma_0 . \quad (4)$$

The measure X_i is the amount the true mean shifted from that specified by the standard, measured in units of the specified standard error of the mean. V_i is the ratio of the true standard deviation to that specified by the standard. Thus equation (2) can be expressed in terms of X_i and V_i as:

$$P_{\bar{x}(1)_i} = \Pr[Z \geq \frac{-X_i+3}{V_i}] + \Pr[Z \leq \frac{-X_i-3}{V_i}] \quad (5)$$

The experimental range of magnitude for X_i and V_i followed essentially the sampling distribution of the mean and variance, respectively. For example, V_i ranged from 0.4 to 1.6, which is illustrated by the following relationships:

$$P[\frac{\chi^2_{1-\alpha/2}\,\sigma^2_0}{(n-1)} < S^2_i < \frac{\chi^2_{\alpha/2}\,\sigma^2_0}{(n-1)}] = 1-\alpha \quad (6)$$

If S^2_i is replaced by σ^2_A and σ^2_0 is divided through, then:

$$P[\frac{\chi^2_{1-\alpha/2}}{(n-1)} < V^2_i < \frac{\chi^2_{\alpha/2}}{(n-1)}] = 1-\alpha \quad (7)$$

from which the approximate confidence interval from 0.4 to 1.6 is obtained for V_i. Shift in mean (X_i) was examined from 0 to 3. Since the subgroups are assumed to come from the same distribution over the entire monitoring process, X_i and V_i can simply be expressed as X and V, respectively.

The probability of rejection of rule 1 at any given subgroup, $P_{\bar{x}(1)_i}$, versus shift in mean (X) and variance (V) is given in Figure 1 in the form of a contour plot. Estimates for probability of rejection can be interpolated for any mean/variance combination. Generally, $P_{\bar{x}(1)_i}$ increases as shift in mean and variance increase, however the gradient with respect to variance(V) is minimal as mean(X) approaches 3. The gradient vector, which is the norm of the contours, could also be estimated numerically from this figure by dividing the change in the response function by the change in X and V.

Note that the definition of X is in terms of standard error of the mean, i.e. it is a function of sample size(n). If the sample size were increased, the variance component would be reduced whereby increasing the standardized shift (X). Since X increases for an increase in n, so does power. The transformation of X when going from a sample of size n to n′ can be expressed as:

$$X' = \sqrt{n'}/\sqrt{n} * X . \qquad (8)$$

Because of this standardization procedure, the same plots are valid for all n. This expression also shows that sample size has no effect on the probability of a false alarm when X equals 0. These findings are consistent with Wheeler's (1983) results.

$\bar{X}$ Chart (Rule 1): Multiple-Subgroup Level

For rule 1, OC on the multiple-subgroup level is easily obtainable from OC on the single-subgroup level because subgroups are independent. The resulting equation is:

$$\begin{aligned} \pi_{\bar{x}(1)} &= \text{Pr[1 or more violations (rule 1) in m subgroups on the } \bar{X} \text{ chart]} \\ &= 1 - \text{Pr[0 violations in m subgroups]} \\ &= 1 - \prod_{i=1}^{m}(1 - \text{Pr[Rejection (rule 1) at subgroup i]}) \qquad (9) \end{aligned}$$

where Pr[Rejection (rule 1) at subgroup i] equals $P_{\bar{x}(1)_i}$ $\forall$ i from equation (5) because each subgroup comes from the same shifted distribution. Therefore, equation (9) can be written as:

$$\begin{aligned} \pi_{\bar{x}(1)} &= \text{Pr[1 or more violations in m subgroups on the } \bar{X} \text{ chart]} \\ &= 1 - (1-P_{\bar{x}(1)_i})^m . \qquad (10) \end{aligned}$$

The contour plot of the probability of rejection on the multiple-subgroup level, $\pi_{\bar{x}(1)}$, is plotted with respect to shifts in mean and variance for m=25 subgroups in Figure 2. $\pi_{\bar{x}(1)}$ does converge to 1 within the given range of X and V. Moreover, Figure 2 describes the gradient of $\pi_{\bar{x}(1)}$ with respect to X as essentially equivalent to the gradient with respect to V. In Figure 3, the probability of rejection at any given subgroup, $P_{\bar{x}(1)_i}$, and the probability of 1 or more violations in 25 subgroups,$\pi_{\bar{x}(1)}$, are compared with respect to shift in the mean. Notice that when X≥2, the probability that "three-sigma" charts can detect at least one out-of-control subgroup in 25 subgroups exceeds 99%. Interestingly, this corresponds to a probability of rejection of only .17 at each subgroup when m equals 25 subgroups.

$\bar{X}$ Chart (Four Rules): Single-Subgroup Level

Although the OC curve has been given considerable attention for rule 1, a direct enumeration for rules (2,3,4) and for (any rule) has not been developed. When all four rules are used in conjunction, the desired characteristic on the single-subgroup level is Pr[violation (any rule) at subgroup i]. Upon inspection of

the sample space for all four rules on the $\bar{X}$ chart, it becomes evident that a single subgroup may violate more than one rule simultaneously. Because the areas associated with each rule are not mutually exclusive to each rule, the four rules are **dependent** on the single-subgroup level. For example, P[2/3 violation at subgroup i] is not equivalent to P[2/3 violation at subgroup i, given 3σ violation at subgroup i].

Therefore, spatial relationships must be used during OC modeling to characterize the dependency relationship between rules. The probability of a violation at subgroup i by (any rule) represents the union of the probabilities associated with the four rules used on the $\bar{X}$ chart. Therefore, the marginal and joint probabilities of these rule violations must be calculated such that:

$$P_{\bar{x}(Any)_i} = P[\text{Rejection (any rule) at subgroup i on the } \bar{X} \text{ chart}] = \tag{11}$$
$$P_{\bar{x}(1)_i} + P_{\bar{x}(2)_i} + P_{\bar{x}(3)_i} + P_{\bar{x}(4)_i} -$$
$$P_{\bar{x}(1,2)_i} - P_{\bar{x}(1,3)_i} - P_{\bar{x}(1,4)_i} - P_{\bar{x}(2,3)_i} - P_{\bar{x}(2,4)_i} - P_{\bar{x}(3,4)_i} +$$
$$P_{\bar{x}(1,2,3)_i} + P_{\bar{x}(1,2,4)_i} + P_{\bar{x}(1,3,4)_i} + P_{\bar{x}(2,3,4)_i} -$$
$$P_{\bar{x}(1,2,3,4)_i}$$

where the subscript (Any) indicates that all four rules were used on the $\bar{X}$ chart, $P_{\bar{x}(j)_i}$ is the marginal probability of a rule j violation at subgroup i, and $P_{\bar{x}(m,n,o,p)_i}$ represents the joint probability of the simultaneous violations of rules m,n,o,p at subgroup i.

__Computation of Marginal Probabilities.__ The marginal probability of rejection of rules (1)-(4) follows the properties of the binomial distribution in that these rules examine the number of successful subgroups falling beyond certain control limits from a particular window of subgroups , e.g. (1/1), (2/3), (4/5), (8/8). The derivations used to compute the marginal probability of rejection of each rule on the $\bar{X}$ chart at subgroup i are given by the following two general formulations:

CASE 1: (k successes out of k) FOR RULE j {j=1,4} (12)

If i<k) $P_{\bar{x}(j)_i} = 0.$

If i=k) $P_{\bar{x}(j)_i} = (PjUP)^k + (PjDN)^k$

CASE 2: (k successes out of l) FOR RULE j {j=2,3} (13)

If i<k) $P_{\bar{x}(j)_i} = 0.$

If i=k) $P_{\bar{x}(j)_i} = (PjUP)^k + (PjDN)^k$

If i>k) $P_{\bar{x}(j)_i} = \binom{l}{k}(PjUP)^k(1-PjUP)^{l-k} + (PjUP)^l + \binom{l}{k}(PjDN)^k(1-PjDN)^{l-k} + (PjDN)^l$

where:

PjUP equals $Pr[\bar{X} \geq \mu_0 + (\mathbf{4-j})\sqrt{\sigma^2{}_0/n}]$ and **PjDN** equals $Pr[\bar{X} \leq \mu_0 - (\mathbf{4-j})\sqrt{\sigma^2{}_0/n}]$

The specific formulations for each rule can be found in Rahn (1990).

__Computation of Joint Probabilities.__ Because rules are dependent on the single-subgroup level, joint probabilities are not equivalent to the product of marginal probabilities. Therefore, using spatial relationships inherent in these rules, appropriate conditional probabilities must be established such that the product of the marginal and conditional probabilities will form the desired joint probabilities. For example, the joint probability of a rule 1 violation and a rule 2 violation $P_{\bar{x}(1,2)_i}$ does not equal $P_{\bar{x}(1)_i} * P_{\bar{x}(2)_i}$, but rather equals $P_{\bar{x}(1)_i} * P_{\bar{x}(2/1)_i}$. Furthermore, just as in the derivation of marginal probability, each derivation of joint probability is also partitioned into two tails, i.e. a positive side and a negative side, since both sides of the $\bar{X}$ chart are used for the rejection of each rule. For instance, the joint probability of the simultaneous occurrence of rules 1 and 2 is equal to:

P[(1) violation on + side] * P[(2) violation, given (1) violation on + side] +
P[(1) violation on - side] * P[(2) violation, given (1) violation on - side]

The probability of the simultaneous rejections of rules 1 and 2 can be written as:

If i<2) $P_{\bar{x}(1,2)_i} = 0.$

If i=2) $P_{\bar{x}(1,2)_i} = P1UP*P2UP + P1DN*P2DN$

If i>2) $P_{\bar{x}(1,2)_i} = P1UP*\{\binom{2}{1}(P2UP)(1\text{-}P2UP) + (P2UP)^2 + (P2DN)^2\} +$

$P1DN*\{\binom{2}{1}(P2DN)(1\text{-}P2DN) + (P2DN)^2 + (P2UP)^2\}$

Because areas associated with rule 1 also are used to reject rule 2 and so on, the formulation of joint probabilities must progress from the outside-in with respect to the distribution, i.e. the derivation must follow the order rule {1, 2, 3, 4}. Thus a general guideline for the joint probability of rules (m,n,o,p) or any subset of these rules, where m<n<o<p , can be expressed as:

$P_{\bar{x}(m,n,o,p)_i} =$ (14)
P[(m) violation on + side] *
P[(n) violation/ (m) violation on + side] *
P[(o) violation/ (n) violation, (m) violation on + side] *
P[(p) violation / (o) violation, (n) violation, (m) violation on + side] +

P[(m) violation on - side] *
P[(n) violation/ (m) violation on - side] *
P[(o) violation/ (n) violation, (m) violation on - side] *
P[(p) violation / (o) violation, (n) violation, (m) violation on - side]

Derivations of the joint probabilities on the $\bar{X}$ chart at subgroup i are listed in Rahn (1990).

Once all the marginal and joint probabilities are estimated at each subgroup, the probability of rejection for (any rule) on the $\bar{X}$ chart can be determined for each subgroup using formula (11). Table 1 lists the probability of rejection of (any rule) at subgroup i, $P_{\bar{x}(Any)_i}$, for {i=1, • • ,10} under the null hypothesis (X=0,V=1), i.e. no shift, and under a shift in mean(X=1,V=1). The first column of table 1 lists α probabilities under no shift, while the second column represents power (1-β) under a specific shift. As expected, the probability of rejection increases with subgroup number because more rules become eligible to reject the null hypothesis. α equals .0184 for subgroups {8-10} compared to .0027 for subgroup 1. Also obtained from this table is the fact that the probability of rejection at subgroup i is identical to other subgroups which display exactly the same rule eligibilities. For example, subgroups {5-7} display the same probability because rules (1-3) have already been fully implemented for each subgroup; consequently the probability does not change until rule 4 becomes eligible at subgroup i=8.

__Average Probability of Rejection.__ The sum of the probabilities of rejection at subgroup i from (i=1, • • , m) yields the expected number of violations in m subgroups (N_m). Consequently, the average probability of rejection for rule j, where j equals {1,2,3,4,Any} can be found by:

$\bar{P}_{\bar{x}(j)}$ = Ave. Pr[Rejection (Rule j) on the $\bar{X}$ chart]

$$= \frac{1}{m}\sum_{i=1}^{m} P_{\bar{x}(j)_i} \qquad (15)$$

In a sense, the average probability of rejection of rule j is a cumulative measure because it is a function of m subgroups. However, it is a point estimate used to characterize single-subgroup performance. Figure 4 depicts the average probability of rejection on the $\bar{X}$ chart for any rule with respect to shifts in mean and variance. Probability of rejections and gradients at specific X-V combinations can be estimated. When compared with the probability of rejection when using strictly rule 1 in Figure 1, the increase in magnitude of probabilities when using all four rules can immediately be observed. Figure 4 also depicts another complex relationship or interaction effect between shifts in mean and variance. For instance, an increase in variance when the mean is small will increase $\bar{P}_{\bar{x}(Any)}$, while an increase in variance when the mean is large will decrease $\bar{P}_{\bar{x}(Any)}$.

Figure 5 compares the average probability of rejection (any rule) with the probability of rejection of each individual rule versus solely a shift in the mean, i.e. V=1 for all cases. This chart indicates that the addition of zone rules (2-4) and the ensuing (any rule) measure from formula (15) had a dramatic effect on power with only a small increase in the probability of rejection when X equals 0. As stated previously, α equals .0027 for rule 1. Using all four rules, the resultant α is .0153. Upon examination of the individual performance of each rule, rule 4 is the most powerful for $\{0<X<1.25\}$, rule 3 is the most powerful for $\{1.25<X<2.75\}$ and rule 2 for $\{2.75<X<?\}$. Interestingly, rule 1 is the least powerful of all rules for the range of mean shifts inspected. When examining the order of importance of the four rules as mean shift increases, it is reasonable to conclude that the ability of each rule to detect mean shifts of certain magnitude is dependent upon the magnitude of the control limit of each rule. Indeed, for this reason, rules 2-4 were introduced. Therefore, to ensure that the most powerful rule is implemented for a specific shift in mean(X), all rules must be incorporated into the control chart because of the uncertainty of X. Because of the distinct pattern of behavior for each rule and (any rule), observed performance values may indicate some pertinent information about the shifted distribution. However, if certain rules are not examined, its corresponding characteristic information is lost.

Table 2 compares derived average probabilities of rejection with 99.73% confidence intervals obtained by simulation for particular shifts in mean. In these simulations, the parent distribution was the standard normal distribution with $\mu_0=0$ and $\sigma^2_0=1$, while the parameters under the alternative hypothesis (μ_A,σ^2_A) were set at prescribed levels. The sample size (n) and number of subgroups (m) were arbitrarily set at 5 and 25, respectively. For each test condition, summary statistics were obtained from 5000 simulated runs. Confidence intervals of these normally distributed summary statistics were constructed from the simulated data in order to certify the accuracy of derived probabilities. For the range of X inspected, the null hypothesis that the derived values from formula (15) are the "true" probabilities cannot be rejected.

$\bar{X}$ Chart (Four Rules): Multiple-Subgroup Level

Using the geometric distribution, Wheeler (1983) successfully modeled the probability of 1 or more violations for (any rule) for up to seven subgroups after a shift in mean. This approach, however, was discontinued due to the complexity of the expression. Once probabilities are derived on the single-subgroup level, measures of power can be assessed on the multiple-subgroup level. In the case of the $\bar{X}$ chart with four rules, the conversion of probabilities from single-subgroup to multiple-subgroup is a highly complex problem that will not be dealt with in this paper. However, the probabilities established using formula (15) on the single-subgroup level are essential to the development of a multiple-subgroup measure of power ($\pi_{\bar{x}(Any)}$).

It was stated previously that the expected number of rule j violations in m subgroups can be obtained by:

$$E[N_{m(j)}] = \sum_{i=1}^{m} P_{\bar{x}(j)_i} \cdot \qquad (16)$$

Although this does not represent a probability of a cumulative nature, it does depict the **location** of the distribution of N_m, which is itself a cumulative measure. For example, $\pi_{\bar{x}(Any)}$, which was not derived in this paper, would equal the probability of 1 or more violations by (any rule) in m subgroups. $\pi_{\bar{x}(Any)}$ represents a probability associated with the distribution of $N_{m(Any)}$ **centered at** $\mathbf{E[N_{m(Any)}]}$.

Consequently, the average probability of rejection obtained in formula (15), $\bar{P}_{\bar{x}(Any)}$, is also a cumulative measure of performance. For example, Table 2 defines the average probability of rejection for (any rule) when X equals 2 as .86315. Therefore, the expected number of violations in 25 subgroups ($E[N_{25}]$) equals 25 times .86315 or 21.578 subgroups. For this distribution of N_{25} centered at 21.578, {f(0), f(1), • •, f(25)} represents the probability of {0, 1, • •, 25} violations in 25 subgroups by any rule. Note that $\pi_{\bar{x}(Any)}$ equals 1-f(0). These measures apply to each individual rule as well.

DEVELOPMENT OF DIAGNOSTIC PROCEDURES

Figure 5 illustrated the behavior of the average probability of rejection for each rule {1,2,3,4,} and for (any rule) with respect to shift in process mean. Figures 1 and 4 depict the average probability for rule 1 and (any rule), respectively, with respect to mean and variance. Similar contour plots could also be produced for rules 2, 3, and 4. For both the one dimensional case (X) and the two dimensional case (X,V), the consistent behavior of these charts demonstrates their ability to characterize the conditional expectation of $\bar{P}_{\bar{x}(1,2,3,4,or Any)}$ given a shift of X or (X,V), respectively. In practical applications of control charts, it may be beneficial to postulate the parameters of a shifted distribution to aid in the identification of faults entering the system. Basically, an inverse function of the model or an inverse plot can estimate the expected mean and variance of the shifted distribution given the behavior in the form of the average probability of a violation for each rule and (any rule).

Diagnostics: One-Dimensional (X)

Upon observation of the derived curve in Figure 5, an estimate of X and V can easily be made graphically once a probability estimate is observed for each rule and (any rule). However, since the derived values are stated in terms of a single parameter (X), a closed form solution is possible. Polynomial models using the method of least squares can easily be fit to the average probability of rejection for each rule and (any rule) in Figure 5 to form:

$$\bar{P}_{\bar{x}(j)} = A_j + B_j X + C_j X^2 + D_j X^3 + E_j X^4 + F_j X^5 + \bullet\bullet \qquad (17)$$

where j designates the rule type. These formulae yield the "best" coefficients of the derived average probability of rejection as a function of X.

When a process has been monitored, and realized probabilities have been attained, a logical step would be to "invert" formula (17) and investigate the expected shift in the mean(X) given the observed performance of each rule and (any rule). Observed probabilities $\{\mathbf{P}_{\bar{x}(1)}, \mathbf{P}_{\bar{x}(2)}, \mathbf{P}_{\bar{x}(3)}, \mathbf{P}_{\bar{x}(4)}, \mathbf{P}_{\bar{x}(Any)}\}$ are readily obtained by dividing the number of violations of a particular rule that occurred in m subgroups by the number of subgroups monitored(m). Therefore, the observed probabilities can replace the theoretical probabilities for each rule. These modified equations of formula (17) can now be utilized as a tool to obtain X as a function of realized probabilities. Theoretically, the roots of the polynomial could be solved for each of the above equations, but five different estimates of X would result. If these five equations are summed, however, the resultant estimate of X represents the combined information of each rule. This method is described in the following manner:

$$\sum_{j=1}^{5} P_{\bar{x}(j)} = \sum_{j=1}^{5} A_j + \sum_{j=1}^{5} B_j X + \sum_{j=1}^{5} C_j X^2 + \sum_{j=1}^{5} D_j X^3 + \sum_{j=1}^{5} E_j X^4 + \sum_{j=1}^{5} F_j X^5 \qquad (18)$$

where 5 is substituted for the "Any" designation. The expected value(s) of the mean, given observed probabilities are revealed as the real root(s) of equation (18). If an odd order polynomial is modeled, at least one real root is guaranteed. The magnitude of the expected shift is estimated from equation (18) , but the direction of the shift is still uncertain. The behavior of control charts is identical for mean shifts in either direction. However, by casual inspection of the frequency of violations on both tails, one should gain insight as to the direction of the shift.

When m equals 25 subgroups, the resultant least squares model for each rule corresponding to equation (17) are:

$$\bar{P}_{\bar{x}(1)} = .0068826 - .018123\ X + .018095\ X^2 + .014657\ X^3 \qquad (19)$$

$$\bar{P}_{\bar{x}(2)} = 0.0 + .056671\ X - .16863\ X^2 + .22054\ X^3 - .046107\ X^4$$

$$\bar{P}_{\bar{x}(3)} = 0.0 + .12\ X - .44395\ X^2 + .78877\ X^3 - .34811\ X^4 + .046988\ X^5$$

$$\bar{P}_{\bar{x}(4)} = .0098921 - .030575\ X + .056013\ X^2 + .2979\ X^3 - .17532\ X^4 + .026577\ X^5$$

$$\bar{P}_{\bar{x}(5)} = .032392 - .1964\ X + .56861\ X^2 + .036086\ X^3 - .13755\ X^4 + .027144\ X^5.$$

All models exhibited an excellent fit with R^2 values equal to 1.0. Equation (18) can be expressed as:

$$\sum_{j=1}^{5} P_{\bar{x}(j)} = .049167 - .068427\ X + .030138\ X^2 + 1.357953\ X^3 - .707087\ X^4 + .100709\ X^5. \qquad (20)$$

Examples. To determine the validity of this diagnostic procedure, two examples are presented. Observed average probabilities of rejection for each rule and (any rule) were obtained from 5000 simulated runs of m=25 subgroups from a **known distribution**. The sum of these probabilities were entered into formula (20) to find the expected shift in mean (X). The known distribution followed an X equal to 0.6 in example 1 and 2.0 in example 2 and resulted in the observed probability of rejections given in Table 3. The sum of the observed probabilities equals 0.222568 in example 1 and 2.803848 in example 2. The singular real root in examples 1 and 2 resulted in X values of 0.59437 and 1.9994, respectively. These estimates are almost identical to the "true" underlying distribution of the simulation.

Diagnostics: Two-Dimensional (X,V)

The method of least squares could also be employed to fit polynomial models to the surface profiles of average probability of rejection, however the model would exhibit tremendous error due to the rather large number of parameters needed to characterize the curvature of the response surface of derived probabilities. Furthermore, equations similar to formulas (17) and (18) but in the two-variable case could not be expressed in a closed form solution because an infinite number of (X-V) combinations exist which satisfy the equation. Therefore, the derived surface response must contain a contour corresponding to the observed probability of rejection. This contour represents the expected (X-V) based on the performance of a particular rule. Hence, a contour can be generated for each rule(1)-(4) and for (any rule). The ideal point would be a region on the X-V plane where the contours intersect, thus satisfying the cumulative information of (any rule) as well as the specific knowledge of each rule.

Examples. This diagnostic procedure is a graphical analysis which estimates the expected X-V given observed probability of rejections obtained during monitoring. If equations (11,12,13,15) engross the underlying nature of behavior for each rule and (any rule), then each rule and their interrelationships provide insight to the location of the true mean and variance. Subsequently, a controlled convergence should emerge which simultaneously satisfies the specific behavior pattern of each rule as well as the more general pattern of (any rule). Figures 6 and 7 demonstrate this method for two known shifted distributions of (X=0.6,V=1) and (X=1.0,V=1.4), respectively, for 25 subgroups. Both figures exemplify a successful convergence to a region in the vicinity of the known distribution. In each example, the contour of rules (2-4) intersected at a region proximal to the known distribution, while rule 1 and (any rule) varied slightly from this point. Rule 1's relative inefficiency stems from its extremely small slope in the region of X examined(see Figure 5). Thus even the smallest uncertainty in probability will yield a sizable uncertainty in the estimate of X. The relative lack of robustness for (any rule) may stem from a combination of two concepts. The first is error propagation, such that any uncertainty of X due to rule 1 {and rules (2-4)} translates into uncertainty of the estimate of X from (any rule). The second theory is that the (any rule) behavior can be manifested in many different specific outcomes pertaining to each rule. For example, (a) subgroup i could be in violation of rule 1 or (b) subgroup i could be in violation of all four rules, but the (any rule) designation would not distinguish between the two cases. The information processed from the (any rule) contour provides a general measure of performance that is not as precise as the information from the contour of each rule. The robustness of these two contours should improve as observed probabilities of rejection increase. In general, this method provides an adequate estimation of the parameters of a shifted distribution based on control chart performance.

APPLICATION EXAMPLE

The previous two examples in the one-dimensional case (X) emphasized the introduction of shifts in mean through simulation of a generic process where observed average probabilities from 5000 simulations were input to project the actual shift in mean. This section (a) applies this one-dimensional diagnostic procedure to a model of AA surface roughness (μ in.) of a bronze bushing as shown in Figure 8 developed from a study of a turning process (Dessouky, 1987) and (b) illustrates the usefulness of this technique in the identification of system faults. This turning process was modeled as a function of ten influencing factors shown in Table 4.

This investigation will concentrate on the manipulation of just one of these variables, tool condition. Process parameter estimates ($\hat{\mu}_0$=70.12, $\hat{\sigma}^2_0$=433.986) were obtained from observing a stable process using a sharp tool. At this point a **dull tool** was introduced into the system and 25 subgroups were collected and plotted with respect to the limits established from $\hat{\mu}_0$ and $\hat{\sigma}^2_0$. Of these 25 subgroups, four subgroups violated rule 3 and two subgroups violated rule 4. Overall, only four different subgroups were violated, which indicates that two subgroups violated both rule 3 and rule 4. Average probabilities can be obtained by simply dividing the number of violations by 25 (m), resulting in .16, .08, .16 for rule 3, 4, and (any rule), respectively. The sum of these probabilities (.40) can then be input into equation (20) to yield a real root of .7605. Since all violations on the $\bar{X}$ chart appeared on the lower tail, the expected standardized shift in the mean(X) is -.7605. Upon conversion back to units of μ in.:

$$\hat{\mu}_A = X * \hat{\sigma}_0/\sqrt{n} + \hat{\mu}_0 \tag{21}$$

The expected surface roughness that manifested this particular chart behavior equals $(-.7605)*\sqrt{\frac{433.986}{5}} + 70.12 = 63.03\ \mu$ in. Therefore, a shift of (63.03-70.12) = -7.09 μ in. is hypothesized from the diagnostic procedure based on control chart behavior. The "true" modeled effect of a dull tool on surface roughness was -8.0 μ in. The first two examples exhibited greater accuracy because average probabilities from 5000 simulations were used. Obviously, more uncertainty exists with a single data set of 25 subgroups. Traditionally, upon evaluation of control chart performance, fault identification methods consist of scatter plots, cause-effect diagrams, or other correlation procedures. The diagnostic procedure developed in this paper provides a more descriptive role of deducing causation in a system.

An estimate of the mean of the underlying shifted distribution in itself does not directly identify the special cause(s). But when this knowledge is used in conjunction with process models expressed in terms of its process parameters, such as in design of experiments(DOE), the process of fault identification becomes methodical. Evaluation of the process model may suggest that certain variables be eliminated from consideration if the sign and magnitude of their gradients at their current level can not account for the observed change. For example, the derived change in surface roughness was -7.09 μ in. In this example, a logical conclusion for the reduction in surface roughness is tool wear(condition). The decision can then be made to require more frequent tool inspection and change as needed.

SUMMARY AND CONCLUSIONS

Although control charts have had a tremendous impact on quality improvement, their inability to precisely measure performance has hindered their full potential of functional performance. The qualifications for effective chart implementation is that performance is both measurable and sufficient. With the addition of rules (2-4), practitioners were willing to sacrifice the ability to measure performance in exchange for a monitoring system which was intuitively more powerful. The work presented in this paper is a major step towards the fulfillment of the control chart's functional capability. The following summarizes the contributions of this paper.

(1) The probability of rejection for each rule and (any rule) were developed using a rule-dependency approach and were analyzed with respect to shifts in mean and variance. These derived probabilities matched well with simulated results. In fact, all derived values fell within their confidence bounds. In comparison with "three-sigma" charts, $\bar{X}$ charts using four rules performed extraordinarily well when a shift occurred with only a minor increase in α under the null hypothesis.

(2) For diagnostic purposes, the expected average probability of rejection was expressed as a function of shift in mean (X) in the one-dimensional case or a function of shift in mean and variance (X,V) in the two-dimensional case. Once this equation or plot is established, the inverse function or inverse plot (in the form of a contour) has been shown to effectively estimate the expected mean and variance based on observed probabilities. In the one-dimensional case, both examples demonstrated that the real root of the resultant polynomial equation was almost an exact match to the mean of the known simulated distribution. Both examples in the two-parameter function provided excellent graphical convergence proximal to the known distribution. However, results indicate that smaller gradients of the probability of rejection for a particular rule will result in greater uncertainty in the estimate of the shifted mean and variance for that rule.

(3) When used in conjunction with process models, this diagnostic procedure can invoke a more scientific approach to fault identification.

(4) The probabilities and diagnostic models in this paper have been developed for the case of 25 subgroups. However, exact probabilities can easily be obtained for any specific number of subgroups (m) monitored using the derived expressions.

(5) This paper only considered shifts prior to monitoring, however, the procedures developed can be applied to shifts which occur during process monitoring. Unfortunately, the distribution of N_m must be developed prior to any such evaluation.

Now that control chart performance is measurable, the question of sufficiency arises along with possible methods to improve control charting performance. The future work of this author will be devoted to the modification of control chart procedures to improve performance, the implementation of charting procedures when the statistical assumptions are no longer valid, and the development of the distribution of N_m for:

1) evaluation of probabilities on the multiple-subgroup level
2) consideration of shifts during the monitoring process
3) construction of confidence intervals for two-dimensional diagnostics.

The growing popularity of $\bar{X}$ control charts over the past 60 years can primarily be attributed to an acceptable perceived level of performance, for their actual performance characteristics have been unknown till now. Because of their statistical soundness and measurable performance characteristics, $\bar{X}$ control charts should gain additional market share in the intensely competitive field of statistical process control.

TABLES AND FIGURES

Table 1. Pr[Rejection at subgroup i on the $\bar{X}$ chart] under no shift (X=0,V=1) and a shift in mean (X=1,V=1)

SUBGROUP	Pr[REJECTION] (X=0,V=1)	Pr[REJECTION] (X=1,V=1)
1	.0026998	.0227821
2	.0036735	.0443459
3	.0056353	.0836677
4	.0068239	.1299576
5	.0109394	.2354709
6	.0109394	.2354709
7	.0109394	.2354709
8	.0184026	.3925588
9	.0184026	.3925588
10	.0184026	.3925588

Table 2. Simulated and derived average probability of rejections for each rule

	X=0.0		X=0.4	
RULE	SIM. CONFIDENCE INT.	DERIVED	SIM. CONFIDENCE INT.	DERIVED
1	.00220 -- .00307	.00269	.00483 -- .00607	.00499
2	.00233 -- .00354	.00285	.00775 -- .00992	.00829
3	.00356 -- .00506	.00469	.01685 -- .02041	.01894
4	.00466 -- .00669	.00562	.02249 -- .02768	.02466
ALL	.01359 -- .01653	.01531	.04981 -- .05645	.05200

	X=0.8		X=1.2	
RULE	SIM. CONFIDENCE INT.	DERIVED	SIM. CONFIDENCE INT.	DERIVED
1	.01254 -- .01450	.01397	.03412 -- .03728	.03594
2	.03235 -- .03681	.03428	.10339 -- .11135	.10817
3	.08536 -- .09401	.08857	.25305 -- .26686	.25824
4	.10320 -- .11479	.10719	.26257 -- .27935	.27077
ALL	.19422 -- .20777	.19960	.45430 -- .47080	.46192

	X=1.6		X=2.0	
RULE	SIM. CONFIDENCE INT.	DERIVED	SIM. CONFIDENCE INT.	DERIVED
1	.07888 -- .08348	.08075	.15619 -- .16230	.15865
2	.24908 -- .26031	.25717	.46754 -- .48045	.47000
3	.48829 -- .50351	.49977	.70358 -- .71566	.70804
4	.44523 -- .46289	.45869	.59070 -- .60465	.59893
ALL	.70404 -- .71724	.71481	.85935 -- .86723	.86315

Table 3. Ave. probability of rejections for each rule

Rules	Ex. 1 X=0.6	Ex. 2 X=2.0
1	.008456	.159248
2	.016648	.474000
3	.040920	.709624
4	.052568	.597680
All	.103976	.863296

Table 4. Potential special causes of variation (Dessouky, 1987)

FACTOR	NATURE OF VARIATION
(1) Depth-to-shoulder dimensions	Varies from (129.532 to 129.547)
(2) Operator	Regular operator or substitute
(3) Set-up person	Two persons may perform setup
(4) Lathe used	Three different lathes were used
(5) Surface measuring	Two different profilometers were used
(6) Tool type	Tools were purchased from three sources
(7) Tool condition	Observed to be relatively sharp or dull
(8) Tool back rake angle	Two different tool geometries were used
(9) Cutting speed	Varies from (5.08-5.588)meters per second
(10) Feedrate	Varies from (0.020-0.023) mm per rev.

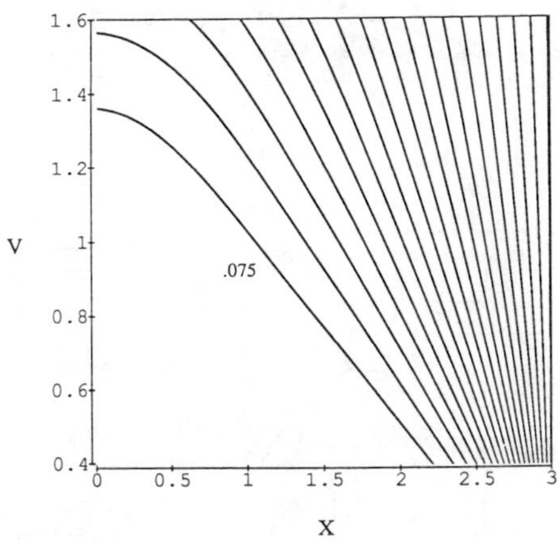

Figure 1. Contour Pot of $P_{\bar{x}(1)i}$

(Intervals of .025)

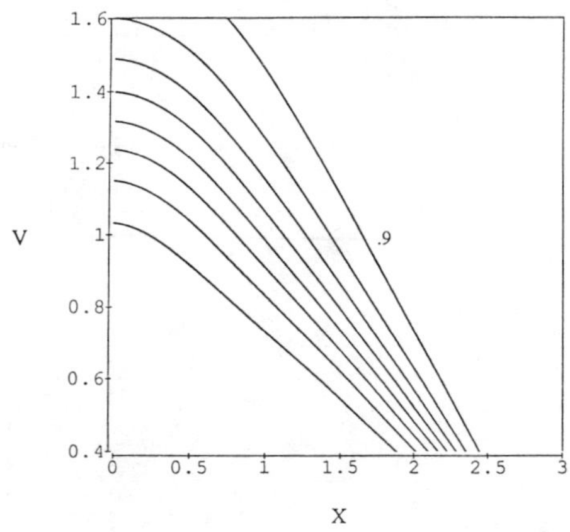

Figure 2. Contour Pot of $\pi_{\bar{x}(1)}$

(Intervals of .1)

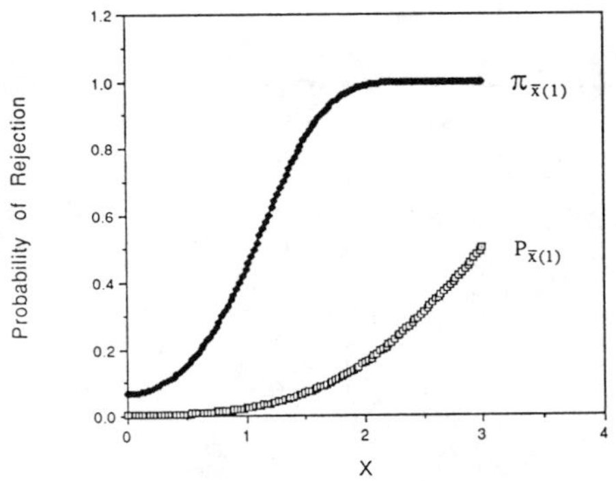

Figure 3. Plot of $P_{\bar{x}(1)i}$ and $\pi_{\bar{x}(1)}$

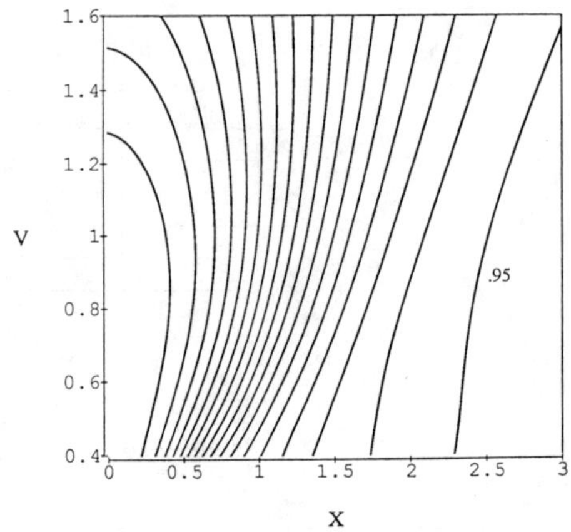

Figure 4. Contour Plot of $\bar{P}_{\bar{x}(Any)}$ (Intervals of .05)

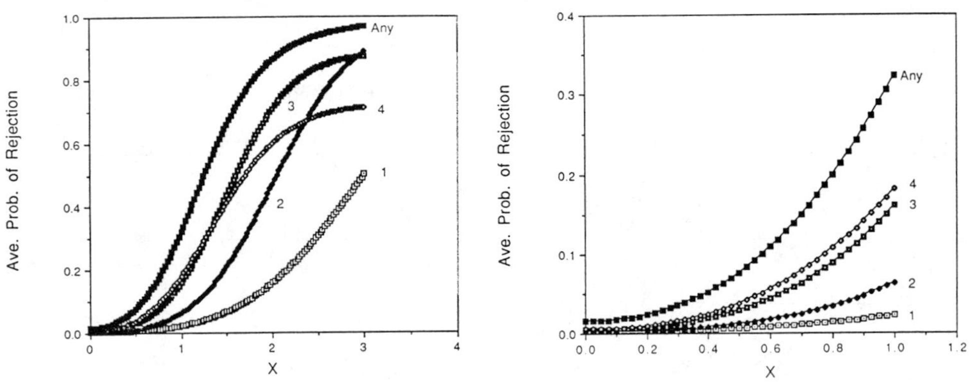

Figure 5. Plot of $\bar{P}_{\bar{x}(j)}$ {j=1,2,3,4,Any}

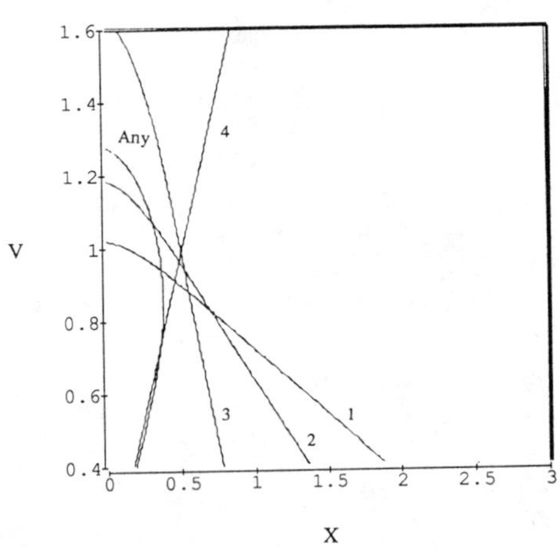

Figure 6. Contours for example 1

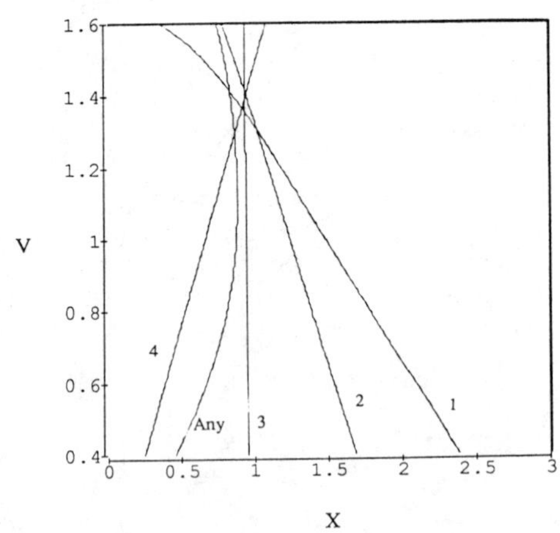

Figure 7. Contours for example 2

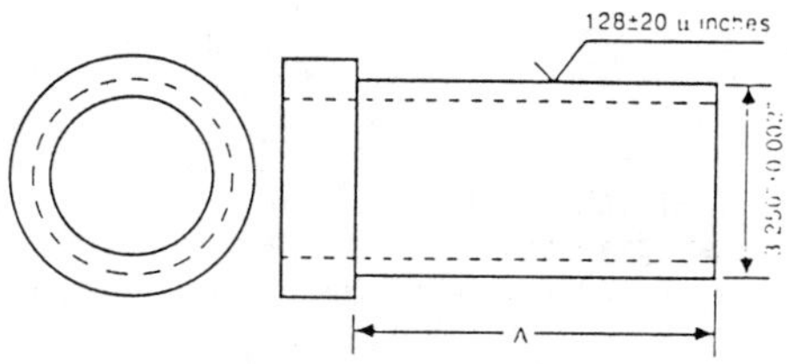

Figure 8. Bushing from (Dessouky, 1987)

REFERENCES

1. Box, George E.P., 1978, William G. Hunter, and J. Stuart Hunter, "Statistics For Experimenters", John Wiley & Sons, Inc., New York.
2. Dessouky, M. I., S. G. Kapoor, and R. E. DeVor, 1987, "A Method for Integrated Quality Systems", Journal of Engineering for Industry, Vol. 109, pp.241-247.
3. DeVor, R. E., T. H. Chang, and J. W. Sutherland, 1991, "Statistical Methods for Quality Design And Control: Contemporary Concepts and Methods" Macmillan Publishing Company, Presented for publication.
4. Ferris, Charles D., Frank E. Grubbs, and Chalmers L. Weaver, 1946, "Operating Characteristics for the Common Statistical Tests of Significance," Annals of Math. Stat., Vol. 17, pp.178-197.
5. Hogg, Robert V. and Allen T. Craig, 1978, "Introduction To Mathematical Statistics," Fourth Edition, Macmillan Publishing Co. Inc., New York.
6. Kanazuka, Takazi, 1986, "The Effect of Measurement Error on the Power of $\bar{X}$ - R Charts," Journal of Quality Technology, Vol. 18, No. 2, pp.91-95.
7. King, Edgar P., 1952_a, "The Operating Characteristics of the Average Chart", Industrial Quality Control, Vol. 9, No. 3, pp.30-32.
8. King, Edgar P., 1952_b , "The Operating Characteristic of the Control Chart for Sample Means," Annals of Math. Statistics, Vol. 23, No. 3, pp.384-395.
9. King, Edgar P., 1954, "Probability Limits for the Average Chart when Process Standards are Unspecified," Industrial Quality Control, Vol. 10, No. 6, pp.62-64.
10. Odeh, R. E. and J. O. Evans, 1974, "Algorithm AS 70 The Percentage Points of the Normal Distribution," Applied Statistics, Vol. 23, pp.96-97.
11. Olds, Edwin G, 1961, "Power Characteristics of Control Charts," Industrial Quality Control, Vol. 18, No. 1, pp.4-10.
12. Proschan F. and I. R. Savage, 1960, "Starting a Control Chart," Industrial Quality Control, Vol. 17, No. 3, pp.12-13.
13. Rahn, Gregory E., 1990, "The Development of a Generalized Control Chart Methodology to Enhance Application and Performance", Ph.D. Dissertation Proposal.
14. Scheffe', Henry, 1959, "Operating Characteristics of Average and Range Charts," Industrial Quality Control, Vol. 5, No. 6, pp.13-18.
15. Shewhart, W. A., 1931, Economic Control of Quality Manufactured Products. Republished in 1981 by American Society for Quality Control Milwaukee, WI.
16. Wheeler, Donald J., 1983, "Detecting a Shift in Process Average: Tables of the Power Function of $\bar{X}$ Charts," Journal of Quality Technology, Vol. 15, No. 4, pp.155-170.
17. Yang, Chung-How, and Frederick S. Hillier, 1970, "Mean and Variance Control Chart Limits Based on a Small Number of Subgroups," Journal of Quality Technology, Vol. 2, No. 1, pp.9-16.

AUTHOR INDEX